D0534484

"I have been doing project management for over 30 years and am considered a subject the *PMBOK Guide*, Third Edition, primarily because I am the project manager who led developed this edition…I can honestly say that *Head First PMP* is by far the best PMP exam preparation book of all I have reviewed in depth. It is the very best basic education and training book that I have read that presents the processes for managing a project, which makes it a great resource for a basic project management class for beginners as well as a tool for practitioners who want to pass the PMP exam. The graphical story format is unique, as project management books go, which makes it both fun and easy to read while driving home the basics that are necessary for preparing someone who is just getting started and those who want to take the exam."

> — **Dennis Bolles, PMP**
> **Project manager for the *PMBOK Guide*, Third Edition, Leadership Team,**
> **DLB Associates, LLC, and coauthor of *The Power of Enterprise-Wide Project***
> ***Management***

"This looks like too much fun to be a PMP study guide! Behind the quirky humor and nutty graphics lies an excellent explanation of the project management processes. Not only will this book make it easier to pass the exam, you'll learn a lot of good stuff to use on the job too."

> — **Carol Steuer, PMP**
> ***PMBOK Guide*, Third Edition, Leadership Team**

"This is the best thing to happen to PMP since, well, ever. You'll laugh, learn, pass the exam, and become a better project manager all at the same time."

> — **Scott Berkun, author of *The Art of Project Management* and *The Myths of***
> ***Innovation***

"I love the brain-friendly approach used by Head First. When was the last time you heard that a PMP prep book was fun to read? This one really is!"

> — **Andy Kaufman, host of the People and Projects Podcast on iTunes**

"*Head First PMP* is the PMP exam prep book for the rest of us: the people who live project management daily and want an exam prep book that is as interesting as the work we live, prepares them for the exam, and helps them become a better project manager. I've taken my copy of the first edition to numerous exam prep classes I have helped teach as a reference book. Students will pick it up, review several pages or topics and say, 'That is how I learn. Can I take your copy?' The impact and satisfaction is immediate."

> — **Ken Jones, PMP and project manager**

"In today's business world, it's not just what you get done, it's how you get it done. To that end, *Head First PMP* has just the right balance of wit and fun that makes learning the Project Management Body of Knowledge engaging and interesting."

> —**Jen Poisson, Director of Production Operations, Disney Online**

Praise for *Head First PMP*

"Wow. In the beginning of March I finished and passed a four-hour adventure called the PMP exam. I can honestly say that though I used a few study guides, without the help of *Head First PMP*, I don't know how I would have done it. Jenny and Andrew put together one of the best 'head smart, brain friendly' training manuals that I have ever seen. I have to say that I am a *huge* fan and *will* be buying their new *Beautiful Teams* book. Anyone I meet who mentions wanting to take the exam, I send them to *http:// www.headfirstlabs.com/books/hfpmp/* to get the sample chapter and free test. Seeing is believing. Thanks, Andrew and Jenny, for putting together an exceptional study guide. Keep up the good work!"

> **—Joe Pighetti Jr., PMP, engineer**

"I think that under the fonts and formalized goofiness, the book has a good heart (intending to cover basic principles in an honest way rather than just to pass the test). *Head First PMP* attempts to educate potential project managers instead of being a mere 'how to pass the PMP exam' book filled with test-taking tips. This is truly something, which sets it apart from the other PMP certification exam books."

> **— Jack Dahlgren, project management consultant**

"I love this format! *Head First PMP* covers everything you need to know to pass your PMP exam. The sound-bite format combined with the whimsical images turns a dry subject into entertainment. The organization starts with the basics, then drills into the details. The in-depth coverage of complex topics like earned value and quality control are presented in an easy-to-understand format with descriptions, pictures, and examples. This book will not only help you pass the PMP [exam], it should be used as a daily reference for practicing project managers. I sure wish I had this when I was studying for the exam."

> **—Mike Jenkins, PMP, MBA**

"It is like an instructor with a blackboard in a book, and the little devil and angel over your shoulder telling you what is right or wrong. I am getting instant results from the first five chapters. An excellent guide/training tool for all those new and somewhat new to project management methodologies."

> **—BJ Moore, PMP**
> **Nashville, TN**
> **Amazon.com reviewer**

"Studying for your PMP exam? Would you like the ability to carry not only an instructor but an entire classroom in your briefcase as you prepare? Then buy this book! The drawings and diagrams are reminiscent of your favorite teacher utilizing the whiteboard to step you through the key points of their lecture. The author's use of redundancy in making the same point in multiple ways, coupled with the "there are no dumb questions" sections, gave the feeling of being in a classroom full of your fellow PMP aspiring peers. At times I actually caught myself feeling relieved that someone else asked such a good question. This book is enjoyable, readable, and most importantly takes the fear out of approaching the subject matter. If you are testing the PMP waters with your big toe, this book will give you the confidence to dive into the deep end."

> **—Steven D. Sewell, PMP**

"With *Head First C#*, Andrew and Jenny have presented an excellent tutorial on learning C#. It is very approachable while covering a great amount of detail in a unique style. If you've been turned off by more conventional books on C#, you'll love this one."

—Jay Hilyard, software developer, coauthor of *C# 3.0 Cookbook*

"I've never read a computer book cover to cover, but this one held my interest from the first page to the last. If you want to learn C# in depth and have fun doing it, this is *the* book for you."

— Andy Parker, fledgling C# programmer

"Going through this *Head First C#* book was a great experience. I have not come across a book series which actually teaches you so well…This is a book I would definitely recommend to people wanting to learn C#"

—Krishna Pala, MCP

"*Head First Web Design* really demystifies the web design process and makes it possible for any web programmer to give it a try. For a web developer who has not taken web design classes, *Head First Web Design* confirmed and clarified a lot of theory and best practices that seem to be just assumed in this industry."

—Ashley Doughty, senior web developer

"Building websites has definitely become more than just writing code. *Head First Web Design* shows you what you need to know to give your users an appealing and satisfying experience. Another great Head First book!"

—Sarah Collings, user experience software engineer

"*Head First Networking* takes network concepts that are sometimes too esoteric and abstract even for highly technical people to understand without difficulty and makes them very concrete and approachable. Well done."

—Jonathan Moore, owner, Forerunner Design

"The big picture is what is often lost in information technology how-to books. *Head First Networking* keeps the focus on the real world, distilling knowledge from experience and presenting it in byte-size packets for the IT novitiate. The combination of explanations with real-world problems to solve makes this an excellent learning tool."

— Rohn Wood, senior research systems analyst, University of Montana

Other related books from O'Reilly

Applied Software Project Management

Making Things Happen

Practical Development Environments

Process Improvement Essentials

Time Management for System Administrators

How to Keep Your Boss from Sinking Your Project (Digital Short Cut)

Other books in O'Reilly's *Head First* series

Head First C#

Head First Java

Head First Object-Oriented Analysis and Design (OOA&D)

Head First HTML with CSS and XHTML

Head First Design Patterns

Head First Servlets and JSP

Head First EJB

Head First SQL

Head First Software Development

Head First JavaScript

Head First Physics

Head First Statistics

Head First Ajax

Head First Rails

Head First Algebra

Head First PHP & MySQL

Head First Web Design

Head First Networking

Head First PMP

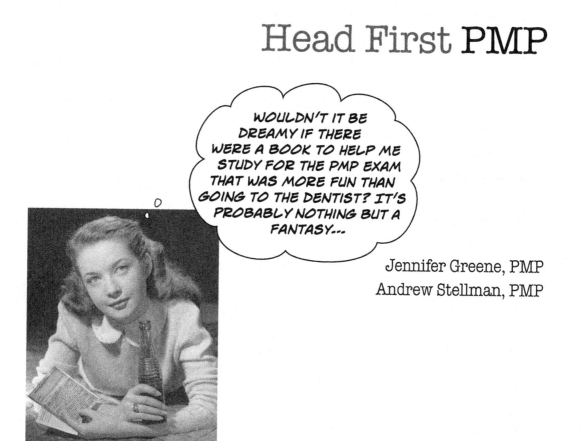

WOULDN'T IT BE DREAMY IF THERE WERE A BOOK TO HELP ME STUDY FOR THE PMP EXAM THAT WAS MORE FUN THAN GOING TO THE DENTIST? IT'S PROBABLY NOTHING BUT A FANTASY...

Jennifer Greene, PMP
Andrew Stellman, PMP

O'REILLY®

Beijing · Cambridge · Köln · Sebastopol · Tokyo

Head First PMP

Third Edition

by Jennifer Greene, PMP and Andrew Stellman, PMP

Published by O'Reilly Media, Inc., 1005 Gravenstein Highway North, Sebastopol, CA 95472.

O'Reilly Media books may be purchased for educational, business, or sales promotional use. Online editions are also available for most titles (*http://my.safaribooksonline.com*). For more information, contact our corporate/institutional sales department: (800) 998-9938 or *corporate@oreilly.com*.

Series Creators:	Kathy Sierra, Bert Bates
Editor:	Courtney Nash
Design Editor:	Louise Barr
Cover Designers:	Karen Montgomery, Louise Barr
Production Editors:	Melanie Yarbrough
Indexer:	Bob Pfahler
Proofreader:	Rachel Monaghan
Page Viewers:	Quentin the whippet and Tequila the pomeranian

Printing History:

March 2007: First Edition.

July 2009: Second Edition.

December 2013: Third Edition.

ISBN: 978-1-449-36491-5

[LSI]   [2014-07-14]

To our friends and family, and the people who make us laugh
(you know who you are)

> THANKS FOR BUYING OUR BOOK! WE REALLY LOVE WRITING ABOUT THIS STUFF, AND WE HOPE YOU GET A KICK OUT OF READING IT...

> ...BECAUSE WE KNOW YOU'RE GOING TO KICK ASS ON THE TEST!

Andrew

Photo by Nisha Sondhe

Jenny

Andrew Stellman, despite being raised a New Yorker, has lived in Pittsburgh *twice*. The first time was when he graduated from Carnegie Mellon's School of Computer Science, and then again when he and Jenny were starting their consulting business and writing their first book for O'Reilly.

When he moved back to his hometown, his first job after college was as a programmer at EMI-Capitol Records—which actually made sense, since he went to LaGuardia High School of Music and Art and the Performing Arts to study cello and jazz bass guitar. He and Jenny first worked together at that same financial software company, where he was managing a team of programmers. He's had the privilege of working with some pretty amazing programmers over the years, and likes to think that he's learned a few things from them.

When he's not writing books, Andrew keeps himself busy writing useless (but fun) software, playing music (but video games even more), experimenting with circuits that make odd noises, studying taiji and aikido, having a girlfriend named Lisa, and owning a pomeranian.

Jennifer Greene studied philosophy in college but figured out pretty soon afterward that she really loved building software. Luckily, she's a great software engineer, so she started out working at an online service, and that's the first time she really got a sense of what good software development looked like.

She moved to New York in 1998 to work on software quality at a financial software company. She's managed a teams of developers, testers, and PMs on software projects in media and finance since then.

She's traveled all over the world to work with different software teams and build all kinds of cool projects.

She loves traveling, watching Bollywood movies, reading the occasional comic book, playing PS4 games, and hanging out with her huge Siberian cat, Sascha.

Jenny and Andrew have been building software and writing about software engineering together since they first met in 1998. Their first book, *Applied Software Project Management*, was published by O'Reilly in 2005. They published their second book in the Head First series, *Head First C#*, in 2009.

They founded Stellman & Greene Consulting in 2003 to build a really neat software project for scientists studying herbicide exposure in Vietnam vets. When they're not building software or writing books, they do a lot of speaking at conferences and meetings of software engineers, architects, and project managers.

Check out their blog, *Building Better Software*, at: http://www.stellman-greene.com.

Table of Contents (Summary)

Table of Contents (the real thing)

Intro

Your brain on PMP.

Here *you* are trying to *learn* something, while here your *brain* is doing you a favor by making sure the learning doesn't *stick*. Your brain's thinking, "Better leave room for more important things, like which wild animals to avoid and whether naked snowboarding is a bad idea." So how *do* you trick your brain into thinking that your life depends on knowing enough to get through the PMP exam?

Introduction

Why get certified?

Tired of facing the same old problems?

If you've worked on a lot of projects, you know that you face the same problems, over and over again. It's time to learn some common solutions to those problems. There's a whole lot that project managers have learned over the years, and passing the PMP exam is your ticket to putting that wisdom into practice. Get ready to change the way you manage your projects forever.

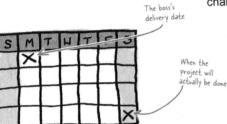

The boss's delivery date

When the project will actually be done

The PMBOK Guide

also known as… A Guide to the Project Management Body Of Knowledge

Organizations, constraints, and projects
In good company

If you want something done right…better hope you're in the right kind of organization. All projects are about teamwork—but how your team works depends a lot on the type of organization you're in. In this chapter, you'll learn about the different types of organizations around—and which type you should look for the next time you need a new job.

Time Cost Resources Quality Risk
Scope

The process framework
It all fits together

All of the work you do on a project is made up of processes. Once you know how all the processes in your project fit together, it's easy to remember everything you need to know for the PMP exam. **There's a pattern** to all of the work that gets done on your project. First you plan it, then you get to work. While you are doing the work, you are always comparing your project to your original plan. When things start to get off-plan, it's your job to make corrections and put everything back on track. And the **process framework**—the **process groups** and **knowledge areas**—is the key to all of this happening smoothly.

Tools

Project integration management

Getting the job done

Want to make success look easy?

It's not as hard as you think. In this chapter, you'll learn about **a few processes** you can use in your projects every day. Put these into place, and your **sponsors** and **stakeholders** will be happier than ever. Get ready for **Integration Management**.

Enterprise environmental factors

Organizational process assets

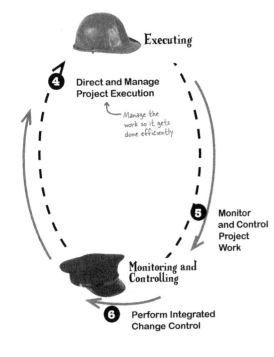

Executing

④ **Direct and Manage Project Execution**

Manage the work so it gets done efficiently.

⑤ **Monitor and Control Project Work**

Monitoring and Controlling

⑥ **Perform Integrated Change Control**

Scope management

Doing the right stuff

Confused about exactly what you should be working on?

Once you have a good idea of what needs to be done, you need to **track your scope** as the project work is happening. As each goal is accomplished, you confirm that all of the work has been done and make sure that the people who asked for it are **satisfied with the result**. In this chapter, you'll learn the tools that help your project team **set its goals** and keep everybody on track.

Updates

Project Management plan

Work breakdown structure

Project scope statement

Time management

Getting it done on time

Time management is what most people think of when they think of project managers. It's where the deadlines are set and met. It starts with **figuring out the work** you need to do, how you will do it, what **resources you'll use**, and how long it will take. From there, it's all about developing and controlling that **schedule**.

Resource calendar

6

Network diagram

IF THE CATERERS COME TOO EARLY, THE FOOD WILL SIT AROUND UNDER HEAT LAMPS! BUT TOO LATE, AND THE BAND WON'T HAVE TIME TO PLAY. I JUST DON'T SEE HOW WE'LL EVER WORK THIS ALL OUT!

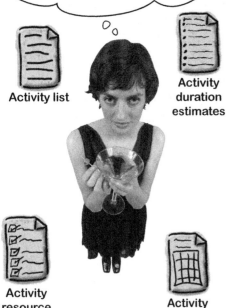

Activity list

Activity duration estimates

Activity resource requirements

Activity attributes

Cost management

Watching the bottom line

7

Every project boils down to money. If you had a bigger **budget**, you could probably get more people to do your project more quickly and deliver more. That's why no project plan is complete until you come up with a budget. But no matter whether your project is big or small, and no matter how many **resources** and **activities** are in it, the process for figuring out the bottom line is *always the same*!

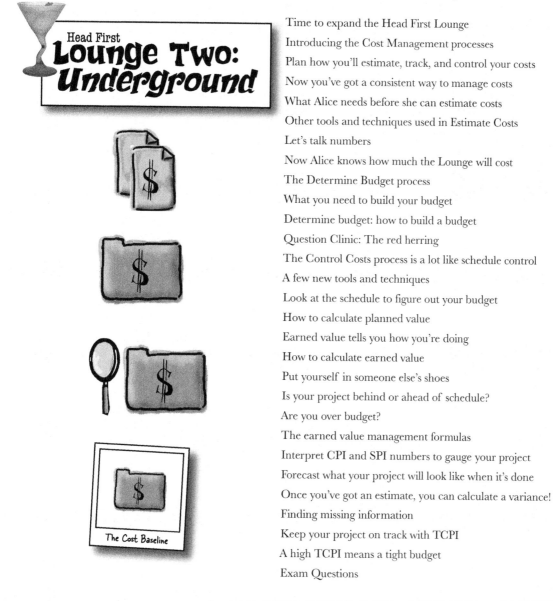

Head First
Lounge Two: Underground

The Cost Baseline

Quality management

Getting it right

It's not enough to make sure you get it done on time and under budget. You need to be sure you make the right product to suit your stakeholders' needs. Quality means making sure that you build what you said you would and that you do it as efficiently as you can. That means trying not to make too many mistakes and always keeping your project working toward the goal of creating the right product!

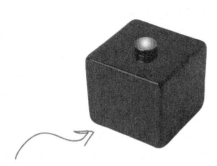

The Black Box 3000™

Lisa also inspected the blueprints for the black box when they were designed.

She looked for defects in the parts as they were being made too.

Human resource management

Getting the team together

9

Behind every successful project is a great team. So how do you make sure that you get—and keep—the best possible team for your project? You need to **plan carefully**, set up a good **working environment**, and negotiate for the **best people** you can find. But it's not enough to put a good team together... If you want your project to go well, you've got to keep the team motivated and deal with any conflicts that happen along the way. **Human resource management** gives you the tools you need to get the best team for the job and lead them through a successful project.

Organization charts

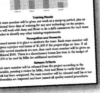

Staffing Management plan

RACI Matrix		Role			
		Mike	Amy	Brian	Peter
Work Package	Project Management	R	I	I	I
	Design	C	R	C	I
	Construction	C	C	R	I
	Testing	C	C	R	I
R = Responsible A = Accountable C = Consult I = Inform					

Roles and responsibilities

10

Communications management

Getting the word out

Communications management is about keeping everybody in the loop. Have you ever tried talking to someone in a really loud, crowded room? That's what running a project is like if you don't get a handle on communications. Luckily, there's **Communications Management**, which is how to get everyone talking about the work that's being done, so that they all **stay on the same page**. That way, everyone has the information they need to **resolve any issues** and keep the project **moving forward**.

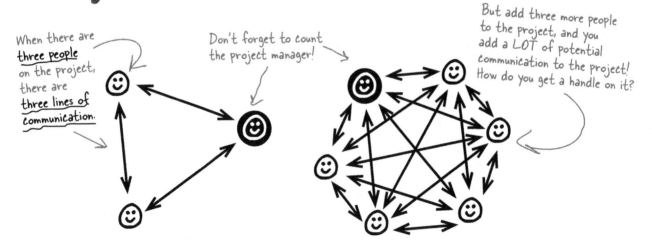

When there are **three people** on the project, there are **three lines of communication.**

Don't forget to count the project manager!

But add three more people to the project, and you add a LOT of potential communication to the project! How do you get a handle on it?

Project risk management

Planning for the unknown

Even the most carefully planned project can run into trouble.

11

No matter how well you plan, your project can always run into **unexpected problems**. Team members get sick or quit, resources that you were depending on turn out to be unavailable—even the weather can throw you for a loop. So does that mean that you're helpless against unknown problems? No! You can use *risk planning* to identify potential problems that could cause trouble for your project, **analyze** how likely they'll be to occur, take action to **prevent** the risks you can avoid, and **minimize** the ones that you can't.

Procurement management

Getting some help

Some jobs are just too big for your company to do on its

own. Even when the job isn't too big, it may just be that you don't have the expertise or equipment to do it. When that happens, you need to use **Procurement Management** to find another company to **do the work for you.** If you find the **right seller**, choose the **right kind of relationship**, and make sure that the **goals of the contract are met**, you'll get the job done and your project will be a success.

Contract

Closed procurements

HELLO, THIS IS TECHNICAL SUPPORT. HOW CAN I HELP YOU?

Stakeholder management

Keeping everyone engaged

Project management is about knowing your audience. If you don't get a handle on the people who are affected by your project, you might discover that they have needs you aren't meeting. If your project is going to be successful, you've got to satisfy your stakeholders. Luckily, there's **Stakeholder Management**, which you can use to understand your stakeholders and figure out what they need. Once you really understand how important those needs are to your project, it's a lot easier to **keep everyone satisfied**.

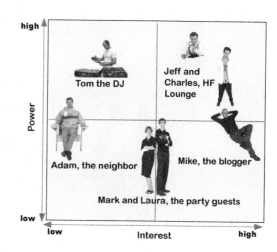

Professional responsibility

Making good choices

It's not enough to just know your stuff. You need to make good choices to be good at your job. Everyone who has the PMP credential agrees to follow the **Project Management Institute Code of Ethics and Professional Conduct**, too. The Code helps you with **ethical decisions** that aren't really covered in the body of knowledge—and it's a big part of the PMP exam. Most of what you need to know is **really straightforward**, and with a little review, you'll do well.

AWESOME. I'VE BEEN WANTING TO GO SHOPPING FOR A WHILE. AND WHAT ABOUT THAT VACATION? ACAPULCO, HERE WE COME!

I WOULD NEVER ACCEPT A GIFT LIKE THAT. DOING A GOOD JOB IS ITS OWN REWARD!

I'M SORRY, I CAN'T ACCEPT THE GIFT. I REALLY APPRECIATE THE GESTURE, THOUGH.

A little last-minute review

Check your knowledge

15

Wow, you sure covered a lot of ground in the last 13 chapters! Now it's time to take a look back and drill in some of the most important concepts that you learned. That'll keep it all fresh and give your brain a final workout for exam day!

Exercise

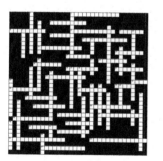

Sharpen your pencil

Pretty soon, this will be YOU!

Practice makes perfect
Practice PMP exam

16

Bet you never thought you'd make it this far! It's been
a long journey, but here you are, ready to review your knowledge and
get ready for exam day. You've put a lot of new information about project
management into your brain, and now it's time to see just how much of it
stuck. That's why we put together this 200-question PMP practice exam
for you. It looks just like the one you're going to see when you take the
real PMP exam. Now's your time to flex your mental muscle. So take a
deep breath, get ready, and let's get started.

how to use this book

Intro

In this section, we answer the burning question:
"So why DID they put that in a PMP exam prep book?"

Who is this book for?

If you can answer "yes" to all of these:

You can also use this book to help you study for the CAPM exam—a lot of the concepts are really similar.

(1) Are you a **project manager**?

(2) Do you want to **learn**, **understand**, **remember**, and *apply* important project management concepts so that you can prepare for **the PMP exam**, and learn to be a better project manager in the process?

We'll help you study for the PMP exam in a way that will definitely make it easier for you to pass.

(3) Do you prefer **stimulating dinner-party conversation** to **dry, dull, academic lectures**?

this book is for you.

Who should probably back away from this book?

If you can answer "yes" to any of these:

(1) Are you **completely new** to project management?

(To qualify to take the PMP exam, you need to show a certain number of hours of experience as a professional project manager.)

But even if you don't have quite enough hours yet, this book can still help you study now, so you can be ready when you've got those hours under your belt! Plus, the ideas will help you on your job immediately...

(2) Are you already PMP certified and looking for a *reference* book on project management?

(3) Are you **afraid to try something different**? Would you rather have a root canal than mix stripes with plaid? Do you believe that a technical book can't be serious if project management concepts are anthropomorphized?

this book is not for you.

[Note from marketing: this book is for anyone with a credit card.]

We know what you're thinking.

"How can *this* be a serious project management book?"

"What's with all the graphics?"

"Can I actually *learn* it this way?"

And we know what your *brain* is thinking.

Your brain craves novelty. It's always searching, scanning, *waiting* for something unusual. It was built that way, and it helps you stay alive.

So what does your brain do with all the routine, ordinary, normal things you encounter? Everything it *can* to stop them from interfering with the brain's *real* job—recording things that *matter*. It doesn't bother saving the boring things; they never make it past the "this is obviously not important" filter.

How does your brain *know* what's important? Suppose you're out for a day hike and a tiger jumps in front of you, what happens inside your head and body?

Neurons fire. Emotions crank up. *Chemicals surge.*

And that's how your brain knows...

This must be important! Don't forget it!

But imagine you're at home, or in a library. It's a safe, warm, tiger-free zone. You're studying. Getting ready for an exam. Or trying to learn some tough technical topic your boss thinks will take a week, 10 days at the most.

Just one problem. Your brain's trying to do you a big favor. It's trying to make sure that this *obviously* unimportant content doesn't clutter up scarce resources. Resources that are better spent storing the really *big* things. Like tigers. Like the danger of fire. Like how you should never again snowboard in shorts.

And there's no simple way to tell your brain, "Hey brain, thank you very much, but no matter how dull this book is, and how little I'm registering on the emotional Richter scale right now, I really *do* want you to keep this stuff around."

Your brain thinks THIS is important.

GREAT. ONLY 800 MORE DULL, DRY, BORING PAGES.

Your brain thinks THIS isn't worth saving.

We think of a "Head First" reader as a <u>learner</u>.

So what does it take to *learn* something? First, you have to *get* it, then make sure you don't *forget* it. It's not about pushing facts into your head. Based on the latest research in cognitive science, neurobiology, and educational psychology, *learning* takes a lot more than text on a page. We know what turns your brain on.

Some of the Head First learning principles:

Make it visual. Images are far more memorable than words alone, and make learning much more effective (up to 89% improvement in recall and transfer studies). It also makes things more understandable. **Put the words within or near the graphics** they relate to, rather than on the bottom or on another page, and learners will be up to *twice* as likely to solve problems related to the content.

You could pound in a nail with a screwdriver, but a hammer is more fit for the job.

THERE HAVE BEEN REPORTS OF BEARS CAUSING PROBLEMS FOR PEOPLE AROUND HERE LATELY. BE CAREFUL OUT HERE.

Use a conversational and personalized style. In recent studies, students performed up to 40% better on post-learning tests if the content spoke directly to the reader, using a first-person, conversational style rather than taking a formal tone. Tell stories instead of lecturing. Use casual language. Don't take yourself too seriously. Which would *you* pay more attention to: a stimulating dinner-party companion, or a lecture?

Get the learner to think more deeply. In other words, unless you actively flex your neurons, nothing much happens in your head. A reader has to be motivated, engaged, curious, and inspired to solve problems, draw conclusions, and generate new knowledge. And for that, you need challenges, exercises, and thought-provoking questions, and activities that involve both sides of the brain and multiple senses.

THIS SUCKS...THEY LOST FRANK'S LUGGAGE, AND IT GOT MY WALLET STOLEN!

Get—and keep—the reader's attention. We've all had the "I really want to learn this, but I can't stay awake past page one" experience. Your brain pays attention to things that are out of the ordinary, interesting, strange, eye-catching, unexpected. Learning a new, tough, technical topic doesn't have to be boring. Your brain will learn much more quickly if it's not.

Touch their emotions. We now know that your ability to remember something is largely dependent on its emotional content. You remember what you care about. You remember when you *feel* something. No, we're not talking heart-wrenching stories about a boy and his dog. We're talking emotions like surprise, curiosity, fun, "what the...?", and the feeling of "I rule!" that comes when you solve a puzzle, learn something everybody else thinks is hard, or realize you know something that "I'm more technical than thou" Bob from engineering *doesn't*.

Metacognition: thinking about thinking

I WONDER HOW I CAN TRICK MY BRAIN INTO REMEMBERING THIS STUFF...

If you really want to learn, and you want to learn more quickly and more deeply, pay attention to how you pay attention. Think about how you think. Learn how you learn.

Most of us did not take courses on metacognition or learning theory when we were growing up. We were *expected* to learn, but rarely *taught* to learn.

But we assume that if you're holding this book, you really want to learn about project management. And you probably don't want to spend a lot of time. And since you're going to take an exam on it, you need to *remember* what you read. And for that, you've got to *understand* it. To get the most from this book, or *any* book or learning experience, take responsibility for your brain. Your brain on *this* content.

The trick is to get your brain to see the new material you're learning as Really Important. Crucial to your well-being. As important as a tiger. Otherwise, you're in for a constant battle, with your brain doing its best to keep the new content from sticking.

So just how *DO* you get your brain to think that the stuff on the PMP exam is a hungry tiger?

There's the slow, tedious way, or the faster, more effective way. The slow way is about sheer repetition. You obviously know that you *are* able to learn and remember even the dullest of topics if you keep pounding the same thing into your brain. With enough repetition, your brain says, "This doesn't *feel* important to him, but he keeps looking at the same thing *over* and *over* and *over*, so I suppose it must be."

The faster way is to do **anything that increases brain activity,** especially different *types* of brain activity. The things on the previous page are a big part of the solution, and they're all things that have been proven to help your brain work in your favor. For example, studies show that putting words *within* the pictures they describe (as opposed to somewhere else in the page, like a caption or in the body text) causes your brain to try to makes sense of how the words and picture relate, and this causes more neurons to fire. More neurons firing = more chances for your brain to *get* that this is something worth paying attention to, and possibly recording.

A conversational style helps because people tend to pay more attention when they perceive that they're in a conversation, since they're expected to follow along and hold up their end. The amazing thing is, your brain doesn't necessarily *care* that the "conversation" is between you and a book! On the other hand, if the writing style is formal and dry, your brain perceives it the same way you experience being lectured to while sitting in a roomful of passive attendees. No need to stay awake.

But pictures and conversational style are just the beginning.

Here's what WE did:

We used ***pictures***, because your brain is tuned for visuals, not text. As far as your brain's concerned, a picture really *is* worth a thousand words. And when text and pictures work together, we embedded the text *in* the pictures because your brain works more effectively when the text is *within* the thing the text refers to, as opposed to in a caption or buried in the text somewhere.

We used ***redundancy***, saying the same thing in *different* ways and with different media types, and *multiple senses*, to increase the chance that the content gets coded into more than one area of your brain.

We used concepts and pictures in ***unexpected*** ways because your brain is tuned for novelty, and we used pictures and ideas with at least *some* ***emotional*** content, because your brain is tuned to pay attention to the biochemistry of emotions. That which causes you to *feel* something is more likely to be remembered, even if that feeling is nothing more than a little ***humor***, ***surprise***, or ***interest.***

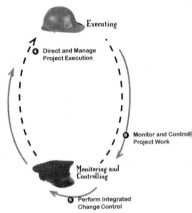

We used a personalized, ***conversational style***, because your brain is tuned to pay more attention when it believes you're in a conversation than if it thinks you're passively listening to a presentation. Your brain does this even when you're *reading*.

We included more than 80 ***activities***, because your brain is tuned to learn and remember more when you ***do*** things than when you *read* about things. And we made the exercises challenging-yet-doable, because that's what most people prefer.

We used ***multiple learning styles***, because *you* might prefer step-by-step procedures, while someone else wants to understand the big picture first, and someone else just wants to see an example. But regardless of your own learning preference, *everyone* benefits from seeing the same content represented in multiple ways.

**BULLET POINTS:
AIMING FOR THE EXAM**

We include content for ***both sides of your brain***, because the more of your brain you engage, the more likely you are to learn and remember, and the longer you can stay focused. Since working one side of the brain often means giving the other side a chance to rest, you can be more productive at learning for a longer period of time.

And we included ***stories*** and exercises that present ***more than one point of view,*** because your brain is tuned to learn more deeply when it's forced to make evaluations and judgments.

Fireside Chats

We included ***challenges***, with exercises, and by asking ***questions*** that don't always have a straight answer, because your brain is tuned to learn and remember when it has to *work* at something. Think about it—you can't get your *body* in shape just by *watching* people at the gym. But we did our best to make sure that when you're working hard, it's on the *right* things. That ***you're not spending one extra dendrite*** processing a hard-to-understand example, or parsing difficult, jargon-laden, or overly terse text.

We used ***people***. In stories, examples, pictures, and so on, because, well, because *you're* a person. And your brain pays more attention to *people* than it does to *things*.

Here's what YOU can do to bend your brain into submission

So, we did our part. The rest is up to you. These tips are a starting point; listen to your brain and figure out what works for you and what doesn't. Try new things.

Cut this out and stick it on your refrigerator.

---

① Slow down. The more you understand, the less you have to memorize.

Don't just *read*. Stop and think. When the book asks you a question, don't just skip to the answer. Imagine that someone really *is* asking the question. The more deeply you force your brain to think, the better chance you have of learning and remembering.

② Do the exercises. Write your own notes.

We put them in, but if we did them for you, that would be like having someone else do your workouts for you. And don't just *look* at the exercises. **Use a pencil.** There's plenty of evidence that physical activity *while* learning can increase the learning.

③ Read the "There are No Dumb Questions"

That means all of them. They're not optional sidebars—*they're part of the core content!* Don't skip them.

④ Make this the last thing you read before bed. Or at least the last challenging thing.

Part of the learning (especially the transfer to long-term memory) happens *after* you put the book down. Your brain needs time on its own, to do more processing. If you put in something new during that processing time, some of what you just learned will be lost.

⑤ Drink water. Lots of it.

Your brain works best in a nice bath of fluid. Dehydration (which can happen before you ever feel thirsty) decreases cognitive function.

⑥ Talk about it. Out loud.

Speaking activates a different part of the brain. If you're trying to understand something, or increase your chance of remembering it later, say it out loud. Better still, try to explain it out loud to someone else. You'll learn more quickly, and you might uncover ideas you hadn't known were there when you were reading about it.

⑦ Listen to your brain.

Pay attention to whether your brain is getting overloaded. If you find yourself starting to skim the surface or forget what you just read, it's time for a break. Once you go past a certain point, you won't learn faster by trying to shove more in, and you might even hurt the process.

⑧ Feel something!

Your brain needs to know that this *matters*. Get involved with the stories. Make up your own captions for the photos. Groaning over a bad joke is *still* better than feeling nothing at all.

⑨ Create something!

Apply this to your daily work; use what you are learning to make decisions on your projects. Just do something to get some experience beyond the exercises and activities in this book. All you need is a pencil and a problem to solve…a problem that might benefit from using the tools and techniques you're studying for the exam.

Read me

This is a learning experience, not a reference book. We deliberately stripped out everything that might get in the way of learning whatever it is we're working on at that point in the book—although we didn't take anything out that you might see on the PMP exam. And the first time through, you need to begin at the beginning, because the book makes assumptions about what you've already seen and learned.

The chapters are ordered the same way as the *PMBOK Guide*

We did this because it makes sense…. The PMP exam focuses on your understanding of the *PMBOK Guide* and the inputs, outputs, tools, and techniques it references. It's a good idea for you to understand the material the way the test organizes it. If you are cross-referencing this book with the *PMBOK Guide*, it will really help that the structure has been pretty much maintained throughout this book, too.

We encourage you to use the *PMBOK Guide* with this book.

This book talks about the practical applications of a lot of the ideas in the *PMBOK Guide*, but you should have a pretty good idea of how the guide talks about the material, too. There's some information that's on the test that isn't in the guide, so we haven't limited this book to a retread of what's in the *PMBOK Guide* at all. But it's a great reference, and you should be cross-referencing the two books as you go. That will help you understand all of the terminology better and make sure that there are no surprises on exam day.

The activities are NOT optional.

The exercises and activities are not add-ons; they're part of the core content of the book. Some of them are to help with memory, some are for understanding, and some will help you apply what you've learned. ***Don't skip the exercises.*** Even crossword puzzles are important—they'll help get concepts into your brain the way you'll see them on the PMP exam. But more importantly, they're good for giving your brain a chance to think about the words and terms you've been learning in a different context.

The redundancy is intentional and important.

One distinct difference in a Head First book is that we want you to *really* get it. And we want you to finish the book remembering what you've learned. Most reference books don't have retention and recall as a goal, but this book is about *learning*, so you'll see some of the same concepts come up more than once.

The Brain Power exercises don't have answers.

For some of them, there is no right answer, and for others, part of the learning experience of the Brain Power activities is for you to decide if and when your answers are right. In some of the Brain Power exercises, you will find hints to point you in the right direction.

We want you to get involved.

Part of being a PMP-certified project manager is getting involved in the community and helping others out. An easy way to start doing this is to head over to the Head First forum, where you'll be able to submit your own Head Libs and see what other people have come up with, too:

http://forums.oreilly.com/forum/73-head-first-pmp/

Check out our free PMP exam simulator online.

The last chapter of this book is a full-length sample PMP exam. But we've also created an exam simulator online so you can see what the test will be like on exam day. It's free and easy to use. By the time you reach the end of this book, you'll have put a lot of new knowledge about project management into your brain, and it'll be time to see just how much of it stuck. The simulator, like the exam in the back of the book, was developed using the official Project Management Professional Exam Specification and has 100% coverage of the exam objectives. Check it out here:

http://www.headfirstlabs.com/PMP/pmp_exam/v1/quiz.html

The technical review team

Lisa Kellner

Jen Poisson

Tequila (the fluffy dog) provided critical input, and this book would not have been possible without her valuable and thorough review.

Joe Pighetti

Technical reviewers:

For the third edition, we had a whole new batch of amazing tech reviewers. They did a great job, and we're really grateful for their incredible contribution.

Jennifer Poisson has more than nine years in technical project management. She is currently the director of production operations at Disney Online. In her spare time, she blows her retirement savings traveling the country in expensive shoes and attending fabulous concerts, while in constant pursuit of maintaining a well-balanced raw diet.

Joe Pighetti has over 11 years of Avionics development experience and is currently an electronic design project manager supporting Boeing 787 Dreamliner products supplied by GE Aviation Systems. He has a master's degree in engineering management from Western Michigan University and a bachelor's degree in electrical engineering from Grand Valley State University. In his spare time, Joe enjoys spending time with the most beautiful girl in the world and their three amazing children as well as helping people to break down walls and find freedom through music.

And, as always, we were lucky to have **Lisa Kellner** return to our tech review team. Lisa was awesome, as usual. Thanks so much, guys!

Acknowledgments

Our editor:

We want to thank our editor, **Courtney Nash**, for editing this book. Thanks!

Courtney Nash

The O'Reilly team:

There are so many people at O'Reilly we want to thank that we hope we don't forget anyone. Special thanks to production editor **Melanie Yarbrough**, indexer **Bob Pfahler**, **Rachel Monaghan** for her sharp proofread, **Ron Bilodeau** for volunteering his time and preflighting expertise, and for offering one last sanity check—all of whom helped get this book from production to press in record time. And as always, we love **Mary Treseler**, and can't wait to work with her again! And a big shout out to our other friends and editors, **Andy Oram**, **Mike Hendrickson**, **Laurie Petrycki**, **Tim O'Reilly**, and **Sanders Kleinfeld**. And if you're reading this book right now, then you can thank the greatest publicity team in the industry: **Marsee Henon**, **Sara Peyton**, and the rest of the folks in Sebastopol.

Safari Books Online

Safari Books Online is an on-demand digital library that delivers expert in both book and video form from the world's leading authors in technology and business.

Technology professionals, software developers, web designers, and business and creative professionals use Safari Books Online as their primary resource for research, problem solving, learning, and certification training.

Safari Books Online offers a range of and pricing programs for organizations, government, and individuals. Subscribers have access to thousands of books, training videos, and prepublication manuscripts in one fully searchable database from publishers like O'Reilly Media, Prentice Hall Professional, Addison-Wesley Professional, Microsoft Press, Sams, Que, Peachpit Press, Focal Press, Cisco Press, John Wiley & Sons, Syngress, Morgan Kaufmann, IBM Redbooks, Packt, Adobe Press, FT Press, Apress, Manning, New Riders, McGraw-Hill, Jones & Bartlett, Course Technology, and dozens more. For more information about Safari Books Online, please visit us online.

1 Introduction

Why get certified?

Tired of facing the same old problems? If you've worked on a lot of projects, you know that you face the same problems, over and over again. It's time to learn some common solutions to those problems. There's a whole lot that project managers have learned over the years, and passing the PMP exam is your ticket to putting that wisdom into practice. Get ready to change the way you manage your projects forever.

Do these problems seem familiar?

Kate's boss promised a delivery date that she couldn't possibly meet.

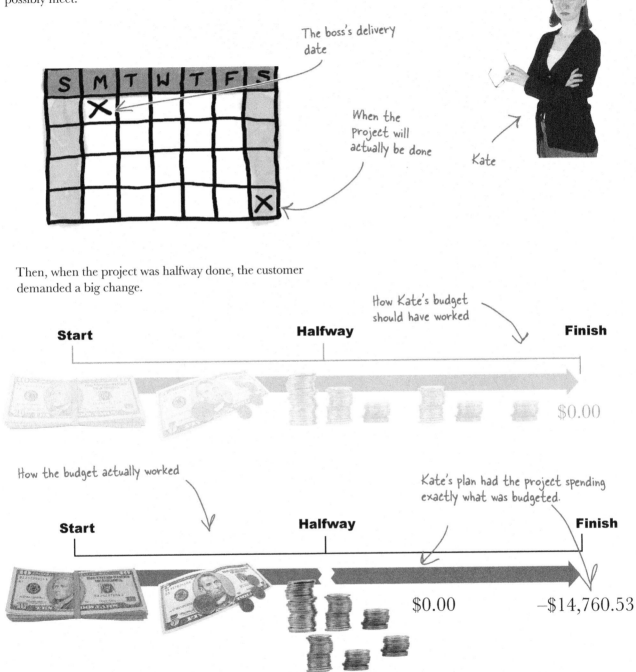

The boss's delivery date

When the project will actually be done

Kate

Then, when the project was halfway done, the customer demanded a big change.

How Kate's budget should have worked

Start **Halfway** **Finish**

$0.00

How the budget actually worked

Kate's plan had the project spending exactly what was budgeted.

Start **Halfway** **Finish**

$0.00 −$14,760.53

Then, just as the project was about to be completed, someone noticed a couple typos, and 10,000 leaflets had to be reprinted.

Two simple typos, and now the project is over budget and dissatisfying to the customer.

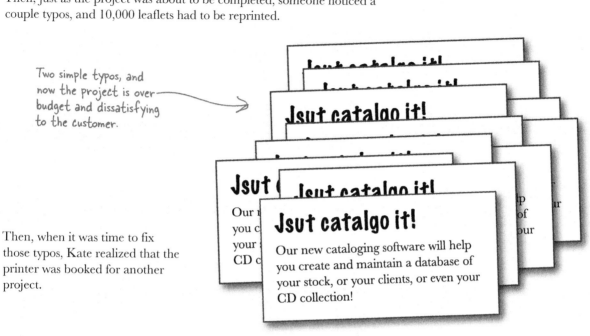

Jsut catalgo it!

Our new cataloging software will help you create and maintain a database of your stock, or your clients, or even your CD collection!

Then, when it was time to fix those typos, Kate realized that the printer was booked for another project.

The short timeframe didn't give Kate enough time to plan for risks.

And even though she knew there was a pretty good chance that someone else might need the printer, she didn't have time to come up with a backup plan.

Now the project's going to be late and over budget, and the customer won't be happy.

Projects don't have to be this way

It may seem like all projects have these types of problems, but there are proven solutions to them…and someone else has already done a lot of the work for you! Realizing that all projects have common problems with solutions, a team of experts put together the *PMBOK Guide* to pass those solutions on to you.

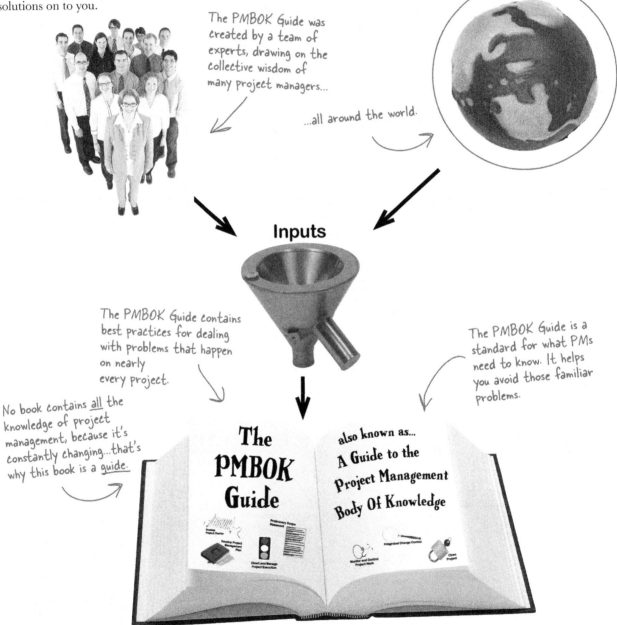

The PMBOK Guide was created by a team of experts, drawing on the collective wisdom of many project managers...

...all around the world.

Inputs

The PMBOK Guide contains best practices for dealing with problems that happen on nearly every project.

The PMBOK Guide is a standard for what PMs need to know. It helps you avoid those familiar problems.

No book contains <u>all</u> the knowledge of project management, because it's constantly changing...that's why this book is a <u>guide</u>.

The
**PMBOK
Guide**

also known as...
**A Guide to the
Project Management
Body Of Knowledge**

Your problems...already solved

Every project eventually runs into the same kinds of issues. But a project manager with good training can spot them early, and quickly figure out the best solutions. The *PMBOK Guide* will help you:

> ✓ Learn from past projects that have run into similar problems to avoid running into them again.

> ✓ Learn a common vocabulary for project management that is used by PMs around the world.

> ✓ Plan and execute your projects to avoid common pitfalls.

Common pitfalls: better avoid these.

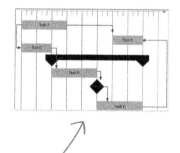

The PMBOK Guide has great ideas on how to estimate your tasks and put them in the right sequence to get your projects done as quickly and efficiently as possible.

It outlines techniques for planning and tracking your costs.

It helps you learn how to plan for and protect against defects in your project.

What you need to be a good project manager

The *PMBOK Guide* is full of practical tools that can help you manage your projects better. But all of that doesn't mean much if you don't have the three core characteristics of a successful project manager. You've got to pay attention to all three if you want to make your project a success. This is what you'll need if you're going to take on the role of project manager:

 1

Knowledge

If you pay attention to what's going on in the field of project management, you can learn from everyone's successes and mistakes so that you can be better at your job.

This means knowing all of the tools and techniques in the PMBOK Guide and how and when to use them.

2

Performance

It's not enough to know what you need to do—you've got to deliver, too. This one is all about keeping your nose to the grindstone and doing good work.

You and your team will have to work hard to deliver a successful project, too.

3

Personal skills

Since you're managing people, you've got to pay attention to what motivates them and what makes things harder on them. Your job as a PM is to make personal connections with your team and help keep everybody on the right track.

As a PM, you've got to lead your team through the project lifecycle, so you need to be skilled at managing people if you're going to be successful, even if they don't report to you directly.

Exercise

Not paying attention to these characteristics is sure to give your project problems. Which of the characteristics of a successful project manager was neglected in the failed projects listed below? Sometimes, more than one will apply; just pick the one that makes the most sense to you.

The project was delivered early, but it didn't have all of the features that the customers asked for. The VP had suggested a new requirements gathering technique, but the PM shot it down because he'd never heard of it.

Neglected characteristic:

....

The project was late because the team couldn't meet the company's standards for productivity. They were always coming into work late and leaving early and taking long lunches. It seemed like the project manager just didn't think the project was important.

Neglected characteristic:

....

The project manager thought his job was to meet the deadline above all else. So he demanded that the product be released on the date it was due, regardless of quality. The team wanted to create a high-quality product, and they fought with the PM throughout the project to try to get him to change his mind. In the end, the team washed their hands of the product after it was released and refused to support it.

Neglected characteristic:

....

The project team had so many conflicts about the project that they couldn't work together. They made decisions that undercut one another, and in the end they couldn't deliver anything at all.

Neglected characteristic:

....

The project was late because the team cut corners that led to sloppy work, and they had to go back and fix all of their mistakes.

Neglected characteristic:

....

The project manager refused to learn to use the scheduling software and templates the company had bought for the team. Instead, he kept track of the schedule in his head and on his whiteboard. Near the end of the project, he realized that he'd forgotten about some important tasks, and his ship date slipped by two months.

Neglected characteristic:

....

Exercise Solution

Not paying attention to these characteristics is sure to give your project problems. Which of the characteristics of a successful project manager was neglected in the failed projects listed below? Sometimes, more than one will apply; just pick the one that makes the most sense to you.

Your project was delivered early, but it didn't have all of the features that the customers asked for. The VP had suggested a new requirements gathering technique, but the PM shot it down because he'd never heard of it

Neglected characteristic:

Knowledge

The project was late because the team couldn't meet the company's standards for productivity. They were always coming into work late and leaving early and taking long lunches. It seemed like the project manager just didn't think the project was important.

Neglected characteristic:

Performance

This could be a knowledge issue too, because the manager didn't learn enough about the stakeholders' expectations.

The project manager thought his job was to meet the deadline above all else. So he demanded that the product be released on the date it was due, regardless of quality. The team wanted to create a high-quality product and they fought with the PM throughout the project to try to get him to change his mind. In the end, the team washed their hands of the product after it was released and refused to support it.

Neglected characteristic:

Personal skills

This could also be a knowledge issue because the manager didn't learn the scheduling software or templates. The differences aren't always 100% clear-cut.

The project team had so many conflicts about the project that they couldn't work together. They made decisions that undercut one another, and in the end they couldn't deliver anything at all.

Neglected characteristic:

Personal skills

The project was late because the team cut corners that led to sloppy work, and they had to go back and fix all of their mistakes.

Neglected characteristic:

Performance

The project manager refused to learn to use the scheduling software and templates the company had bought for the team. Instead, he kept track of the schedule in his head and on his whiteboard. Near the end of the project, he realized that he'd forgotten about some important tasks, and his ship date slipped by two months.

Neglected characteristic:

Performance

there are no
Dumb Questions

Q: How can the *PMBOK Guide* claim to be the entire body of knowledge for project management?

A: Actually, it doesn't claim that at all. That's why the *PMBOK Guide* is called "A **Guide** to the Project Management Body of Knowledge." It's a reference book that organizes a lot of information about how project managers do their jobs—but it doesn't claim to have all the information itself. Instead, it provides you with a framework for managing projects and tells you what information you need to know.

A lot of people are surprised to find out that there are a bunch of things on the PMP exam that are never explicitly mentioned in the *PMBOK Guide*. (Don't worry: we'll cover that stuff in the rest of this book.) There's a whole lot of information that modern project managers should know about risk management and time management and cost and quality...and you're expected to learn more about the knowledge areas as you move forward in your career. That's why you should never limit your study to just what's in the *PMBOK Guide*. It's meant only as a guide to all of the knowledge areas that project managers use on the job.

Q: What if I don't do all of this stuff in my job?

A: The *PMBOK Guide* isn't necessarily meant to be followed like a recipe for every project. It's a broad collection of many tools and processes that are used across the project management profession, and project managers have a lot of discretion about how they run their projects. So you shouldn't throw out all that you're doing at work and replace it with every single one of the tools in this book immediately. But you'll notice as you go that some of the tools you're learning about will solve problems

for you on the job. When you find places where these tools can help, you really should start using them. Seriously, it's the best way to learn. You might find that your projects go better after you start using a new concept that you learn while you study.

Q: I've heard that there are a whole bunch of formulas you have to memorize for the PMP exam. Will I have to do that?

A: Yes, but it won't be that bad. The formulas are actually really useful. They help you understand how your project is doing so you can make better decisions. When you read about them later in the book, you'll focus on how to use them and why. Once you know that, it's not about memorizing a bunch of useless junk. The formulas will actually make sense, and you'll find them intuitive and helpful in your day-to-day work.

Q: Aren't certification exams just an excuse that consultants use so that they can charge their clients more money?

A: Some consultants charge more money because they are certified, but that's not the only reason to get certified. The best reason to get PMP Certification is because it helps you understand all of the project management concepts available to help you do your job better. If you learn these tools and apply them to your job, you will be a better project manager. And hey, if it turns out you can make more money too, that's great.

What's more, it's worth keeping in mind that for a project manager, being PMP-certified is a requirement for a large amount of contracting work, especially in government, and it's increasingly seen in job postings of all kinds. Some employers won't even interview project managers who don't have a PMP Certification!

Q: Doing all of the stuff in the *PMBOK Guide* seems like it will take a long time. How much of this really applies to me?

A: That's a great question. You might find that there are documents that are mentioned in the *PMBOK Guide* that you're not used to writing or creating for your projects, and some planning steps that you've never taken before. That's because the *PMBOK Guide* is a framework, not a recipe for a successful project.

When you get your certification, it means that you have a solid understanding of all of the tools and techniques that are typically used by project managers to plan projects, track them, and deal with problems that come up along the way. It doesn't mean that you follow the exact same recipe for project success every time you lead a project.

Q: But I work for a company that always runs projects on really tight deadlines. You can't honestly expect me to write a bunch of project documents and use all of these formulas for my projects.

A: One of the useful things that you'll learn in Chapter 8 of this book is that sometimes the processes that seem like a lot of work up front actually end up saving you time in the end.

If you find a problem in a two-hour planning meeting that would've cost you two weeks to fix, then that two hours you spent planning actually saved your project two weeks of time. A lot of the planning and documentation that you'll be tested on is there to help you head off problems before they derail your project. So in the end, doing all of that work up front and writing it all down can actually make your project go faster and be cheaper than not doing it would've been!

The PMBOK Guide is just a guide, but if you understand all the material in it, then you'll ultimately be a better project manager.

Understand your company's big picture

Your project is an important part of the work your company is doing, but you need to understand how it fits in to the higher-level strategy your company is executing, too. That's where programs and portfolios come in.

Portfolios might include programs and projects.

Programs are groups of projects that should be managed together.

Projects have a beginning and an end and produce something specific.

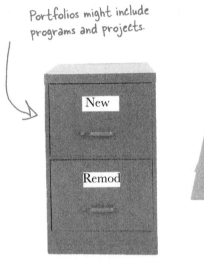

New Construction:
Residential Homes

Portfolio

A portfolio is a group of projects or programs that are linked together by a business goal. If an architecture firm was venturing into remodelling existing buildings as well as designing new ones, it might split its efforts into separate New Construction and Remodelling portfolios, since the goals for each are quite different.

Program

A program is a group of projects that are closely linked, to the point where managing them together provides some benefit. The firm knows from experience that creating huge skyscrapers is dramatically different than building residential homes, so residential home construction would be its own separate program.

Project

A project is any work that produces a specific result and is temporary. Projects always have a beginning and an end. Building a house is a classic example of a project. Projects can be part of programs or portfolios, but portfolios and programs can't be part of a project.

Projects in a program are often dependent on each other. Program management focuses on these interdependencies.

Your project has value

Think about the projects you've worked on in your career. Each one of them did something beneficial for your company. You might've created a product and sold it to customers to make money directly. You might've made someone's job easier by automating work that would've taken time and effort to do. No matter how you count the benefit you created when you completed your project, that benefit is the real reason that your company decided to do the work in the first place. That benefit has an impact on the overall **business value** of the company you work for. Sometimes it can be easy to spend so much time dealing with your project's issues that you lose track of the goals you set out to achieve. It's important to think about the value of your project with every decision you make.

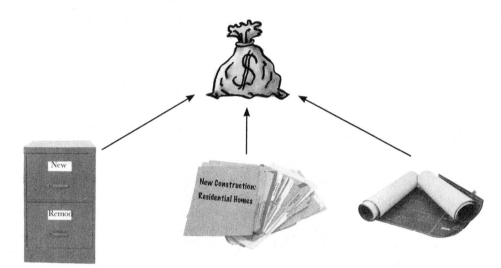

Business value is the sum of all of the things your company is made of, from desks and chairs to people and the intellectual property they produce.

Portfolio
Portfolio managers divide up the projects, programs, and operations your company is doing so that they align with business goals. That way, they can be sure that projects get the most out of managing their timelines, budgets, and resource commitments so that the company's goals are met.

Program
Program managers focus on the places where projects depend on each other and coordinate activities to make sure that the work gets done in the most direct way possible.

Project
Project managers keep the team focused on the business value the project is providing. By constantly helping each team member to understand how his or her work impacts the project's value, project managers make sure that everyone on the team makes the best decisions to keep the project on track.

Portfolios, programs, and projects have a lot in common

We've talked about the differences between portfolio management, program management, and project management, but there are a lot of similarities between them too.

Professions with proven processes

Portfolio managers and program managers have a set of proven processes, tools, and techniques that have been used to manage many successful programs and portfolios. Like the PMP, the Project Management Institute offers certifications in Portfolio (PfMP) and Program Management (PgMP) too.

Business value

Portfolio managers prioritize work to meet a company's strategic goals. Managing a program is all about keeping track of resources and other constraints affecting groups of projects so that all of those projects can achieve some shared benefit. Projects, on the other hand, are about managing the work to achieve some result. A portfolio's strategic goals, a program's shared benefit, and the result of that project add business value to your company.

Deal with constraints

Portfolio managers need to prioritize work in environments with limited resources, budgets, risk tolerance, and many other constraints that set up the environment for their portfolio's success. Program managers need to manage groups of projects that are being produced by the same resource pool or from the same budget. Project managers often have predefined schedules, resource constraints, and scope requirements to manage. All three need to use the processes, tools, and techniques that have worked for other managers to balance all of the constraints in their business environments.

Portfolios, programs, and projects all use charters

All of the work you do to initiate, plan, execute, control, and close your project helps your program and portfolio managers understand how your project is doing and keep it on track. Following all of the processes in the *PMBOK Guide* will ensure that the programs and portfolios of which your project is a part always know how you're doing and what you'll accomplish. While there are many differences in the documents that are used in portfolio and program management, all three use a **charter** to define their objectives.

Charter

Portfolios, programs, and projects all use a charter to define their goals and initiate work. A charter lists any known constraints and goals and gives the manager authority to get the work started.

Portfolio charter

A portfolio charter will lay out the strategic benefits that a portfolio is going to accomplish. It will list all of the programs and projects included in the portfolio.

Program charter

A program charter will define the shared benefit that the program is achieving as well as the list of projects it includes.

Project charter

A project charter gives a project description, summary schedule, and business case, and assigns a project manager.

Relax

> **We'll talk about the project charter in depth in Chapter 4.**
>
> While the charter is used in projects, programs, and portfolios, you really only need to focus on the project charter for the exam. We'll spend a lot of time on the project charter later in the book.

Projects, Programs, and Portfolios Way Up Close

Let's take a look at a charter for a portfolio, a program, and a project for a software company called Ranch Hand Games to get a better understanding of how they break down.

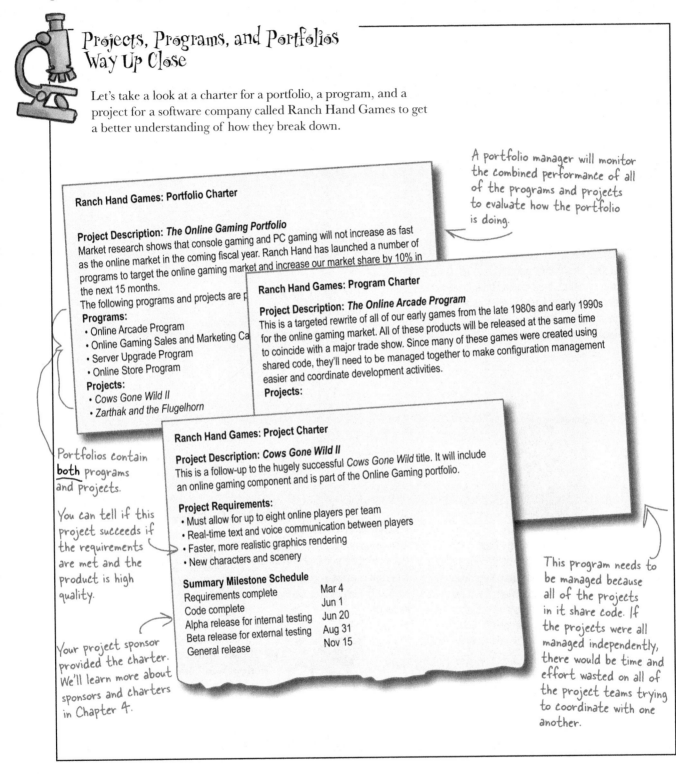

A portfolio manager will monitor the combined performance of all of the programs and projects to evaluate how the portfolio is doing.

Ranch Hand Games: Portfolio Charter

Project Description: *The Online Gaming Portfolio*
Market research shows that console gaming and PC gaming will not increase as fast as the online market in the coming fiscal year. Ranch Hand has launched a number of programs to target the online gaming market and increase our market share by 10% in the next 15 months.
The following programs and projects are p

Programs:
• Online Arcade Program
• Online Gaming Sales and Marketing Ca
• Server Upgrade Program
• Online Store Program

Projects:
• *Cows Gone Wild II*
• *Zarthak and the Flugelhorn*

Ranch Hand Games: Program Charter

Project Description: *The Online Arcade Program*
This is a targeted rewrite of all of our early games from the late 1980s and early 1990s for the online gaming market. All of these products will be released at the same time to coincide with a major trade show. Since many of these games were created using shared code, they'll need to be managed together to make configuration management easier and coordinate development activities.

Projects:

Ranch Hand Games: Project Charter

Project Description: *Cows Gone Wild II*
This is a follow-up to the hugely successful *Cows Gone Wild* title. It will include an online gaming component and is part of the Online Gaming portfolio.

Project Requirements:
• Must allow for up to eight online players per team
• Real-time text and voice communication between players
• Faster, more realistic graphics rendering
• New characters and scenery

Summary Milestone Schedule

Requirements complete	Mar 4
Code complete	Jun 1
Alpha release for internal testing	Jun 20
Beta release for external testing	Aug 31
General release	Nov 15

*Portfolios contain **both** programs and projects.*

You can tell if this project succeeds if the requirements are met and the product is high quality.

Your project sponsor provided the charter. We'll learn more about sponsors and charters in Chapter 4.

This program needs to be managed because all of the projects in it share code. If the projects were all managed independently, there would be time and effort wasted on all of the project teams trying to coordinate with one another.

Exercise

You'll need to know the difference between a portfolio, a program, and a project on the exam. Which one of those does each story below describe?

A consulting company wanted to increase the amount of billable time for each consultant, so it started several company-wide programs to help consultants to get more productivity out of each year.

....

A company wanted to switch from a paper-based Human Resources group to a software-based one. It spent some time looking into the best software packages for the job, and decided to manage all of the HR functions together since it needed the same people to help with all of the work.

....

A software game company wanted to build up its online presence. It started several marketing and sales initiatives, created some new games, and rewrote some old ones in order to reach more gamers online.

....

A university wanted to build admissions websites for all of its departments. It realized that all of the sites would be feeding into the same registration interface and decided to manage all of them together in order to save time.

....

A company wanted to build a better reporting interface so that it could have more accurate data on year-end goals.

....

A construction company bid on several parking garage projects at the same time. It won one of the bids, and built the garage a month under schedule and $5,000 under budget.

....

Exercise Solution

You'll need to know the difference between a portfolio, a program, and a project on the exam. Which one of those does each story below describe?

A consulting company wanted to increase the amount of billable time for each consultant, so it started several company-wide programs to help consultants to get more productivity out of each year.

Portfolio

A company wanted to switch from a paper-based Human Resources group to a software-based one. It spent some time looking into the best software packages for the job, and decided to manage all of the HR functions together since it needed the same people to help with all of the work.

Program

A software game company wanted to build up its online presence. It started several marketing and sales initiatives, created some new games, and rewrote some old ones in order to reach more gamers online.

Portfolio

A university wanted to build admissions websites for all of its departments. It realized that all of the sites would be feeding into the same registration interface and decided to manage all of them together in order to save time.

Program

A company wanted to build a better reporting interface so that it could have more accurate data on year-end goals.

Project

A construction company bid on several parking garage projects at the same time. It won one of the bids, and built the garage a month under schedule and $5,000 under budget.

Project

What a project I**S**...

Temporary

Projects always have a start and a finish. They start when you decide what you are going to do, and they end when you've created the product or service you set out to create. Sometimes they end because you decide to stop doing the project. But they are never ongoing.

> Operations are ongoing. If you're building cars on an assembly line, that's an operation. If you're designing and building a prototype of a specific car model, that's a project.

Creating a unique result

When you create the product of your project, it is measurable. If you start a project to create a piece of software or build a building, you can tell that software or that building from any other one that has been produced.

> You might also see the word "process" instead of "operation." A team might run a project to build software, but the company might have an ongoing process for keeping the servers that run the software from going down. In fact, the group that keeps those servers running is often called "IT Operations." Get it?

Progressively elaborated

You learn more and more about a project as it goes on. When you start, you have goals and a plan, but there is always new information to deal with as your project progresses, and you'll always have to make decisions to keep it on track. While you do your best to plan for everything that will happen, you know that you will keep learning more about your project as you go.

... and what a project is N**OT**

Projects are N**OT**: always strategic or critical

Projects are N**OT**: ongoing operations (or processes)

Projects are N**OT**: always successful

Sharpen your pencil

Which of these scenarios are operations, and which are projects?

1. Building an extension on a house

☐ Operation   ☐ Project

2. Shelving books at the library

☐ Operation   ☐ Project

3. Baking a wedding cake

☐ Operation   ☐ Project

4. Stapling programs for a play

☐ Operation   ☐ Project

5. Watering your plants twice a week

☐ Operation   ☐ Project

6. Walking the dog every day

☐ Operation   ☐ Project

7. Knitting a scarf

☐ Operation   ☐ Project

8. Making a birdhouse

☐ Operation   ☐ Project

9. Changing your air filters every six months

☐ Operation   ☐ Project

10. Running an assembly line in a toy factory

☐ Operation   ☐ Project

11. Organizing a large conference

☐ Operation   ☐ Project

12. Going to the gym three times a week

☐ Operation   ☐ Project

➞ Answers on page 20.

BULLET POINTS: AIMING FOR THE EXAM

- **Knowledge**, **performance**, and **personal skills** are the three areas that project managers focus on to get better at their jobs.

- A **project charter** is a document that describes a project's requirements and high-level schedule, assigns a project manager, and authorizes the project.

- A **program** is a collection of projects that should be managed together in order to achieve a specific goal or benefit to the company.

- A **portfolio** is a collection of projects or programs.

- A **project** gathers a team together to do work that's **temporary**, creates a **unique result**, and is **progressively elaborated**.

- An **operation** (or **process**) is work that's done in a way that's repeatable and ongoing, but is not a project.

A day in the life of a project manager

You probably already know a lot of what a project manager does: gets a project from concept to completed product. Usually a project manager works with a team of people to get the work done. And PMs don't usually know a lot about the project when they start. When you think about it, you can categorize pretty much everything a project manager does every day into three categories.

Gather product requirements

Being a project manager almost always means figuring out what you're going to build. It's one of the first things you do when you start to plan the project! But as you go, you are always learning more and more. Sometimes that can mean changes to your product, while other times it's just more detail on what you already knew.

One of the most important stakeholders is the sponsor. That's the person who provides financial and political support for the project.

Manage stakeholder expectations

There are a lot of people involved in making most projects happen: the team that actually does the work, the people who pay for it, everybody who will use the product when you're done, and everybody who might be impacted by the project along the way. Those people are called your **stakeholders**. And a big part of the PM's job is communicating with everybody and making sure their needs are met.

You'll need to use your interpersonal skills to keep everyone on the same page.

Deal with project constraints

Sometimes there will be **constraints** on the project that you'll need to deal with. You might start a project and be told that it can't cost more than $200,000. Or it absolutely MUST be done by the trade show in May. Or you can do it only if you can get one specific programmer to do the work. Or there's a good chance that a competitor will beat you to it if you don't plan it well. It's constraints like these that make the job more challenging, but it's all in a day's work for a project manager.

Even though you're constantly gathering requirements, managing stakeholders, and working within constraints, different situations can call for different tools for dealing with all of those challenges. When you think about it, all of the tools and techniques that are discussed in the *PMBOK Guide* are there to help you do those three things at different points in your project's lifecycle. That's why the *PMBOK Guide* divides up the work you do on a project into the five process groups. The groups help you organize all of the work you do as your project progresses and keep your role in the project straight.

You'll learn all about the process groups in Chapter 3!

Sharpen your pencil
Solution

Which of these scenarios are operations, and which are projects?

1. Building an extension on a house

   ☐ Operation   ☒ Project

2. Shelving books at the library

   ☒ Operation   ☐ Project

3. Baking a wedding cake

   ☐ Operation   ☒ Project

4. Stapling programs for a play

   ☐ Operation   ☒ Project

5. Watering your plants twice a week

   ☒ Operation   ☐ Project

6. Walking the dog every day

   ☒ Operation   ☐ Project

7. Knitting a scarf

   ☐ Operation   ☒ Project

8. Making a birdhouse

   ☐ Operation   ☒ Project

9. Changing your air filters every six months

   ☒ Operation   ☐ Project

10. Running an assembly line in a toy factory

    ☒ Operation   ☐ Project

11. Organizing a large conference

    ☐ Operation   ☒ Project

12. Going to the gym three times a week

    ☒ Operation   ☐ Project

there are no
Dumb Questions

Q: Do project constraints just mean restrictions on time and cost?

A: No. A project constraint is any limitation that's placed on your project before you start doing the work. It's true that project managers are really familiar with time and cost constraints, because those are really common. But there are lots of other kinds of constraints, too.

Here's an example. Let's say that some of your team members won't be available for three weeks because they have to attend a mandatory training session. That's called a **resource constraint**, because some of your project resources (people you need) are restricted.

There are lots of other kinds of constraints, too: risk constraints, scope constraints, and quality constraints.

Q: Wait a minute—a quality constraint? Shouldn't I always run my project to build high-quality products?

A: Of course. But quality is more important for some projects than it is for others, and as a project manager, you need to be realistic about it.

If you're running a project to build a playground, quality is important. You don't want to build unsafe playground equipment, because children could get hurt. Does that mean that you spend the highest possible portion of your budget on quality? Take a minute and think about how you'd approach quality for that project, as compared to, say, a project to build a heart monitor for a medical device company. It's likely that quality is a much more important constraint for the heart monitor than it is for the playground.

How project managers run great projects

There are plenty of ways that you can run a project; people have been running projects for about as long as civilization has been around. But some project managers run their projects really effectively, while others consistently come in late, over budget, and with poor quality. So what makes the difference between a great project and one that faces challenges?

That's exactly the question that the folks at the Project Management Institute asked when they started putting together their *Guide to the Project Management Body of Knowledge*. They surveyed thousands of project managers and analyzed tens of thousands of successful and unsuccessful projects to come up with a structured way of thinking about how to effectively run a project.

One goal of the *PMBOK Guide* is to give you a repeatable way to run your projects. It does so by breaking the work down into 47 processes that describe different, specific kinds of work that project managers do. To help you understand how those processes fit together, they came up with two different ways to think about them. Each process falls into one of the five process groups, which tell you the sequence that the processes are performed on a project. But the *PMBOK Guide* is also a tool for organizing knowledge about project management, so each process also falls into one of 10 knowledge areas. The *PMBOK Guide* is organized around these knowledge areas…and so is this book!

The PMBOK Guide describes 47 processes your project will go through from start to finish.

There are also 10 knowledge areas that help organize the processes to make them easier to learn and understand.

The PMBOK Guide

also known as...
A Guide to the Project Management Body Of Knowledge

Head First PMP has one chapter per knowledge area…and so does the PMBOK Guide.

It has five process groups that show you the order that the processes happen on a project, and how they interact with one another.

Each process is assigned to a process group, and it's also in a knowledge area.

Project management offices help you do a good job, every time

Every project your company completes can teach you a lot about what works and what doesn't within your company's culture. **Project management offices** (PMOs) help you to learn from all of the work that's been done in the past. They'll give you the templates and the guidance you need to make sure your project takes the right approach and makes sense to everyone you work with. There are three different kinds of PMOs that you might run into in your career.

Supportive
PMOs that play a supportive role provide all of the templates you need to fill out while your project is under way. They'll lay out the standards for how you should communicate your project's scope, resources, schedule, and status as your project progresses from its initial stages through to delivery and closing.

Controlling
PMOs that control the way project management is done in a company will be able to check that you're following the processes they prescribe. Like supportive PMOs, they'll tell you what templates you should fill out and prescribe a framework for doing project management in your company. They'll also periodically review the work that you're doing on your project to make sure you're following their guidelines.

Directive
PMOs that take a directive approach actually provide project managers to project teams. In a directive PMO, the project manager usually reports to the PMO directly. That reporting structure makes sure that the project managers follow the frameworks and templates prescribed by the PMO, because their job performance depends on it.

Directive PMOs have a lot of control over the way things are done on projects.

Good leadership helps the team work together

It's not enough to have a good plan and all the resources you need to make your project a success. You need to think about your **interpersonal skills** if you want to keep your project on track. Here are a few examples:

Leadership

A good leader gets the team to see the end goal and focus on getting there. With the right leadership, team members feel like they can take control of the work they're doing, and make good decisions to help the team achieve its goals as directly as possible.

Team building and trust building

Everything you do to help team members feel like they can rely on one another is part of team building. When a team feels like they're all working together to achieve the project's goals, they're able to do so much more than each of them can do individually. Trust building is all about sharing information with all of your stakeholders so that they know they can trust one another.

Motivation

Some people are motivated by the kind of work they do, some are looking for experience that will help their résumés, and others are hoping for a promotion or a pay raise. Understanding what motivates your team members and helping them to achieve their personal goals will help your project, too.

Some teams can be so motivated by the value of the project that their personal goals are secondary to making the project a success. When that happens, they can be really productive as a team.

Project teams are made of people

Keeping your team motivated and helping them to feel included are just a couple of the interpersonal skills that make your project a success. You also need to help your team members work through problems and maintain an environment where it's easy for everyone to get along. Here are a few more **interpersonal skills** that a good project manager uses to keep the team on track:

Influencing

Sometimes you need to collaborate with others to get your work done. When you influence people, you focus on the shared benefit of the work with them and share power toward a common goal. You're probably using your influence to make things happen every day on the job.

You usually can't just tell people to do what you want and have them do it.

Conflict management

When people work together, there will always be disagreements. A good project manager works to find positive solutions when conflicts pop up.

We'll talk more about conflict management in Chapter 13.

There are a few more interpersonal skills discussed in the PMBOK Guide that we'll talk about along the way:

* Communication (Chapter 10 is all about communication).

* Negotiation and decision making (we'll talk more about those in Chapter 9).

Coaching

As the people on your team grow and take on new responsibilities, they might want some help developing new skills. That's where coaching comes in. When you coach people, you help them develop their skills and get better at what they do. Sometimes this means helping them find training, while other times it's just acting as a sounding board to help your team members sort through the problems they run into.

A good PM is always looking for ways to help the team get better at what they do.

Political and cultural awareness

It's important to make sure that everybody on the team feels included. You need to be aware of the topics that might alienate people or make them feel uncomfortable, so that you can maintain an open and inclusive environment on your team.

BRAIN POWER

Can you think of a time when one of these skills helped your team succeed on a past project?

Q: Directive and controlling PMOs seem pretty similar to me. What's the difference again?

A: That's a good question. Both of them are pretty active in managing the projects they govern, but there are some differences.

Controlling PMOs tend to review the work a project team does at various points in the project to make sure that they are following the company's agreed-upon process. The people who work in the controlling PMO are like auditors who take a look at a project team's work products to make sure they're complying with the company's project management rules.

Directive PMOs actually manages the projects on their own. The people who work in the PMO take on the role of project manager for all of the projects the PMO is responsible for.

Q: My company doesn't have a PMO at all. Do the 47 processes apply only to companies that have PMOs? How much of what's in the *PMBOK Guide* applies to plain old project managers who aren't in a PMO?

A: Yes! The *PMBOK Guide* is all about project management, and all of the processes it talks about are meant to be used on projects to help them succeed. If your company has a PMO, following the 47 processes will help you work with that PMO. But if you don't have a PMO where you work, it's still good for a project to follow all 47 of them.

Q: Does the PMP exam test you on your interpersonal skills?

A: Yes. You need to know what all of the interpersonal skills are and when you would use them in managing your project. Interpersonal skills are an important part of managing a team. If your team members each feel like their opinion is valued and they are motivated to do good work, you'll have a much better chance at success.

Q: Hold on, how are team building and coaching different?

A: Another good question. Team building usually means going out, having lunch, doing group exercises, and in general socializing as a team so that the group gets more familiar with one another and can have a better internal set of relationships. Coaching is about identifying your team members' talents and helping them develop them.

Say a member of your team is really good at explaining technical concepts to other people. You might want to coach that person to write project documents that will help the whole team understand what they're doing better. That would let the team member get better at using her talent, and also help with team communication.

Operations management handles the <u>processes</u> that make your company tick

Think about all of the processes your company goes through every day to keep things running smoothly. You probably have worked with operations teams on many of your projects—from accounting departments to support teams to infrastructure teams who maintain software environments. Each operations team is staffed with specialists in the aspect of your company that they run. Sometimes, you'll work with them to build out parts of your project; sometimes, you'll depend on them to do work before your project can proceed. Operations teams are almost always <u>stakeholders</u> in your projects.

Operations management will direct, oversee, and control the business processes you work with on your project every day.

You'll learn more about stakeholders when we talk about stakeholder management in Chapter 13.

✏️ Sharpen your pencil

Which kind of project management office is being described?

1. Provides the process documents and templates for your project.

☒ Supportive   ☐ Controlling   ☒ Directive

2. You meet with them once a month to go over project docs and make sure you're following the right process.

☒ Supportive   ☒ Controlling   ☐ Directive

3. Provides a knowledge base of common project problems and lessons learned for you to use.

☒ Supportive   ☐ Controlling   ☐ Directive

4. A centralized group of project managers who are assigned to manage projects.

☐ Supportive   ☒ Controlling   ☒ Directive

5. When a project gets started, this group makes sure that they've followed all of the initiating processes and have the right approvals to start working.

☒ Supportive   ☒ Controlling   ☐ Directive

6. When you sit down to do your risk planning for your project, you go to them to find a good example of a risk plan that's been useful on other projects.

☐ Supportive   ☒ Controlling   ☐ Directive

7. This team audits your project work at regular intervals to confirm the status reports you're giving and guide you when you run into trouble.

☐ Supportive   ☒ Controlling   ☒ Directive

→ Answers on page 28.

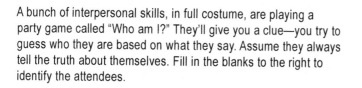

A bunch of interpersonal skills, in full costume, are playing a party game called "Who am I?" They'll give you a clue—you try to guess who they are based on what they say. Assume they always tell the truth about themselves. Fill in the blanks to the right to identify the attendees.

Any of the charming skills you've seen so far just might show up!

Name

I get everybody on the team to understand the goals of the project so that they can get behind them.

leadership

I work to maintain an open and inclusive environment by paying attention to the things that team members might be sensitive to.

Political/cul awareness

I try to figure out what each team member wants out of the project, and then I help him or her get it.

motivation

I share power with other people in order to get some shared benefit.

leadership

I help team members get better at doing project work.

coaching

When arguments or disagreements happen, I try to solve them in a way that benefits the team as much as possible.

conflict Mgt

I help everybody on the team feel like they can rely on one another.

Team bld/trust bld

I'm transparent with all of the stakeholders in my project so that everyone has all of the information they need to make good decisions.

Trust bld

———→ Answers on page 29.

Sharpen your pencil
Solution

Which kind of project management office is being described?

1. Provides the process documents and templates for your project.

☒ Supportive ☐ Controlling ☐ Directive

2. You meet with them once a month to go over project docs and make sure you're following the right process.

☐ Supportive ☒ Controlling ☐ Directive

3. Provides a knowledge base of common project problems and lessons learned for you to use.

☒ Supportive ☐ Controlling ☐ Directive

4. A centralized group of project managers who are assigned to manage projects.

☐ Supportive ☐ Controlling ☒ Directive

5. When a project gets started, this group makes sure that the team has followed all of the initiating processes and have the right approvals to start working.

☐ Supportive ☒ Controlling ☐ Directive

6. When you sit down to do your risk planning for your project, you go to them to find a good example of a risk plan that's been useful on other projects.

☒ Supportive ☐ Controlling ☐ Directive

7. This team audits your project work at regular intervals to confirm the status reports you're giving and guide you when you run into trouble.

☐ Supportive ☒ Controlling ☐ Directive

It's true that all of the types of PMOs provide process documents, but only the supportive PMO provides them as its main function.

A bunch of interpersonal skills, in full costume, are playing a party game called "Who am I?" They'll give you a clue—you try to guess who they are based on what they say. Assume they always tell the truth about themselves. Fill in the blanks to the right to identify the attendees.

Any of the charming skills you've seen so far just might show up!

Name

I get everybody on the team to understand the goals of the project so that they can get behind them.

Leadership

I work to maintain an open and inclusive environment by paying attention to the things that team members might be sensitive to.

Political and cultural awareness

I try to figure out what each team member wants out of the project and then I help him or her get it.

Motivation

I share power with other people in order to get some shared benefit.

Influencing

I help team members get better at doing project work.

Coaching

When arguments or disagreements happen, I try to solve them in a way that benefits the team as much as possible.

Conflict resolution

I help everybody on the team feel like they can rely on one another.

Team building

I'm transparent with all of the stakeholders in my project so that everyone has all of the information they need to make good decisions.

Trust building

A PMP certification is more than just passing a test

Getting your PMP certification means that you have the **knowledge to solve** most **common project problems.**

It proves that **you know your stuff.**

Once you're certified, **your projects are more likely to succeed** because:

you have the skills and knowledge to make them successful.

Meet a real-life PMP-certified project manager

PMs have demonstrated that they understand the tools it takes to be successful at leading projects. They know what it means to juggle their project priorities and still have their projects come out on top. Being certified doesn't mean you won't have problems on your projects anymore, but it does mean that you'll have the wisdom of many experienced and smart project managers behind you when you make decisions about how to solve these problems.

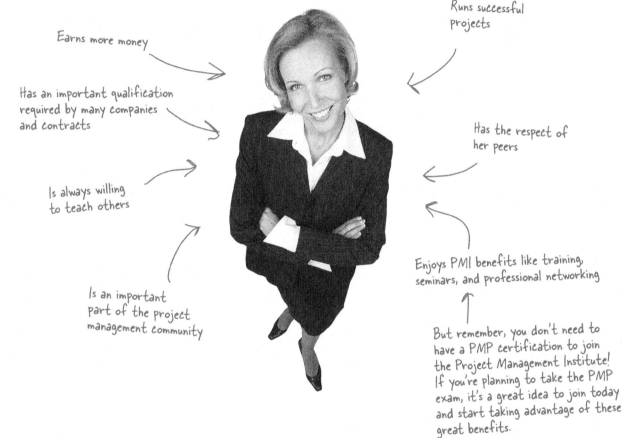

Runs successful projects

Earns more money

Has an important qualification required by many companies and contracts

Has the respect of her peers

Is always willing to teach others

Is an important part of the project management community

Enjoys PMI benefits like training, seminars, and professional networking

But remember, you don't need to have a PMP certification to join the Project Management Institute! If you're planning to take the PMP exam, it's a great idea to join today and start taking advantage of these great benefits.

This could be <u>YOU</u>!

Exam Questions

1. Which of the following is NOT a type of project management office?

   A. Directive

   B. Value-driven

   C. Supportive

   D. Controlling

2. Which of the following is NOT a characteristic of a project?

   A. Temporary

   B. Strategic

   C. Specific result

   D. Progressively elaborated

3. An energy company is investing in a series of initiatives to look for alternative energy sources so that the company can be competitive in 10 years. The initiatives are tracked and managed together because this goal is vital to the success of the company. This is an example of:

   A. A portfolio

   B. A program

   C. A project

   D. A enterprise environmental factor

4. Which of the following is NOT a responsibility of a project manager?

   A. Managing stakeholder expectations

   B. Managing project constraints

   C. Gathering product requirements

   D. Sponsoring the project

5. Which of the following is NOT an interpersonal skill?

   A. Motivation

   B. Brainstorming

   C. Team building

   D. Coaching

Exam Questions

6. Which of the following is NOT true about interpersonal skills?

- A. Coaching means helping your team to get more exercise.
- B. Motivation means helping team members get what they want out of the project.
- C. Influencing means sharing power with people to get something done.
- D. Conflict management means finding positive solutions to conflicts during the project.

7. Which of the following is NOT true about portfolio management?

- A. The portfolio manager judges the success of the portfolio by combining data from all of its programs and projects.
- B. A portfolio can contain projects and programs.
- C. A portfolio is organized around a business goal.
- D. A portfolio is always a group of programs.

8. You're managing a project to remodel a kitchen. You use earned value calculations to figure out that you're going to run $500 over budget if your project continues at the current rate. Which of the following core characteristics of a project manager are you using to find the problem?

- A. Knowledge
- B. Performance
- C. Personal
- D. None of the above

9. At the beginning of a project, a software team project manager is given a schedule with everyone's vacations on it. She realizes that because the software will be delivered to the QA team exactly when they have overlapping vacations, there is a serious risk of quality problems, because there won't be anyone to test the software before it goes into production. What BEST describes the constraint this places on the project?

- A. Quality constraint
- B. Time constraint
- C. Resource constraint
- D. Risk constraint

10. A project manager is having trouble with his project because one of his team members is not performing, which is causing him to miss an important date he promised to a stakeholder. He discovers that the team member knew about the project problem, but didn't tell him because the team members are all afraid of his bad temper. Which BEST describes how the project manager can avoid this situation in the future?

- A. Increasing his knowledge of the *PMBOK Guide*
- B. Measuring personal performance
- C. Improving his personal skills
- D. Managing stakeholder expectations

Answers

Exam ~~Questions~~

1. Answer: B

Although PMOs are usually value-driven, that's not a valid type of PMO. The three types of PMOs are supportive, controlling, and directive. Supportive PMOs provide templates and guidelines for running projects, controlling PMOs audit projects to ensure adherence to processes and standards, and directive PMOs provide project managers to manage projects.

2. Answer: B

A project doesn't have to be strategic or critical. It only needs to be temporary, have a specific result, and be progressively elaborated.

Look out for questions like this one on the exam. Common sense might tell you that a project should be important for a company to want to do it, but that's not what the question is asking.

3. Answer: A

Since the initiatives are being managed together because of a strategic business goal, you can tell that this is a portfolio.

Portfolios are organized around business goals, and programs are organized around a shared benefit in managing them together.

4. Answer: D

The sponsor is the person who pays for the project. The project manager doesn't usually play that role.

In fact, you'll learn more about its role in defining project requirements in a couple of chapters.

5. Answer: B

Brainstorming is an activity that you do with other people, but it's not an interpersonal skill that you need to hone to help manage all of the stakeholders on your project.

6. Answer: A

Coaching is really about helping your team members to get better at what they do. Anything you do to challenge them to develop their skills is coaching.

7. Answer: D

Since a portfolio can be a group of programs and projects, option D is the one that's not true. It *can* be a group of programs, but it doesn't *have to* be.

Exam ~~Questions~~ Answers

8. Answer: A

Your knowledge of earned value management techniques is how you can predict that the project will be over budget. Knowing that could help you plan ahead to avoid further cost overruns. Minimally, it can help you to reset expectations with your stakeholders so they have a better idea of what's coming.

9. Answer: C

This is a resource constraint, because the project manager's resources—in this case, the people who will be testing the software—are not going to be available to her when she needs them. Yes, this will cause problems with the quality, introduce risks, and cause schedule problems. But they're not schedule, time, or risk constraints, because there's no outside limitation placed on the project quality, schedule, or risks. The only outside limitation is the resource availability. If they were available, there wouldn't be a problem!

10. Answer: C

The way that the project manager interacts with the people on his team interfered with his work getting done. This is a good example of how a lack of personal skills can lead directly to major project problems down the line, and it's why this particular project manager needs to work on his personal skills.

The project manager's temper led to a disappointed stakeholder, but that doesn't mean that the stakeholder's expectations were out of line. This was an avoidable project problem, and better personal skills would have fixed it.

YOU MADE IT THROUGH YOUR FIRST CHAPTER AND EXAM QUESTIONS! HOW'D IT GO?

2 Organizations, constraints, and projects

In good company

If you want something done right...better hope you're in the right kind of organization. All projects are about teamwork—but how your team works depends a lot on the type of organization you're in. In this chapter, you'll learn about the different types of organizations around—and which type you should look for the next time you need a new job.

A day in Kate's life

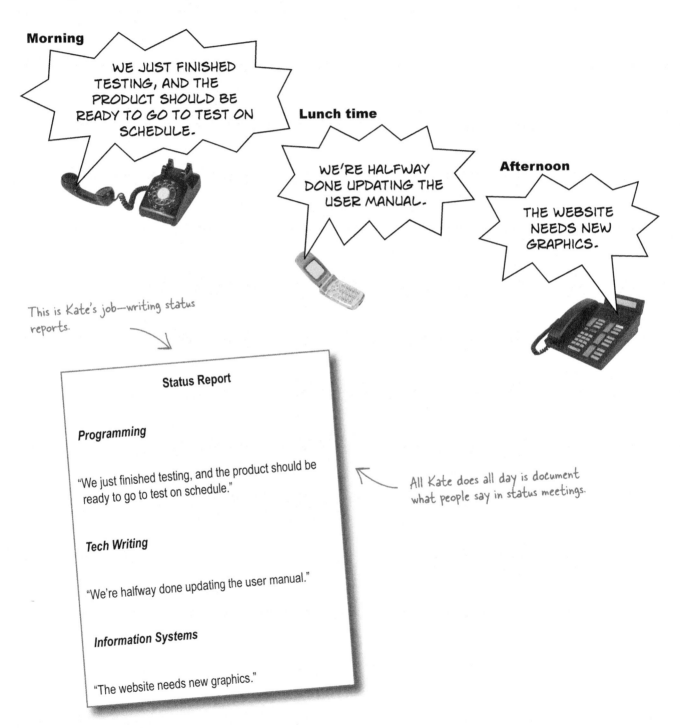

Morning

WE JUST FINISHED TESTING, AND THE PRODUCT SHOULD BE READY TO GO TO TEST ON SCHEDULE.

Lunch time

WE'RE HALFWAY DONE UPDATING THE USER MANUAL.

Afternoon

THE WEBSITE NEEDS NEW GRAPHICS.

This is Kate's job—writing status reports.

Status Report

Programming

"We just finished testing, and the product should be ready to go to test on schedule."

Tech Writing

"We're halfway done updating the user manual."

Information Systems

"The website needs new graphics."

All Kate does all day is document what people say in status meetings.

Kate wants a new job

Now that she's working on getting her PMP certification, Kate's learning a whole load of new skills. And she's even started to look for a new job—one where she does more than write down what other people say all day…

> ALL I DO ALL DAY IS COLLECT STATUS. EVEN IF I HAVE IDEAS ABOUT HOW TO IMPROVE THE PROJECT, IT'S NOT LIKE I HAVE THE POWER TO ACTUALLY CHANGE ANYTHING.

Kate's not responsible for the success or failure of her project. She just keeps everybody informed of its progress.

Kate is a <u>project expediter</u> right now.

Kate may have the job title of "project manager," but even though that's what's printed on her business cards, that's not really her job. Kate's job is to document what's happening on a project, but she doesn't have the authority to make decisions on it. The *PMBOK Guide* calls this role a **project expediter**. She may *work* on projects, but she's certainly not *managing* anything.

BRAIN POWER

When Kate checks out Monster.com, what types of things do you think she should look for in a new organization?

Exercise

Kate spilled a hot cup of Starbuzz half-caf nonfat latte on her job-hunting checklist. Can you match the notes she scribbled at the bottom of the page to what's covered up by coffee stains?

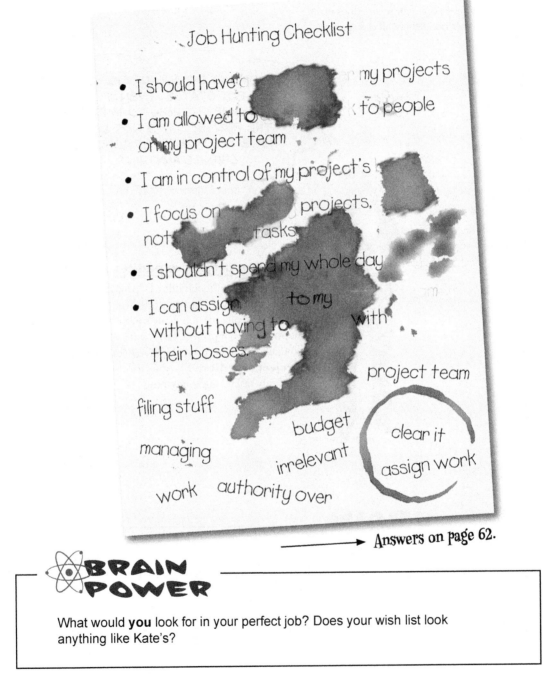

> Answers on page 62.

Answers on page 62.

BRAIN POWER

What would **you** look for in your perfect job? Does your wish list look anything like Kate's?

Organization Magnets

In a **functional organization**, which is what Kate works in, project managers don't have the authority to make major decisions on projects. **Projectized organizations**, on the other hand, give all of the authority to the PM.

Can you work out which description goes with which organization type?

In this kind of company, the team reports to the project manager, who has a lot more authority.

Functional organization

In a functional organization, the teams working on the project don't report directly to the PM. Instead, the teams are in departments, and the project manager needs to "borrow" them for the project.

Projectized organization

1. ~~PM's don't set budget~~

2. ~~PM - dept mgrs~~

3. ....

1. ~~PM's set + track bu~~

2. ....

3. ....

Teams are organized around projects.

Project managers estimate and track budget and schedule.

PMs don't set the budget.

Project managers choose the team members, and release them when the project is over.

PMs spend half their time doing admin tasks.

Project managers need to clear major decisions with department managers.

⟶ Answers on page 63.

There are different types of organizations

Kate's got three major options when looking at the kinds of organizations she can work for. **Functional organizations** are set up to give authority to functional managers, **projectized organizations** give it to the PM, and **matrix organizations** share responsibility and authority between the two.

Functional

In this kind of organization, the project team members always report to a functional manager who calls all the shots.

- Project management decisions need to be cleared with functional managers.

- Project managers are assistants to the functional managers in getting the work done.

- Project managers spend a lot of time doing administrative tasks and often only work as PMs part of the time.

- You're likely to find project expediters in functional organizations.

All of the project work typically happens within a particular department, and that department's manager is completely in charge of everything.

Weak Matrix

Balanced Matrix

Matrix organizations

- PMs have some authority, but they aren't in charge of the resources on a project.

- Major decisions still need to be made with the functional manager's cooperation or approval.

- Project expediters (like Kate) and project coordinators can work in weak matrix organizations, too.

Project coordinators are like expediters, except that coordinators typically report to higher-level managers and have some decision-making ability. Expediters have no authority at all.

- Project managers share authority with the functional managers.

- PMs run their people-management decisions by the functional manager, but the functional manager runs his project decisions by the PM, too.

Folks who work in a balanced matrix organization report to a project manager AND a functional manager equally.

The project manager has the most authority and power in a projectized organization.

⟶

Strong Matrix

Projectized

For the PMP exam, most questions assume that you work in a matrix organization unless they say otherwise.

If you've worked with a contractor or consulting company, they are usually organized like this.

- Project managers have more authority than functional managers, but the team still reports to both managers.

- The team might be judged based on performance on their projects, as well as on their functional expertise. In a strong matrix, delivery of the project is most important.

- Teams are organized around projects. When a project is done, the team is released, and the team members move on to another project.

- The project manager makes all of the decisions about a project's budget, schedule, quality, and resources.

- The PM is responsible for the success or failure of the project.

> WAIT A SECOND. NOT ALL COMPANIES WILL FIT INTO ONE OF THESE FIVE CATEGORIES, WILL THEY?

Good point.

Sometimes companies will use multiple organization types to get different kinds of projects done. Those organizations are called **composite organizations**.

there are no
Dumb Questions

Q: I'm still not clear on the difference between a project *coordinator* and a project *expediter*.

A: They're actually pretty similar. A project expediter is somebody who keeps track of status but has no decision-making authority on a project at all. A project coordinator is someone who does pretty much the same thing, but does get to make some of the minor decisions on the project without having to run them by the functional manager. Coordinators usually report to somebody who is pretty high up in the organization, while expediters are more like assistants to the functional manager. Both of them usually exist in weak-matrix or functional organizations.

Q: What's the difference between the way teams are run in a functional organization and a projectized one?

A: Think of a major bookkeeping project being run by the Admin department. Usually the head of Admin is the one who is ultimately responsible for what happens to it. If a project manager is called in to help out, she's just there to keep things straight for the Admin department manager. The team is made up of people who already report to the Admin manager, so nobody questions his authority. That's an example of a functional organization.

Contrast that with the way the bookkeeping project would be run if a consulting company that specialized in bookkeeping were contracted to do it. The company would assemble a team of bookkeepers and assign a project manager to lead them. When the project was over, the team would dissolve, and the team members would go join other teams working for other project managers. That's how a projectized organization works. The team is organized around a project and not around a job function.

Q: Can I be an effective PM in a functional organization?

A: Since project managers don't have much authority in a functional organization, it's hard to have as much impact in a functional organization as you would in a matrixed or projectized one.

Of course, you can be good at your job in any kind of organization. But, for your company to really get the most out of having project managers on staff, it really pays for it to look into changing the way it balances power. The project managers who are accountable for project success or failure should also have the chance to influence the team, budget, and schedule for those projects.

Q: Does the PMP exam favor any kind of organization?

A: When you're taking the PMP exam, if you see a question that mentions a PM, then you should assume that the question is asking about a matrix organization if it doesn't say up front which kind of organization is being described. Functional organizations are usually painted in a negative light because they tend to give less authority to project managers.

BULLET POINTS: AIMING FOR THE EXAM

- Functional managers have all the power in a functional organization. Project managers have the power in a projectized organization.

- If a question on the exam doesn't state an organization type, assume it's referring to a matrix organization. That means the PM is responsible for making budgets, assigning tasks to resources, and resolving conflicts.

- Project coordinators and expediters don't exist in a projectized organization.

- A project expediter keeps track of project status only. A project coordinator has some authority, and usually reports to someone higher up in the company. Neither role has as much power or authority as a real project manager, even though expediters or coordinators may have "Project Manager" printed on their business cards.

Exercise

Here are a few excerpts from some of Kate's job interviews. Can you figure out what kind of organization each interviewer is representing?

Interviewer #1: We're looking for someone who can work with our development manager to deliver our products on time. We have a good programming team; they just need a little encouragement to meet their deadlines. You'll be expected to keep really good status meeting notes. If you run into any trouble with the team, just kick it back to the Dev Manager, and she'll address the problem.

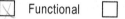

☒ Functional ☐ Matrix ☐ Projectized

Interviewer #2: We need someone who can manage the whole effort, start to finish. You'll need to work with the client to establish goals, choose the team, estimate time and cost, manage and track all of your decisions, and make sure you keep everybody in the loop on what's going on. We expect the project to last six months.

☐ Functional ☐ Matrix ☒ Projectized

Interviewer #3: We have a project coming up that's needed by our customer service team. The project is a real technical challenge for us, so we've assembled a team of top-notch programmers to come up with a good solution. We need a project manager to work with the programming manager on this one. You would be responsible for the schedule, the budget, and managing the deliverables. The programming manager would have the personnel responsibilities.

☐ Functional ☒ Matrix ☐ Projectized

Interviewer #4: Most of the work you'll be doing is contract work. You'll put together three different teams of software engineers, and you'll need to make sure that they build everything our customer needs. And don't forget: you've got to stay within budget, and it's got to be done on time! It's a big job, and it's your neck on the line if things go wrong. Can you handle that?

☐ Functional ☐ Matrix ☒ Projectized

Exercise Solution

Here are a few excerpts from some of Kate's job interviews. Can you figure out what kind of organization each interviewer is representing?

This is just like the job Kate wants to leave. Just gathering status sounds pretty boring.

Interviewer #1: We're looking for someone who can work with our development manager to deliver our products on time. We have a good programming team; they just need a little encouragement to meet their deadlines. You'll be expected to keep really good status meeting notes. If you run into any trouble with the team, just kick it back to the Dev Manager, and she'll address the problem.

☒ Functional   ☐ Matrix   ☐ Projectized

Interviewer #2: We need someone who can manage the whole effort, start to finish. You'll need to work with the client to establish goals, choose the team, estimate time and cost, manage and track all of your decisions, and make sure you keep everybody in the loop on what's going on. We expect the project to last six months.

☐ Functional   ☐ Matrix   ☒ Projectized

Everybody moves from project to project in this organization.

Interviewer #3: We have a project coming up that's needed by our customer service team. The project is a real technical challenge for us, so we've assembled a team of top-notch programmers to come up with a good solution. We need a project manager to work with the programming manager on this one. You would be responsible for the schedule, the budget, and managing the deliverables. The programming manager would have the personnel responsibilities.

☐ Functional   ☒ Matrix   ☐ Projectized

Shared authority between the PM and the functional manager.

Most contractors are projectized: the PM builds the team and makes sure the work gets done.

Interviewer #4: Most of the work you'll be doing is contract work. You'll put together three different teams of software engineers, and you'll need to make sure that they build everything our customer needs. And don't forget: you've got to stay within budget, and it's got to be done on time! It's a big job, and it's your neck on the line if things go wrong. Can you handle that?

☐ Functional   ☐ Matrix   ☒ Projectized

Kate takes a new job

WELCOME ABOARD, KATE! I'M BEN, THE FUNCTIONAL MANAGER.

Ben

Kate: Hi, Ben. I'm excited to be here. It's such a relief to be hired as a project manager, and not just a project expediter anymore.

Ben: We're excited too, since you'll be taking care of our main software development project. It's in maintenance mode right now.

Kate: Sounds great. How do we handle that here?

Ben: Well, we're constantly getting business reports from the field, and when people think of new ideas, we just add them to the project.

Kate: Umm…so how do you know when you're done?

Ben: We're never really done; we try to release new versions as often as possible.

Kate's being asked to do <u>operational work</u>

Kate's spent a lot of time studying for the PMP exam, and the first thing she learned was that a project is temporary. When she sees ongoing work that doesn't really have a start or a finish, it's not a project at all. Ben asked Kate to do **operational work**, which has no beginning and no end. Since there's no way for Kate to know when she's done, it will be harder for her to be successful at her job. And that makes her nervous!

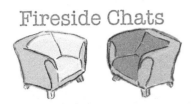

Fireside Chats

Tonight's talk: **Operational Work and A Project spar over who's more valuable.**

Operational Work

I've been meaning to sit down and talk to you for a while.

Hey, don't knock that day-to-day work! It's your bread and butter. If I weren't here keeping the lights on, there'd be no chance for you to go out and build all of the flashy stuff you do. It's thankless work keeping the business running, I'll give you that, but where would you be without me?

Don't patronize me. I know you think you're pretty hot because everybody wants to know when you'll be done and how much you'll cost. But remember, you're temporary. When you're done producing your product, you close down and I'll be left to maintain the systems you create. Not only are you nothing without me, you have an expiration date. Still feel like a star?

A Project

Really? I thought you were too busy doing your day-to-day business to care too much about us projects.

There you go again..."thankless work." Give me a break. Eveybody knows that you represent all of the work the business does on an ongoing basis. You're the work that keeps the computers running, and the paychecks flowing. You're the systems upgrades and the maintenance...all of the work that has no definite beginning and end.

You make it possible for me to break new ground. You're a great supporting player. You make it easy for me to be a star! While I'm out there expanding the business, you're making sure we can take care of our old stuff. Er, I mean, you're maintaining our **core** work.

I'm out there every day making progress, building new products, and changing the way that you do business. Yeah, I do feel like a star! I'm sorry you can't see how important I am. I guess we'll just never agree on this one.

Stakeholders are impacted by your project

Anyone who will be affected by the outcome of your project is a **stakeholder**. It's usually pretty easy to come up with the first few people on the list of affected people. The sponsor who's paying for the project, the team who's building it, and the people in management who gave the project the green light are all good examples. But it can get a little tricky as your project gets going. You might start with that core group of people and find that the number keeps growing as time goes on. It's your job as a project manager to find all of the stakeholders who are influential in your project and keep them updated on where your project is going. Making sure that their expectations are managed can be the difference between your project succeeding and failing.

> I HOPE KATE DOESN'T DO ANYTHING TO SLOW DOWN MY DEVELOPMENT TEAM.

Kate might not think about Ben's goals when she starts planning her project, and that could cause problems for her.

Negative stakeholders

Not all of the people you're working with are rooting for your project to succeed. Sometimes, the people you're working with think that your project might bring negative consequences for them. Ben's worried that bringing any kind of planning into his company will slow down his team. Kate's going to have to **manage his expectations** and work with him to set goals that make sense to him if she's going to bring him around to supporting her work. You need to know what's motivating all of your project stakeholders if you're going to understand the influence they'll have over your project.

It may sound like having a negative stakeholder is a really bad thing, but if Kate can turn Ben around then he'll be a great ally in the future.

Identify stakeholders

One of the first things you'll do when you start a project is figure out who your stakeholders are and write down their goals and expectations in a **stakeholder register**. That's part of the Identify Stakeholders process that you'll learn more about in Chapter 13 of this book. Even though you do that work up front, you'll find that new stakeholders are always popping up, and you'll need to make changes to your stakeholder register to include them as you learn about them.

You'll learn more about how the Identify Stakeholders process helps you understand their goals and expectations in Chapter 13.

More types of stakeholders

Take a minute to think about all of the stakeholders who've ever had something to do with the projects you've worked on in the past. Making sure that they are all informed and helping your project to succeed is the point of **Project Stakeholder Management**, one of the 10 knowledge areas that are covered in the *PMBOK Guide*. Let's take a look at some of the kinds of stakeholders that will impact your project.

> Project Stakeholder Management is covered in depth in Chapter 13 of this book.

Sponsor

The sponsor is the person who pays for the project. Without the sponsor's help, there's no way the project can be a success.

Consumers or users

Usually, you build a product or service so that someone can buy or use it. You have to make sure your project meets the customer's needs if you want to call it successful.

Seller

You might license software or contract consultants to help you build your product. The companies you work with to help you deliver it are stakeholders in your project's success too.

Organizational Groups

You might not think about it at first, but there are many ways that groups outside your team can be affected by your project. Your sales team, your internal support teams, all kinds of groups inside your company will have a stake in your project's success

Business partner

Your company might contract a company to provide training or other materials that affect your project. They're important stakeholders of your project too.

Functional manager

If you're building an accounting software package, you're going to need accounting expertise to understand what you're building. Functional managers provide the subject matter expertise to make things run smoothly on your project.

Your project team has lots of roles too

When you think about it, most of the roles we just talked about have a place on your project team too. When you think about your team, it's easy to focus on the PM and the project staff. But your team can be comprised of roles from many different stakeholder organizations. Think about all of the stakeholder organizations that might help you deliver the product of your project by actually assigning people to your project team.

Human Resource Management is covered in depth in Chapter 9 of this book.

Exercise

Can you name the project team role from the list of stakeholder roles on the opposite page for each of the examples below?

1. The project was nearly complete and it was time to have it tested for acceptance by the people who would need to use it everyday.

Project Team Role:

consumer

3. The company contracted an external vendor to build a big component of the product. The vendor met with the team everyday so that they could make sure there were no integration problems between the part they were building and the part the rest of the team was working on.

Project Team Role:

Seller

2. The company had an external company come in and train all of the project staff on how to deploy the product of they project using a new software package.

Project Team Role:

Biz part

4. The company just purchased a new accounting software package. The company that makes the accounting software assigned a representative to answer any questions the team might have as they built the product.

Project Team Role:

biz partner

Answers: 1. Consumers or Users, 2. Business Partners, 3. Seller, 4. Business Partners

Back to Kate's maintenance nightmare

Let's figure out how things are working in Kate's new organization...
and start to think about how we can improve things.

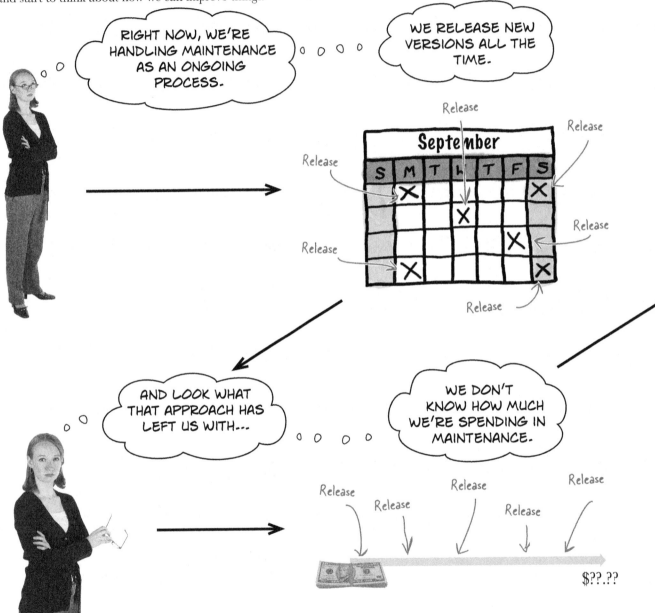

The customer service department's report on user feedback.

BRAIN POWER

What would you do to fix this problem?

Managing project constraints

When Kate thinks about solutions, she's going to have to deal with the project's constraints. Every project, regardless of what is being produced or who is doing the work, is affected by the **constraints** of time, scope, cost, quality, resources, and risk. These constraints have a special relationship with one another, because doing something to deal with one of the constraints always has an effect on the others.

Your project will need to get done on schedule.

Your project will always have to stay within a budget.

You need to manage the scope of work you do for the project.

You have to have the people and materials to get the work done.

If your product doesn't do what it's supposed to do, you wont succeed.

Unexpected obstacles can wreck your project if you don't deal with this one.

Time   Cost   Scope   Resources   Quality   Risk

If you don't manage all six constraints at the same time, you risk managing in favor of just one constraint.

For Kate's project to succeed, she needs to think about the project **constraints**. If she doesn't manage these six constraints at the same time, she'll find that her project is either late, over budget, or unacceptable to her customers.

Any time your project changes, you'll need to know how that change affects all of the constraints.

Can you figure out the constraint that's causing the biggest headache for the project manager in each of these scenarios?

The project was running late, so the project manager decided to release it on time even though it was missing some of its features.

Constraint affected:

. . . .

The team wanted to add more testers to find defects, but the project manager overruled them.

Constraint affected:

. . . .

A construction project manager assumed that the weather would cooperate with the plans to complete the job, but thunderstorms have derailed the project.

Constraint affected:

. . . .

The company didn't have enough money to invest in the project, so they had to draft people from other departments to work part time to get the job done.

Constraint affected:

. . . .

About halfway through the project, the PM realized that the money was running out faster than expected. She went through the schedule to try to find ways to move up the deadline.

Constraint affected:

. . . .

The project manager didn't take software license fees into account, which caused the budget to balloon out of control.

Constraint affected:

. . . .

Exercise Solution

Can you figure out the constraint that's causing the biggest headache for the project manager in each of these scenarios?

The project was running late, so the project manager decided to release it on time even though it was missing some of its features.

Constraint affected:

scope ⟵

The PM stuck to the original budget and schedule, but released a product that wasn't complete. That means the scope was affected.

The team wanted to add more testers to find defects, but the project manager overruled them.

Constraint affected:

quality

⟵ *Any time you're talking about tests and defects, you're talking about quality.*

A construction project manager assumed that the weather would cooperate with the plans to complete the job, but thunderstorms have derailed the project.

Constraint affected:

risk

Whenever you make assumptions about a project, you're introducing risk.

The company didn't have enough money to invest in the project, so they had to draft people from other departments to work part time to get the job done.

Constraint affected:

resources ⟵

Resources are people or materials that you need for your project, and when you cut corners you end up straining them.

About halfway through the project, the PM realized that the money was running out faster than expected. She went through the schedule to try to find ways to move up the deadline.

Constraint affected: ⟵

time

There are lots of ways to change how long it'll take to do your project, but sometimes there simply isn't enough time.

The project manager didn't take software license fees into account, which caused the budget to balloon out of control.

It's the project manager's job to always look after the bottom line.

Constraint affected:

⟶ cost

These are the answers we thought fit best!

Did you get different answers? That's okay! For this exercise, a good case can be made for almost any of the constraints. Don't worry about which answer is "right" for now – it's more important to get some practice thinking about projects in terms of constraints. And when you get to the actual exam questions, there will always be a clear, correct, BEST answer.

Q: I've heard project constraints referred to as the triple constraint. But there are six of them here. What gives?

A: Some project managers focus on Cost, Scope, and Time as the main constraints of a project. But just thinking about those three constraints doesn't give a clear picture of all of the constraints you need to account for when planning a project. The important thing here is to understand that Cost, Time, Scope, Quality, Risk, and Resources are all related to each other. You need to pay attention to all of them and if you manage your project in favor of one of them, it will affect the others.

Q: I've heard of an old saying: "Faster, cheaper, better—pick two," but doesn't that mean that there are only two constraints that you can manage at any given time?

A: No, that's an old (and somewhat cynical) project management saying. When a project manager says it to a customer or stakeholder, what he is saying is that there's no way to reduce cost, shorten the schedule, and increase quality all at the same time. At least one of those things absolutely has to give... but the saying is a little disingenuous! We already know that all six of the constraints are related to each other, and there's almost never an easy, obvious trade-off where you can sacrifice one to improve the others.

Q: What if I know that a change will impact just scope, but not schedule or cost or any of the other constraints. Can I go ahead and make it?

A: Whenever you are making a change that affects the project constraints, you need to be sure that the change is acceptable to **your stakeholders**. They're the people who will be impacted by your project. The term applies to your team, your customer, your sponsor, and anybody else who is affected by the change.

A lot of project management is about evaluating what a change is going to do to your project constraints, and using that impact analysis to help stakeholders make choices about what to do when changes come up. Sometimes a change that affects the quality of your product is completely unacceptable to your stakeholders, and they would rather delay the project than sacrifice the product's quality.

Q: In my organization, we have some projects that would sound like they're functional and some that are fully projectized. Where does that kind of organization fit in?

A: Sometimes a company that mostly runs their projects in a functional way will create a special team that gives the project manager more authority. When a company manages using many different types of organizational structure, it's called a **composite organization**.

Q: I don't quite get this whole negative stakeholder thing. Why do I care about people who aren't helping me with my project?

A: Think of it this way: sometimes a project might have really good overall outcome for your company, but it might make some of the people who are impacted by it uncomfortable. (Here's a quick example: think about another project manager who won't get to use the resources he planned on because they're taken up by your project.) Change can be really hard for people to adapt to, and sometimes your stakeholders are not going to be happy about changes that your project is making. It's important to know how negative stakeholders feel, and understand why they're resistant to your project.

You need to identify and manage the expectations of all of the stakeholders who have influence over your project if you're going to succeed. So don't take it personally if there are people out there who aren't as enthusiastic about your project as you are. Use it as an opportunity to find out what your project can do to get buy-in from the negative stakeholders out there.

Q: You mentioned that it's possible to "turn around" a negative stakeholder. How does that work?

A: Today's negative stakeholders can become tomorrow's **advocates** if you make sure their needs are met. By listening to them, taking their needs into account, and making changes to your project so that those needs are satisfied, those previously negative stakeholders will feel good about what you're doing... and they'll often become your closest allies in the future.

A stakeholder is anyone who is affected either positively or negatively by your project.

You can't manage your project in a vacuum

Even the best project managers can't control everything that affects their projects. The way your company is set up, the way people are managed, the processes your team needs to follow to do their jobs... they all can have a big impact on how you manage your project. On the exam, all of those things are called Enterpise Environmental Factors.

It's easy to fall into the trap of thinking that these factors only apply to big companies. In fact, they apply to all organizations of all sizes and types... and you need to understand them about YOUR company if you want your projects to be successful! Your company's culture is one of its most important Enterprise Environmental Factors.

Enterprise Environmental Factors

People

The skills and organizational culture where you work.

Risk Tolerance

Some companies are highly tolerant of risk and some are really risk averse.

Market

The way your company is performing in the market can affect the way you manage your project.

Databases

Where your company stores its data can make a big difference in the decisions you make on your project.

Standards

Some companies depend on government standards to run their business and when they change, it can have a big impact

Kate's project needs to follow company processes

HERE'S A PROJECT GOVERNANCE DOCUMENT FROM OUR PAST PROJECTS. MAYBE THIS WILL HELP YOU AS YOU START TO GET YOUR PLAN TOGETHER.

Kate: This will really come in handy. There's an organization chart that describes all of the teams and people that will rely on our project. Wow. There's also a whole process for escalating issues that come up.

Ben: I wanted to make sure we didn't re-invent the wheel when we drew up the plans for our development project.

Kate: This will make my life a lot easier. One question though: I don't see any guidelines for project acceptance. How do we know if a project is a success?

Ben: Usually our sales team takes the new features out in the field and they let us know what the response is. If our customers like what we've done, the project is a success. If not, well... you get the picture.

Kate: Umm... that sounds a little hard to manage.

Briefcase with project governance documents inside.

Kate's project needs clear acceptance criteria

It's a lot harder for a project to be successful if the team doesn't have a clear goal in mind. Kate's project needs to follow all of the company governance guidelines, but she also needs to write down the goal her team is shooting for. That way, it will be clear that the project has met its goals when it completes. Most projects aim to finish within the constraints we talked about in the last chapter (time, cost, resources, quality, risk, and scope). It helps to write down concrete goals for those constraints as acceptance criteria up front. That way there are no surprises when the project ends.

Kate makes some changes...

1 She divides maintenance into releases, each with a well-defined start *and* a finish.

... and once Ben saw that Kate's careful planning made the project go faster, not slower, he stopped being a negative stakeholder and started being a project <u>advocate</u>!

One BIG release.

2 She follows the company's project governance guidelines and works with stakeholders to set scope goals that each release must meet.

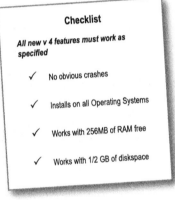

Checklist

All new v 4 features must work as specified

✓ No obvious crashes

✓ Installs on all Operating Systems

✓ Works with 256MB of RAM free

✓ Works with 1/2 GB of diskspace

All these acceptance criteria must be completed before the project can end.

3 She manages the budget for each release and keeps the costs contained.

Did you expect this to be zero? Real projects rarely come in on budget to the dollar.

Start   **1/2 Way**   **Finish**

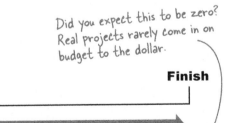

$3.68

...and her project is a success!

Now the company knows when their products will be done, how much they will cost, and that the products will satisfy their customers...

...and that earns Kate and Ben big bonuses!

Exercise Solution

Kate spilled a hot cup of Starbuzz half-caf nonfat latte on her job-hunting checklist. Can you match the notes she scribbled at the bottom of the page to what's covered up by coffee stains?

Job Hunting Checklist

- I should have authority over my projects

- I am allowed to assign work to people on my project team

- I am in control of my project's budget

- I focus on managing projects, not irrelevant tasks

- I shouldn't spend my whole day filing stuff

- I can assign work to my project team without having to clear it with their bosses.

Organization Magnets Solutions

In a functional organization, which is what Kate works in, project managers don't have the authority to make major decisions on projects. Projectized organizations give all of the authority to the PM.

Can you work out which description goes with which organization type?

In a functional organization, the teams working on the project don't report directly to the PM. Instead, the teams are in departments, and the project manager needs to "borrow" them for the project.

In this kind of company, the team reports to the project manager, who has a lot more authority.

Functional Organization

1. Project managers need to clear major decisions with department managers.

2. PMs don't set the budget.

3. PMs spend half their time doing admin tasks.

Projectized Organization

1. Teams are organized around projects.

2. Project managers choose the team members, and release them when the project is over.

3. Project managers estimate and track budget and schedule.

Exam Questions

1. Which of the following is NOT a Project Constraint?

   A. Quality

   B. Scale

   C. Time

   D. Cost

2. A project manager is running a data center installation project. He finds that his stakeholder is angry because he's run over his budget because the staff turned out to be more expensive than planned. The stakeholder's unhappy that when the project is over, the servers won't have as much drive space as he needs. Which of the following constraints was not affected by this problem?

   A. Quality

   B. Resource

   C. Time

   D. Cost

3. Which of the following is NOT an example of operational work?

   A. Building a purchase order system for accounts payable

   B. Submitting weekly purchase orders through a purchase order system

   C. Deploying weekly anti-virus software updates

   D. Yearly staff performance evaluations

4. You're managing a project to build a new accounting system. One of the accountants in another department really likes the current system and is refusing to be trained on the new one. What is the BEST way to handle this situation?

   A. Refuse to work with him because he's being difficult

   B. Appeal to the accountant's manager and ask to have him required to take training

   C. Get a special dispensation so that the accountant doesn't have to go to the training

   D. Work with him to understand his concerns and do what you can to help alleviate them without compromising your project

5. Which of the following is used for identifying people who are impacted by the project?

   A. Resource List

   B. Stakeholder Register

   C. Enterprise Environmental Factors

   D. Project Plan

Exam Questions

6. Your manager asks you where to find a list of projects that should be managed together. What is the BEST place to find this information?

    A.   Project Plan

    B.   Project Charter

    C.   Portfolio Charter

    (D.)   Program Charter

7. You want to know specifically which business goal a group of projects and programs are going to accomplish. Which is the best place to look for this information?

    A.   Project Plan

    B.   Project Charter

    (C)   Portfolio Charter

    D.   Program Charter

8. A project coordinator is having trouble securing programmers for her project. Every time she asks her boss to give a resource to the project he says that they are too busy to help out with her project. Which type of organization is she working in?

    (A)   Functional

    B.   Weak Matrix

    C.   Strong Matrix

    D.   Projectized

9. A project manager is having trouble securing programmers for her project. Every time she asks the programming manager for resources for her project, he says they're all assigned to other work. So she is constantly having to go over his head to overrule him. Which type of organization is she working for?

    A.   Functional

    B.   Weak Matrix

    (C)   Strong Matrix

    D.   Projectized

10. The project manager for a construction project discovers that a new water line is being created in the neighborhood where he's managing a project. Company policy requires that a series of forms for city environmental changes need to be filled out before his team can continue work on the project. This is an example of:

    A.   A portfolio

    B.   A program

    (C)   An enterprise environmental factor

    D.   A project

Answers

Exam ~~Questions~~

1. Answer: B

Scale is not a project constraint. The constraints are Scope, Time, Cost, Quality, Resource, and Risk.

2. Answer: C

There is no mention of the project being late or missing its deadlines in the example. The project was over budget, which affects the project's cost. The project won't meet the stakeholder's requirements, which is a quality problem. And the staff was more expensive than planned, which is another cost problem.

3. Answer: A

Building a purchase order system for accounts payable is a project. It's a temporary effort that has a unique result.

4.Answer: D

When a stakeholder is negatively impacted by your project, you need to manage his expectations and help him to buy into your project.

5. Answer: B

The stakeholder register is where you identify all of the people who are impacted by your project.

6. Answer: D

A program is a group of projects that should be managed together because of interdependencies. A program charter fits the description in this question.

Usually there's some benefit to the company by managing them together.

7. Answer: C

A portfolio charter will give the business goal that a group of projects and programs will accomplish as part of a portfolio.

Exam Questions
Answers

8. Answer: A

Since the project manager has to ask permission from the functional manager and can't overrule him, she's working in a functional organization.

9. Answer: C

The Project Manager in this scenario can overrule the functional manager, so she's working in a Strong Matrix organization. If it were a projectized organization, she wouldn't have to get permission from the functional manager at all because she'd be the person with authority to assign resources to projects.

10. Answer: C

Since the project manager is filling out the form because of a company policy. This is a good example of an enterprise environmental factor.

3 The process framework

It all fits together

WE'RE USING PROCESS GROUPS AND KNOWLEDGE AREAS TO DO OUR PART!

All of the work you do on a project is made up of processes.

Once you know how all the processes in your project fit together, it's easy to remember
everything you need to know for the PMP exam. **There's a pattern** to all of the work that
gets done on your project. First you plan it, and then you get to work. While you are doing
the work, you are always comparing your project to your original plan. When things start
to get off-plan, it's your job to make corrections and put everything back on track. And the
process framework—the **process groups** and **knowledge areas**—is the key to all of
this happening smoothly.

Cooking up a project

When you cook something from a recipe for the first time, there are certain steps you always follow:

1 First figure out what you're going to make.

2 Then make all your plans.

Make a shopping list of everything you need.

3   Next, it's time to start cooking!

Set the oven to 375°F.

Cook 'em until they're golden (around 8–10 minutes).

4   Finally, you can give the cookies to a loved one.

SHE'S GOING TO LOVE THESE. TONIGHT, I'LL BASK IN THE WARM GLOW OF HER LOVE (AND THE KNOWLEDGE OF A PROJECT WELL DONE).

Projects are like recipes

All projects, no matter how big or small, break down into
process groups. **Process groups** are like the steps you use
when following a recipe.

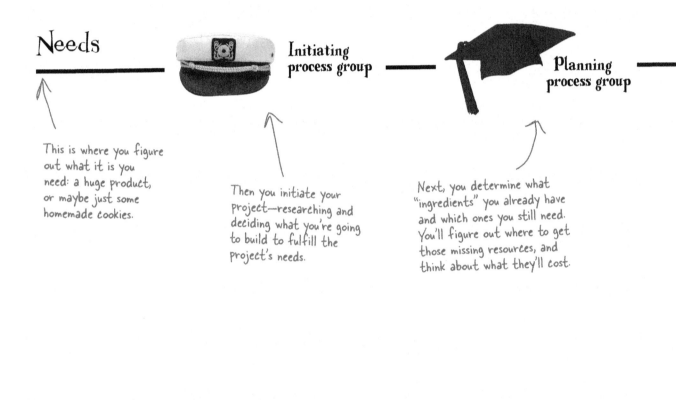

Needs

This is where you figure
out what it is you
need: a huge product,
or maybe just some
homemade cookies.

**Initiating
process group**

Then you initiate your
project—researching and
deciding what you're going
to build to fulfill the
project's needs.

**Planning
process group**

Next, you determine what
"ingredients" you already have
and which ones you still need.
You'll figure out where to get
those missing resources, and
think about what they'll cost.

This is where the bulk of the project work is done.

Executing process group

This is where you actually mix the ingredients, put the dough on a cookie sheet, pop the sheet into the oven...

Closing process group

Success!

Closing out a project means making sure you get paid...and closing out a recipe means making sure you get to eat good food!

Monitoring and Controlling process group

Another large part of project management is keeping an eye on everything that's happening, and adjusting processes as needed. So as you're mixing, you check that the consistency is right, and you keep an eye on the oven temperature while baking.

If your project's really big, you can manage it in <u>phases</u>

A lot of project managers manage projects that are big or complex, or simply need to be done in stages because of external constraints, and that's when it's useful to approach your project in **phases**. Each phase of the project *goes through all five process groups*, all the way from Initiating to Closing. The end of a phase is typically a natural point where you want to assess the work that's been done so that you can hand it off to the next phase. When your project has phases that happen one after another and don't overlap, that's called a **sequential relationship** between the phases.

You're managing a large web development project...

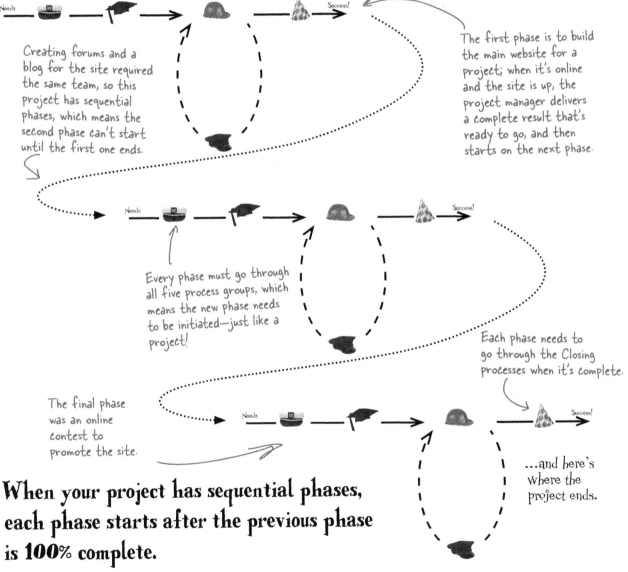

Creating forums and a blog for the site required the same team, so this project has sequential phases, which means the second phase can't start until the first one ends.

The first phase is to build the main website for a project; when it's online and the site is up, the project manager delivers a complete result that's ready to go, and then starts on the next phase.

Every phase must go through all five process groups, which means the new phase needs to be initiated—just like a project!

Each phase needs to go through the Closing processes when it's complete.

The final phase was an online contest to promote the site.

...and here's where the project ends.

When your project has sequential phases, each phase starts after the previous phase is 100% complete.

Phases can also overlap

Sometimes you need teams to work independently on different parts of the project, so that one team delivers their results while another team is still working. That's when you'll make sure that your phases have an **overlapping relationship**. But even though the phases overlap, and may not even start at the same time, they still need to go through all five process groups.

> The first phase is set to deliver while the second is still executing... but for some projects you might have an overlapping phase that ends before the previous phase. As a result, overlapping phases can get pretty complicated to manage! That's why overlapping phases can increase risk, because your team might have to do a lot of rework.

Needs → ▮ ▸ → ⬛ ··· 🍕 Success!

> This project has two overlapping phases. In this case, they don't start at the same time—the first phase's team needs to get started before the team for the second phase.

Needs → ▮ ▸ → ⬛ ··· 🍕 Success!

> When the second phase begins, it needs to go through the Initiating process group independently, even though the first phase is already in the Executing processes.

> Since one team is planning one phase while executing another, this means the whole team (INCLUDING designers, testers, etc.) is usually working at the same time.

<u>Iteration</u> means executing one phase while planning the next

There's a third approach to phased projects that's partway between sequential and overlapping. When your phases have an **iterative relationship**, it means that you've got a single team that's performing the Initiating and Planning processes for one phase of the project while also doing the Executing processes for the previous phase. That way, when the processes in the Executing and Closing process groups are finished, the team can jump straight into the next phase's Executing processes.

Iteration is a really effective way to run certain kinds of software projects. **Agile software development** is an approach to managing and running software projects that's based on the idea of iterative phases.

> This is a really good way to deal with an environment that's very uncertain, or where there's a lot of rapid change. Does this sound like any of the projects you've worked on?

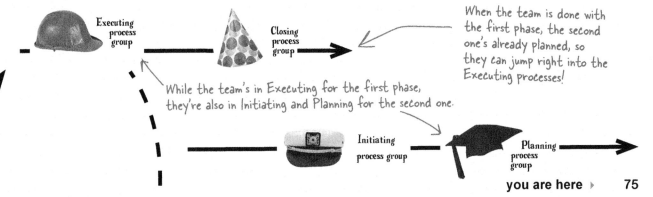

Executing process group → Closing process group →

> When the team is done with the first phase, the second one's already planned, so they can jump right into the Executing processes!

> While the team's in Executing for the first phase, they're also in Initiating and Planning for the second one.

Initiating process group → Planning process group →

Break it down

Within each process group are several individual **processes**, which is how you actually do the work on your project. The *PMBOK Guide* breaks every project down into 47 processes—that sounds like a lot to know, but don't start looking for the panic button! In your day-to-day working life, you actually use most of them already…and by the time you've worked your way through this book, you'll know all of them.

Relax

Taking a vacation is simple, even though there are several steps.

Forty-seven processes might seem like a lot to remember, but once you've been using them for a while, they'll be second nature—just like all the things you do without thinking when you go on a trip.

Stuff you do when you take a vacation

The PMBOK Guide Processes that those steps correspond to

1 Figure out how much time you have off, how much money you can spend, and where you want to go.

1 Develop Project Charter.

2 Find your flights and hotel information and put together an itinerary using a travel website.

2 Develop Project Management Plan.

Simultaneous

3 Take your flight, stay in the hotel, see the sights. Enjoy yourself.

3 Direct and Manage Project Work.

4 Make sure you get the seat you want on the plane, your hotel room is clean, and the sightseeing tours are worth your money. If not, complain, correct any problems that come up, and try to get better service.

4 Monitor and Control Project Work.

Don't worry about memorizing these process names now… you'll see a lot more of each of them throughout the book.

5 Come home, pay all the bills, and write up your reviews of the trip for the hotel feedback website.

5 Close Project.

Process Magnets

Below are several of the 47 processes. Try to guess which process group each process belongs to just from the name. We've done the first two for you.

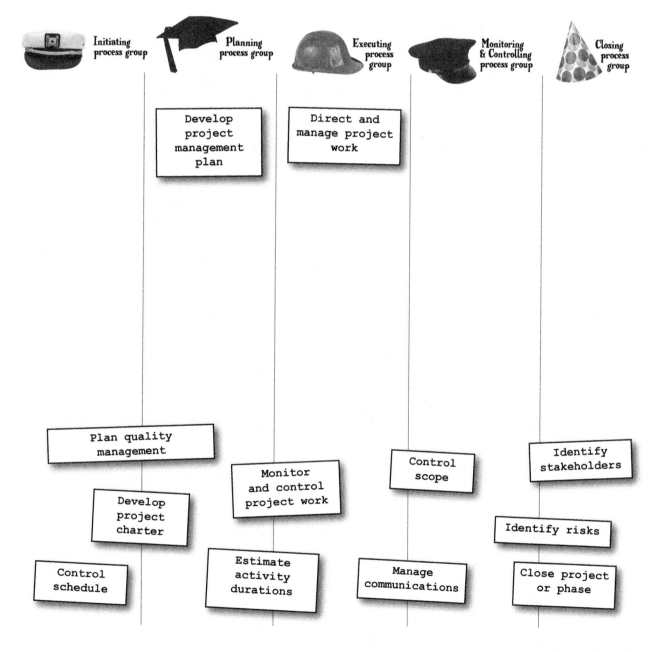

Initiating process group

Planning process group

Executing process group

Monitoring & Controlling process group

Closing process group

Develop project management plan

Direct and manage project work

Plan quality management

Monitor and control project work

Control scope

Identify stakeholders

Develop project charter

Identify risks

Control schedule

Estimate activity durations

Manage communications

Close project or phase

Process Magnets

Below are several of the 47 processes. Try to guess which process group each process belongs to just from the name!

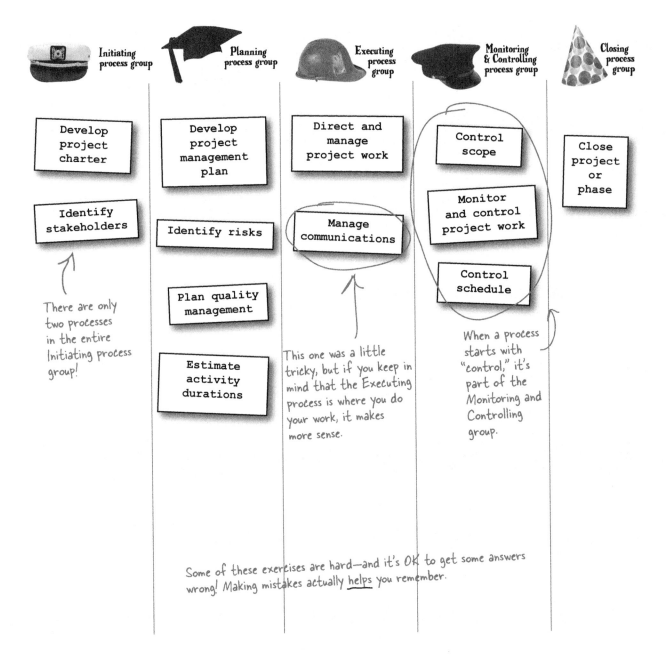

Initiating process group

Planning process group

Executing process group

Monitoring & Controlling process group

Closing process group

Develop project charter

Develop project management plan

Direct and manage project work

Control scope

Close project or phase

Identify stakeholders

Identify risks

Manage communications

Monitor and control project work

Plan quality management

Control schedule

Estimate activity durations

There are only two processes in the entire Initiating process group!

This one was a little tricky, but if you keep in mind that the Executing process is where you do your work, it makes more sense.

When a process starts with "control," it's part of the Monitoring and Controlling group.

Some of these exercises are hard—and it's OK to get some answers wrong! Making mistakes actually <u>helps</u> you remember.

Anatomy of a process

You can think of each process as a little machine. It takes the **inputs**—information you use in your project—and turns them into **outputs**: documents, deliverables, and decisions. The outputs help your project come in on time, within budget, and with high quality. Every single process has inputs, **tools and techniques** that are used to do the work, and outputs.

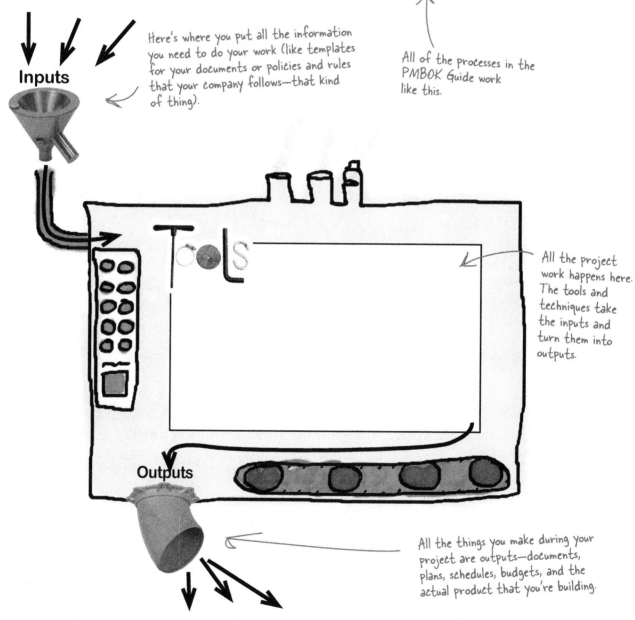

Inputs

Here's where you put all the information you need to do your work (like templates for your documents or policies and rules that your company follows—that kind of thing).

All of the processes in the PMBOK Guide work like this.

Tools

All the project work happens here. The tools and techniques take the inputs and turn them into outputs.

Outputs

All the things you make during your project are outputs—documents, plans, schedules, budgets, and the actual product that you're building.

Sharpen your pencil

Think of the vacation we talked about on page 76 as a project, and each of its steps as a process. Here are some inputs, tools, and outputs that could be used in each of the vacation steps. Can you look at each of the underlined words and figure out if the words represent an input, tool, or output? (Here's a hint: some of them are an output from one process and an input for another.)

Look at each of these underlined things, and figure out if it's an input, output, and/or tool.

1. You log in and check your company's <u>vacation calendar</u> to see how much vacation time you have for your trip.

    ☒ Input   ☒ Tool   ☐ Output

2. You create <u>an itinerary</u> on a travel website. You'll use the itinerary when you board your flight.

    ☐ Input   ☐ Tool   ☐ Output

3. You have some <u>hotel reservation documents</u> you created on the travel website, too. You'll use those when you check in to your hotel.

    ☐ Input   ☐ Tool   ☐ Output

4. You use a <u>travel website</u> to book the plane, hotel, and sights you'll see on your trip.

    ☐ Input   ☐ Tool   ☐ Output

5. You verify your <u>bank account balance to</u> make sure you have enough money to pay for everything.

    ☐ Input   ☐ Tool   ☐ Output

6. You use a <u>hotel feedback website</u> to review your stay in the hotel once you get back home.

    ☐ Input   ☐ Tool   ☐ Output

⟶ Answers on page 90.

OK, I UNDERSTAND HOW THIS WORKS FOR VACATIONS AND COOKIES, BUT 47 PROCESSES ON EVERY PROJECT? YOU'VE GOT TO BE KIDDING ME...

These processes are meant to work on <u>any</u> type of project.

The processes are there to help you organize how you do things. But they have to work on small, medium, and large projects. Sometimes that means a lot of processes—but it also ensures that what you're learning here will work on <u>all</u> your projects.

there are no Dumb Questions

Q: Can a process be part of more than one process group?

A: No, each process belongs to only one process group. The best way to figure out which group a process belongs to is to remember what that process does. If the process is about defining high-level goals of the project, it's in Initiating. If it's about planning the work, it's in Planning. If you are actually doing the work, it's in Executing. If you're tracking the work and finding problems, it's in Monitoring and Controlling. And if you're finishing stuff off after you've delivered the product, that's Closing.

Q: Do you do all of the processes in every project?

A: Not always. Some of the processes apply only to projectized organizations or subcontracted work, so if your company doesn't do that kind of thing, then you won't need those processes. But if you want to make your projects come out well, then it *really does make sense* to use the processes. Even a small project can benefit from taking the time to plan out the way you'll handle all of the knowledge areas. If you do your homework and pay attention to all of the processes, you can avoid most of the big problems that cause projects to run into trouble!

Q: Can you use the same input in more than one process?

A: Yes. There are a lot of inputs that show up in multiple processes. For example, think about a schedule that you'd make for your project. You'll need to use that schedule to build a budget, but also to do the work! So that schedule is an input to at least two processes. That's why it's really important that you write down exactly how you use each process, so you know what its inputs and outputs are.

Your company should have records of all of these process documents, and the stuff the PMs learned from doing their projects. We call these things "organizational process assets," and you'll see a lot of them in the next chapter.

Combine processes to complete your project

Sometimes the output of one process becomes an input of the next
process. In the cookie project, the raw ingredients from the store are
the outputs of the planning process, but they become the inputs for the
executing process, where you mix the ingredients together and bake them:

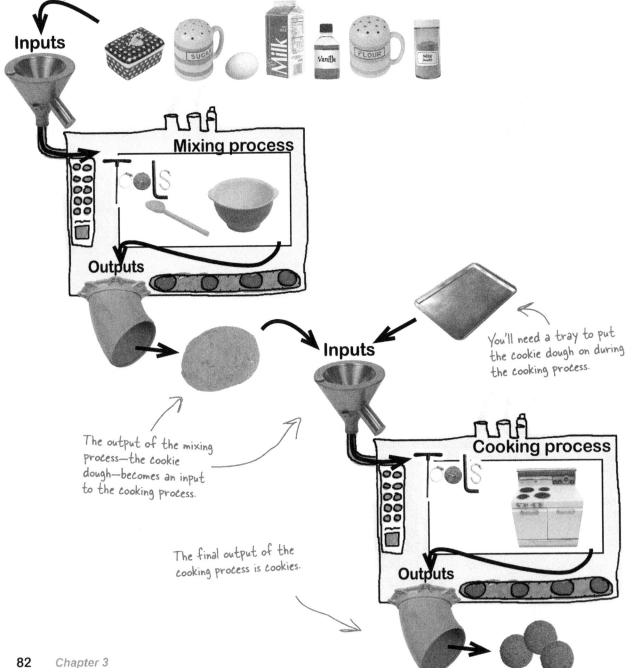

Inputs

Mixing process

Outputs

Inputs

You'll need a tray to put
the cookie dough on during
the cooking process.

The output of the mixing
process—the cookie
dough—becomes an input
to the cooking process.

Cooking process

The final output of the
cooking process is cookies.

Outputs

Knowledge areas organize the processes

The **process groups** help you organize the processes by the *kind of work* you do. The **knowledge areas** help you organize by the *subject matter* you're dealing with. The following 10 elements of the cookie process are the *PMBOK Guide* knowledge areas.

The processes are organized in two ways—the process groups are about how you do the work, and the knowledge areas are there to help you categorize them and help you learn.

Integration

Making sure all the right parts of the project come together in the right order, at the right time

Scope

Could you have decorated the cookies? Or made more batches?

Time

Preparation and cooking time

Cost

Budgeting for the cookie project

Quality

Checking that the cookies look and taste right

Human Resource

Making sure your schedule is clear, and your honey is going to be home on time

Risk

Could you burn the cookies or yourself on the range? Are the eggs fresh?

Communications

Making sure you're not mixing metric and imperial measurements

Stakeholder

Does your sweetheart like chocolate chip cookies, peanut butter, or oatmeal?

Procurement

Selecting the right store to supply your ingredients

Knowledge Area Magnets

Match the knowledge areas to each description. We've filled in a couple for you.

Time Management

Coordinating all of the work so that it happens correctly. Making sure changes are approved before they happen.	Figuring out what work needs to be done for your project. Making sure your end product has everything you said it would.	Figuring out the time it will take to do your work and the order you need to do it in. Tracking your schedule and making sure everything gets done on time.	Knowing how much you're able to invest in the project and making sure you spend it right.	Making sure you work as efficiently as you can and don't add defects into the product.

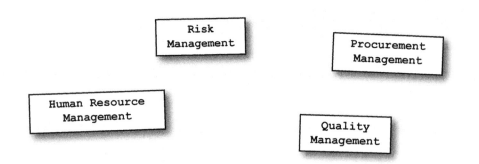

Risk Management

Procurement Management

Human Resource Management

Quality Management

Communications
Management

Getting the people to work on the team and helping them stay motivated. Rewarding them for a job well done and resolving conflicts that come up.

Making sure that everybody knows what they need to know to do their job right. Tracking how people talk to each other and dealing with misunderstandings or miscommunications if they happen.

Figuring out how to protect your project from anything that could happen to it. Dealing with the unexpected when it does happen.

Finding contractors to help you do the work. Setting the ground rules for their relationships with your company.

Identifying the group of people who might have an impact on your project or who your project will affect. Understanding what they need and making sure your project delivers it.

Scope
Management

Stakeholder
Management

Integration
Management

Cost
Management

Knowledge Area Magnets Solutions

Match the knowledge areas to each description.

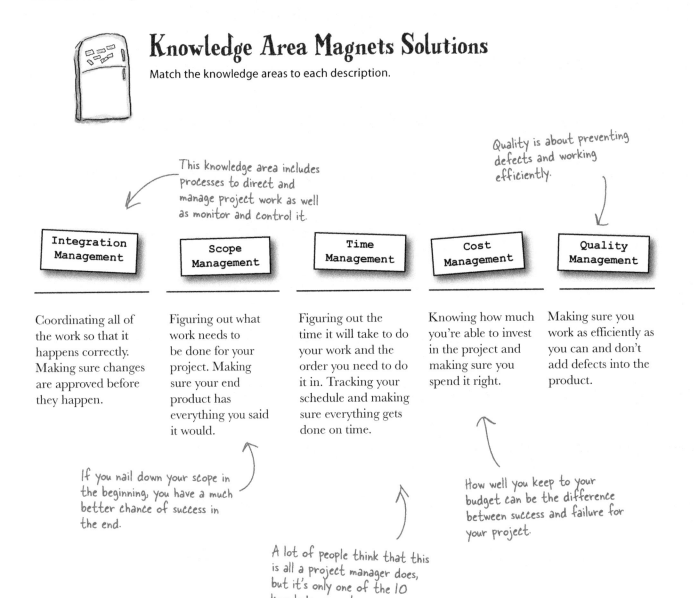

Quality is about preventing defects and working efficiently.

This knowledge area includes processes to direct and manage project work as well as monitor and control it.

Integration Management	Scope Management	Time Management	Cost Management	Quality Management
Coordinating all of the work so that it happens correctly. Making sure changes are approved before they happen.	Figuring out what work needs to be done for your project. Making sure your end product has everything you said it would.	Figuring out the time it will take to do your work and the order you need to do it in. Tracking your schedule and making sure everything gets done on time.	Knowing how much you're able to invest in the project and making sure you spend it right.	Making sure you work as efficiently as you can and don't add defects into the product.

If you nail down your scope in the beginning, you have a much better chance of success in the end.

How well you keep to your budget can be the difference between success and failure for your project.

A lot of people think that this is all a project manager does, but it's only one of the 10 knowledge areas!

Since the PMBOK Guide covers projectized organizations, it talks about actually acquiring your team as a process, too. People in most organizations don't get a chance to do that. The team is often determined by the time you get assigned to it.

This one is another area that a lot of PMs don't have much experience with. It's all about selecting suppliers, contractors, and vendors, and setting up contracts with them.

Human Resource Management

Communications Management

Risk Management

Procurement Management

Stakeholder Management

Getting the people to work on the team and helping them stay motivated. Rewarding them for a job well done and resolving conflicts that come up.

Making sure that everybody knows what they need to know to do their job right. Tracking how people talk to each other and dealing with misunderstandings or miscommunications if they happen.

Figuring out how to protect your project from anything that could happen to it. Dealing with the unexpected when it does happen.

Finding contractors to help you do the work. Setting the ground rules for their relationships with your company.

Identifying the group of people who might have an impact on your project or who your project will affect. Understanding what they need and making sure your project delivers it.

Communication is a really important part of the project manager's job.

Risk Management can also be about making sure that you are in the right position to take advantage of the opportunities that come your way.

Once you know who your stakeholders are, you can keep a constant check on how well your project is meeting their expectations.

Watch it!

Even though all of these knowledge areas are important throughout your project, the *PMBOK Guide* covers them in the order above.

All of the knowledge areas are used throughout every project to keep your project on track.

there are no
Dumb Questions

Q: So what's the difference between process groups and knowledge areas?

A: The process groups divide up the processes by function. The knowledge areas divide up the same processes by subject matter. Think of the process groups as being about the *actions* you take on your project, and the knowledge areas as the things you need to *understand*.

In other words, the knowledge areas are more about helping you understand the *PMBOK Guide* material than about running your project. But that doesn't mean that every knowledge area has a process in every process group! For example, the Initiating process group has only two processes. The Risk Management knowledge area has only Planning and Monitoring and Controlling processes. So the process groups and the knowledge areas are two different ways to think about all of the processes, but they don't really overlap.

Q: Is every knowledge area in only one process group?

A: Every process belongs to exactly one process group, and every process is in exactly one knowledge area. But a knowledge area has lots of processes in it, and they can span some, or all, of the groups. Think of the processes as the core information in the *PMBOK Guide*, and the process groups and knowledge areas as two different ways of grouping these processes.

Q: It seems like the Initiating and Planning process groups would be the same. How are they different?

A: Initiating is everything you do when you first start a project. You start by writing down (at a very high level) what the project is going to produce, who's in charge of it, and what tools are needed to do the work. In a lot of companies, the project manager isn't even involved in much of this. Planning just means going into more detail about all of that as you learn more about it, and writing down specifically how you're going to do the work. The Planning processes are where the project manager is really in control and does most of the work.

> **Process groups and knowledge areas are two different ways to organize the processes...but they don't really overlap each other! Don't get caught up trying to make them fit together.**

The benefits of successful project management

Take a moment to digest all this new knowledge, because you're going to start putting it all into practice when we take a look at Project Integration Management in Chapter 4.

Sharpen your pencil
Solution

Think of the vacation we talked about on page 70 as a project, and each of its steps as a process. Here are some inputs, tools, and outputs that could be used in each of the vacation steps. Can you look at each of the underlined words and figure out if the words represent an input, tool, or output? (Here's a hint: some of them are an output from one process and an input for another.)

1. You log in and check your company's vacation calendar to see how much vacation time you have for your trip.

   ☒ Input   ☐ Tool   ☐ Output

2. You create an itinerary on a travel website. You'll use the itinerary when you board your flight.

   ☒ Input   ☐ Tool   ☒ Output   ← *The itinerary was an output of the Develop Project Management Plan process but an input to the Direct and Manage Project Work process.*

3. You have some hotel reservation documents you created on the travel website, too. You'll use those when you check in to your hotel.

   ☒ Input   ☐ Tool   ☒ Output

4. You use a travel website to book the plane, hotel, and sights you'll see on your trip.

   ☐ Input   ☒ Tool   ☐ Output   ← *This one was the tool you used to book your tickets and hotel reservations.*

5. You verify your bank account balance to make sure you have enough money to pay for everything.

   ☒ Input   ☐ Tool   ☐ Output   *You had to know this to know how much you could spend on your trip. It's an input.*

6. You use a hotel feedback website to review your stay in the hotel once you get back home.

   ☐ Input   ☒ Tool   ☐ Output

   Here's the tool you used to give feedback about your hotel in the Close Project process.

Exam Questions

1. You're a project manager working on a software engineering project. The programmers have started building the software, and the testers have started to create the test environment. Which process group includes these activities?

A. Initiating

B. Planning

C. Executing

D. Closing

2. Which of the following is not a stakeholder?

A. The project manager who is responsible for building the project

B. A project team member who will work on the project

C. A customer who will use the final product

D. A competitor whose company will lose business because of the product

3. A project manager runs into a problem with her project's contractors, and she isn't sure if they're abiding by the terms of the contract. Which knowledge area is the BEST source of processes to help her deal with this problem?

A. Cost Management

B. Risk Management

C. Procurement Management

D. Communications Management

4. You're a project manager for a construction project. You've just finished creating a list of all of the people who will be directly affected by the project. What process group are you in?

A. Initiating

B. Planning

C. Executing

D. Monitoring and Controlling

5. You're a project manager working in a weak matrix organization. Which of the following is NOT true?

A. Your team members report to functional managers.

B. You are not directly in charge of resources.

C. Functional managers make decisions that can affect your projects.

D. You have sole responsibility for the success or failure of the project.

Exam Questions

6. Which of the following is NOT a project?

    A.   Repairing a car

    B.   Building a highway overpass

    C.   Running an IT support department

    D.   Filming a motion picture

7. A project manager is running a software project that is supposed to be delivered in phases. She was planning on dividing the resources into two separate teams to do the work for two phases at the same time, but one of her senior developers suggested that she use an agile methodology instead, and she agrees. Which of the following BEST describes the relationship between her project's phases?

    A.   Sequential relationship

    B.   Iterative relationship

    C.   Constrained relationship

    D.   Overlapping relationship

8. Which of the following is NOT true about overlapping phases?

    A.   Each phase is typically done by a separate team.

    B.   There's an increased risk of delays when a later phase can't start until an earlier one ends.

    C.   There's an increased risk to the project due to potential for rework.

    D.   Every phase must go through all five process groups.

9. You're the project manager for an industrial design project. Your team members report to you, and you're responsible for creating the budget, building the schedule, and assigning the tasks. When the project is complete, you release the team so they can work on other projects for the company. What kind of organization do you work in?

    A.   Functional

    B.   Weak matrix

    C.   Strong matrix

    D.   Projectized

10. Which process group contains the Develop Project Charter process and the Identify Stakeholders process?

    A.   Initiating

    B.   Executing

    C.   Monitoring and Controlling

    D.   Closing

Answers

Exam ~~Questions~~

1. Answer: C

The Executing process group is the one where the team does all the work. You'll get a good feel for the process groups pretty quickly!

2. Answer: D

One of the hardest things that a project manager has to do on a project is figure out who all the stakeholders are. The project manager, the team, the sponsor (or client), the customers and people who will use the software, the senior managers at the company—they're all stakeholders. Competitors aren't stakeholders, because even though they're affected by the project, they don't actually have any direct influence over it.

3. Answer: C

The Procurement Management knowledge area deals with contracts, contractors, buyers, and sellers. If you've got a question about a type of contract or how to deal with contract problems, you're being asked about a Procurement Management process.

4. Answer: A

People who will be directly affected by the project are stakeholders, and when you're creating a list of them you're performing the Identify Stakeholders process. That's one of the two processes in the Initiating process group.

5. Answer: D

In a weak matrix, project managers have very limited authority. They have to share a lot of responsibility with functional managers, and those functional managers have a lot of leeway to make decisions about how the team members are managed. In an organization like that, the project manager isn't given a lot of responsibility.

That's why you're likely to find a project expediter in a weak matrix.

Exam ~~Questions~~ Answers

6. Answer: C

The work of an IT support department doesn't have an end date—it's not temporary. That's why it's not a project. Now, if that support team had to work over the weekend to move the data center to a new location, then that would be a project!

I SEE—SO EVEN WHEN SOMETHING IS A PROCESS AND NOT A PROJECT, THERE COULD BE PROJECTS RELATED TO IT.

7. Answer: B

Agile development is a really good example of an iterative approach to project phases. In an agile project, the team will typically break down the project into phases, where they work on the current phase while planning out the next one.

Answers

Exam Questions

8. Answer: B

If there's an increased risk of a project because one phase can't start until another one ends, that means your project phases aren't overlapping. When you've got overlapping phases, that means that you typically have multiple teams that start their phases independently of one another.

Also, take another look at answer C, because it's an important point about overlapping phases. When your phases have an overlapping relationship, there's an increased risk of rework. This typically happens when one team delivers the results of their project, but made assumptions about what another team is doing as part of their phase. When that other team delivers their work, it turns out that the results that both teams produced aren't quite compatible with each other, and now both teams have to go back and rework their designs. This happens a lot when your phases overlap, which is why overlapping phases have an increased risk of rework.

9. Answer: D

In a projectized organization, the project manager has the power to assign tasks, manage the budget, and release the team.

10. Answer: A

The first things that are created on a project are the charter (which you create in the Develop Project Charter process) and the stakeholder register (which you create in the Identify Stakeholders process). You do those things when you're initiating the project.

Getting the job done

Want to make success look easy? It's not as hard as you think. In this chapter, you'll learn about **a few processes** you can use in your projects every day. Put these into place, and your **sponsors** and **stakeholders** will be happier than ever. Get ready for **Integration Management**.

Time to book a trip

Everyone in the Midwestern Teachers' Association has gotten together and planned a trip—a tour of Asia and Europe, starting with Mumbai, India, and ending up in Paris, France.

> WE WANT TO GO IN FEBRUARY, AND WE NEED TO STAY IN BUDGET. CAN YOU GET US A GOOD DEAL? GOTTA RUN; WE'VE GOT CLASSES IN FIVE.

> YES, I GOT THAT: YOU WANT WINDOW, NOT AISLE...

Larry, the teachers' travel agent

Joanne and Frank were "volunteered" to organize the trip by the other eight teachers in the group.

Larry's cutting corners

Larry wants to dive into the project and make his clients happy. When he sees an opportunity to save them money, he takes it! But sometimes the cheapest way to do things isn't the way that will end up satisfying everyone.

> HMM... IF I BOOK THEIR TICKETS IN JUNE INSTEAD OF FEBRUARY, THAT'LL REDUCE THE FARE. I'LL BET THAT CONVINCES THEM TO CHANGE THEIR PLANS. WHO WOULDN'T LOVE THOSE SAVINGS?!

The teachers are thrilled...for now

Larry convinces the teachers to travel in June because of the great price he got on tickets. But he's not really planning for the results of that decision—and neither are the teachers.

One of the keys of project management is thinking a project through *before* starting the work, so problems that could arise down the line are anticipated ahead of time. That's why so much of project management is spent **planning**.

Larry may think this itinerary is a plan, but it doesn't detail any of the problems that could arise on the teachers' vacation.

Acme Travel

```
TRAVEL ITINERARY FOR
MIDWESTERN TEACHERS ASSN.

Record Locator     HF184-Z              Agent ID      Larry
Trip ID            189435163                          Acme Travel
Travel Details

Flight Information

Leg 1
Airline            Econo Airlines       Departing     1:45PM
Flight             8614                 Arriving      1:00AM
Origin             St. Paul, MN         Terminal      1
Destination        Mumbai, INDIA        Arriving      June 13
Est Time           17 hours 45 Minutes  Distance      7942mi
```

Larry changed the date to June, and now the project's coming in way under budget.

These clients are definitely <u>not</u> satisfied

When the clients arrived in Mumbai, they found out why the fare was so low: June is monsoon season in India! Larry may have saved them a bundle, but it didn't keep him from soaking his clients.

LARRY NEVER MENTIONED THE CHEAPER FARES WERE BECAUSE OF THE RAINY SEASON!

A LITTLE WARNING WOULD'VE BEEN NICE. WE COULD HAVE PACKED RAINCOATS!

Even though the itinerary got them there exactly when Larry specified, there were no plans to deal with the pouring rain.

Larry's been let go

From the minute they got off the plane, the clients were extremely unhappy. The senior managers at Acme Travel don't want to lose the teachers' business…so they've appointed YOU as the new travel agent.

It's your job to finish planning the trip, and make sure that the teachers leave their vacation satisfied.

The day-to-day work of a project manager

Project managers make projects run well. They plan for what's going to happen on the project. A big part of the job is watching closely to make sure the plan is followed, and when things go wrong, making sure they're fixed. And sometimes the plan itself turns out to be inadequate! Project managers look for those kinds of problems, and fix them too. That day-to-day work is what the **Integration Management** processes are all about.

A bird's-eye view of a project

Every project follows the same kind of pattern. First it gets initiated, then planned, then executed (and monitored), and finally closed. That's why the process groups are so useful—they're a good way to think about how you do the work.

For a large project, you'll often see this pattern repeated several times. Each major chunk of deliverables is treated as its own **subproject** that goes through all of the process groups and processes on its own.

Here's where every project or subproject begins.

First you get assigned to a project.

This is done by the process in the Initiating group.

Then you plan out all the work that will get done.

So you always need to use the processes in the Closing group, even when it's a subproject of a larger project.

Three Executing and Monitoring and Controlling processes make sure the project runs smoothly.

Then you make sure the work is done properly, dealing with changes along the way.

And once it's finished, you close out the project.

The six Integration Management processes

The *PMBOK Guide* divides Integration Management into six processes that you need to understand for the exam. They're what people usually think of as a project manager's "core" responsibilities.

Without the project charter, you don't have the authority to tell your team what to do and when to do it.

❶ Develop Project Charter

The very first thing that's done on a new project is the development of the project charter. That's the document that authorizes you to do your work. But you're not always involved in making it—oftentimes it's handed to you by the sponsor.

The sponsor is the person who pays for the project.

Develop Project Charter

Develop Project Management Plan

❷ Develop Project Management Plan

The Project Management plan is the most important document in the entire *PMBOK Guide* because it guides everything that happens on the project. It spans all of the knowledge areas.

A big part of the Project Management plan is that it tells you how to handle changes when problems come up.

❸ Direct and Manage Project Work

After you're done planning, it's time to do the work. Your job is to make sure that everybody is doing what they should be doing, and that the products or services your project creates meet the needs of the stakeholders.

Here's where the work gets done. It's where all of the planning you'll do in all of the other knowledge areas comes together so that you can actually make stuff. It's the day-to-day work that you help your team do, and make sure gets done.

❹ Monitor and Control Project Work

Keep everyone satisfied by catching problems as early as possible.

A good project manager is constantly monitoring every single thing that goes on in the project. Remember, the later you find a problem, the harder and more expensive it usually is to fix.

❺ Perform Integrated Change Control

Once you catch problems, this is where you figure out how to fix them—or if they should be fixed at all.

Once you've found problems on your project, you've got to work with your stakeholders and sponsors to figure out how to deal with those problems. You should also update your Project Management plan to reflect any extra steps you'll need to take to complete the project. Updating the Project Management plan also makes sure everyone working on the project stays on the same page.

Keep an eye out for <u>potential</u> changes. Part of your job is helping the people around you anticipate changes, and maybe even prevent them.

❻ Close Project or Phase

The last thing you do on the project is close it out. Make sure you document everything…especially the lessons you and your team have learned along the way. You can never tell when these lessons may help you out on your *next* project.

Sharpen your pencil

Here are a few of the things you might have to deal with in working on the teachers' vacation trip. Figure out which of the six Integration Management processes you'd use in each situation, and write down the process name in the blank.

1 It turns out that one of the teachers is a vegetarian, so some of the restaurant reservations will need to be canceled, and new reservations will need to be made at restaurants that can accommodate him.

_____

2 You come up with a detailed description of everything that you plan to do to get the teachers where they want to be.

_____

3 The CEO of Acme Travel sends you a document that assigns you to the project.

_____

4 You check in with the teachers at each destination to make sure everything is going according to plan.

_____

5 When the teachers get back, you write up everything you learned while handling the trip so other travel agents can learn from your experience.

_____

6 You book the tickets and hotel accommodations.

_____

Develop Project Charter

Develop Project Management Plan

Direct and Manage Project Work

Monitor and Control Project Work

Perform Integrated Change Control

Close Project or Phase

——————→ Answers on page 147.

Start your project with the Initiating processes

All you need to get your project started are the only two processes in the
Initiating process group. First, the **Develop Project Charter** process
tells everyone in the company why the project is needed, and gives you
the authority you need to make it happen. Then you use the **Identify
Stakeholders** process to figure out who is affected by the project and
how to communicate with them.

These are the only
two processes in the
Initiating process group.

Initiating
process group

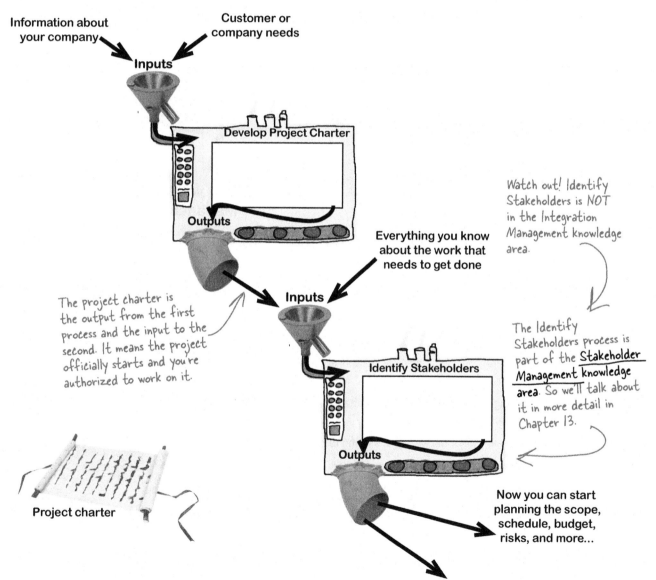

Information about
your company

Customer or
company needs

Inputs

Develop Project Charter

Outputs

Everything you know
about the work that
needs to get done

Watch out! Identify
Stakeholders is NOT
in the Integration
Management knowledge
area.

The project charter is
the output from the first
process and the input to the
second. It means the project
officially starts and you're
authorized to work on it.

Inputs

Identify Stakeholders

The Identify
Stakeholders process is
part of the Stakeholder
Management knowledge
area. So we'll talk about
it in more detail in
Chapter 13.

Outputs

Project charter

Now you can start
planning the scope,
schedule, budget,
risks, and more...

Integration Management and the process groups

Here is how the process groups all fit into this whole Integration Management thing. The process groups show you the order in which these things happen, and how the processes interact.

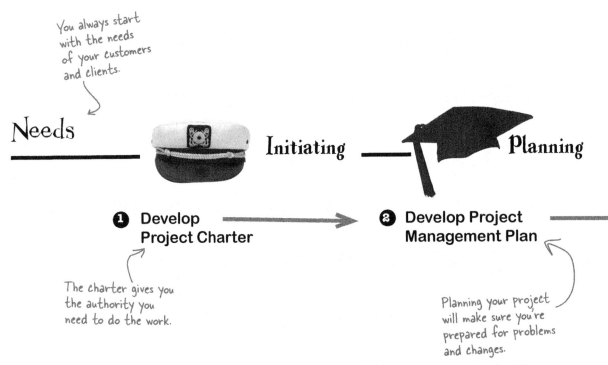

You always start with the needs of your customers and clients.

Needs ———— **Initiating** ———— **Planning**

① Develop Project Charter ——————→ **② Develop Project Management Plan**

The charter gives you the authority you need to do the work.

Planning your project will make sure you're prepared for problems and changes.

The Integration Management knowledge area brings all of the process groups together. A project manager has to integrate the work of everyone on the team through all of these major activities to keep the project on track:

1. Being authorized by the project charter to control the budget and assign resources

2. Planning all of the work that's going to happen throughout the project

3. Directing the work once it gets started

4. Monitoring the way the work progresses and looking for potential problems

5. Looking out for changes, understanding their impacts, and making sure they don't derail the project

6. Closing out the project and making sure that there are no loose ends when it's over

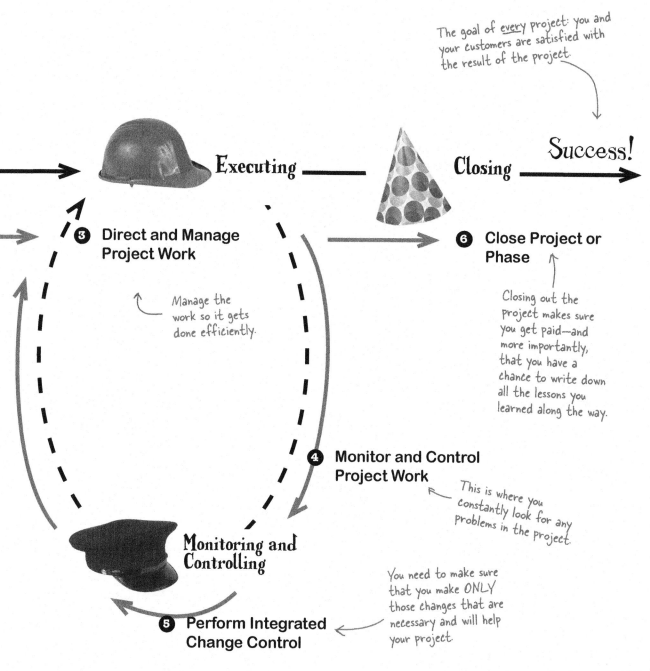

The goal of <u>every</u> project: you and your customers are satisfied with the result of the project.

Executing

Success!

Closing

❸ Direct and Manage Project Work

Manage the work so it gets done efficiently.

❻ Close Project or Phase

Closing out the project makes sure you get paid—and more importantly, that you have a chance to write down all the lessons you learned along the way.

❹ Monitor and Control Project Work

This is where you constantly look for any problems in the project.

Monitoring and Controlling

You need to make sure that you make ONLY those changes that are necessary and will help your project.

❺ Perform Integrated Change Control

The Develop Project Charter process

Initiating
process group

If you work in a matrix organization, then your team doesn't report to you. They report to functional managers, and might have other work to do. But when they're on your project, you're effectively their boss. So how do you make that happen? Well, you need some sort of **authorization**, and that's what the project charter is for. It says exactly what you're authorized to do on the project (like assign work to the team members and use the company's resources), and why you've been assigned to it. But the charter isn't just important for matrix companies. In any kind of company, it's really important to know who's in charge, and what resources you have available to you when you manage a project.

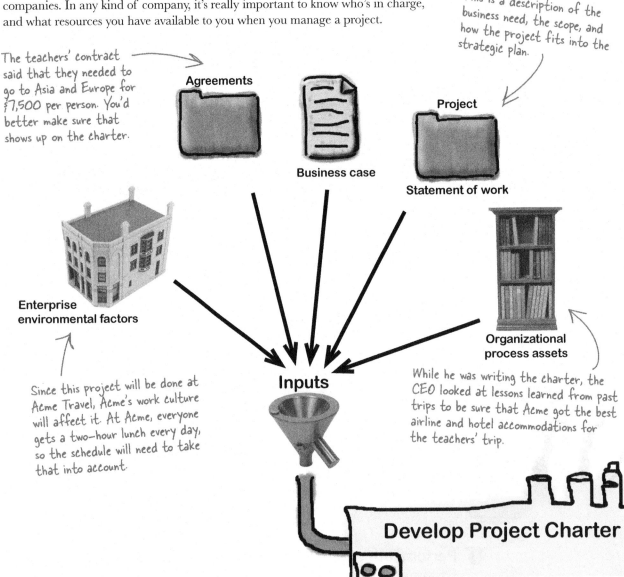

The teachers' contract said that they needed to go to Asia and Europe for $7,500 per person. You'd better make sure that shows up on the charter.

This is a description of the business need, the scope, and how the project fits into the strategic plan.

Agreements

Business case

Project

Statement of work

Enterprise environmental factors

Since this project will be done at Acme Travel, Acme's work culture will affect it. At Acme, everyone gets a two-hour lunch every day, so the schedule will need to take that into account.

Organizational process assets

While he was writing the charter, the CEO looked at lessons learned from past trips to be sure that Acme got the best airline and hotel accommodations for the teachers' trip.

Inputs

Develop Project Charter

Make the case for your project

The Midwestern Teachers' Association contract wasn't the only one that Acme could have taken. The company's got more work than it can handle right now, and occasionally it needs to turn away a client. That's where a **business case** comes in handy. If a project is too risky, won't make enough money, isn't strategic, or isn't likely to succeed, then the senior managers at Acme could choose to pass on it.

But to figure all that out, you need to do some thinking about what makes taking on this project a good idea for Acme Travel. Preparing a business case means thinking about the value of the project to business. Is there a big market for world travel packages that Acme can break into if it does this project? Should Acme do it just because the customer requested it? Will it help the company in other ways?

BUSINESS CASE DOCUMENT

Acme Travel

Midwestern Teachers' Association World Tour

Project Description: A group of teachers from Minnesota wants to take a trip around the world, starting with Mumbai, India, and ending somewhere in Europe.

Strategic Analysis: Taking on this project would give Acme Travel Agency an edge over most of the other travel agencies in town that don't offer travel packages to southeast Asia. The only travel agencies in the area that offer this kind of package charge about $500 more for the package than our clients are willing to pay. By offering the package at the cost the Teachers' Association has suggested, we'll make around $700 profit on the trip and still be able to undercut the closest competition.

Intangible Benefits: The agents who work on this trip will gain experience booking travel in Asia, and that will help us with some other prospects that have expressed interest in similar trips.

Related Projects: This project is similar to the 2007 Handbell Enthusiasts European Tour we managed. If all goes well, we should be able to use the outcome of this project as leverage to win the travel planning job for the Midwestern High School Horticulture Club World Tour that's coming up in 2011.

Conclusion: It's in Acme's best interest to do this project.

This project will make the company money.

Acme needs more Asia Travel specialists. This project will help train them.

Doing this project will not only profit the company, but also might win it further business.

A business case document says why it's worth it to spend money on the project.

Use expert judgment and facilitation techniques to write your project charter

When you think about it, a lot of different people's opinions can help your company come to a good decision about whether or not to start a project. Sometimes project sponsors will call on experts to help them decide which projects to do. At Acme Travel, the CEO called a meeting with the VP of Asia Travel to make sure that the teachers' trip was worth doing. The VP of Asia Travel had set up trips like this one before and he knew where things could go wrong in planning them. Together, they looked at all of the project documentation to make sure that this project looked like it would make Acme enough money to be worth doing.

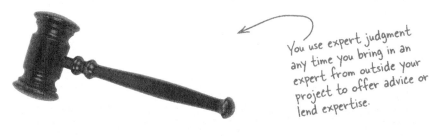

You use expert judgment any time you bring in an expert from outside your project to offer advice or lend expertise.

Your company might need to talk to subject matter experts from a bunch of different departments to decide if a project will be beneficial to it. It might rely on outside consultants or industry groups to tell it how other companies have solved the same problem. All of those different opinions are called **expert judgment**.

If the experts agree that the project's business case, contract, and statement of work all add up to a product that's going to do good things for your company, they'll usually give the green light to write the charter.

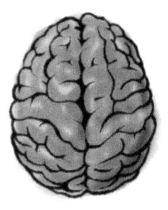

Facilitation techniques help everyone understand the goal of your project

When you sit down to write your project charter, you'll need to get your stakeholders on the same page about what your project team will do. You might set up meetings with your stakeholders to brainstorm project goals or work with them to resolve conflicts around how your project will run. All of the approaches you take to get everybody on the same page are called **facilitation techniques**.

Exercise

Here are a bunch of ways Acme evaluated the inputs for the Develop Project Charter process. Try to figure out which ones involve expert judgment and which are facilitation techniques.

1. Acme Travel creates a committee to review all of the business case documents that have been submitted for possible projects and compare them to figure out which projects should be funded in the next quarter.

A. Expert judgment   B. Facilitation technique

2. Acme hires an outside consultant to help it figure out whether or not its current strategic goals are the right ones for the company.

A. Expert judgment   B. Facilitation technique

3. Acme asks the VP of Asia Travel to review the business case for the Midwest Teachers' Association trip and decide whether or not the projected costs and schedule look right.

A. Expert judgment   B. Facilitation technique

4. Acme has a big meeting with all of the project stakeholders to help it evaluate all of its project proposals and decide which ones are most likely to benefit the company.

A. Expert judgment   B. Facilitation technique

5. The travel agent who is assigned to the project holds a brainstorming session with all of the other travel agents to propose a new goal for the project.

A. Expert judgment   B. Facilitation technique

Answers on page 114.

A closer look at the project charter

The charter is the **only output** of the Develop Project Charter process. We know that it makes sense to do the project—that's what we did with the business case. And we know that it assigns authority so that you can do your job. But what else does a charter have in it?

Outputs

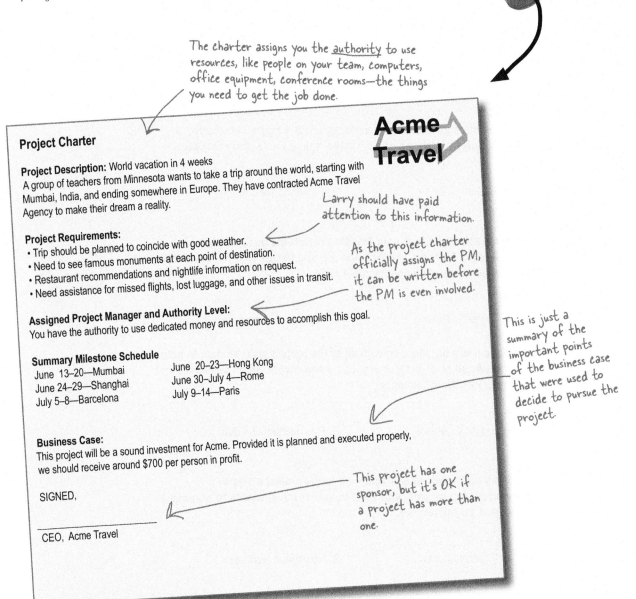

The charter assigns you the authority to use resources, like people on your team, computers, office equipment, conference rooms—the things you need to get the job done.

Project Charter

Acme Travel

Project Description: World vacation in 4 weeks
A group of teachers from Minnesota wants to take a trip around the world, starting with Mumbai, India, and ending somewhere in Europe. They have contracted Acme Travel Agency to make their dream a reality.

Larry should have paid attention to this information.

Project Requirements:
• Trip should be planned to coincide with good weather.
• Need to see famous monuments at each point of destination.
• Restaurant recommendations and nightlife information on request.
• Need assistance for missed flights, lost luggage, and other issues in transit.

As the project charter officially assigns the PM, it can be written before the PM is even involved.

Assigned Project Manager and Authority Level:
You have the authority to use dedicated money and resources to accomplish this goal.

Summary Milestone Schedule
June 13–20—Mumbai
June 24–29—Shanghai
July 5–8—Barcelona
June 20–23—Hong Kong
June 30–July 4—Rome
July 9–14—Paris

This is just a summary of the important points of the business case that were used to decide to pursue the project.

Business Case:
This project will be a sound investment for Acme. Provided it is planned and executed properly, we should receive around $700 per person in profit.

SIGNED,

_____
CEO, Acme Travel

This project has one sponsor, but it's OK if a project has more than one.

Sharpen your pencil

Take a look at the charter for the teachers' trip, and write down what you think each of the following sections of a typical project charter is used for.

Project Description:

Project Requirements:

Assigned Project Manager and Authority Level:

Summary Milestone Schedule:

Business Case:

Sharpen your pencil
Solution

Take a look at the charter for the teachers' trip, and write down what you think each of the following sections of a typical project charter is used for.

Project Description:

The purpose of the project ← This is a high-level description of the goals of your project. It's usually a few sentences that describe the project's main purpose.

Project Requirements:

Describes the product your project has to make ← Anything you know that the customer, stakeholder, or sponsor expects to get out of the project should go here.

Assigned Project Manager and Authority Level:

Who the project manager is and what he has to do ← This is where you're assigned to the project. If it's known who is going to be the project manager, the name of that person is noted. Otherwise, you may just have a department listed that you know the PM will come from. This is also where any specific decision-making authority you might need can be described.

Summary Milestone Schedule:

A list of dates that your project needs to meet

Business Case:

Why your company has decided to do this project ← This section lists the reasons why it makes sense for your business to do this project. You might note the return on investment, building infrastructure, goodwill with clients, or anything else that will help people understand why this project is important.

Exercise

Here are a bunch of ways Acme evaluated the inputs for the Develop Project Charter process. Try to figure out which ones involve expert judgment and which are facilitation techniques.

Expert judgment always refers to people using their experience to make decisions on your project.

1. (A. Expert judgment)   B. Facilitation technique

2. (A. Expert judgment)   B. Facilitation technique

3. (A. Expert judgment)   B. Facilitation technique

4. A. Expert judgment   (B. Facilitation technique)

5. A. Expert judgment   (B. Facilitation technique)

Facilitation techniques are the meetings and sessions that are used to get everybody to agree on major project decisions.

Two things you'll see over and over and over...

There are two inputs that you'll see repeatedly for a bunch of different processes throughout the rest of the book. **Enterprise environmental factors** are anything that you need to know about how your company does business. And **organizational process assets** have information about your projects: how people in your company are supposed to perform them, and how past projects have gone.

Enterprise environmental factors tell you about how your company does business.

There's a lot of information about your company that will be really useful to you when you're planning your project. You need to know how each of the different departments operates, the market conditions you're working in, the company's overall strategy, any policies you need to work with, your company's culture, and all about the people who work at the company.

Enterprise environmental factors

One of the enterprise environmental factors you'll use in the Integration Management processes is **the work authorization system**, which determines how your company assigns work to people and ensures that tasks are done properly and in the right order.

Organizational process assets tell you about how your company normally runs its projects.

Every company has standards for how to run its projects. There are guidelines and instructions for managing projects, procedures you need to follow, categories for various things you need to keep track of, and templates for all of the various documents that you need to create. These things are usually stored in some sort of library.

The Project Management plan template is an organizational process asset, too.

Organizational process assets

One of the most important organizational process assets is called **lessons learned**, which is how you keep track of valuable historical information about your project. At the end of every project, you sit down with the project team and write down everything you learned about the project. This includes both positive and negative things. That way, when you or another project manager in your company plans the next project, you can take advantage of the lessons you learned on this one.

Can you think of how these would be useful for starting and planning your project?

there are no
Dumb Questions

Q: I've never had a project charter. Is it really necessary?

A: Yes, definitely. Have you ever been on a project where you didn't feel like you had enough authority to do your job? The project charter gives you the authority to manage your project. Every project should have a charter, and writing the charter is the first thing that should happen on any project.

Q: Wait a minute! How can I be the one writing the charter, when it's what gives me all of my authority and I might not even be assigned to the project yet?

A: Right, you're not usually going to write a charter. The charter is usually handed to you. The project sponsor usually writes the charter. And it's always easy to tell who the project sponsor is: the sponsor is the person who pays for the project, and comes up with the project's overall goals.

Q: I'm still not sure I get the idea behind a business case document. How's that different from the project charter?

A: The business case is a description of what your company is trying to get out of the project—like how much money you're planning on making from the project, how it will benefit parts of your organization, and future business you might gain from the project.

The project charter is a high-level description of your project. It tells you—and anyone else who needs to know about your project—what you'll be delivering, including a really high-level description of what it is that you'll build.

A really important difference between them is that the project charter is what authorizes the project manager to do the work, while the business case helps give justification for the project. You can think of the business case as the background research that had to be done in order to make sure the project was worth doing, and the project charter as the thing that formally announces the decision to do it.

Q: I'm still not clear on who the sponsor is. How's that different than the customer?

A: The sponsor is the person (or people) paying for the project. The customer is the person who uses the product of the project. Sometimes the customer is the same person as a sponsor. This is often true in consulting companies. For the teachers' project, the two sponsors are the CEO and VP of Asia Travel, and the customers are the teachers. But it's possible that in another travel agency, the teachers themselves would be the sponsors. This happens a lot in contracted work.

For the exam, you'll need to be careful about this. Sometimes you'll see the word *customer* in a question that's asking you about the sponsor. You might even see the word *client*—a word that appears in the *PMBOK Guide* only four times! (It's usually used when you're talking about procurement.) When you see this, you should assume that the question is asking you about a consulting situation, where the sponsor, customer, and client are all the same person.

The CEO and VP of Asia Travel are paying for this project in the sense that they're providing funding for the project team at the travel agency and cutting checks to the airlines, hotels, tour groups, etc. The customers are definitely paying Acme Travel, but they're not paying out the budget for the specific work that has to be done to complete the project.

Q: Hold on. My project sponsors are really important people in my company. I can't imagine them actually typing up a project charter.

A: Good point. That's why the project sponsor will often delegate the actual creation of the charter to the project manager. For the exam, though, keep in mind that the sponsor is ultimately responsible.

The sponsor of a project is responsible for creating the project charter.

The sponsor of a project pays for the project. The PM manages the project.

BULLET POINTS: AIMING FOR THE EXAM

- The **project charter** officially sanctions the project. Without a charter, the project cannot begin.

- The **sponsor** is the person (or people) responsible for paying for the project and is part of all important project decisions.

- **Develop Project Charter** is the very first process performed in a project.

- The project charter gives the project manager authority to **do the project work**, and to **assign work** or take control of project resources for the duration of the project. It also gives the project manager authority to **spend money** and **use other company resources**.

- **Facilitation techniques** (like brainstorming) are ways to get all of your stakeholders on the same page about your project goals and your approach to meeting them.

- The **business case** tells everyone why the company should do the project. It's an input to building the **project charter** that tells everyone that the project actually started, explains what it's going to deliver, and authorizes the project manager to do the work.

- The project charter does not include details about what will be produced or how. Instead, it contains the **summary milestone schedule**.

- Two additional inputs to Develop Project Charter are the **agreements** and the **statement of work**. The contract is what you agreed to do, although not all projects have a contract. The statement of work lists all of the **deliverables** that you and your team need to produce.

- **Enterprise environmental factors** tell you how your company does business. An important one is the **project management information system**, which determines how work is assigned, and makes sure that tasks are done in the right order.

- **Organizational process assets** tell you how your company normally runs projects. One of the most important assets is **lessons learned**, which is where you write down all of the valuable historical information that you learn throughout the project to be used later.

In matrix organizations, your team doesn't report to you, so the charter gives you the authority to put them to work.

The project charter shouldn't be too detailed. You shouldn't have to update the charter every time you change something about your project for it to stay accurate.

Agreements are sometimes referred to as contracts on the exam.

At Acme, the CEO and VP of Asia Travel were the sponsors. But at another travel agency, Frank and Joanne could just as easily sponsor the project, since they're the customers.

Watch it!

When you're taking the PMP exam, be careful when you see a question that asks you about the customer or client.

There's a good chance that the question is asking you about a consulting or procurement situation where the customer or client is also the sponsor.

Plan your project!

Planning the project is when you really take control. You write a plan that says exactly how you're going to handle everything that goes on in the project. The **Develop Project Management Plan** process is where you organize all of the information about your project into one place, so everyone knows exactly what needs to happen when they do the project work—no matter what their jobs are.

Planning
process group

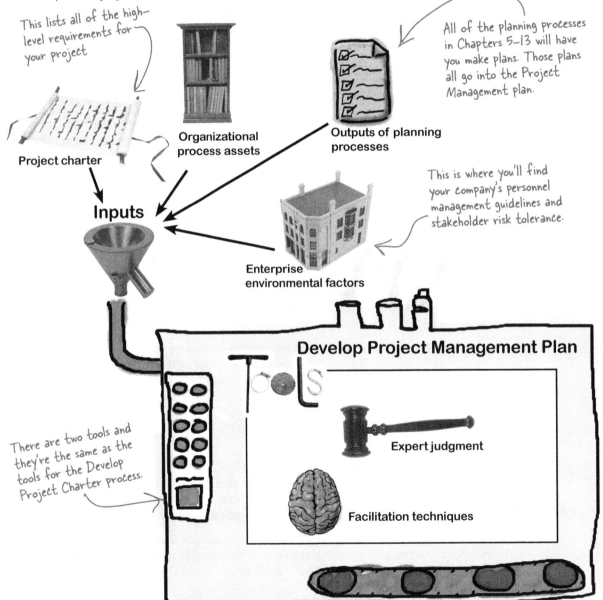

This lists all of the high-level requirements for your project

Organizational
process assets

All of the planning processes in Chapters 5–13 will have you make plans. Those plans all go into the Project Management plan.

Outputs of planning
processes

Project charter

This is where you'll find your company's personnel management guidelines and stakeholder risk tolerance.

Inputs

Enterprise
environmental factors

Develop Project Management Plan

T⬤⬤LS

There are two tools and they're the same as the tools for the Develop Project Charter process.

Expert judgment

Facilitation techniques

The Project Management plan lets you plan ahead for problems

The **Planning** process group is where you figure out how you're going to do the project—because you need to come up with a plan before you bring the team in to do the work. This is where you think about everything that will happen on your project, and try to plot a course to completing it with as few errors as possible.

And it's where you figure out how you'll handle changes—because every project has plenty of problems, but not all of those problems mean that you need to change course. If you plan well, your project will make only the right changes.

The Project Management plan is a collection of other plans

The Project Management plan is a single document, but it's broken into a bunch of pieces called **subsidiary plans**. There's one subsidiary plan for each of the other knowledge areas: Scope Management, Time Management, Cost Management, Quality Management, Human Resource Management, Communications Management, Risk Management, Procurement Management, and Stakeholder Management.

Outputs

The Project Management plan is the only output of the Develop Project Management Plan process.

The Project Management plan is actually a whole bunch of documents called "subsidiary plans," each dedicated to a knowledge area and the problems that could happen related to that area.

The Project Management plan is all about planning for problems, and having the information you need to correct those problems when they occur.

Communications Management is another knowledge area. In that plan, we've got important numbers for the trip.

Project Management plan

If you take over a project that's already under way, but there isn't a Project Management plan or it's out of date, the first thing you need to do is get a current, accurate plan written up.

There's a subplan for Risk Management. We used it when we took out traveler's insurance for the teachers' trip. That means if bags or cash are stolen, we'll have a plan for dealing with it.

The Project Management plan also has baselines. A baseline is a snapshot of the scope, schedule, or budget that you can use for planning. You'll learn all about baselines in the next three chapters!

ISN'T A PROJECT PLAN JUST SOMETHING I GET OUT OF MICROSOFT PROJECT?

No. The Project Management plan is not the same thing as a project schedule.

You'll use a tool like Microsoft Project when you're doing Time Management to build the project schedule. (It's useful for other knowledge areas as well.) But you'll use your Project Management plan as a guide to help you develop that schedule. It will tell you what tools to use when you develop it, and how changes will be handled.

Relax

Don't worry about memorizing all of the subsidiary plans.

You're going to learn about all of the knowledge areas throughout the book, so don't worry about memorizing all of these subsidiary plans right now. Just know that the Project Management plan has plans within it that map to each of the knowledge areas.

A quick look at all those subsidiary plans

You'll be learning about each of the knowledge areas throughout this book, and you'll learn all about the subsidiary plan that goes with each area. But let's take a quick look at what each subsidiary plan focuses on.

Project Management Plan—Subsidiary Plans and Baselines

The **Scope Management** plan describes how scope changes are handled—like what to do when someone needs to add or remove a feature for a service or product your project produces.

The **Requirements Management** plan describes how you'll gather, document, and manage the stakeholders' needs, and how you'll meet those needs with the project deliverables.

The **Schedule Management** plan shows you how to deal with changes to the schedule, like updated deadlines or milestones.

The **Cost Management** plan tells you how you'll create the budget, and what to do when your project runs into money problems.

The **Quality Management** plan deals with problems that could arise when a product doesn't live up to the customer or client's standards. The **Process Improvement** plan tells you how you can change the processes you're using to build your product to make them better.

You use the **Human Resource Management** plan to deal with changes in your staff, and to identify and handle any additional staffing needs and constraints you might have in your specific project.

The **Communications Management** plan lists all of the ways that you communicate with your project's team, stakeholders, sponsors, and important contacts related to the project.

The **Risk Management** plan is about detailing all the bad things that might happen and coming up with a plan to address each risk when and if it occurs.

The **Procurement Management** plan focuses on dealing with vendors outside of your company.

The **Stakeholder Management** plan focuses on managing the expectations of all of the people who are affected by the project.

There are three **baselines** in the Project Management plan. The **scope baseline** is a snapshot of the scope, which helps you keep track of changes to the work that you'll be doing and the planned deliverables you'll be building. The **schedule baseline** does the same for the project schedule, and the **cost performance baseline** does the same for the budget.

The Project Management plan is the core of Integration Management. It's your main tool for running a project.

Exercise

Below is a whole crop of problems that the teachers are running into. Write down which subsidiary plan you'd look in to get some help. If you're not sure, just reread the descriptions of each subsidiary plan on the previous page, and take your best guess.

1. The teachers want to go Bali, but Acme Travel doesn't book flights there so you need to subcontract one leg of the travel to another travel agency.

   _____

2. The teachers are having so much fun that they want to stay at a better hotel. They tell you to increase their budget by 15% to do that.

   _____

3. Just as you're about to mail off the teachers' tickets, you notice they've been printed incorrectly.

   _____

4. The teachers might run into more bad weather, and you've got to figure out what contingencies you can put into place if that happens.

   _____

5. The teachers are concerned that they won't be able to get in touch with you when they're away.

   _____

6. One of the teachers realizes that he needs to come back earlier, and you want to make sure the budget reflects his lessened costs.

   _____

7. You find out that you need to get the tickets out earlier than expected, because the teachers' contract requires that all trips be preapproved by the superintendent of their school district.

   _____

———————➤ Answers on page 149.

there are no
Dumb Questions

Q: How far should I go when trying to anticipate every possible problem and list it in the Project Management plan?

A: It's really important to think about what could go wrong on your project, so that you can have plans for what to do when problems crop up. An unexpected change can sometimes derail a project, and doing some planning up front can keep issues like that to a minimum. Planning can help you avoid problems in the first place, which is a lot better for everyone than reacting to them when they happen. So think of everything you can; the extra time you spend planning could be what keeps your project a success.

Q: Does the project manager create the Project Management plan all by himself?

A: No, it should be a group effort between the PM and the stakeholders. Everyone on the project team and all of the stakeholders need to agree that the plan is acceptable.

Q: What about things that I don't think about? And sometimes, I know there could be problems in a certain area, but I'm not sure what they'll be until the project gets going.

A: You're never going to think of everything that could go wrong. To help keep your plan flexible, you should add an Open Issues section to the plan. You can write down any open issues or concerns in this section, and deal with them as they come down the line. However, you have to have all your project requirements complete before starting the project—you should *never* have any requirements in your Open Issues section.

Q: I still don't get what enterprise environmental factors are.

A: Your company's enterprise environmental factors are all of the information you have about its policies, processes, departments, and people. You need to know how your company does business in order to do a project. For example, you need to know about the different departments in your company if you're managing a project that will be used by people in them.

BULLET POINTS: AIMING FOR THE EXAM

- Remember that the Project Management plan is **formal**—which means that it's **written down** and **distributed** to your team.

- You may get a question on the exam that asks what to do when you encounter a change. You *always* begin dealing with change by **consulting the Project Management plan.**

- The **project management information system** is a part of your company's enterprise environmental factors, and it's generally part of any change control system. It defines how **work is assigned to people.**

- The Project Management plan includes **baselines**: snapshots of the scope, schedule, and budget that you can use to keep track of them as they change.

Question Clinic: The "just-the-facts-ma'am" question

A great way to prepare for the exam is to learn about the different kinds of questions, and then try writing your own. Each of these Question Clinics will look at a different type of question, and give you practice writing one yourself.

Take a little time out of the chapter for this Question Clinic. It's here to let your brain have a break and _think about something_ different.

> A LOT OF QUESTIONS ON THE EXAM ARE PRETTY STRAIGHTFORWARD—BUT IT'S THE ANSWERS TO THOSE QUESTIONS THAT CAN REALLY HANG YOU UP. HERE, TAKE A LOOK:

27. Which of the following can be found in the project charter?

A. Business case document

Some answers will clearly be wrong. The business case document is one of the inputs to the Develop Project Charter process.

B. Expert judgment

Some answers are a little misleading! This is part of the Develop Project Charter process—but it's from the tools and techniques, not a part of the project charter itself.

C. Authorization for the project manager

Here's the right answer! The project manager's authorization is included in the project charter.

D. Project management information system

You haven't seen this one yet—it's part of enterprise environmental factors, an input to the Develop Project Charter process, but not in the charter itself.

> WHEN YOU SEE A "JUST-THE-FACTS-MA'AM" QUESTION, READ THE QUESTION REALLY CAREFULLY! IF YOU DON'T, IT'S EASY FOR A WRONG ANSWER TO LOOK RIGHT.

HEAD  LIBS

Fill in the blanks to come up with your own "just-the-facts-ma'am" question.

You are managing a _____ project. You are using
(an industry)

_____ and _____
(an input) _(an input)_
to create a _____ . What process are you performing?
(an output)

A. _____
(the name of the wrong process)

B. _____
(the name of the right process)

C. _____
(a made-up process that sounds like a real process)

D. _____
(the name of a tool and technique from the right process)

LADIES AND GENTLEMEN,
WE NOW RETURN YOU
TO CHAPTER FOUR

The Direct and Manage Project Work process

Once you have a Project Management plan, your project is ready to begin. And as the project unfolds, it's your job to direct and manage each activity on the project, every step of the way. That's what happens in the **Direct and Manage Project Work** process: you simply follow the plan you've put together and handle any problems that come up.

Executing process group

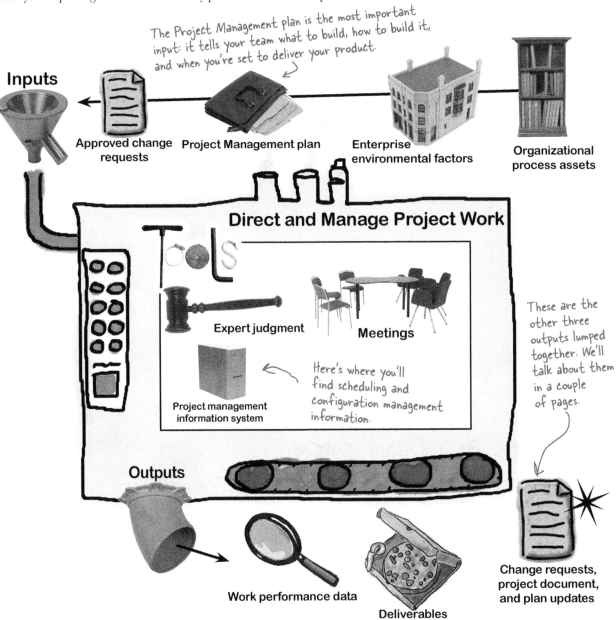

The Project Management plan is the most important input: it tells your team what to build, how to build it, and when you're set to deliver your product.

Inputs

Approved change requests   Project Management plan   Enterprise environmental factors   Organizational process assets

Direct and Manage Project Work

Tools

Expert judgment   Meetings

Project management information system

Here's where you'll find scheduling and configuration management information.

These are the other three outputs lumped together. We'll talk about them in a couple of pages.

Outputs

Work performance data   Deliverables   Change requests, project document, and plan updates

The project team creates <u>deliverables</u>

The work you're doing on the teachers' project creates lots of things: airline reservations, hotel reservations, invoices, defect reports, and customer comments (to name a few). These things are all your **deliverables**, and they are one of the *five* outputs of the **Direct and Manage Project Work** process.

Another output is **Work Performance Data**, and that's what we call the reports Acme's running on the project. These reports track how many negative versus positive customer comments the project gets, and how well the project is doing at meeting its cost estimates. In fact, a project manager should figure out a way to use the work performance data that is generated from each knowledge area during the Direct and Manage Work process to understand how well the processes are being performed.

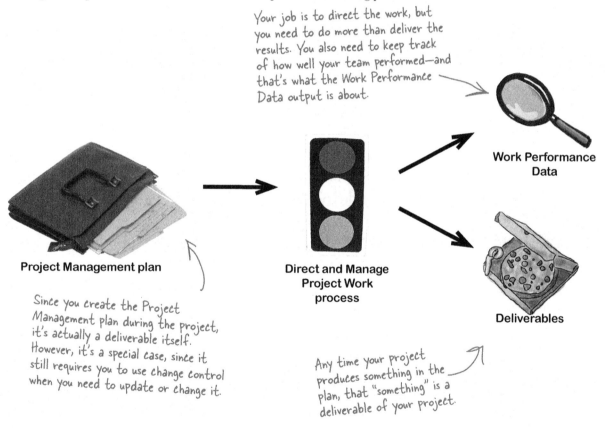

Your job is to direct the work, but you need to do more than deliver the results. You also need to keep track of how well your team performed—and that's what the Work Performance Data output is about.

Work Performance Data

Project Management plan

Since you create the Project Management plan during the project, it's actually a deliverable itself. However, it's a special case, since it still requires you to use change control when you need to update or change it.

Direct and Manage Project Work process

Any time your project produces something in the plan, that "something" is a deliverable of your project.

Deliverables

You create <u>work performance data</u> by measuring how the processes from each knowledge area are being performed.

Executing the project includes repairing defects

The Direct and Manage Project Work process has a bunch of inputs and outputs—but most of them have to do with implementing changes, repairs, and corrective action. If there's a defect repair that's been approved, this is where it happens. Once the defect is repaired, the result is an **implemented defect repair.** The same is true for changes and corrective actions; once they're approved, they become process inputs, and then they can be implemented and become process outputs.

Any time you have to correct a mistake or make a repair in a deliverable, you're fixing a defect.

The three components of the Direct and Manage Project Work process:

Deliverables are anything you produce in the course of doing your project activities.

1. Use the plan to create deliverables.

2. Repair defects in deliverables.

Your Quality Management plan focuses on catching defects as you go, so you can repair them as soon as possible.

3. As the project plan changes, make sure those changes are reflected in the deliverables.

This is different from fixing defects. A defect means that the plan was right, but your deliverable was built wrong.

Deliverables include everything that you and your team produce for the project

The word **deliverable** is pretty self-explanatory. It means anything that your project **delivers**. The deliverables for your project include all of the products or services that you and your team are performing for the client, customer, or sponsor.

But deliverables include more than that. They also include every single document, plan, schedule, budget, blueprint, and anything else that gets made along the way…including all of the project management documents that you put together.

Deliverables can be either internal to your company or to the customer.

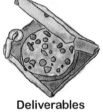

Deliverables

> # The Direct and Manage Project Work process is where you and your team actually do the work to produce the <u>deliverables</u>.

Sharpen your pencil

Here's a list of things produced by some typical projects. Some of them are deliverables, and others are work performance data produced by running reports. There's also a list of changes, some of which affect the Project Management plan, and some of which just affect the project deliverables. It's up to you to figure out which is which.

1. The software project team builds software.

☑ Deliverable   ☐ Work performance data   ①

2. A builder hangs a door.

☑ Deliverable   ☐ Work performance data

3. A wedding photographer sends the photo proofs to the client.

☑ Deliverable   ☒ Work performance data

4. The cable repair technicians take an average of four hours per job.

☐ Deliverable   ☑ Work performance data

5. The construction crew worked 46 hours of overtime in March.

☐ Deliverable   ☑ Work performance data

Sometimes something that looks like a defect in a deliverable is really a change that you need to make to the plan.

6. The construction crew built the six houses required by the plan.

☑ Deliverable   ☐ Work performance data

7. A software test team finds bugs in the software.

☑ Defect in deliverable   ☐ Change to Project Management plan

8. A bride asks the photographer to stop asking her mother for permission to make changes.

☐ Defect in deliverable   ☑ Change to Project Management plan

9. A construction crew used the wrong kind of lumber in a house.

☑ Defect in deliverable   ☐ Change to Project Management plan

10. A photographer's prints are grainy.

☑ Defect in deliverable   ☐ Change to Project Management plan

Answers on page 150.

Eventually, things **<u>WILL</u>** go wrong...

Even if you work through all the processes you've seen so far, things can still go wrong on your project. In fact, the teachers are already letting you know about some issues they're having.

...but if you keep an eye out for problems, you can stay on top of them!

It's a good thing you've been monitoring the project. Otherwise, you might not have found out about their problems in time to help.

Sometimes you need to change your plans

Take a minute and flip back to page 107. Notice how there's a loop between the Executing and the Monitoring and Controlling processes? That's because when your team is executing the plan and working on the deliverables, you need to keep a constant lookout for any potential problems. That's what the **Monitor and Control Project Work** process is for. When you find a problem, you can't just make a change…because what if it's too expensive, or will take too long? You need to look at how it affects the project constraints—time, cost, scope, resources, risks, and quality—and figure out if it's worth making the change. That's what you do in the **Perform Integrated Change Control** process.

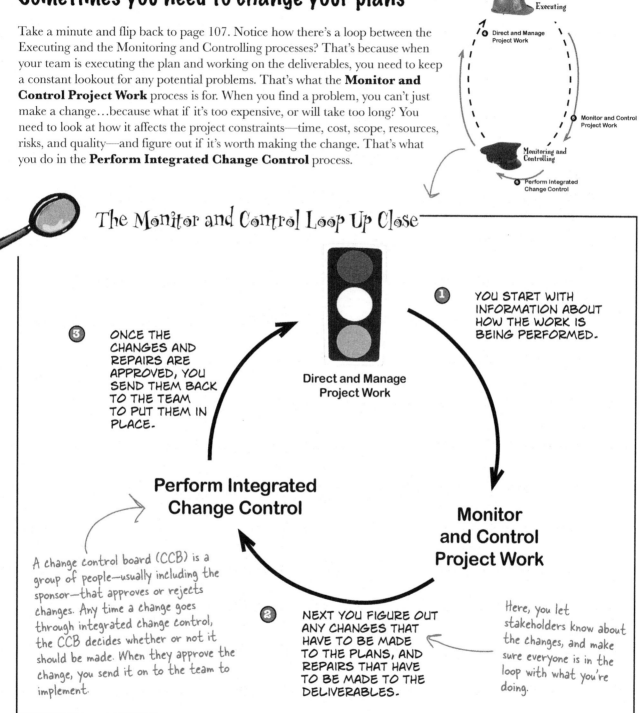

The Monitor and Control Loop Up Close

① YOU START WITH INFORMATION ABOUT HOW THE WORK IS BEING PERFORMED.

Direct and Manage Project Work

③ ONCE THE CHANGES AND REPAIRS ARE APPROVED, YOU SEND THEM BACK TO THE TEAM TO PUT THEM IN PLACE.

Perform Integrated Change Control

A change control board (CCB) is a group of people—usually including the sponsor—that approves or rejects changes. Any time a change goes through integrated change control, the CCB decides whether or not it should be made. When they approve the change, you send it on to the team to implement.

Monitor and Control Project Work

② NEXT YOU FIGURE OUT ANY CHANGES THAT HAVE TO BE MADE TO THE PLANS, AND REPAIRS THAT HAVE TO BE MADE TO THE DELIVERABLES.

Here, you let stakeholders know about the changes, and make sure everyone is in the loop with what you're doing.

Look for changes and deal with them

Monitoring
& Controlling
process group

You need to stay on top of any possible changes that happen throughout your project, and that's what the **Monitor and Control Project Work** process is for. Usually the work is progressing just fine. But sometimes you find out that you need to change something, and that's when you use the **Perform Integrated Change Control** process to see if the change is worth the impact it will have on your project.

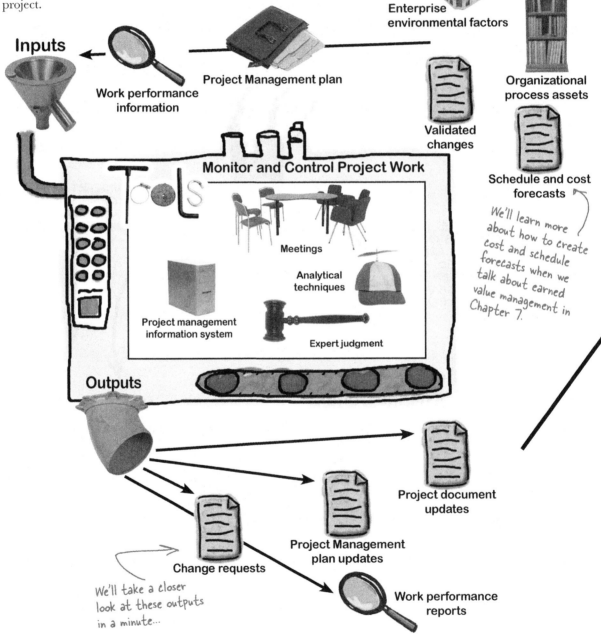

Inputs

Work performance information

Project Management plan

Enterprise environmental factors

Validated changes

Organizational process assets

Schedule and cost forecasts

We'll learn more about how to create cost and schedule forecasts when we talk about earned value management in Chapter 7.

Monitor and Control Project Work

Tools

Meetings

Analytical techniques

Project management information system

Expert judgment

Outputs

Change requests

We'll take a closer look at these outputs in a minute...

Project Management plan updates

Project document updates

Work performance reports

Make only the changes that are right for your project

The Monitor and Control Project Work process is where you find the changes that you may want to make. The **Perform Integrated Change Control** process is where you decide whether or not to make them. But you're not the one actually making that decision—a big part of Perform Integrated Change Control is that you **need to get your changes approved by the change control board**.

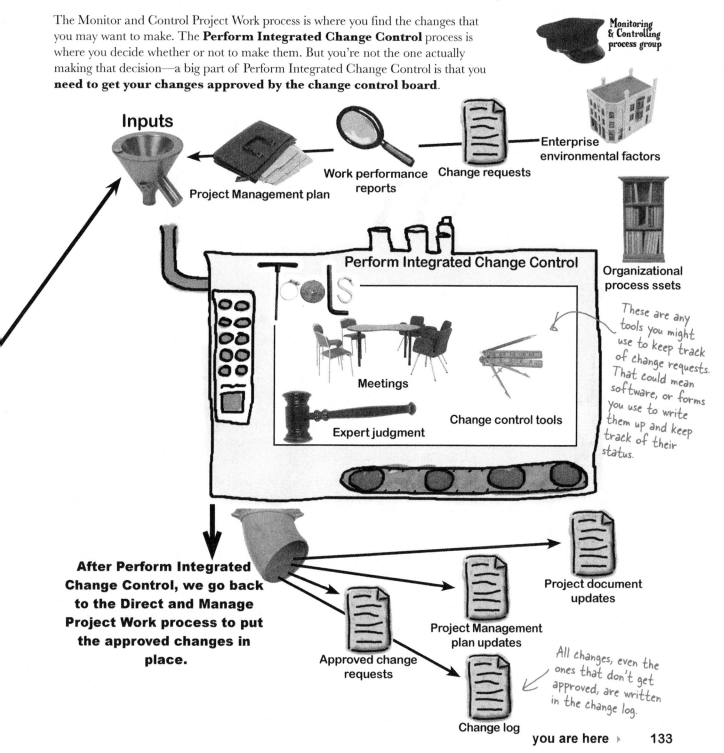

Monitoring & Controlling process group

Inputs

Project Management plan

Work performance reports

Change requests

Enterprise environmental factors

Perform Integrated Change Control

T**oo**ls

Meetings

Expert judgment

Change control tools

Organizational process ssets

These are any tools you might use to keep track of change requests. That could mean software, or forms you use to write them up and keep track of their status.

After Perform Integrated Change Control, we go back to the Direct and Manage Project Work process to put the approved changes in place.

Approved change requests

Project Management plan updates

Project document updates

Change log

All changes, even the ones that don't get approved, are written in the change log.

Changes, defects, and corrections

You've already seen how a project can change as it goes along. When the teachers asked for their hotel to be upgraded, you took the request through the **change control** process at Acme, and when the change control board approved the change, you directed the agents to make the booking for the group.

But sometimes, things go wrong with what you intended to have happen in the first place. When your quality department told you that you had booked the teachers on the flight to Rome without putting them in the same row, you quickly fixed the reservation. But you intended for the teachers to sit together in the first place, so that's not a change, it's a **defect**.

In the process, you realized that your team wasn't reading your documentation carefully, which is why they screwed up the airline reservations. To fix the way your team is working, you need to take corrective action. That's when you need to change the way you're doing the work on your project. Got all that?

When the team is repairing defects to deliverables, they <u>still</u> need to go through change control.

Decide your changes in change control meetings

Sometimes a change you make will have a direct impact on other teams and projects, and it's a good idea to be sure that everybody who will be impacted knows that it's coming and thinks that it's worth it before you make the change. You can't always know everything that might happen as a result of a change, and that's why it's a good idea to get buy-in from key people in your company before you go through with it. And that's what a change control meeting is all about!

Usually, a change control meeting will be a regularly scheduled thing, where people representing the affected areas of the company will get together to review proposed changes and decide whether or not to make them. A change control board is never made up of just the people on your team. A change control meeting is all about getting people with different perspectives together to talk about the pros and cons of changes before deciding whether to approve or reject them.

It's your job as a project manager to know the impact of requested changes to your project and prioritize them for the change control board. Once you've done that, the change control board can make informed decisions about whether or not to approve them.

How the processes interact with one another

While monitoring the teachers' trip, you notice that they all ask for nonsmoking rooms every time they check into a hotel. But some hotels don't have enough nonsmoking rooms available, and the teachers aren't too thrilled about that.

After you talk it over with the teachers it's clear that it's worth splitting up the group over multiple hotels to make sure they all are in nonsmoking rooms—and some hotels are more expensive than you'd planned. The cost change will put you over budget, so the cost management plan needs to be updated. Time to take the request to change control:

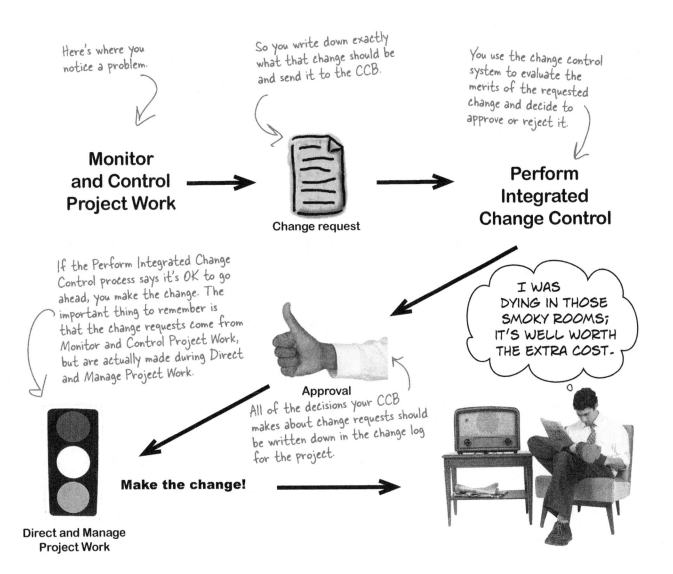

Here's where you notice a problem.

So you write down exactly what that change should be and send it to the CCB.

You use the change control system to evaluate the merits of the requested change and decide to approve or reject it.

Monitor and Control Project Work

Change request

Perform Integrated Change Control

If the Perform Integrated Change Control process says it's OK to go ahead, you make the change. The important thing to remember is that the change requests come from Monitor and Control Project Work, but are actually made during Direct and Manage Project Work.

Approval

All of the decisions your CCB makes about change requests should be written down in the change log for the project.

I WAS DYING IN THOSE SMOKY ROOMS; IT'S WELL WORTH THE EXTRA COST.

Make the change!

Direct and Manage Project Work

Control your changes: use <u>change</u> <u>control</u>

> THERE'S A BEAUTIFUL HOTEL ACROSS THE STREET, AND WE WANT TO TRANSFER. INCREASE THE BUDGET BY 15%, AND BOOK US THERE.

Project Management plan

Your Project Management plan should detail how you deal with changes that happen during your project.

Any time you need to make a change to your plan, you need to start with a **change request**. This is a document that either you or the person making the change needs to create. Any change to your project needs to be <u>documented</u> so you can figure out what needs to be done. Once you have a change request, that then kicks off your project's set of change control procedures.

The key here is PROCEDURE— change control is about how your company handles changes. You may use a computer system to monitor and document changes, but that's just one part of your change control system.

This means you need to write down exactly what needs to be changed and put it in something called a change request. That's a form that you fill out to send a change through change control.

Change control is how you deal with changes to your Project Management plan.

A change control system is the set of procedures that lets you make those changes in an organized way.

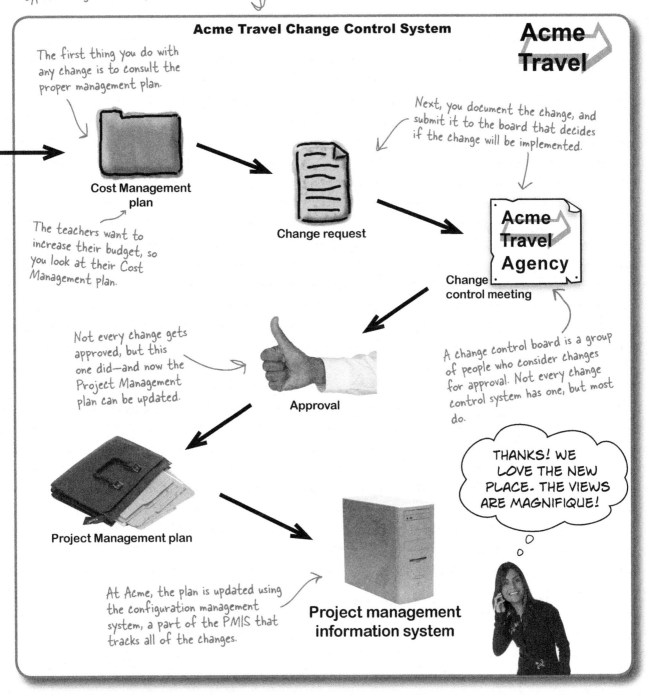

This is Acme's change control system. It's specific to the company, but it contains all of the steps you'd see in a typical change control system.

Acme Travel Change Control System

Acme Travel

The first thing you do with any change is to consult the proper management plan.

Cost Management plan

The teachers want to increase their budget, so you look at their Cost Management plan.

Change request

Next, you document the change, and submit it to the board that decides if the change will be implemented.

Acme Travel Agency

Change control meeting

A change control board is a group of people who consider changes for approval. Not every change control system has one, but most do.

Not every change gets approved, but this one did—and now the Project Management plan can be updated.

Approval

THANKS! WE LOVE THE NEW PLACE. THE VIEWS ARE MAGNIFIQUE!

Project Management plan

At Acme, the plan is updated using the configuration management system, a part of the PMIS that tracks all of the changes.

Project management information system

Preventing or correcting problems

When you monitor your project, you might be checking the actual time it's taking you to do scheduled work versus the amount of time you planned, or you might be gathering information on the number of defects you have found versus the number you expected. In both cases, it's possible that you might find problems. If you do, you have to change the way you do your work and keep your project from being dragged down. When you make a course change on your project, that's taking **corrective action**.

It's also possible that you might see problems that are going to occur even though they haven't happened yet. If you do, you will want to take **preventive action**, or steps that you take to avoid potential problems.

When people predict problems on projects before they happen, it's called a forecast. A forecast can be a good reason to make a change too!

In both corrective and preventive action, you always need to submit your proposed change and put it through the Perform Integrated Change Control process—and only if it is approved will you implement it. If your recommended action makes it through, you need to change the plan and any of your **baselines** to include it.

The documented scope, schedule, and cost baselines in the Project Management plan are called the **performance measurement baseline.**

We'll learn more about the performance measurement baseline in upcoming chapters.

Exercise

Here is a list of actions that are recommended by a project manager. Which are preventive and which are corrective?

1. A software project is running late, so a software project manager looks to find slack time and reassign resources to get things done more quickly.

- ☒ Preventive action
- ☑ Corrective action

2. A caterer notices that the crudités are all gone and assigns a chef to make more.

- ☐ Preventive action
- ☑ Corrective action

3. A photographer brings an extra camera body to a shoot, in case one breaks down.

- ☑ Preventive action
- ☐ Corrective action

4. A consulting company assigns extra resources to a project to compensate for possible attrition.

- ☑ Preventive action
- ☐ Corrective action

→ Answers on page 148.

there are no
Dumb Questions

Q: Sometimes my team members come to me and tell me that the project could have problems later. What do I do with that?

A: For some project managers, it seems natural to dismiss these "negative Nellies" who seem concerned with problems that could go wrong in the future. But working with them instead is one of the best ways you can satisfy your stakeholders.

When someone makes an estimate or prediction of a future condition that could lead to trouble, it's called a **forecast**, and that's very valuable information. You should distribute it along with your work performance information, and try to think of ways to avoid the problem—which is what preventive action is all about.

A big part of your job as a project manager is to figure out how to prevent changes. This might seem a little weird—how can you prevent changes before the project is implemented? One way to do this is plan as well as possible, because a lot of changes happen because of a lack of planning. But it also means talking to stakeholders throughout the project and keeping an eye out for potential problems. When you take the PMP exam, if you see the phrase **"influencing factors that cause change,"** this is what it's referring to.

Q: Who approves changes?

A: Usually there's a **change control board (CCB)** that approves changes. That's a group of people, most often including the stakeholders and sponsor, who look at the benefits of a change and figure out if it's worth the cost. If there's a CCB, your change control system will include a procedure that mentions it. But not every company has a CCB, and there is no requirement in the *PMBOK Guide* that you have one.

Q: What if there's a problem outside my project, and I'm not sure it affects me?

A: You should still consider its potential impact when you're monitoring your project's work. It's important that you're always on the lookout for potential problems. If you're not sure whether something could impact your project, it's your responsibility as a project manager to bring it to the attention of your stakeholders. And if you can make a change on your own that doesn't impact the project constraints (scope, cost, time, quality, risk, or resources), then **it's completely within your rights as a project manager to do it**.

Q: Once a change is approved, what do I do with it?

A: You change your Project Management plan to incorporate the change. This can mean that you create a new baseline with the new Project Management plan. For example, say you forgot to add a stakeholder to the change control board, so your project plan now describes the wrong process for making changes. You'll need to fix that, and you'll need to go through change control to do it.

Every time a change is reviewed by the change control board, you keep a record of it in your change log. So whether the change was approved or rejected, the change request and the decision the CCB makes about it should be documented.

Q: What about changes that don't affect the project constraints?

A: If you evaluate the impact of a change and find that it won't have an impact on the project constraints, then you can make the change without going through change control. Sometimes you need to change resources or move tasks around, and you

can make those changes without affecting the bottom line or the end product. In these cases, change control wastes time and resources, rather than helping your project.

Q: Now, what's a performance baseline again, and what do I do with it?

A: A performance baseline is a snapshot of your project's scope, schedule, and cost. When you plan out the work you'll do on a project, you write down all of the activities you'll need to do and save that understanding as your scope baseline. You'll do the same with your understanding of the project's schedule and its cost. That way, you can always compare your actual performance to your plan.

Every time a change is approved, that means the plan has changed. So you have to update your baseline to include the new work (or cost, or schedule).

> You always have the authority to make changes to your project if they don't affect cost, schedule, or scope.

Finish the work, close the project

You can't finish the project until you get paid! Most projects start with contracts, and when they do you need to make sure the terms are met. Acme signed a contract with the Midwestern Teachers' Association when the project started, and now it's time to make sure all of the parts of that contract are met. And that's part of what you do in the **Close Project or Phase** process. But an even more important part of this process is that you create the **lessons learned** and add them to your company's organizational process assets. That way you and other project managers can learn from this **historical** information in the future. The inputs to the Close Project or Phase process include the project management plan, organizational process assets, and accepted deliverables. And you use the same familiar **tools and techniques** that you've seen throughout the chapter.

Close project or phase

The most important output of the Close Project or Phase process is the final product that you deliver to the customer!

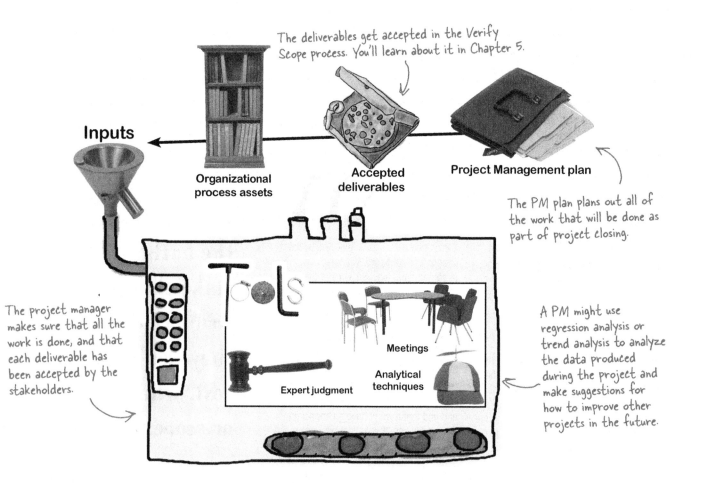

The deliverables get accepted in the Verify Scope process. You'll learn about it in Chapter 5.

Inputs

Organizational process assets

Accepted deliverables

Project Management plan

The PM plan plans out all of the work that will be done as part of project closing.

The project manager makes sure that all the work is done, and that each deliverable has been accepted by the stakeholders.

Tools

Meetings

Analytical techniques

Expert judgment

A PM might use regression analysis or trend analysis to analyze the data produced during the project and make suggestions for how to improve other projects in the future.

You don't have to go home, but you can't stay here

Closing
process
group

The teachers have gone through their entire itinerary. They're now on their way to Paris, which is the final leg of their tour. They've had a great time, and now it's time for you to finish up.

Every project needs to end, and that's what the Close Project or Phase process is all about. You want other travel agents at Acme to learn from anything new you've discovered. Remember how you had to scramble with the nonsmoking room request? Maybe your friends at Acme can learn from that, and ask new clients up front what they want! That's why you write down your lessons learned, and that's a big part of closing the project.

Even if your project ends early, you still need to follow the Close Project or Phase process.

Outputs

You've seen the organizational process assets input a bunch of times now. But where does it really come from? It turns out that it comes from other project managers just like you. Every time a project is closed, you update those assets so that you can use them later. And new project managers will be able to learn from everything that's happened on your project.

Lessons learned are finished in Close Project or Phase, but written down throughout the entire project. And it's not just by the project manager—the whole team writes down lessons learned.

**Organizational
process asset
updates**

The final product of your project is the thing your customers will remember most.

**Final product,
service, or result
transition**

⚛️ BRAIN POWER

Think about a major project you've heard of that did not end well, like one that was shut down before the work was done. What lessons could have been learned from that project?

How can the project manager use the Close Project or Phase process to make sure that something good comes out of early termination?

So why INTEGRATION Management?

The Integration Management knowledge area has all of the processes that you do in your day-to-day work as a project manager. So why are they called "Integration Management" processes? Well, think about what it takes to run a project: you need people and other resources from all around your company; knowledge about how your company does its business; standards, templates, and other assets that you've gathered from other projects; and the ability to put it all together—that's what a project manager does. And that's where the "integration" part comes in.

This is especially important when you need to work with consultants, because your job is to procure services for the project. And you need to plan for all of it at the beginning—which is when you **integrate** all of these things together into a single plan. It's your job to make sure that every one of the 47 processes in the *PMBOK Guide* is addressed in the plan, even if you're not going to use it (for example, if you don't need contractors or consultants, you won't use Procurement processes).

> I GET IT! WHEN I'M PUTTING TOGETHER MY PROJECT MANAGEMENT PLAN, I NEED TO LOOK AT EVERY SINGLE PROCESS AND FIGURE OUT HOW IT'S INTEGRATED INTO MY PROJECT. SO THAT'S WHAT "INTEGRATION MANAGEMENT" MEANS.

Integration Management means making sure that all of the processes work together seamlessly to make your project successful.

What else is there?

Huh…it seems like we covered the whole project, right? You got authorized to do the work, you planned the project, you executed it, you corrected problems along the way, and you closed it out. Isn't that everything?

Well, of course not! There's a whole lot more planning that you have to do, and many skills that you need to have. Luckily, we've got the *PMBOK Guide* to help us figure out exactly what we need to know to manage projects effectively.

That's what the rest of the book is about.

Project Integration Management Magnets

These inputs, outputs, and processes are all scrambled up on the fridge.
Can you reconstruct them so that the processes go under the correct process groups, and the inputs, outputs, and tools go in the right categories?

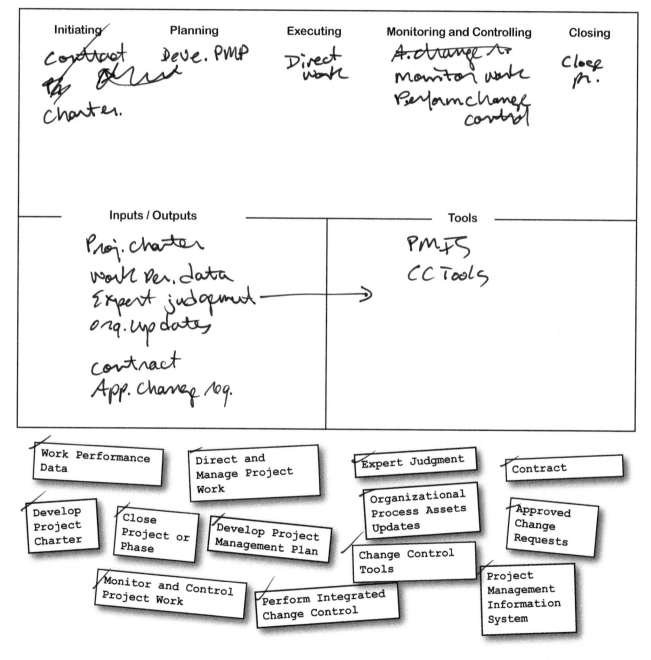

Initiating	Planning	Executing	Monitoring and Controlling	Closing
Contract ~~the first~~ Charter.	Deve. PMP	Direct work	A. change ↑ Monitor work Perform change control	Close pr.

Inputs / Outputs

Proj. charter
Work Per. data
Expert judgment ————→
Org. updates

Contract
App. Change log.

Tools

PMIS
CC Tools

Work Performance Data

Direct and Manage Project Work

Expert Judgment

Contract

Develop Project Charter

Close Project or Phase

Develop Project Management Plan

Organizational Process Assets Updates

Approved Change Requests

Change Control Tools

Monitor and Control Project Work

Perform Integrated Change Control

Project Management Information System

Project Integration Management Magnets Solution

These inputs, outputs, and processes are all scrambled up on the fridge.
Can you reconstruct them so that the processes go under the correct process groups, and the inputs, outputs, and tools go in the right categories?

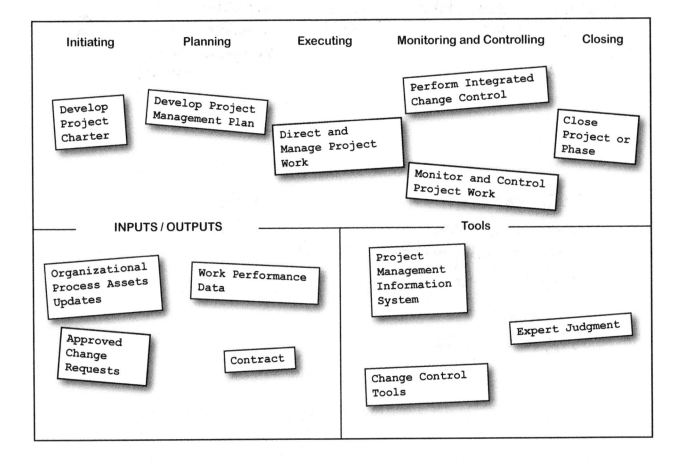

Initiating	Planning	Executing	Monitoring and Controlling	Closing

Develop Project Charter

Develop Project Management Plan

Direct and Manage Project Work

Perform Integrated Change Control

Monitor and Control Project Work

Close Project or Phase

INPUTS / OUTPUTS

Organizational Process Assets Updates

Work Performance Data

Approved Change Requests

Contract

Tools

Project Management Information System

Expert Judgment

Change Control Tools

Integration Management kept your project on track, and the teachers satisfied

By using all of the Integration Management processes, you kept the project on track. You handled all of the problems that came up, made some important changes in the process, and the teachers got to all of their destinations on time and on budget.

THANKS SO MUCH! WE HAD A GREAT TRIP, AND WE'LL DEFINITELY BE USING ACME AGAIN NEXT YEAR.

AH, MON DIEU! QUEL PROJET MAGNIFIQUE! HEY, JOANNE'S RIGHT...ACME REALLY DID US RIGHT ON THIS TRIP.

Integrationcross

Take some time to sit back and give your right brain something to do. It's your standard crossword; all of the solution words are from this chapter.

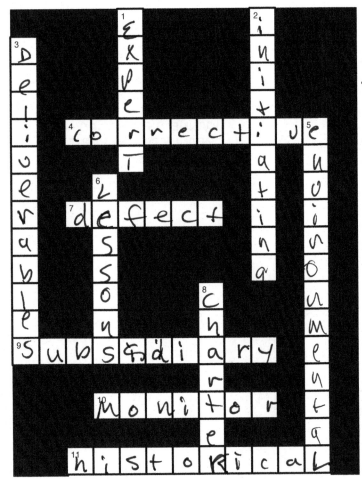

Across

4. Fixing problems that have already happened is called **Corrective** action.

7. A problem in a deliverable that shows that it does not do what you meant for it to do.

9. The Project Management plan is a collection of _____ plans.

10. When you watch what's happening in your project to look for changes, corrective actions, and preventive actions, you are in the **Monitor** and Control Project Work process.

11. _____ information is an important organizational process asset that comes from documenting lessons learned.

Down

1. When you ask someone who has experience to help you figure something out, you are using the **expert** judgment tool and technique.

2. The **___** process group contains the processes that help you start your project.

3. The things your project produces. **deliverables**

5. Work culture and company policies are called enterprise **environ** factors.

6. A record of all of the decisions you have made and their consequences that you write when you close your project is called **lessons** learned.

8. The project **charter** is a document that gives the project manager authority over the team.

———➤ Answers on page 151.

Sharpen your pencil
Solution

Here are a few of the things you might have to deal with in working on the teachers' vacation trip. Figure out which of the seven Integration Management processes you'd use in each situation, and write down the process name in the blank.

① It turns out that one of the teachers is a vegetarian, so you need to change your plans to include vegetarian meals on the airlines and find restaurants that accommodate him.

Perform Integrated Change Control

② You come up with a detailed description of everything that you plan to do to get the teachers where they want to be.

Develop Project Management Plan

③ The CEO of Acme Travel sends you a document that assigns you to the project.

Develop Project Charter

④ You check in with the teachers at each destination to make sure everything is going according to plan.

Monitor and Control Project Work

⑤ When the teachers get back, you write up everything you learned while handling the trip so other travel agents can learn from your experience.

Close Project or Phase

⑥ You book the tickets and hotel accommodations.

Direct and Manage Project Work

Develop Project Charter

Develop Project Management Plan

Direct and Manage Project Work

Monitor and Control Project Work

Perform Integrated Change Control

Close Project or Phase

Exercise Solution

Here is a list of actions that are recommended by a project manager. Which are preventive and which are corrective?

1. A software project is running late, so a software project manager looks to find slack time and reassign resources to get things done more quickly.

☐ Preventive action   ☑ Corrective action

2. A caterer notices that the crudités are all gone and assigns a chef to make more.

☐ Preventive action   ☑ Corrective action

3. A photographer brings an extra camera body to a shoot, in case one breaks down.

☑ Preventive action   ☐ Corrective action

4. A consulting company assigns extra resources to a project to compensate for possible attrition.

☑ Preventive action   ☐ Corrective action

Exercise Solution

Below is a whole crop of problems that the teachers are running into. Write down which subsidiary plan you'd look in to get some help. If you're not sure, just reread the descriptions of each subsidiary plan on the last page, and take your best guess.

1. The teachers want to go Bali, but Acme Travel doesn't book flights there so you need to subcontract one leg of the travel to another travel agency.

   Procurement Management plan

2. The teachers are having so much fun that they want to stay at a better hotel. They tell you to increase their budget by 15% to do that.

   Cost Management plan

3. Just as you're about to mail off the teachers' tickets, you notice they've been printed incorrectly.

   Quality Management plan

4. The teachers might run into more bad weather, and you've got to figure out what contingencies you can put into place if that happens.

   Risk Management plan

5. The teachers are concerned that they won't be able to get in touch with you when they're away.

   Communications Management plan

6. One of the teachers realizes that he needs to come back earlier, and you want to make sure the budget reflects his lessened costs.

   Cost Management plan

7. You find out that you need to get the tickets out earlier than expected, because the teachers' contract requires that all trips be preapproved by the superintendent of their school district.

   Schedule Management plan

Sharpen your pencil
Solution

Here's a list of things produced by some typical projects. Some of them are deliverables, and others are work performance data produced by running reports. There's also a list of changes, some of which affect the Project Management plan, and some of which just affect the project deliverables. It's up to you to figure out which is which.

1. The software project team builds software.

☑ Deliverable   ☐ Work performance data

2. A builder hangs a door.

☑ Deliverable   ☐ Work performance data

3. A wedding photographer sends the photo proofs to the client.

☑ Deliverable   ☐ Work performance data

4. The cable repair technicians takes an average of four hours per job.

☐ Deliverable   ☑ Work performance data

5. The construction crew worked 46 hours of overtime in March.

☐ Deliverable   ☑ Work performance data

6. The construction crew built the six houses required by the plan.

☑ Deliverable   ☐ Work performance data

7. A software test team finds bugs in the software.

☑ Defect in deliverable   ☐ Change to Project Management plan

8. A bride asks the photographer to stop asking her mother for permission to make changes.

☐ Defect in deliverable   ☑ Change to Project Management plan

9. A construction crew used the wrong kind of lumber in a house.

☑ Defect in deliverable   ☐ Change to Project Management plan

10. A photographer's prints are grainy.

☑ Defect in deliverable   ☐ Change to Project Management plan

Integrationcross Solution

Take some time to sit back and give your right brain something to do. It's your standard crossword; all of the solution words are from this chapter.

Exam Questions

1. You've just received a change request. This means:

    A.   The project charter is complete, but the work cannot begin yet because you need to make a change to the scope baseline.

    B.   You are in the Direct and Manage Project Work process, and you can implement the change now.

    C.   The change needs to be approved before it can be implemented.

    D.   There is a defect in a deliverable that must be repaired.

2. Which of these is not an input to Develop Project Charter?

    A.   Enterprise environmental factors

    B.   Project Management plan

    C.   Agreements

    D.   Project statement of work

3. What is the output of Direct and Manage Project Work?

    A.   Approved change requests

    B.   Project Management processes

    C.   Deliverables

    D.   Forecasts

4. You're managing a graphic design project. One of your team members reports that there is a serious problem, and you realize that it will cause a delay that could harm the business of the stakeholders. Even worse, it will take another two days for you to fully assess the impact—until then, you won't have the whole story. What is the BEST way to handle this situation?

    A.   Create a change request document and submit it to the change control meeting.

    B.   Pull out the project charter and show them that you have authority to make decisions.

    C.   Meet with the stakeholders and tell them that there's a problem, and you need two more days to get them the information they need.

    D.   Update the lessons learned and add it to your organizational process assets.

5. You're a project manager on a construction project. The electrician has started laying out the wiring, when the client comes to you with a change request. He needs additional outlets, and you think that will increase the cost of the electrical work. What is the first thing you do?

    A.   Refuse to make the change because it will increase the cost of the project and blow your budget.

    B.   Refer to the Project Management plan to see how the change should be handled.

    C.   Consult the contract to see if there is a clause.

    D.   Make the change, since the client requested it.

Exam Questions

6. The work authorization system:

A. Ensures that every work package is performed at the right time and in the proper sequence

B. Authorizes the project manager to spend money on work

C. Is a set of processes and tools that aids project manager in effectively guiding the project to completion

D. Is a formalized, written description of how to carry out an activity

7. You're the project manager at a telecommunications company. You recently had stakeholders approach you with changes. You figured out that the changes would cost additional time and money. The stakeholders agreed, you were given additional time and budget, and the changes were approved. Now you have to incorporate the changes into the project. What do you do next?

A. Modify the project charter to include the changes.

B. Use the project management information system to make sure the work is performed.

C. Make sure to track your changes against the project's baseline so you know how much they eventually cost.

D. Incorporate the changes into the baseline so you can track the project properly.

8. You are a project manager on a software project. When you planned the project, your enterprise environmental factors included a policy that all changes that cost over 2% of the budget need to be approved by the CFO, but smaller changes could be paid for by a management contingency fund. One of your stakeholders submitted a change request that requires a 3% increase in the budget. Your company has an outsourcing effort, and you believe that a small change to the way that the change is requested could allow you to take advantage of it and cut your costs in half. What is the BEST way to handle this situation?

A. Work with the stakeholder to figure out how to reduce the cost of the change by a third.

B. Request approval from the CFO.

C. Refuse the change because it is over 2% of the budget.

D. Document the change request, since all changes must be documented.

9. You're on the project selection committee. You're reviewing a document that describes the strategic value of a potential project and its benefits to the company. What's this document called?

A. Project charter

B. Business case

C. Benefit measurement method

D. Contract

Exam Questions

10. One of your team members has discovered a defect in a deliverable and has recommended that it be repaired. Which of the following is NOT true:

   A.   The project charter has authorized you to perform the work.

   B.   Your project is in Monitor and Control Project Work process.

   C.   The defect repair must be approved before the deliverable can be repaired.

   D.   You must update the Project Management plan to document the defect.

11. You are holding a formal, approved document that defines how the project is executed, monitored, and controlled. You are holding:

   A.   The Project Management plan

   B.   The performance measurement baseline

   C.   The project charter

   D.   The work breakdown structure

12. You are the project manager for a software project, when the sponsor pulls the plug and cancels the project. What do you do?

   A.   Give the team the day off to recuperate from the bad news.

   B.   Create a budget summary for the remaining unspent budget.

   C.   Follow project closure procedures to close the project and update lessons learned.

   D.   Find new assignments for any people previously assigned to your project.

13. You are managing a software project, when you find out that a programming team whom you were supposed to have access to has been reassigned to another project. What is the first thing that you should do?

   A.   Figure out the impact that this will have on your project.

   B.   Bring a copy of your project's charter to the other manager, and explain that you need that team for your own project.

   C.   Go to your sponsor and demand the team.

   D.   Figure out a way to compress the project schedule so that you can work with the team if they become available.

14. You are a project manager on a software project. There are several changes that need to be made, and you need to decide how to apply project resources in order to implement them. What do you do?

   A.   Decide the priority of the changes and announce them to the team.

   B.   You should call a team meeting and invite the stakeholders, so that everyone can reach a consensus on the priority.

   C.   Deny the changes because they will delay the project.

   D.   Consult the Change Prioritization plan for guidance on prioritizing new changes.

Exam Questions

15. You're a project manager on a software project. Your team is busy executing the project and creating the deliverables, but there have been several changes requested by stakeholders over the past few weeks. Each time you got one of these changes, you called a meeting with your team and the stakeholders to discuss it. Why did you do this?

    A. Every change needs to be evaluated by a change control board.

    B. You're delegating the work of evaluating changes.

    C. You do not have a good change control system in place.

    D. You are using a project management information system to assign the work.

16. You are the project manager on a construction project, and you have just received a change request. You consulted the Project Management plan, and followed the procedures laid out in the change control system. You are in the process of reviewing the change and documenting its impact. Your manager asks you why you are doing this. Which process are you doing by reviewing the change and documenting its impact?

    A. Perform Integrated Change Control

    B. Monitor and Control Project Work

    C. Manage Requested Changes

    D. Direct and Manage Project Work

17. Which of the following is NOT true about the project charter?

    A. The project charter defines the requirements that satisfy customer needs.

    B. The project charter defines the work authorization system.

    C. The project charter makes the business case that justifies the project.

    D. The project charter includes the milestone schedule.

18. You have just verified that all of the work on your project is completed. Which of these things is NOT part of the Closing process?

    A. Update historical information by documenting lessons learned.

    B. Document the work performance information to show the deliverables that have been completed and record the lessons learned.

    C. Verify that all of the deliverables have been accepted by the stakeholders.

    D. Follow the project closure procedure.

19. Which of the following is NOT true about the Project Management Plan?

    A. The Project Management plan contains the Scope Management plan.

    B. The Project Management plan gives authority to the project manager.

    C. The Project Management plan contains the schedule baseline.

    D. The Project Management plan contains the performance baseline.

Exam Questions

20. Which of the following is NOT an output of the Direct and Manage Project Work process?

A.   Work performance information

B.   Deliverables

C.   Implemented change requests

D.   Forecasts

21. You are a project manager starting a new project. Your manager warns you that previous projects ran into trouble. Which of the following would be BEST for you to rely on to help plan your project:

A.   Our project management expertise

B.   Historical information

C.   The change control system

D.   Forecasts

22. Which is NOT true about the project charter:

A.   The project manager must be consulted before the charter is finalized.

B.   The charter is issued by the project sponsor.

C.   The project manager's authority to manage the project is granted by the charter.

D.   The charter gives a summary milestone schedule.

23. Which of the following is NOT an input to the Develop Project Management Plan process?

A.   Outputs of the planning processes

B.   Project charter

C.   Expert judgment

D.   Enterprise environmental factors

24. You are the project manager on a network engineering project. Two weeks ago, your team began executing the project. The work has been going well, and you are now a day ahead of schedule. Two stakeholders just approached you to tell you that they have an important change that needs to be made. That change will put you behind schedule. What do you do?

A.   Implement the change because you're ahead of schedule.

B.   Refuse to make the change because the stakeholders did not take it to the change control board.

C.   Refuse to make the change until the stakeholders document it in a change request.

D.   Make sure the stakeholders know that you're open to change, and tell them to talk to the project sponsor.

Exam Questions

25. Diane is a project manager at a software company. She just got a change request from one of her stakeholders, but is concerned that it will cause a serious problem with her schedule. She called a meeting with the project team, and decided that there was a real change, and now they need to start change control. Which of the following is NOT an output of the Perform Integrated Change Control process?

A.   Project document updates

B.   Change request status updates

C.   Project Management plan updates

D.   Change requests

Start thinking about the kinds of questions you're seeing. Some have extraneous details—we call them "red herrings." Others are about inputs and outputs. That will definitely make the exam more familiar and easier.

OH, I SEE. SOMETIMES THE DETAILS OF THE QUESTION DON'T MATTER. THEY'RE JUST THERE TO THROW YOU OFF TRACK.

Watch out for those red herrings.

Take some time to go over the answers to these questions and if they did throw you off track, reread the question to understand why.

Just remember...if you get something wrong now, that means you're actually **MORE** likely to remember it on the exam! That's why practice exams are so useful.

Answers

Exam ~~Questions~~

1. Answer: C

This is really a question about inputs and outputs. There's only one process that takes "change requests" as an input, and that's Perform Integrated Change Control. That's where your changes get approved. The other answers all refer to other processes: A is about building a baseline (which is part of Develop Project Management Plan), while B and D are both about Direct and Manage Project Work.

This is a "which-is-not" question. When you see a question asking you to choose which input or output is not associated with a process, one good strategy is to try to think of what it is that process does.

2. Answer: B

The Project Management plan is created in the Develop Project Management Plan process, which happens after Develop Project Charter. Develop Project Charter is the very first process on any project, and the inputs in answers A, C, and D exist before the project started. The Project Management plan is created during the project.

3. Answer: C

The whole reason for the Direct and Manage Project Work process is to actually do the project work, and the deliverables are the products or services that are created by the project. Don't get fooled by answer D—even though the work is performed in Direct and Manage Project Work, the information about how that work is performed is turned into forecasts in Monitor and Control Project Work.

That makes sense. You need to monitor the work to figure out how well it's being performed.

4. Answer: C

When you get a question about communication, look for the answer that provides the most complete, honest, and up-front information, even if that information won't necessarily solve the problem or make everyone happy.

5. Answer: B

All changes must be handled using the change control system, which is a set of procedures that is contained in the Project Management plan. There is no way to tell from the question what specific steps will be in that change control system—answers A, C, and D are all possible ways to deal with changes, depending on the situation. The only way to know for sure what to do is to follow the change control procedures in the Project Management plan.

Answers

Exam ~~Questions~~

6. Answer: A

This is a "just the facts" question, and answer A is the actual definition of the Work Authorization System from the *PMBOK Guide*. After you're done with these questions, look it up—it's on page 567. Underline or highlight it, and then read it out loud. Once you've read about it in the chapter, answered this question about it, and then looked up the definition, you'll never forget it!

7. Answer: D

The first thing you do after a change is approved is to update the baseline. If you chose answer C, don't feel bad—it's easy to get a little mixed up about what a baseline is used for. The whole purpose of the baseline is to figure out whether your project has deviated from the plan. But a change isn't a deviation from the plan! A deviation is accidental, while a change is done on purpose. That's why it's so important to get the change approved: that way, everyone knows about it, which means that you can plan for it. And updating the baseline is how you do that planning.

> You use the baseline to protect yourself from nasty surprises... and an approved change is not a surprise.

8. Answer: B

When your company has a policy, you need to follow it and not try to work around it. Also, don't get fooled by answer D—the question said that a change request was submitted, so it's already documented. The exam could contain tricks like that!

> The important stuff in this question is all in the second and third sentences. The outsourcing detail is a red herring.

9. Answer: B

This is a business case—it describes the benefits of doing a project and can be used to decide whether it's worth it for your company to do the work. Sometimes the benefits will be about gaining capabilities, not just money.

> There will be questions on the exam where there are two valid answers but only one BEST answer.

10. Answer: D

Defects do not need to be documented in the Project Management plan. Take a look at the other answers—do you understand why they are correct? Answer A is simply the definition of the project charter; it doesn't have anything to do with the defect, but it's still true. When you're performing the Monitor and Control Project Work process, you need to make sure defect repairs are approved before you change the deliverables, so answer B is true as well. And as far as answer C goes, that's the whole purpose of the Perform Integrated Change Control process: to approve defect repairs, changes, and preventive and corrective actions!

Answers

Exam ~~Questions~~

11. Answer: A

This is the definition of the Project Management plan!

12. Answer: C

Even when a project is terminated, you still need to close it out.

A question like this needs you to actually think about what you'd do—it's not just about applying a rule that you've learned.

13. Answer: A

If a resource is not available to you, it doesn't matter what's in your project charter or what your sponsors and stakeholders want. You need to figure out how to move forward from here, and the first step in doing that is evaluating the impact that this new problem will have on your project.

There's no such thing as a Change Prioritization plan! Keep an eye out for fake artifacts and processes.

14. Answer: A

The project manager must decide the priority of the changes. If the changes need to be made, that means that they were approved. So you can't simply deny them. And you can't call the team in for a meeting, because they need to do the work. Some people may think that the stakeholders need to be involved—but since the change was already approved, you've gotten their buy-in. Now it's up to you to decide the order in which they're implemented.

This is NOT a good change control board because a change control meeting doesn't usually include the whole team!

15. Answer: C

When you get a change request, you need to consult the Project Management plan and follow the procedures defined in the change control system. It is generally not a good idea to involve the entire team in evaluating each change that comes in—there may be many changes, and if you pull your team off the job for each one, they'll never get their job done!

Doesn't C seem like the right answer? Too bad it's not a real process!

16. Answer: A

Once a change is requested, all of the work that you do with it falls under Perform Integrated Change Control, right up until it's approved and you can implement it.

Exam ~~Questions~~ Answers

17. Answer: B

The work authorization system is defined by the company, and it's external to the project. You can think about it as the rules that you are told to follow in order to assign work in your company. They are part of the enterprise environmental factors, an input to Develop Project Charter.

Remember that lessons learned are documented throughout the project, not just at the end! That's why they're part of Work Performance Information.

When you close a process or phase, you need to make sure each deliverable has been accepted by the stakeholders.

18. Answer: B

The work performance information is documented as part of Direct and Manage Project Work. By the time the project closes, it's too late to use the work performance information! That's why it's an input to Monitor and Control Project Work—so you can take corrective action if the work is not being performed well.

We'll learn about Scope Management in the next chapter.

19. Answer: B

The project charter authorizes the project manager.

20. Answer: D

You'll learn about forecasts in Chapter 7—they're used to help predict whether the project will come in on time and within budget. If not, preventive or corrective actions will be needed! But you don't need to know that to know they're not an output of Direct and Manage Project Work.

> IT SEEMS LIKE HISTORICAL INFORMATION IS AN IMPORTANT CONCEPT. I'LL BET THERE WILL A QUESTION OR TWO ABOUT IT ON THE EXAM.

21. Answer: B

Historical information is an important input into Develop Project Charter, which is the first process that you perform when you start a new project. Historical information is very important, because it's how you learn about past projects' successes and failures. It's not actually listed as its own input. It's a part of organizational process assets—and it really is a huge asset to any organization!

When you add lessons learned to your organizational process assets, you're recording important historical information that other project managers can use later.

Answers

Exam ~~Questions~~

22. Answer: A

The project manager may be consulted when the project charter is created, but that's not always the case. It's possible that the project manager for a project is not even known when the charter is created!

23. Answer: C

While you may employ good judgment in developing your project management plan, expert judgment is not an input. It's a tool/technique used in the various processes.

Didn't D look like a good answer?

24. Answer: C

The first step in handling any change is to document it. That's why change requests are an input to Perform Integrated Change Control: the change control process cannot begin until the change is written down!

25. Answer: D

If you're having trouble remembering what the inputs and outputs are for Monitor and Control Project Work and Perform Integrated Change Control, one way to think about it is that change control is all about deciding whether or not to do something. Monitor and Control Project Work is where you spot the problems—that's why all of the REQUESTED changes are outputs of it, and inputs into Perform Integrated Change Control.

Perform Integrated Change Control is where those recommendations get evaluated and turned into APPROVED actions and changes. The ones that are not approved are REJECTED. Then they go back to Direct and Manage Project Work, where they are IMPLEMENTED, because that's project work and all project work happens in that process.

REJECTING CHANGES MEANS THAT SOMETIMES YOU NEED TO SAY NO TO PEOPLE TO MAKE THEM HAPPY IN THE END—THEY MIGHT NOT LIKE IT, BUT THEY'LL END UP SATISFIED WHEN THE PROJECT GOES WELL.

Remember, this is how you handle changes: Find it...evaluate it...fix it.

So how did you do?

5 Scope management

Doing the right stuff

Confused about exactly what you should be working on?

Once you have a good idea of what needs to be done, you need to **track your scope** as
the project work is happening. As each goal is accomplished, you confirm that all of the
work has been done and make sure that the people who asked for it are **satisfied with
the result**. In this chapter, you'll learn the tools that help your project team **set its goals**
and keep everybody on track.

Out of the frying pan...

The people at Ranch Hand Games have been working hard for over a year on the sequel to their most successful title, *Cows Gone Wild*. It seemed like the project would never end…

I NEED A BREAK! EVERY TIME I THOUGHT WE WERE DONE, SOMEONE WOULD COME UP WITH SOMETHING ELSE TO ADD.

I HAVEN'T SEEN MY GIRLFRIEND IN TWO WEEKS...AND NOW THIS GAME IS ENORMOUS.

Amy's the creative director. She's in charge of the story and the artwork.

Brian's the development manager. His job is managing the team that builds and tests the software.

...and right back into the fire

Since it took so long to get this version out, it's already time to start working on the next version. But nobody wants to see that project spin out of control the way it did last time.

WE CAN'T LOSE CONTROL OF THE PROJECT THIS TIME AROUND. CAN'T WE HIRE SOMEONE TO HELP US KEEP CGW III ON TRACK?

They're wondering what they can do to get this new project started off on the right foot.

The *Cows Gone Wild II* team ran into a lot of changes throughout the project. Could they have done something to avoid that problem?

Cubicle conversation

IT TOOK WAY TOO LONG TO GET COWS GONE WILD II OUT THE DOOR.

Brian: The project rocked in the beginning. We brought in some really talented programmers so that we could handle all of the technical challenges that might come up. We spent all that time whiteboarding and working our way through the technical issues in design. It really felt like this game was going to be amazing and fun to build. What went wrong?

Amy: We got sidetracked all over the place. Remember what happened with the website? We spent months making that site look just like the game. It got to the point where it actually looked a lot better than the game did.

Brian: Yeah, you're right. And there were all these changes along the way—the story got updated like a thousand times. It was nuts.

Amy: I remember that. What a mess.

Brian: Totally. Oh man, and that time we realized you had to redraw all the artwork for the Haymaker level? We all slept in the office for like a week!

Amy: Right...um, so what's gonna keep that from happening this time?

Maybe the Cows Gone Wild II project would have gone better if they'd had a project manager on board...

Exercise

How would you solve these problems that happened in *Cows Gone Wild II* so they don't cause the same kind of trouble on *CGW III*?

Just write down a short sentence for each of these.

1. The website got larger and larger and took almost as much time to build as the game itself.

....

The team had to rework a bunch of artwork because the game story changed.

2. Last-minute story changes. ....

....

3. Artwork changes that caused rewrites at the last minute. ....

....

4. The game was over a year late. ....

....

Here are some answers that are good for dealing with these scenarios.

Exercise Solution

1. The website got larger and larger and took almost as much time to build as the game itself.

Keep the team from doing unnecessary work.

You can't depend on the team to figure out what to do along the way. You need to scope out the work from the very beginning.

Luckily, if you nail down the scope up front, your team won't waste time doing unnecessary work later.

2. Last-minute story changes. **Plan ahead and avoid late-breaking changes.**

Writing down all of the work and the effort required to do it will help everyone understand the impact of their changes.

If the creative team figured out earlier that they'd need to make changes, the programmers could have worked on parts that weren't going to change. That would have been a lot more efficient.

3. Artwork changes that caused rewrites at the last minute.

Get started on the artwork changes sooner.

It's easier to figure out what's going to have to change if everyone is in sync on the scope.

Sounds like this game was late because the scope kept changing. Better planning could have fixed this.

4. The game was over a year late. **Start planning (sooner.) Figure out what the team is going to do before they start.**

Knowing what you're going to build BEFORE you build it means you can do a better job predicting how long it will take.

Doing more planning at the start of the project helps you prioritize so that the most important work gets done efficiently.

It looks like we have a scope problem

All of the major problems on *Cows Gone Wild II* were **scope problems**. The website was bloated with features that were added on late in the project. The creative team kept realizing that they had to do a lot more work. These are classic scope problems.

The product scope is all about the final product—its features, components, pieces.

When people talk about scoping out their products, a lot of times they're talking about figuring out the features of the product, not the work that goes into it.

Product scope means the features and functions of the product or service that you and your team are building.

When we talk about scoping out a project, we mean figuring out all of the work that needs to be done to make the product.

Project scope is all of the **work** that needs to be done to make the product.

THIS is a big part of what the project manager is concerned with...the work the team has to do.

This means changes that just went in without anyone bothering to figure out what effect they'd have on the project's time, cost, scope, quality, risk, or resources.

Scope creep means uncontrolled changes that cause the team to do extra work.

For the exam, you need to understand both product and project scope.

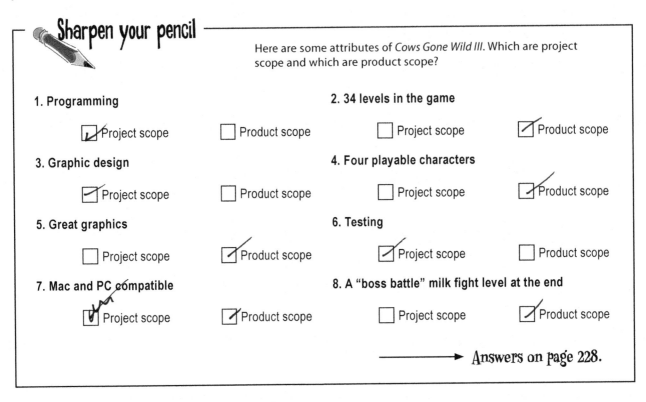

Sharpen your pencil

Here are some attributes of *Cows Gone Wild III*. Which are project scope and which are product scope?

1. Programming
☑ Project scope   ☐ Product scope

2. 34 levels in the game
☐ Project scope   ☑ Product scope

3. Graphic design
☑ Project scope   ☐ Product scope

4. Four playable characters
☐ Project scope   ☑ Product scope

5. Great graphics
☐ Project scope   ☑ Product scope

6. Testing
☑ Project scope   ☐ Product scope

7. Mac and PC compatible
☑ Project scope   ☐ Product scope

8. A "boss battle" milk fight level at the end
☐ Project scope   ☑ Product scope

⟶ Answers on page 228.

there are no Dumb Questions

Q: Does the scope include all of the stuff that I make, like a project schedule or a budget? What about things that are used to build the product but not actually delivered to the people who use it?

A: Yes, the project scope includes every single thing made by you and the team, and that includes the project plan and other project management documents. There are plenty of things on a project that are deliverables, but which the people who use the product will never see...like a project schedule, specifications, blueprints, and budgets. And while some of these things are made by the project manager, there are a lot of them that aren't, and it's not your job to figure out what goes into them. You just need to make sure they get done.

Q: Won't the team care more about what they are making than how they are making it?

A: Yes, definitely. It's your job as project manager to worry about all of the work the team does to build the product, so that they can focus on actually building it. But that doesn't mean you don't need their cooperation to make sure you've written down all of the work, and nothing else.

Q: Does that mean the project manager doesn't care about the product scope at all, just the project scope?

A: No, you still need to think about your project's final product. You can never ignore product scope, because most projects have changes to the product scope along the way. You'll have to change your project scope to include the work that's caused by product scope changes. Changes like that will probably have an impact on time and cost, too.

Here's an example: if somebody asks for a new feature in *Cows Gone Wild III*, the first thing the team needs to do is understand how much work is involved to accommodate it, and what that scope change will do to the cost and schedule.

As a project manager, your main concern is understanding that impact, and making sure everyone is OK with it before the change gets made. It's not your job to decide which is the best feature for the product, just to help everybody involved keep their priorities in mind and do what's best for the project.

You've got to know what (and how) you will build <u>before</u> you build it

You always want to know exactly what work has to be done to finish your project *before* you start it. You've got a bunch of team members, and you need to know exactly what they're going to do to build your product. So how do you write down the scope?

That's the goal of the **six Scope Management processes**. They're about figuring out how you will identify all of the work your team will do during the project, coming up with a way to make sure that you've written down what work will be done (and nothing else!) and making sure that when things change on your project, you keep its scope up to date so that your team is always building the right product.

Scope Management means figuring out what's OUT OF scope, not just what's part of it.

> CAN'T I JUST HAVE MY TEAM SPEND A DAY BRAINSTORMING A LIST OF EVERY POSSIBLE THING THEY MIGHT HAVE TO DO?

Scope Management plan

You need to write down exactly how you're going to do all of those things in the Scope Management plan.

That's a good idea. But what happens if they miss something?

It often seems like you should just be able to get everyone in the same room when the project starts and just hash all this stuff out. But it's really easy to miss something, and it's even easier for a team to get sidetracked.

It's way too easy for people to go off track and start doing things that don't really contribute to the project—like building the website for a video game instead of building the game itself.

This is why the Scope Management plan needs to say how you're going to keep unnecessary work out of the project.

The Scope Management plan describes how you write down the scope, make sure it's right, and keep it up to date.

The power of Scope Management

When you take control of your project's scope, you're doing more than just planning. It turns out that when projects have scope problems, the results are actually pretty predictable. Take a look at these problems that the Ranch Hand team ran into. Do any of these sound familiar to you? Many project managers run into similar problems on their own projects.

1 **The team had trouble getting the project off the ground.** Everyone on the team was good at their individual jobs, but it seemed like nobody knew how to get the project started.

They'd sit around in meetings talking about what they wanted to build, but it seemed like weeks before anything started getting done.

2 **There were a lot of false starts.** Just when they thought they were getting the project under way, it seemed like something would shift and they'd be back to square one.

3 **The sponsor and stakeholders were unpredictable.** There were three different times that Amy and Brian thought they were done. But each time, a stakeholder found a problem that sent them back to the drawing board.

The worst part about this was that there was no way to know when they were done with the project without asking for the sponsor's opinion...and it seemed like that opinion was always changing.

4 **There were a whole lot of changes.** They were always scrambling to keep up with shifting priorities and ideas, and they never knew for sure what they'd be working on each week.

The team was tempted to lay down the law and forbid any changes...but a lot of those changes were necessary, and good ideas.

The six Scope Management processes

Each of the Scope Management processes was designed to help you avoid the kinds of scope problems that cause a lot of projects to go off track. One of the best ways to remember these processes for the exam is to understand why they're useful, and how they solve the kinds of problems that you've seen on your own projects.

Project Management plan

Requirements documentation

Project scope statement

Work breakdown structure

Change requests

Accepted deliverables

Plan Scope Management

Here's where you write down the subsidiary plan for the project management plan that we talked about in the last chapter. You plan out all of the work you'll do to define your scope, make sure the team is planning to do the right work, and control it.

Collect Requirements

In this process, you find out all of the stakeholders' needs and write them down so that you know what to build and your requirements can be measured and tracked.

Define Scope

Here's where you write down a detailed description of the work you'll do and what you'll produce.

When you do this right, the stakeholders are never unpredictable because you already understand their needs.

Create WBS

The work breakdown structure (WBS) organizes all of your team's work into work packages—or discrete pieces of work that team members do—so that you can keep the momentum of the project going from the start.

Control Scope

Pay attention to the WBS—there will be a lot of questions about it on the exam.

We already know how important it is to control changes on your project. When scope changes aren't controlled, it leads to the most frustrating sort of project problems. Luckily, you already know about change control, and now you can use it to manage your project's scope.

Remember integrated change control from Chapter 4? Now you'll see it in action.

Validate Scope

Once the work is complete, you need to make sure that what you're delivering matches what you wrote down in the project scope statement. That way, the team never delivers the wrong product to the customer.

On the exam, "customer" can mean the same thing as "client" and "sponsor."

Plan your scoping processes

Here's where you figure out how you'll approach defining and validating the scope of your project. The **Plan Scope Management** process is where you lay out your approach to figuring out what work you'll do and what's out of scope. All of the other processes in the Scope Management knowledge area are defined and described in this document. It's the blueprint you'll use for everything else you'll do to manage scope through the project.

Planning process group

Since you're in the process of creating these plans for the other knowledge areas at the same time as this one, there's a good chance that you can use some of the same ideas in this plan that you've uncovered in the process of creating your Time Management plan, your Cost Management plan, or any of the other subsidiary plans.

The charter already includes a high-level description of the scope of the project. So it's a good place to start.

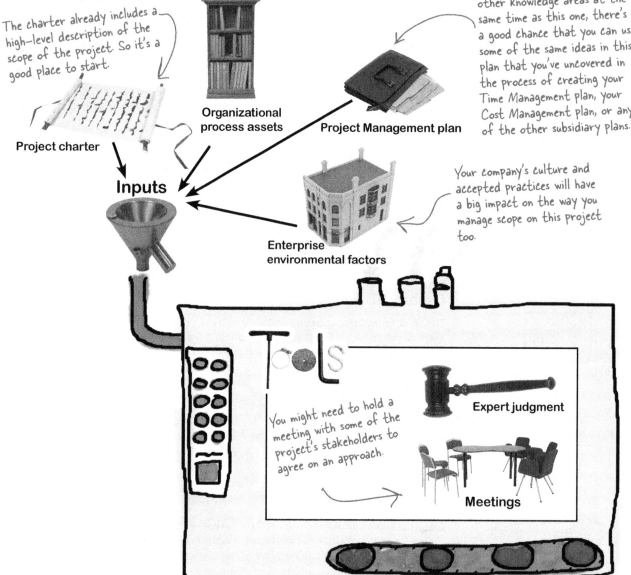

Project charter

Organizational process assets

Project Management plan

Inputs

Enterprise environmental factors

Your company's culture and accepted practices will have a big impact on the way you manage scope on this project too.

Tools

You might need to hold a meeting with some of the project's stakeholders to agree on an approach.

Expert judgment

Meetings

Now you've got a roadmap for managing scope

There are two outputs of the Plan Scope Management process: the Scope Management plan and the Requirements Management plan. Both of them help you define the scope of your project and make sure that you and your team are focused on only the work that will help you satisfy your customers' needs. The Scope Management plan keeps you on track by detailing the processes you and your team will follow as you document your scope, figure out your work breakdown structures, and validate and control your scope for the rest of the project. The Requirements Management plan details the process you'll use to collect requirements and how you'll manage them once they've been written down.

Outputs

Your Requirements Management plan will describe all of the processes your team will use to document your requirements and maintain that document throughout he project.

The Scope Management plan isn't just about writing a scope document; it details the process you use to come up with your work breakdown structure too.

Requirements Management plan

Here's where you'll find a description of the approach the team will take to planning, tracking, and reporting on requirements. You'll use this document to describe the prioritization process for requirements, and how you'll build a traceability matrix for your requirements as well.

Scope Management plan

Here's where you write down the subsidiary plan for the Project Management plan that we talked about in Chapter 4. You plan out all of the work you'll do to define your scope, with the right work planned for the team, and control it.

The Plan Scope Management process helps you think through everything you'll need to do to keep your project focused on the right work from beginning to end.

Cubicle conversation

Meet Mike, the new project manager at Ranch Hand Games.

IT LOOKS LIKE I GOT HERE JUST IN TIME.

Brian: So we finally hired a project manager. Welcome aboard!

Amy: I'm glad they brought you in to help fix this mess.

Brian: So what are you gonna do to help us? Because I don't see what you can really change.

Mike: Thanks for the vote of confidence. Look, I might not be able to fix everything, but we should be able to keep this scope under control.

Brian: Sure, you say that now. But we all thought the last project would go fine too, and that one was a real pain!

Mike: Well, did you gather the requirements for your last project?

Amy: No, but we've built video games before and we knew basically what we needed to do when we started out.

Mike: It sounds like that wasn't enough.

BRAIN POWER

What's the first thing Mike should do to make sure that *Cows Gone Wild III* goes well?

Collect requirements for your project

Planning
process group

Gathering requirements is all about sitting down with all of the stakeholders for your project and working out what their needs are, and that's what you do in the **Collect Requirements** process. If your project is going to be successful, you need to know what it will take for all of your stakeholders to agree that your project has met its goals. You need to have a good idea of what's required of your project up front, or you'll have a tough time knowing whether or not you're doing a good job as you go. That's why you need to write down all of your project and product requirements with enough detail that you can measure your team's progress.

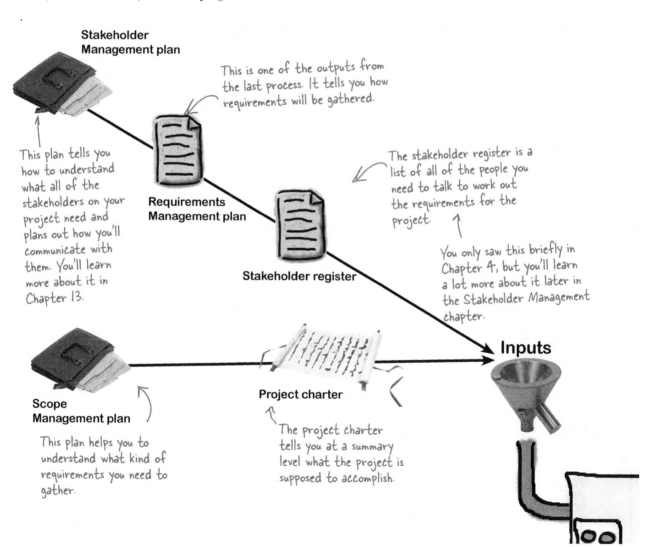

Stakeholder Management plan

This is one of the outputs from the last process. It tells you how requirements will be gathered.

This plan tells you how to understand what all of the stakeholders on your project need and plans out how you'll communicate with them. You'll learn more about it in Chapter 13.

Requirements Management plan

The stakeholder register is a list of all of the people you need to talk to work out the requirements for the project.

You only saw this briefly in Chapter 4, but you'll learn a lot more about it later in the Stakeholder Management chapter.

Stakeholder register

Scope Management plan

This plan helps you to understand what kind of requirements you need to gather.

Project charter

The project charter tells you at a summary level what the project is supposed to accomplish.

Inputs

Talk to your stakeholders

Tools

The **Collect Requirements** process involves talking to the people who are affected by your project to find out what they need. All of the tools in this process are focused on getting your stakeholders to tell you about the problem that the project is going to solve. Sometimes that means sitting down with each of them one-on-one, and other times you can do it in a group setting. One of the most important things to understand about requirements is that every requirement fulfills a specific stakeholder need. Lucky for you, a lot of those needs are already written down—in your business case document.

But that's not the only place you'll find requirements, so here are three really useful tools and techniques to help you gather requirements:

Interviews are important ways to get your stakeholders to explain how they'll use the product or service your project is creating. By talking to people one-on-one, you can get them to explain exactly what they need so that you can be sure that your project can meet its goals.

Focus groups are another way to get a group of people to discuss their needs with you. By letting a group discuss the end product together, you can get them to tell you requirements that they might not have thought of by themselves.

Facilitated workshops are more structured group conversations where a moderator leads the group through brainstorming requirements together. In facilitated workshops, misunderstandings and issues can get reconciled all at once because all of the stakeholders are working together to define the requirements.

If you've ever done a joint application design (JAD) session where users and the development team work together to define requirements, it's considered a facilitated workshop.

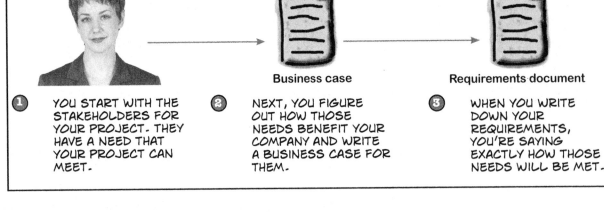

Requirements Up Close

GAMERS HAVE BEEN ASKING FOR AN UNDERWATER LEVEL.

All your requirements fill stakeholders' needs, and many start with a need that you identified in your business case document.

Your requirements are a direct line from your stakeholder's needs to your project.

Business case

Requirements document

1. YOU START WITH THE STAKEHOLDERS FOR YOUR PROJECT. THEY HAVE A NEED THAT YOUR PROJECT CAN MEET.

2. NEXT, YOU FIGURE OUT HOW THOSE NEEDS BENEFIT YOUR COMPANY AND WRITE A BUSINESS CASE FOR THEM.

3. WHEN YOU WRITE DOWN YOUR REQUIREMENTS, YOU'RE SAYING EXACTLY HOW THOSE NEEDS WILL BE MET.

Make decisions about requirements

A big project usually has a lot of stakeholders, and that means a lot of opinions. You'll need to find a way of making decisions when those opinions conflict with each other. There are four major decision-making techniques you can choose from. These are referred to as **group decision-making techniques** on the test.

Unanimity means everyone agrees on the decision.

Majority means that more than half the people in the group agree on the decision.

Plurality means that the idea that gets the most votes wins.

Dictatorship is when one person makes the decision for the whole group.

Exercise

You'll need to know the difference between the four different decision techniques for the exam. Here are the minutes from a facilitated workshop that the CGW team held with all of its stakeholders. Identify which decision-making technique was used in each case.

1. The group voted on the CCG (cud-chewer gun) five times, but decided not to include it because they couldn't get everyone to agree on it.

☐ Unanimity   ☐ Plurality

☐ Majority   ☐ Dictatorship

2. The VP of Engineering told everyone that they had to come up with a new character for Team Guernsey. Since he's the highest-ranking person in the room, nobody argued with him.

☐ Unanimity   ☐ Plurality

☐ Majority   ☐ Dictatorship

3. There were 10 new scenery suggestions up for approval, but only 5 could make it into the game. The team chose the top 5 in a general vote.

☐ Unanimity   ☐ Plurality

☐ Majority   ☐ Dictatorship

4. Over half the group wanted to see a new story that involved Farmer Ted. So that requirement was recorded as an absolute necessity.

☐ Unanimity   ☐ Plurality

☐ Majority   ☐ Dictatorship

Answers: 1. Unanimity 2. Dictatorship 3. Plurality 4. Majority

Help your team to get creative

Getting your team to think creatively can help you create a better product from the start. **Group creativity techniques** are all about getting those creative juices flowing while you gather your requirements.

Idea/mind maps are a good way to visualize the way your ideas relate to each other. When you've finished working through an idea, it sometimes helps to create a map of how you got there and show which ideas can be grouped together.

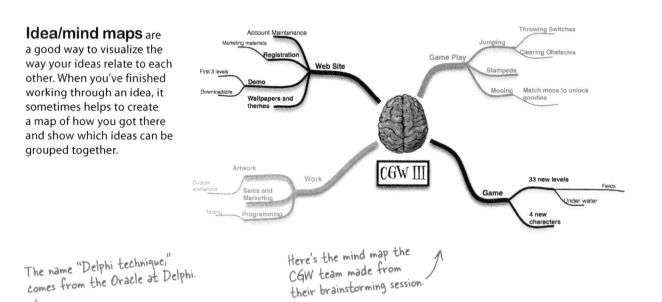

The name "Delphi technique," comes from the Oracle at Delphi.

Here's the mind map the CGW team made from their brainstorming session.

The Delphi technique is a way of letting everyone in the group give their thoughts about what should be in the product while keeping them anonymous. When you use the Delphi technique, everybody writes down their answers to the same questions about what the product needs to do and then hands them into a moderator. The questions could be about specific features that the product should have.

When the CGW team used the Delphi technique, here were a few of their questions:

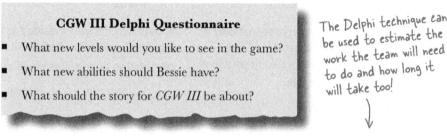

CGW III Delphi Questionnaire

- What new levels would you like to see in the game?

- What new abilities should Bessie have?

- What should the story for *CGW III* be about?

The Delphi technique can be used to estimate the work the team will need to do and how long it will take too!

The moderator keeps everybody's names to himself , but shares the ideas so that everyone can learn from them and think of new ones. After everybody discusses those ideas, they're given a chance to adjust their original answers to the questions and hand them back in to the moderator. These iterations continue a few times until the group settles on a list of requirements for the product.

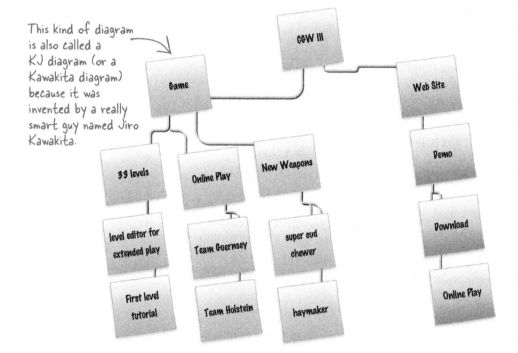

This kind of diagram is also called a KJ diagram (or a Kawakita diagram) because it was invented by a really smart guy named Jiro Kawakita.

Affinity diagrams are great when you have a lot of ideas and you need to group them so you can do something with them. A lot of people make affinity diagrams using Post-it notes on walls. That way, you can move the ideas around and change the groupings when you think of new areas to explore. Sometimes just putting requirements in categories will help you to find new ones.

The nominal group technique is a form of brainstorming where you write down the ideas as you find them and have the group vote on which ones they like the best. You then use the votes to rank all of the ideas and separate the ones that aren't important from the ones you want to delve into deeper.

Context diagrams help your team show the way all of the processes and features in your product scope relate to each other. It's a picture of the scope of your product that shows how users will interact with it.

Brainstorming is one of the most commonly used ways of collecting requirements. Whenever you sit a group of people down to think of new ideas, you're brainstorming.

Benchmarking is a way of comparing the processes and practices used in building your software with the practices and processes in other organizations so you can figure out the best ideas for improvement.

Document analysis is a way of collecting requirements by reading through all of the existing documents for your product.

Use a questionnaire to get requirements from a bigger group of people

The *Cows Gone Wild* development team needed to talk to the people who play their games to figure out what would make the gamers happy in the next version. The team obviously couldn't go around to every customer's house asking questions, so they wrote a questionnaire about new possible features for the game that they sent to gamers who had registered the game.

When it was time to start collecting requirements for the new version, the team started with all of the data they'd gathered from those surveys and did some analysis to figure out which features were most important to the gaming community. Here's an excerpt from their survey results:

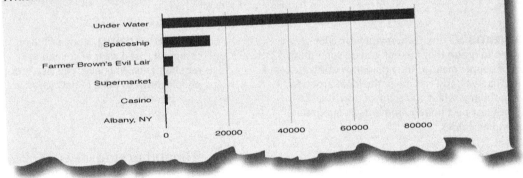

Cows Gone Wild II Registration Survey Results

The *Cows Gone Wild* series released *Cows Gone Wild II* three months ago. Since then, we've sold 500,000 copies of the game. Of those sales, 350,000 have been registered and 100,000 have responded to the *CGW III* requirements collection survey. Here are the results:

Artwork:
Which new environments would you like to see included in a follow-up to the game?

Observation can help you see things from a different point of view

Sometimes observing the people who will use your product while they work with it will give you a better idea of how to solve their problems. People don't always know what to say when you ask them for requirements, so watching them deal with the problem your product is going to address can help you to find requirements that they might not tell you about on their own.

A prototype shows users what your product will be like

Sometimes the best way to get your stakeholders to give you an opinion on what your product should be is to show it to them in a **prototype**. Prototypes are models of the product that you're going to build that give your stakeholders a better idea of what your team is thinking. Sometimes users who are experimenting with a prototype will come up with a brand new requirement that they never thought of before. If you can get them to find it in the prototype, it's a lot easier to deal with than if you wait until the end of the project to show to them. When you're making a really complicated product, it can make sense to prototype it as part of the requirements collection process so that you can find changes that your users will ask for early on.

Prototypes are a great tool if you're developing your project using iterative techniques. If you're using agile software development processes or defining requirements in phases, prototypes are a great way to keep your stakeholders involved in the project and get their feedback on changes that might be needed.

there are no Dumb Questions

Q: In my company, business analysts collect the requirements, not project managers. Why do I need to know all this stuff?

A: Good point. A lot of project teams will have a business analyst who will work on gathering requirements for the project and writing specifications for it. As the project manager, though, you are responsible for making sure that the needs of all of the stakeholders are met. So it's a good idea for you to stay on top of the requirements collection process, and be an active participant in it.

Some organizations even divide up the requirements-gathering activities into project requirements and product requirements. The project requirements would be things like staying within the budget, meeting specific deadlines, and using a certain number of resources, while product requirements would be about features of the product. Even if you are lucky enough to have a business analyst on your project to help you gather requirements, you'd better understand both the project and product requirements if you're going to keep your project on track.

Q: Can I just skip these requirements-gathering tools and jump straight into code? We do iterative development where I work. That means I can jump right in and plan the work as it's happening, right?

A: The short answer is no. The more you know up front, the easier it's going to be for you to plan out your project. Even iterative projects must plan out their requirements for each phase up front. Now, it's true that you should be able to get through the Collect Requirements process more quickly if you're only gathering requirements for a small phase of your project, but it doesn't mean that you can skip requirements altogether.

Q: How do I know when I'm done collecting requirements?

A: That's a good question. Your requirements need to be measurable to be complete. So it's not enough to write down that you want good performance in your product. You need to be able to tell people what measurement counts as good performance for you. You have to be able to confirm that all of your requirements are met when you close out your project, so you can't leave requirements up to interpretation.

You know your requirements are complete when you've got a way to verify each of them once they're built.

Now you're ready to write a requirements document

The outputs of the Collect Requirements process are the requirements document and a requirements traceability matrix, which allows you to follow the requirements from the document through implementation and verification.

Outputs

> This requirement is measurable. If the end product has puzzles that involve swimming, the requirement will pass its test. If not, it will fail.

CGW III Requirements Document

1. Introduction
CGW II was a huge hit. We've done some market research and some internal brainstorming and compiled these requirements for *Cows Gone Wild III: The Milkening*, which will be released next year in time for the holidays.

2. Organizational Impact
This product will have an impact on many departments at Ranch Hand Games, including Research and Development, Marketing, Distribution, Shipping, Administration, Finance, and Customer Service.

3. Functional requirements

Name	RU001—Include Underwater levels.
Summary	The cows will need to be able to move around under water.
Rationale	Underwater environment was the single biggest request from polled gamers.
Requirement	Cows will need to be able to swim, and underwater puzzles will need to be developed that require swimming.

...

4. Nonfunctional requirements

> Here, you can load the levels and time it to figure out if the product meets its requirements.

Name	RNF001—Performance as good or better than *CGW II*.
Summary	The new functionality cannot slow down game play.
Rationale	Gamers were very happy with the performance upgrades in *CGW II*. We cannot be seen as losing that improvement in the next version.
Requirement	All levels must load in under 15 seconds. All online levels must load in under 25 seconds over a cable connection at 256K.

The requirements document needs to list all of the functional and nonfunctional requirements of your product. **Functional requirements** are most of the kinds of things that you think of right away: new features, bug fixes, and new or different behavior. **Nonfunctional requirements** are sometimes called *quality attributes* because they're things that you expect from your deliverables, but aren't specific features. Some examples of nonfunctional requirements are performance, reliability, error handling, and ease of use.

CGW III Requirements Traceability Matrix

Origin codes: Business case - BC, Survey-S1, Internal- I
Requirement Nos: Cross-reference with requirements document
Work Module: Where implemented, cross-reference with WBS
Test: Where verified, cross reference with design of experiments

Requirements for Underwater Levels

Origin	Requirement	Module	Test
S1	RU001	3.3.1	TC01-TC57
BC1	RU002	3.4.1	TC101-TC350
S3	RU003	3.6.2, 3.7.1	TC2

This document shows where the requirements come from, where they get implemented, and how they get verified. It's a great way to take a quick high-level look at all your requirements and make sure they're mapped to specific test cases.

We'll be talking more about what a WBS is and how to build one in just a few pages.

BULLET POINTS — AIMING FOR THE EXAM

- **Product scope** means the features and functions of the product or service being built. **Project scope** means the work that's needed to build the product.

- **Functional requirements** are the behavior of the product. **Nonfunctional requirements** are implicit expectations about the product.

- **Scope Management** is about figuring out all of the work that's going to be needed for the project, and making sure only that work is done—and nothing else.

- The **Scope Management plan** is created as part of the Project Management plan. It defines the process you'll use for defining scope and managing changes to it.

- You'll need to know the **order of processes** for the exam. A good way to remember them is to understand how the output of one process is used as the input for another.

CGW II Requirements Management

The Requirements Collection Process:
The following techniques will be used for requirements elicitation:
1. Questionnaires and surveys
2. Facilitated workshops
3. Delphi technique
4. Focus groups
5. Interviews
6. Observation

They will be prioritized based on strategic alignment with *CGW III*'s business case document.

Requirements will be managed as part of integrated change control once approved.

The Requirements Management plan tells how requirements will be gathered and analyzed.

Once the requirements document is approved by the stakeholders, any changes to it need to be approved using integrated change control.

Exercise

Write down the Collect Requirements tool or technique that's being used in each one of these scenarios.

1. The team got together to come up with ideas for the game. As they thought of them, they grouped them on different colored index cards and used thumb tacks to arrange them on a bulletin board by type.

....

2. Ranch Hand Games listed questions for people visiting the website to answer in exchange for a game promo coupon.

....

3. The team got together to brainstorm and periodically voted to rank requirements and separate the least important from the most important.

....

Answers: 1. Affinity diagram, 2. Questionnaire, 3. Nominal group technique

BRAIN POWER

Now that Mike's gathered the requirements, what do you think he should do with them? How can he make sure they actually get implemented in the game?

Define the scope of the project

Now that the Ranch Hand team has a project manager, everything will go smoothly, right? Well, not exactly. Just assigning a project manager isn't enough to get the scope under control. That's why you need the **Define Scope** process. Even the best project managers need to rely on things from the company and the people around them. That's why the inputs to Define Scope are so important. They contain everything you need to know before you can begin to break the project down into the work that the team members will do.

Planning
process group

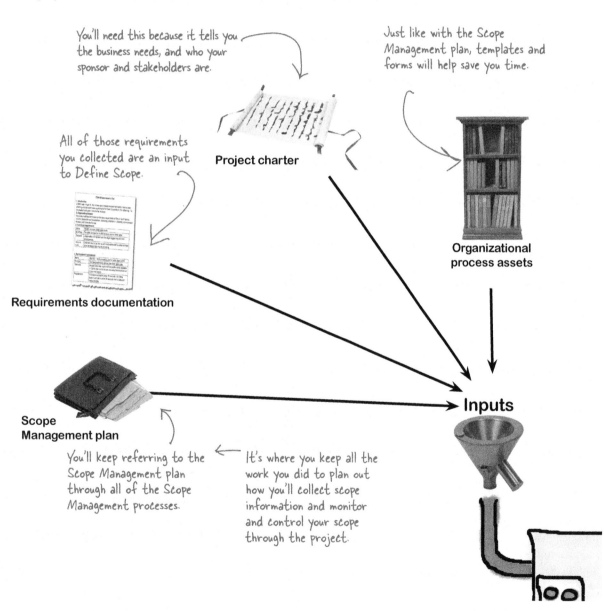

You'll need this because it tells you the business needs, and who your sponsor and stakeholders are.

Just like with the Scope Management plan, templates and forms will help save you time.

All of those requirements you collected are an input to Define Scope.

Project charter

Requirements documentation

Organizational process assets

Scope Management plan

You'll keep referring to the Scope Management plan through all of the Scope Management processes.

It's where you keep all the work you did to plan out how you'll collect scope information and monitor and control your scope through the project.

Inputs

How do you define the scope?

You already got a head start on defining the project scope when you wrote down the requirements. But now you need to go a lot further and write down all of the work that you and your team are going to do over the course of the project. Luckily, the **Define Scope process tools and techniques** are there to help guide you through creating the project scope statement (which you'll learn about in a minute).

These are the four tools and techniques of Define Scope.

You need to figure out what the stakeholders need so you can deliver it to them.

Facilitated workshops

When you do facilitated workshops with your stakeholders, figure out what they need, and write it all down. The reason you do this is because you need to make sure that what you're delivering really meets the needs of the stakeholders. This keeps the team from delivering a poor product.

An important part of stakeholder analysis is doing your best to set **quantifiable goals**. That means writing down specific project goals that you can measure, which makes it a lot easier for the team to plan for the work they have to do.

> WE NEED TO IMPROVE CUSTOMER SATISFACTION.

> WE NEED TO REDUCE SUPPORT CALLS BY 15%

That's a great goal, but it's not quantifiable.

VS.

Everybody can shoot for that.

Product analysis

Remember product versus project scope? People naturally think about the product they are making when they start to define the scope. This tool is all about turning those things into project work that needs to be done.

Once the work is complete, you're going to have to make sure that what you're delivering matches what you put in your requirements. The better your product analysis is at the start of the project, the happier your stakeholders will be with the product, and the less likely it is that you'll discover painful, last-minute problems at the end.

COWS GONE WILD III
The Milkening

storyboards
scenery
great graphics

The game needs this...

...so Amy does this.

storyboarding sessions
drawing the scenery
designing the graphics

Alternatives generation

Think of other ways that you could do the work. Exploring different ways to do the work will help you find the one that is most efficient for the project. It's always possible that you might find a better way of doing things and need to change your original plan.

Designing the graphics: alternatives

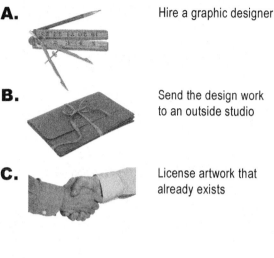

A. Hire a graphic designer

B. Send the design work to an outside studio

C. License artwork that already exists

Expert judgment

You've seen this one before! Bring in an expert to help you figure out what work needs to be done.

Expert judgment

there are no
Dumb Questions

Q: Is product analysis the same as requirements gathering?

A: Not exactly. When people gather requirements, they're trying to understand what needs the product should fill. Requirements are the contents of the product. When you use product analysis to define the scope of the work to be done, you're figuring out what deliverables the team needs to work on in order to build your project scope statement. So product analysis is concerned with how the work will be done, not what's in it.

Q: What if there is only one way to do something? Do I still need to do alternatives identification?

A: There aren't too many things out there that can only be done one way, but if you happen across one, then you don't have to spend much time on alternative identification because there aren't any alternatives to identify.

Q: What if a stakeholder can't tell me how to measure his needs?

A: That can get kind of tricky. Sometimes stakeholders know that they want things to get better, but they don't know how to tell when they've succeeded. You need to work with them to find something that can be measured in their ideas about project success. Without a way to measure your success, you won't know whether or not you are accomplishing your goals.

The project scope statement tells you what you have to do

After you have done your scope planning, figured out as much as you could using stakeholder and product analysis, and identified all of the possible ways of doing the work, you should be ready to add any new findings to the project scope statement.

Outputs

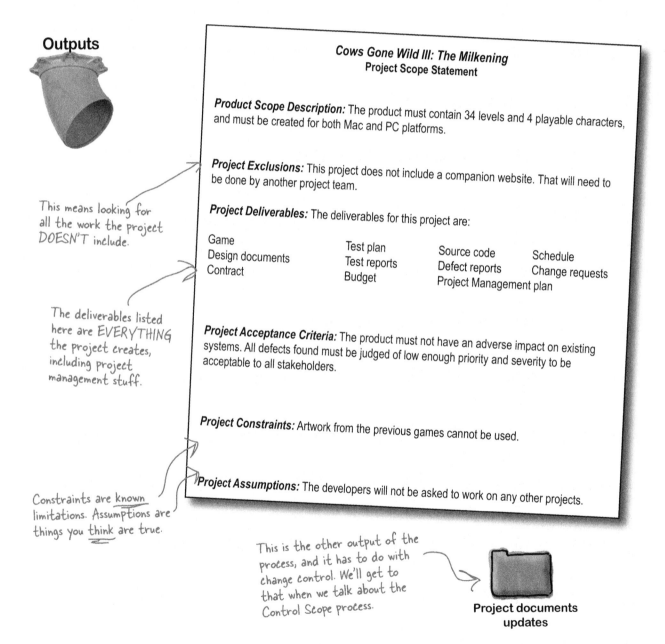

Cows Gone Wild III: The Milkening
Project Scope Statement

Product Scope Description: The product must contain 34 levels and 4 playable characters, and must be created for both Mac and PC platforms.

Project Exclusions: This project does not include a companion website. That will need to be done by another project team.

Project Deliverables: The deliverables for this project are:

Game	Test plan	Source code	Schedule
Design documents	Test reports	Defect reports	Change requests
Contract	Budget	Project Management plan	

Project Acceptance Criteria: The product must not have an adverse impact on existing systems. All defects found must be judged of low enough priority and severity to be acceptable to all stakeholders.

Project Constraints: Artwork from the previous games cannot be used.

Project Assumptions: The developers will not be asked to work on any other projects.

This means looking for all the work the project DOESN'T include.

The deliverables listed here are EVERYTHING the project creates, including project management stuff.

Constraints are known limitations. Assumptions are things you think are true.

This is the other output of the process, and it has to do with change control. We'll get to that when we talk about the Control Scope process.

Project documents updates

WHAT'S MY PURPOSE

Here are a few things that Mike left out of the *CGW III* project scope statement. Can you figure out where each of them should go?

100%

1. The game must have fewer than 15 defects ~~F~~ per 10,000 lines of code.

2. There will be four graphic designers D reporting to the art director, and six programmers and four testers reporting to the development manager.

3. No more than 15 people can be allocated C to work on the game at any time.

4. Scenery artwork. B

5. The product will not include bug fixes for A the previous version.

6. The game needs to run on a machine with E 1 GB of memory or less.

A. Project exclusions

B. Project deliverables

C. Project constraints

D. Project assumptions

E. Project requirements

F. Acceptance criteria

⟶ Answers on page 229.

The <u>project scope statement</u> tells what <u>work</u> you are— and are not—going to do in the project.

Fireside Chats

Tonight's talk: **Requirements Documentation and Project Scope Statement spar over what's important in Scope Management**

Requirements Documentation:

I'm glad we're finally getting a chance to chat in person.

I wouldn't say that! It's just that, well, I think it's not hard to see why I'm such a critical part of Scope Management.

Well, it ought to be. I mean, you wouldn't even exist if it weren't for me.

There's no work to do if there's no product and without me, nobody knows what to build. So without me, really, who needs you?

But they still need me to tell them what to build. I tell everybody what the product needs to be.

That's true. And it's no wonder that so many projects have problems. But the more you know up front, the easier it is to plan for what might happen along the way.

Project Scope Statement:

Really? I never got the impression that you had much respect for me.

Typical. Everything's about you.

How do you figure?

Now that's just not fair. You think that just because people get together and talk about you in focus groups and brainstorming sessions, you're something special. Without me, people would be arguing over your requirements forever. I'm the one who puts limits on all of this stuff.

That may be true, but think about it for a minute. Your requirements almost always change from the time you start the project until it ends. You're so high-maintenance. You hardly ever hear of a project where the team gets all of the requirements right from the beginning. And when you change, I have to change too. It's so obnoxious.

Requirements Documentation:

I guess that means that knowing the project scope up front is pretty important too, now that I think about it.

But you're so broad. I mean, if you really want to know what's getting done on a project, you have to look at me. I represent the need the project is filling; without me, it never would've happened in the first place.

That seems almost as important as my job to me.

Project Scope Statement:

That's exactly my point. We're both useful, but I'm the one everyone thinks of first when they think about managing scope.

Here we go again with your attitude. Trust me, without me, no one would know how those needs were going to be met. I'm just as important as you are.

I guess we're never going to see eye-to-eye on this.

Exercise

You'll need to know the difference between defining the scope and collecting the project's requirements for the exam. Which of these things is part of the project scope statement, and which is part of the requirements document?

1. The work required to create the graphics

☐ Requirements document   ☑ Project scope statement

2. New characters in the game

☑ Requirements document   ☐ Project scope statement

3. 33 new levels

☑ Requirements document   ☐ Project scope statement

4. The performance requirements for the product

☑ Requirements document   ☐ Project scope statement

5. A description of how the WBS is created

☐ Requirements document   ☑ Project scope statement

6. How the software will be tested

☐ Requirements document   ☑ Project scope statement

7. How the stakeholders will verify the deliverables

☐ Requirements document   ☑ Project scope statement

8. A list of all artwork that will be created

☑ Requirements document   ☐ Project scope statement

Answers: Requirements doc: 2, 3, 4, 8 Project scope statement: 1, 5, 6, 7

Question Clinic: The "which-is-BEST" question

WHEN YOU'RE TAKING ANY SORT OF EXAM, THE MORE FAMILIAR YOU ARE WITH IT, THE MORE RELAXED YOU'LL BE. AND ONE WAY TO GET FAMILIAR WITH THE PMP EXAM IS TO GET TO KNOW THE DIFFERENT KINDS OF QUESTIONS YOU'LL SEE. ONE IMPORTANT SORT IS THE "WHICH-IS-BEST" QUESTION.

The which-is-BEST question sometimes starts with a sentence or two talking about a particular situation.

This is one of those questions where "customer" is used in place of "sponsor."

OK, now you have enough information to answer the question. What do you do when you find out that certain deliverables need to change?

36. You are the project manager for a building contracting project. You schedule a meeting with your customer and stakeholders to give them an update on the progress of the project. At that meeting, they tell you that certain deliverables need to be changed before they can be accepted. Which is the BEST way for you to proceed?

A. Inform the stakeholders that they have no authority to decide what deliverables are acceptable.

B. Consult the project charter and use it to show the stakeholders that you are the authorized project manager.

C. Figure out what needs to be fixed so that you can tell the team how to make the deliverables acceptable.

D. Document the requested changes so that you can put them through change control.

Some of the answers will simply be wrong. You should be able to eliminate them first.

This one sounds good... That's what the project charter is for, right? But wait a minute! What does the charter have to do with the scope of the work?

OK, this actually seems right—you do need to do that. But is it really the BEST answer?

THE WHICH-IS-BEST QUESTION MAY HAVE MORE THAN ONE GOOD ANSWER, BUT IT ONLY HAS ONE **BEST** ANSWER.

Aha! Here's the BEST answer! Even though C was technically correct, D is a much better description of how change control actually works.

The BEST answer

HEAD LIBS

Fill in the blanks to come up with your own "which-is-BEST" question.

You are the project manager for _____ . At the end
(an industry or the name of a project)
of _____ , you ran into a problem. You find
(a Scope Management process)
out that _____ was not performed by
(a tool or technique that is part of that process)
_____ correctly. Which is the BEST way for you to proceed?
(the team member or person who is supposed to do that tool or technique)

A. _____
(an obviously wrong answer where the person or project manager uses the tool or technique incorrectly)

B. _____
(an answer that sounds correct, but isn't the BEST answer)

C. _____
(the BEST answer that describes exactly how to use the process properly)

D. _____
(an answer that says something that's true about an irrelevant process, like one from Chapter 4)

Create the work breakdown structure

The **Create WBS** process is the most important process in the Scope Management knowledge area because it's where you actually figure out all the work you're going to do. It's where you create the **work breakdown structure** (or WBS), which is the main Scope Management output. Every single thing that anyone on the project team—including you—will do is written down in the WBS somewhere.

Planning process group

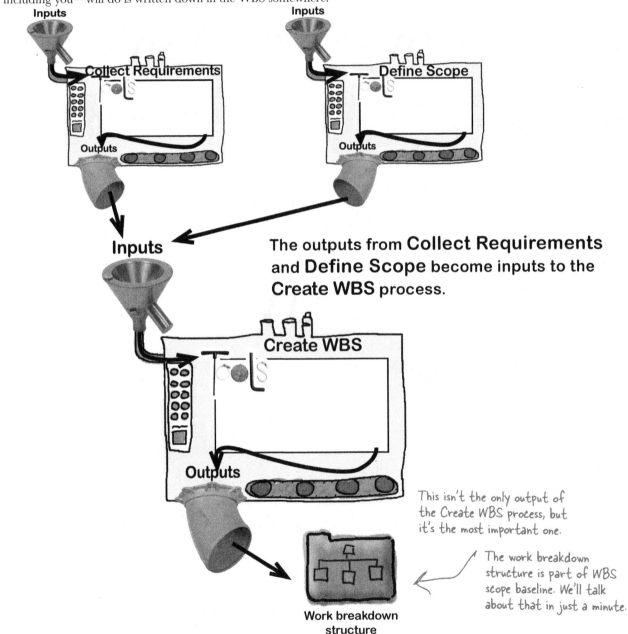

The outputs from **Collect Requirements** and **Define Scope** become inputs to the **Create WBS** process.

This isn't the only output of the Create WBS process, but it's the most important one.

The work breakdown structure is part of WBS scope baseline. We'll talk about that in just a minute.

Work breakdown structure

The inputs for the WBS come from other processes

You've already seen all of the inputs that you need to create the WBS. It shouldn't be too surprising that you need the requirements document, project scope statement, and organizational process assets before you create the WBS. When you're developing these things, you're learning what you need to know in order to decompose the project work.

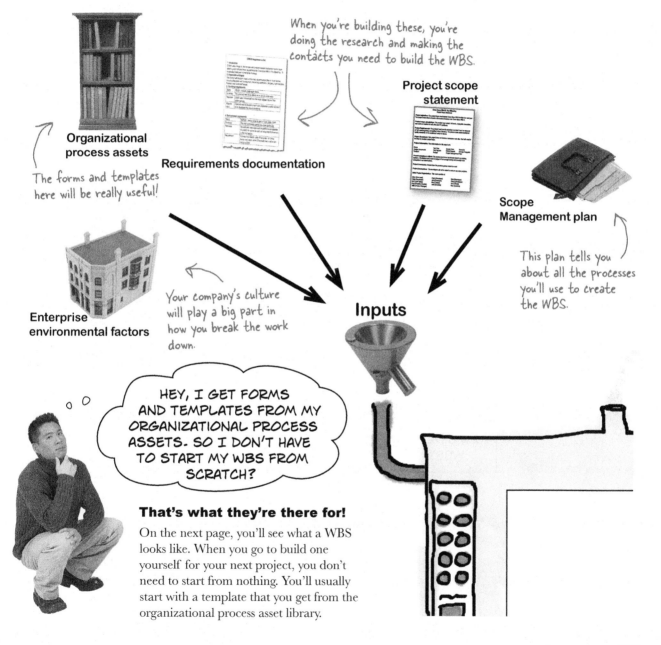

Organizational process assets

The forms and templates here will be really useful!

When you're building these, you're doing the research and making the contacts you need to build the WBS.

Requirements documentation

Project scope statement

Scope Management plan

This plan tells you about all the processes you'll use to create the WBS.

Enterprise environmental factors

Your company's culture will play a big part in how you break the work down.

Inputs

HEY, I GET FORMS AND TEMPLATES FROM MY ORGANIZATIONAL PROCESS ASSETS. SO I DON'T HAVE TO START MY WBS FROM SCRATCH?

That's what they're there for!

On the next page, you'll see what a WBS looks like. When you go to build one yourself for your next project, you don't need to start from nothing. You'll usually start with a template that you get from the organizational process asset library.

Breaking down the work

One way to get a clear picture of all of the work that needs to be done on a project is to create a work breakdown structure. The WBS doesn't show the order of the work packages or any dependencies between them. Its only goal is to show the work involved in creating the product.

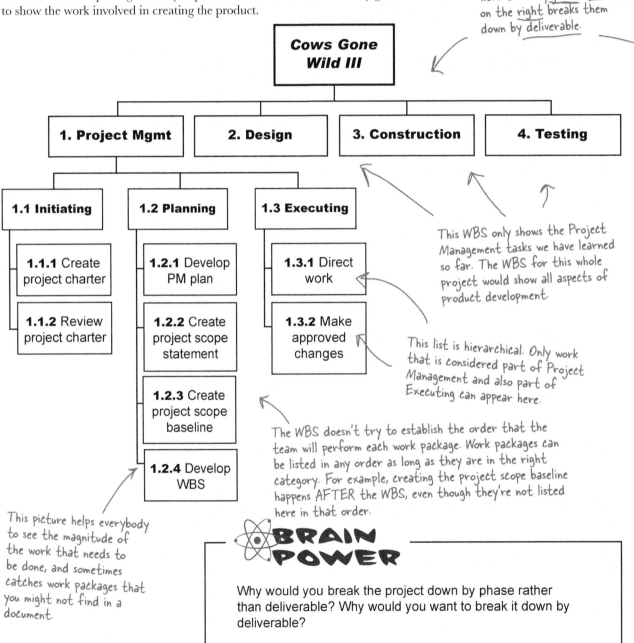

This WBS breaks the project work down by phase; the one on the right breaks them down by deliverable.

This WBS only shows the Project Management tasks we have learned so far. The WBS for this whole project would show all aspects of product development.

This list is hierarchical. Only work that is considered part of Project Management and also part of Executing can appear here.

The WBS doesn't try to establish the order that the team will perform each work package. Work packages can be listed in any order as long as they are in the right category. For example, creating the project scope baseline happens AFTER the WBS, even though they're not listed here in that order.

This picture helps everybody to see the magnitude of the work that needs to be done, and sometimes catches work packages that you might not find in a document.

⚛ BRAIN POWER

Why would you break the project down by phase rather than deliverable? Why would you want to break it down by deliverable?

Break it down by project or phase

A WBS can be structured any way it makes the most sense to you and your project team. The two most common ways of visualizing the work are by deliverable or by phase. Breaking down the work makes it easier to manage, because it means you are less likely to forget work packages that need to be included. This is the same project as the one on the left, but this time, it's broken down by <u>deliverable</u>.

These are the same phases we talked about in Chapter 3.

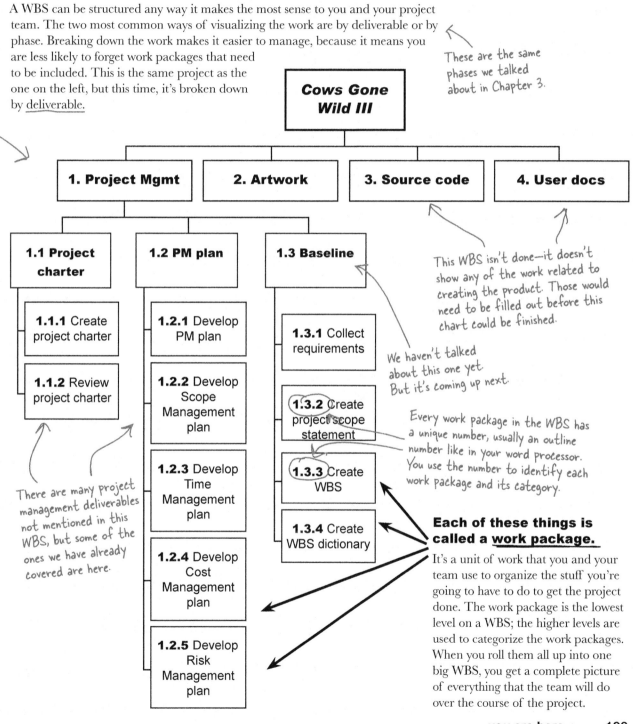

Cows Gone Wild III

1. Project Mgmt | **2. Artwork** | **3. Source code** | **4. User docs**

1.1 Project charter

1.1.1 Create project charter

1.1.2 Review project charter

1.2 PM plan

1.2.1 Develop PM plan

1.2.2 Develop Scope Management plan

1.2.3 Develop Time Management plan

1.2.4 Develop Cost Management plan

1.2.5 Develop Risk Management plan

1.3 Baseline

1.3.1 Collect requirements

1.3.2 Create project scope statement

1.3.3 Create WBS

1.3.4 Create WBS dictionary

This WBS isn't done—it doesn't show any of the work related to creating the product. Those would need to be filled out before this chart could be finished.

We haven't talked about this one yet. But it's coming up next.

Every work package in the WBS has a unique number, usually an outline number like in your word processor. You use the number to identify each work package and its category.

There are many project management deliverables not mentioned in this WBS, but some of the ones we have already covered are here.

Each of these things is called a <u>work package</u>.

It's a unit of work that you and your team use to organize the stuff you're going to have to do to get the project done. The work package is the lowest level on a WBS; the higher levels are used to categorize the work packages. When you roll them all up into one big WBS, you get a complete picture of everything that the team will do over the course of the project.

Decompose deliverables into work packages

Creating the WBS is all about taking deliverables and coming up with work packages that will create them. When you do that, it's called **decomposition**, and it's the main tool you use to create a WBS.

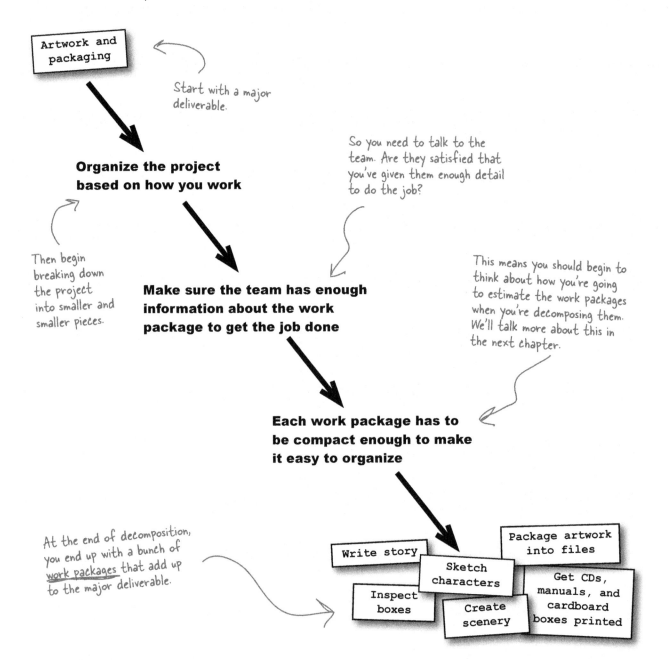

Artwork and packaging

Start with a major deliverable.

Organize the project based on how you work

So you need to talk to the team. Are they satisfied that you've given them enough detail to do the job?

Then begin breaking down the project into smaller and smaller pieces.

Make sure the team has enough information about the work package to get the job done

This means you should begin to think about how you're going to estimate the work packages when you're decomposing them. We'll talk more about this in the next chapter.

Each work package has to be compact enough to make it easy to organize

At the end of decomposition, you end up with a bunch of <u>work packages</u> that add up to the major deliverable.

Write story

Package artwork into files

Sketch characters

Inspect boxes

Get CDs, manuals, and cardboard boxes printed

Create scenery

You won't find any solutions for this, because there aren't any right or wrong answers! It's your chance to take a minute to think things through—that'll get it into your brain.

Sharpen your pencil

You'll need to understand decomposition for the exam. Here are a few deliverables from *Cows Gone Wild III*. Based on what you've seen so far, decompose them into work packages. There are no right or wrong answers—this is practice for thinking about decomposition.

Software ....
....
....
....

Artwork ....
....
....
....

Marketing materials ....
....
....
....

Throwing a party for the team ....
....
....
....

Online play promotional events ....
....
....
....

Game add-ons ....
....
....
....

Support forums and message boards. ....
....
....
....

Project Scope Management Magnets

Understanding how to build a work breakdown structure is very important for the exam—it's one of the most important parts of the Scope Management knowledge area. Here's your chance to create a WBS for *Cows Gone Wild III: The Milkening*. There are two ways you can break down the work. See if you can use decomposition to do it!

On this page, create a work breakdown structure broken down by <u>project phase</u>.

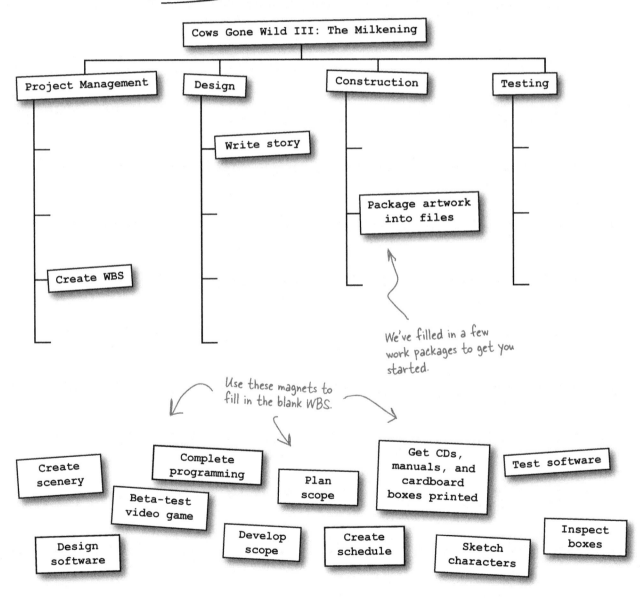

Cows Gone Wild III: The Milkening

Project Management | Design | Construction | Testing

Write story

Package artwork into files

Create WBS

We've filled in a few work packages to get you started.

Use these magnets to fill in the blank WBS.

Create scenery

Complete programming

Plan scope

Get CDs, manuals, and cardboard boxes printed

Test software

Beta-test video game

Develop scope

Create schedule

Sketch characters

Inspect boxes

Design software

More Magnets

Oops! Looks like the magnets fell off the fridge. Here's your chance to practice breaking down the work to create a different WBS using the same magnets as before. But this time, instead of decomposing project phases into work packages, break the project down by deliverable.

On this page, create a work breakdown structure broken down by <u>deliverable.</u>

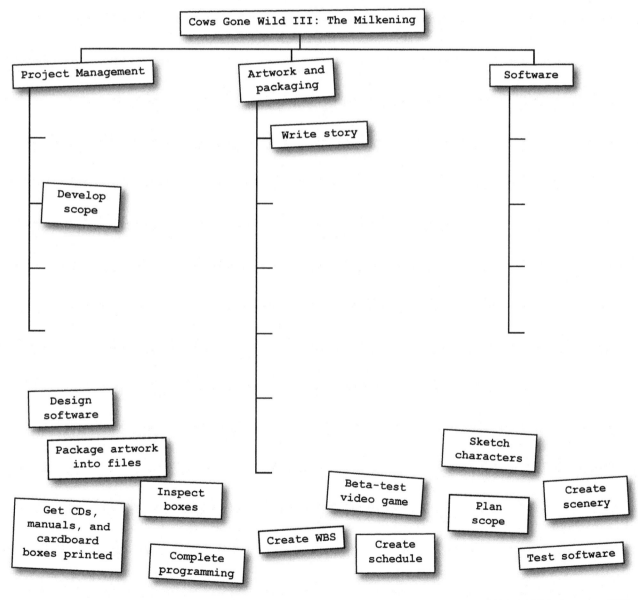

Cows Gone Wild III: The Milkening

Project Management

Artwork and packaging

Software

Write story

Develop scope

Design software

Package artwork into files

Sketch characters

Inspect boxes

Beta-test video game

Create scenery

Get CDs, manuals, and cardboard boxes printed

Create WBS

Plan scope

Complete programming

Create schedule

Test software

Project Scope Management Magnets Solutions

You can break down the work for a project in any number of ways.

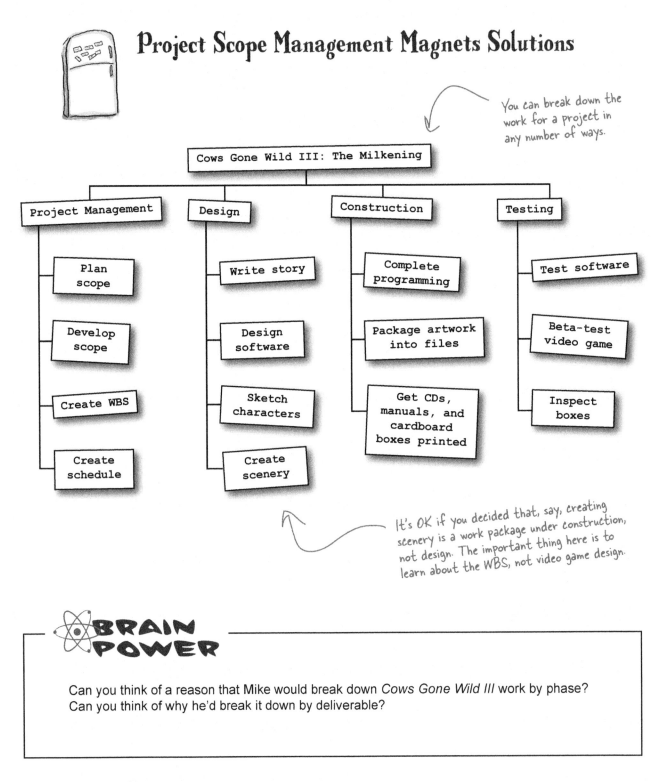

Cows Gone Wild III: The Milkening

Project Management
- Plan scope
- Develop scope
- Create WBS
- Create schedule

Design
- Write story
- Design software
- Sketch characters
- Create scenery

Construction
- Complete programming
- Package artwork into files
- Get CDs, manuals, and cardboard boxes printed

Testing
- Test software
- Beta-test video game
- Inspect boxes

It's OK if you decided that, say, creating scenery is a work package under construction, not design. The important thing here is to learn about the WBS, not video game design.

BRAIN POWER

Can you think of a reason that Mike would break down *Cows Gone Wild III* work by phase? Can you think of why he'd break it down by deliverable?

This WBS has the same work packages, but they're broken down differently.

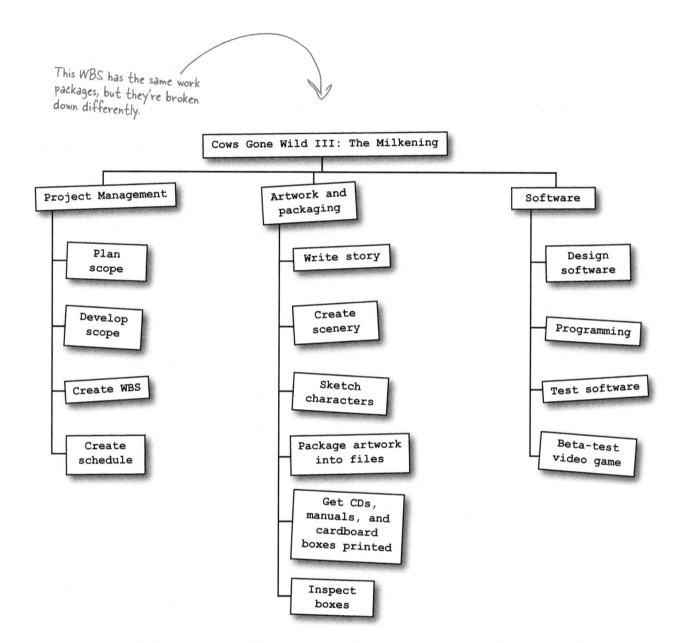

Cows Gone Wild III: The Milkening

Project Management
- Plan scope
- Develop scope
- Create WBS
- Create schedule

Artwork and packaging
- Write story
- Create scenery
- Sketch characters
- Package artwork into files
- Get CDs, manuals, and cardboard boxes printed
- Inspect boxes

Software
- Design software
- Programming
- Test software
- Beta-test video game

Did you notice how the project management work packages are the same in both WBSes? You could break them down into more detailed project management deliverables, and then you'd see a difference.

Inside the work package

You've probably noticed that the work breakdown structure only shows you the name of each work package. That's not enough to do the work! You and your team need to know a lot more about the work that has to be done. That's where the **WBS dictionary** comes in handy. It brings along all of the details you need to do the project work. The WBS dictionary is an important output of the Create WBS process—the WBS wouldn't be nearly as useful without it.

The WBS dictionary contains the details of every work package. It's a separate output of the Create WBS process.

This is one of the WBS Dictionary entries for the Cows Gone Wild III project. It goes with the "Test Software" work package in the WBS.

Each work package has a name, and in many WBSes the work packages will also have ID numbers.

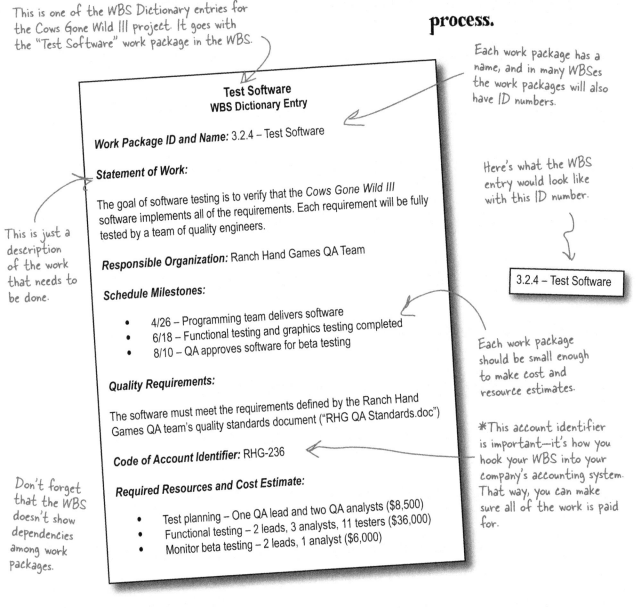

Test Software
WBS Dictionary Entry

Work Package ID and Name: 3.2.4 – Test Software

Statement of Work:

The goal of software testing is to verify that the *Cows Gone Wild III* software implements all of the requirements. Each requirement will be fully tested by a team of quality engineers.

Responsible Organization: Ranch Hand Games QA Team

Schedule Milestones:

- 4/26 – Programming team delivers software
- 6/18 – Functional testing and graphics testing completed
- 8/10 – QA approves software for beta testing

Quality Requirements:

The software must meet the requirements defined by the Ranch Hand Games QA team's quality standards document ("RHG QA Standards.doc")

Code of Account Identifier: RHG-236

Required Resources and Cost Estimate:

- Test planning – One QA lead and two QA analysts ($8,500)
- Functional testing – 2 leads, 3 analysts, 11 testers ($36,000)
- Monitor beta testing – 2 leads, 1 analyst ($6,000)

This is just a description of the work that needs to be done.

Here's what the WBS entry would look like with this ID number.

3.2.4 – Test Software

Each work package should be small enough to make cost and resource estimates.

*This account identifier is important—it's how you hook your WBS into your company's accounting system. That way, you can make sure all of the work is paid for.

Don't forget that the WBS doesn't show dependencies among work packages.

Here's another chance for you to think things through. Putting it down on paper helps the cognitive process.

scope *management*

Sharpen your pencil

It will help you on the exam to know why all of the outputs are important, and the WBS is one of the most important ones. Write down as many reasons for using a WBS as you can think of.

....

....

....

....

....

there are no Dumb Questions

Q: Does the work breakdown structure need to be graphical? It looks like a lot of work. Can't I just write out a list of tasks?

A: Yes, the WBS has to be graphical. The WBS needs to show all of the work packages, and how they decompose into phases or deliverables. When you look at a simple WBS, it might seem like you could manage your work packages just as efficiently using a simple list. But what if you have a large team with dozens, hundreds, or even thousands of work packages? That's when you'll be really happy that you know how to decompose deliverables into a hierarchy.

Q: What if one work package depends on another one?

A: There are definitely dependencies among work packages. For example, the Ranch Hand QA team can't begin to test the software until the programming team has finished building it. But while this information is important, the WBS isn't where you figure out the dependencies.

The reason is that you need to figure out what work needs to be done before you start to figure out how the work packages depend on each other.

Q: What if I don't know enough to estimate the cost of a work package? What do I add to the WBS dictionary?

A: The WBS dictionary should contain only information that you can fill in when you create it. A lot of the time, you'll know all of the information that needs to go into it. If you have an estimate and know the resources that should be used, then put it in. But if all you have is a statement of work and an account code, then that's all the information you'll be able to add to the entry.

The project scope baseline is a snapshot of the plan

As the project goes on, you will want to compare how you are doing to what you planned for. So, the **project scope baseline** is there to compare against. It's made up of the project scope statement, the WBS, and the WBS dictionary. When work gets added to the scope through change control, you need to change the baseline to include the new work packages for that work, so you can always track yourself against the plan.

The project scope baseline is a snapshot of the plan, and it's an important output of Create WBS.

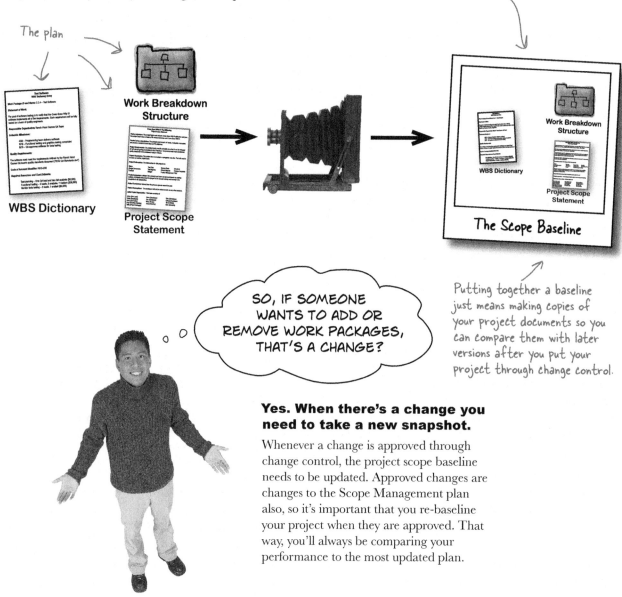

The plan

Work Breakdown Structure

WBS Dictionary

Project Scope Statement

The Scope Baseline

Putting together a baseline just means making copies of your project documents so you can compare them with later versions after you put your project through change control.

> SO, IF SOMEONE WANTS TO ADD OR REMOVE WORK PACKAGES, THAT'S A CHANGE?

Yes. When there's a change you need to take a new snapshot.

Whenever a change is approved through change control, the project scope baseline needs to be updated. Approved changes are changes to the Scope Management plan also, so it's important that you re-baseline your project when they are approved. That way, you'll always be comparing your performance to the most updated plan.

there are no
Dumb Questions

Q: What happens if I need to change the scope?

A: You need to put it through change control—just like a change to the product scope. As you're building the product, it's always possible that some work will pop up in an unexpected place.

It could be that the initial technical design is inadequate or buggy. Or maybe you just think of a better way to do things while you're working. In either case, you have to determine the impact to the schedule, the budget, the scope, and the quality of the product, and put the proposed change through change control. That's what it means to look at the project constraints every time there's a change.

Once everyone understands the impact and approves the change, you need to go back and adjust your project scope baseline to include the new work. If your budget or schedule is affected, you'll need to change those baselines too and integrate all of them into the Project Management plan. But we'll talk more about that in later chapters.

Q: Do I really need to create a project scope baseline?

A: Yes. It might seem like a formality in the beginning, but the baseline is a really useful tool. As you are building your project, you will need to refer back to the baseline if you want to know how you are tracking against stakeholders' expectations.

Let's say you said it would take you 12 months to build *Cows Gone Wild III*, and a wrong technical decision creates a two-week delay. You can use the project scope baseline to figure out the impact of that change to all of the different plans you have made, and then explain to everybody the impact of the change.

You can think of the baseline as a way of keeping track of the project team's understanding of their goals and how they are going to meet them. If the goals change, then the understanding of them needs to change too. By telling everyone who needs to approve the two-week delay about it, you make sure that the goals change for the team as well. Then you change the baseline, so you can measure your team against the new deadline of 12 and a half months.

Q: Wait a minute. Doesn't that mean I need to do change control and update the baseline every time I make any change to the document while I'm writing it? That's going to make it really hard to write the first version of anything!

A: Don't worry, you don't have to go through change control until the baseline is approved. And that goes for ANY document or deliverable. Once it's accepted and approved by all of the stakeholders, only then do the changes need to go through change control. Until it's approved, you can make any changes you want. That's the whole reason for change control—to make sure that once a deliverable is approved, you

run all of the changes by a change control board to make sure that they don't cause an unacceptable impact to the schedule, scope, cost, or quality.

Q: How can you know all of this up front?

A: You can't. Even the best planned projects have a few surprises. That's why the scope planning cycle is iterative. As you find out something new about your scope of work, you put it through change control. When it's approved, you need to add it to your Scope Management plan, your project scope statement, your WBS, and your WBS dictionary.

It's also possible that you might find new things that the team should do when you're making your WBS or your project scope statement. So all of the scope planning documents are closely linked and need to be kept in sync with one another.

Q: What if I come up with new work for the team later on?

A: You use change control to update the baseline. Your project can change at any time, but before you make a change you need to figure out how it will affect the project constraints—and make sure your sponsors and stakeholders are OK with that impact. That's what change control does for you.

Any time you make a change, you need to get it approved, and then update the baseline.

The outputs of the Create WBS process

The Create WBS process has three major outputs: the **work breakdown structure**, the **WBS dictionary,** and the **baseline**. But there are others as well. When you create the WBS, you usually figure out that there are pieces of the scope that you missed, and you may realize that you need to change your plan. That's what the project document updates are for.

Outputs

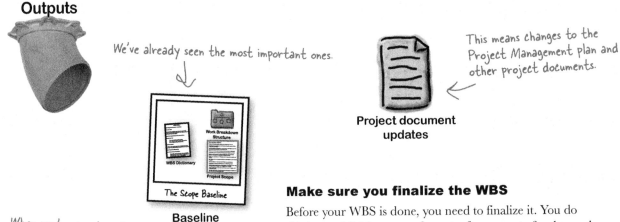

We've already seen the most important ones.

This means changes to the Project Management plan and other project documents.

Project document updates

The Scope Baseline

Baseline

When you're creating the WBS, you often discover missing pieces of the scope. You'll need to go back and plan for them. That kicks off the planning cycle again.

Make sure you finalize the WBS

Before your WBS is done, you need to finalize it. You do this by establishing a set of **control accounts** for the work packages. A control account is a tool that your company's management and accountants use to track the individual work packages. For example, Mike gets a list of control accounts from Ranch Hand Games' accounting department, so they know how to categorize the work for tax purposes.

BULLET POINTS: AIMING FOR THE EXAM

- The **Create WBS** process is a really important process on the PMP exam.

- You create the WBS by **decomposing** large work products into **work packages**.

- To finalize the WBS, **control accounts** are established for the work packages.

- The **WBS dictionary** is a description of each work package listed in the WBS.

- The inputs to WBS creation are the outputs to the Define Scope and Collect Requirements processes: the requirements document and the project scope statement.

- As you decompose the work, you find new information that needs to be added to the requirements document and the project scope statement. That information is treated as a change and goes through change control. Once it's approved, it can be added into the document, and that kicks off the planning cycle again.

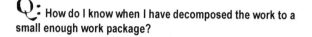

there are no
Dumb Questions

Q: How do I know if I should use phases or deliverables for my WBS?

A: It really depends on the project. You want to present the information so that it allows the management in your organization the ability to visualize and control your project. So, if most people in your organization divide it by phases, then you should, too.

If people do it different ways from project to project where you work, then you might make your decision based on how people think about the work you are about to do.

The point behind the WBS is to help other people see the work that is necessary to get the project done, so if your management thinks of projects in terms of phases and understands them best that way, then it's better to divide your project work along those lines.

It could be that the work you are doing is anxiously awaited by a lot of people who will look at the WBS to understand the project, and, in that case, it probably makes sense to divide your work up by deliverable.

> WE'VE PLANNED THE SCOPE AND WE'VE BROKEN DOWN THE WORK FOR COWS GONE WILD III. NOW WE'RE READY TO BEGIN THE PROJECT!

Q: How do I know when I have decomposed the work to a small enough work package?

A: The short answer is that you should decompose that work until it is manageable.

You need to be careful when you come up with the work packages for your WBS. If you decompose to the most granular level, you could end up wasting everybody's time trying to figure out exactly how much effort goes into, say, writing up meeting minutes for each and every meeting in your project.

So, you should break down the work to small enough packages that everybody can understand what's being done and describe it in the dictionary…and no further.

Q: I know how to make scope changes during Planning. What do I do if I run into scope changes during Execution?

A: Any time you run into a change to your scope, regardless of where you are in the process, you put it through change control. Only after examining the impact and having the change approved can you incorporate the change.

Q: Can you back up a minute and go over the difference between the Scope Management plan and the Project Management plan one more time?

A: Remember how the Project Management plan was divided into subsidiary plans? The Project Management plan tells you how to manage all of the different knowledge areas, and it has baselines for the scope, schedule, and budget.

The Scope Management plan is one of those subsidiary plans. It has really specific procedures for managing scope. For example, Mike's Scope Management plan tells him which stakeholders he needs to talk to when he's gathering requirements. It lists what tools and techniques he's planning to use when he uses scope definition to define the scope (for example, it says that he needs to consult with specific experts when he does alternatives analysis). And when there's an inevitable change—because even the best project manager can't prevent every change—it gives him procedures for doing Scope Management. So even though the Scope Management plan is created in the Develop Project Management Plan process, it's used throughout all of the Scope Management processes. So definitely expect questions about it on the exam!

Cubicle conversation

Everything is great. The project is rolling along, and there are no problems with the scope…until something goes wrong.

SOMETHING'S NOT RIGHT
WITH THE ARTWORK…

Brian: At first I thought we could use the same five backgrounds over and over, but it's starting to look really stale.

Amy: Huh, I guess you're right. It looks like we need to create more scenery.

Mike: Why were we trying to limit the backgrounds in the first place?

Amy: I think they were worried about disk space.

Brian: Yeah, but that's not so much a concern right now.

Amy: Great! Let's just change the artwork, then.

Mike: Not so fast, Amy. There are a couple of things we need to do first…

This is work that was not planned for, and isn't in the WBS. That means it's a scope change.

What homework do you need to do before you make a change to the scope by adding or removing project work? Why?

Why scope changes

Sometimes something completely unexpected happens. Say, a really important customer asks for a new feature that nobody saw coming and demands it right away. Or a design for a feature just isn't working, and you need to rethink it. Or new stakeholders come on board and ask for changes.

The scope can change while you are working for a lot of reasons. Some changes are good for your project, while others will definitely reduce your chance of success. Change control is there to help you to see which is which.

Good change

A good change makes the product better with very little downside. It doesn't cost more time in the schedule or more money from the budget, and it doesn't destabilize the product or otherwise threaten its quality.

Good changes happen pretty rarely, and nearly EVERY change has some impact that should be fully explored before you go forward.

Bad change

A bad change is one that might seem from the outside like a good idea but ends up making an impact on the project constraints. Here are a couple of examples:

Scope creep

This happens when you think you know the impact of a change so you go ahead, but it turns out that *that* change leads to another one, and since you are already making the first change, you go with the next. Then another change comes up, and another, and another, until it's hard to tell what the scope of the project is.

The way to avoid scope creep is to plan your changes completely.

Gold plating

Sometimes people think of a really great improvement to the product and go ahead and make it without even checking the impact. In software, this can happen pretty easily. A programmer thinks of a way to make a feature better, for example, and just implements it, without talking it over with anybody. This may sound good, but it's not—because now you have to pay for these features you never asked for.

Be on the lookout for examples of scope creep and gold plating on the exam. Both are considered very bad and should never be done.

Exercise

Here's the WBS that you created for the *Cows Gone Wild III* project, and below that are some changes that the team has asked Mike to make since the work started. All of them are bad changes. Check either scope creep or gold plating for each one.

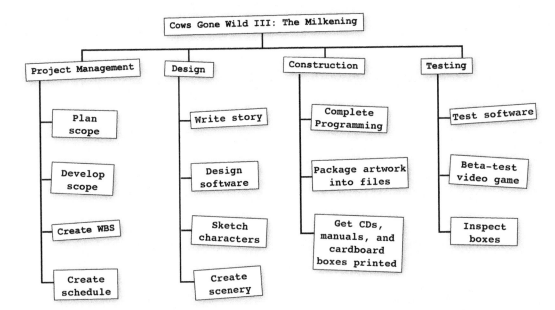

Cows Gone Wild III: The Milkening

Project Management
- Plan scope
- Develop scope
- Create WBS
- Create schedule

Design
- Write story
- Design software
- Sketch characters
- Create scenery

Construction
- Complete Programming
- Package artwork into files
- Get CDs, manuals, and cardboard boxes printed

Testing
- Test software
- Beta-test video game
- Inspect boxes

1. We need to create a screensaver to market the game. Let's kill two birds with one stone and test out a brand new graphics engine on it. Oh, and we'll need a story for the screensaver, so we should write that too. Of course we have to recruit some killer voice talent for the screensaver. Memorable names sell more games.

   ☐ Scope creep   ☐ Gold plating

2. Testing the most recent build, I just noticed that if the player presses x–x–z–a–Shift–Shift–Space in that order, Bessie does the Charleston—it's really funny.

   ☐ Scope creep   ☐ Gold plating

3. We should add a calculator for tracking gallons of milk collected in the game. It will be really easy. We could even release the calculator as a separate add-in, and we could probably make it full-featured enough for the folks developing the game down the hall to use it too.

   ☐ Scope creep   ☐ Gold plating

4. The printer just told us that she could also do silk screen T-shirts for everybody as a ship gift. Let's get our design team to do some special artwork for them. We can have everybody's names written in cows!!! Then we could use the same artwork on posters that we put around the office—oh, and coffee mugs for new people, too

   ☐ Scope creep   ☐ Gold plating

Answers: 1, 3, and 4 are Scope creep. 2 is Gold plating

The Control Scope process

There's no way to predict every possible piece of work that you and your team are going to do in the project. Somewhere along the way, you or someone else will realize that a change needs to happen, and that change will affect the baseline. That's why you need the **Control Scope** process. It's how you make sure that you make only those changes to the scope that you need to make, and that everyone is clear on what the consequences of those changes are.

You'll also use organizational process assets as an input here.

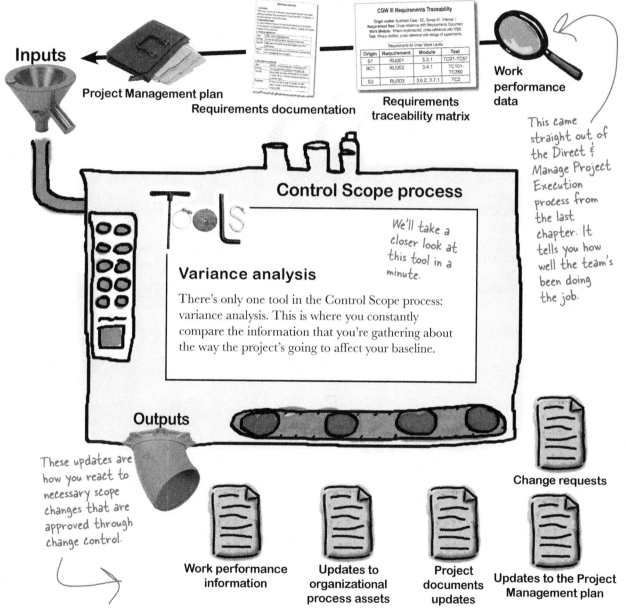

Inputs

Project Management plan

Requirements documentation

Requirements traceability matrix

Work performance data

This came straight out of the Direct & Manage Project Execution process from the last chapter. It tells you how well the team's been doing the job.

Control Scope process

We'll take a closer look at this tool in a minute.

Variance analysis

There's only one tool in the Control Scope process: variance analysis. This is where you constantly compare the information that you're gathering about the way the project's going to affect your baseline.

Outputs

These updates are how you react to necessary scope changes that are approved through change control.

Work performance information

Updates to organizational process assets

Project documents updates

Change requests

Updates to the Project Management plan

Anatomy of a change

Let's take a closer look at what happens when you need to make a change. You can't just go and change the project whenever you want—the whole reason that you have a baseline is so you can always know what work the team is supposed to do. If you make changes, then you need to change the baseline...which means you need to make sure that the change is *really* necessary. Luckily, you have some powerful tools to help you manage changes:

❶ A change is needed.
Every change starts the same way. Someone realizes that if the project sticks with the plan, then the outcome will lead to problems.

❷ Create a change request.
Before a change can be made, it needs to be approved. That means that it needs to be documented as a requested change. The only way to get a handle on a change is to write it down and make sure everyone understands it.

> I KNOW WE'RE UNDER PRESSURE TO GET THE GAME OUT THE DOOR, BUT WE NEED TO MAKE A CHANGE. WE ONLY PLANNED ON MAKING FOUR MEADOWS ON LEVEL 3 AND REUSING THEM FOR LEVEL 6, BUT IT'S JUST NOT WORKING FOR US. WE'VE GOT TO CHANGE THIS IF WE WANT THE GAME TO SELL.

A change can come from anywhere—the project manager, a team member, even a stakeholder!

> WE'LL WRITE UP A CHANGE REQUEST, AND THEN PUT IT THROUGH INTEGRATED CHANGE CONTROL TO GET IT APPROVED.

❸ Get the change approved.
Remember integrated change control from Chapter 4? That's the process where the project manager takes a requested change and works with the sponsor and stakeholders to get approval to put it in place.

Think of integrated change control as a kind of machine that converts requested changes into approved changes.

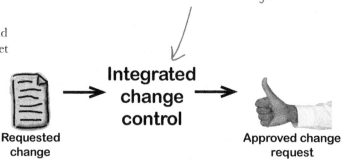

| Requested change | → | **Integrated change control** | → | Approved change request |

4 **Do variance analysis.**

Take a look at the baseline and see how the change will affect it. This is where you decide whether you need to take some sort of corrective action. You compare the baseline against the change that you want to make, and figure out just how big the change really is.

You're weighing the change against the baseline to see if it's going to require a big change to your plan.

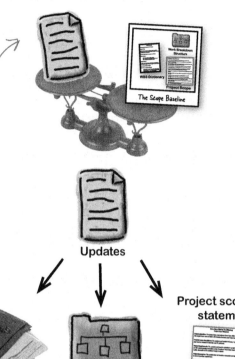

The Scope Baseline

5 **Replan the work.**

Now it's time to go back to the scope documentation and update it to reflect the change.

Updates

**Project
Management plan**

**Work breakdown
structure**

**Project scope
statement**

Don't forget to update the WBS dictionary, too.

6 **Create a new baseline.**

Now that you've figured out that you need to change the scope, it's time to update the baseline. Go back to the scope statement, WBS, and WBS dictionary, and update them so that they reflect the change that needs to be made.

The change is done!

Now you can move on with the project using the new baseline that you saved and distributed to the team.

This will come in handy when you go back to put together the lessons learned!

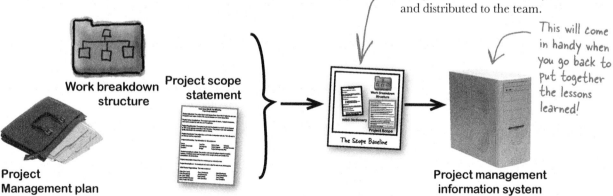

**Work breakdown
structure**

**Project scope
statement**

**Project
Management plan**

The Scope Baseline

**Project management
information system**

A closer look at the change control system

One of the most important tools in any Monitoring and Controlling process is the **change control system**. Let's take a closer look at how it works.

Since the folks at Ranch Hand need a change to add more scenery to *Cows Gone Wild III*, Mike takes a look at the Scope Management plan to understand the impact before forwarding it to the change control board. Once they approve the change, he updates the Project Management plan, checks it into the configuration management system, and changes the WBS and WBS dictionary to include the new work packages.

Remember this from Chapter 4? It's exactly the same change control system tool that we already learned about.

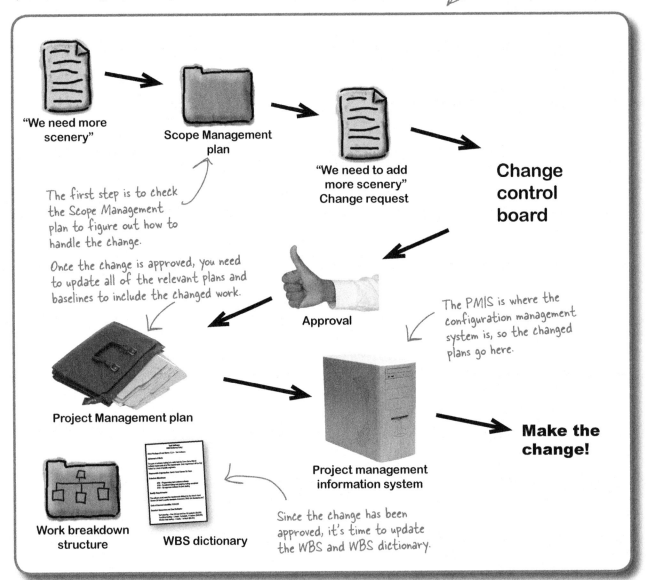

"We need more scenery"

Scope Management plan

"We need to add more scenery" Change request

Change control board

The first step is to check the Scope Management plan to figure out how to handle the change.

Once the change is approved, you need to update all of the relevant plans and baselines to include the changed work.

Approval

The PMIS is where the configuration management system is, so the changed plans go here.

Project Management plan

Make the change!

Work breakdown structure

WBS dictionary

Project management information system

Since the change has been approved, it's time to update the WBS and WBS dictionary.

Just one Control Scope tool/technique

There's just one tool/technique in the Control Scope process. It's pretty intuitive: take a minute and think of what you would need to do if you had to make a change to your project's scope. You'd need to figure out how big the change is, and what needs to change. And when you do that, it's called **variance analysis**.

Variance analysis

This means comparing the data that can be collected about the work being done to the baseline. When there is a difference between the two, that's variance.

This tool of Control Scope is all about analyzing the difference between the baseline and the actual work to figure out if the plan needs to be corrected. If so, then you recommend a corrective action and put that recommendation through change control.

The goal of Control Scope is updating the scope, plan, baseline, and WBS info.

There's no "right order" for the Control Scope and Validate Scope processes

If you've got a copy of the *PMBOK Guide* handy, take a look at how it presents the Scope Management processes. Did you notice how the section on the Validate Scope process comes before Control Scope? We're putting these processes in this book in a different order, and it's the only time we deviate from the order of the *PMBOK Guide*. That's not because the *PMBOK Guide* is wrong! We could do this because there is no "right" order: Control Scope can happen at any time, because project changes can happen at any time. Validate Scope (the next process you'll learn about) is *usually* the last Scope Management process that you'll do in a project. The trick is that sometimes you'll find a scope problem while you're verifying the scope, and you'll need to do Control Scope and then go back and gather new requirements, rebuild the WBS, etc. So the Control Scope process can happen *either before or after* Validate Scope.

So why did we change the order? Because thinking about how the two processes relate to each other will help you remember this for the exam!

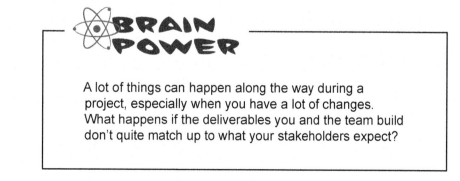

BRAIN POWER

A lot of things can happen along the way during a project, especially when you have a lot of changes. What happens if the deliverables you and the team build don't quite match up to what your stakeholders expect?

there are no Dumb Questions

Q: Is Control Scope always about work and *project* scope? Can it ever be about deliverables and *product* scope?

A: No. The Control Scope process is **always about the work that the team does**, because the whole Scope Management knowledge area is about the project scope, not the product scope. In other words, as a project manager, you manage the work that the team is doing, not the things that they're making. Now, that doesn't mean you should never pay attention to deliverables. You still need to pay attention to the scope of the product, too, since the two are pretty closely related. For example, in the *CGW III* project, any time somebody wants to add a new feature to the game, a programmer will need to program it, an artist will need to make new artwork, and a tester will have to test it. Any time you make changes to the project scope, it affects the product scope, and vice versa.

Q: What if a change is really small? Do I still have to go through all of this?

A: Yes. Sometimes what seems like a really small change to the scope—like just adding one tiny work package—turns out to be really complex when you take a closer look at it. It could have a whole lot of dependencies, or cause a lot of trouble in other work packages. If you don't give it careful consideration, you could find yourself watching your scope creep out of control. Each and every change needs to be evaluated in terms of impact. If there is any impact to the project constraints—time, cost, scope, quality, resources, or risk—you HAVE to put it through change control.

Q: How can you do variance analysis without knowing all of the changes that are going to happen?

A: You do variance analysis as an ongoing thing. As information comes in about your project, you constantly compare it to how you planned. If you're running a month behind, that's a good indication that there are some work packages that took longer than your team estimated—or that you missed a few altogether. Either way, you need to take corrective action if you hope to meet your project objectives.

Waiting until all possible changes are known will be too late for you to actually meet your goals. So you need to constantly check your actuals versus your baseline and correct where necessary (after putting your recommended actions through change control, of course!).

Q: I thought the configuration management system was part of the project management information system from Chapter 4. What does that have to do with change control?

A: When you write and modify documents throughout your project, you need to make sure that everybody is working with the same version of them. So you check them into a configuration management system, and that way everybody always knows where to go for the latest version.

Since you are checking all of your documents in, that's where you will keep your work performance information also. The most recent version of the schedule, any reports you have gathered on defects, and individual work performance should all be there. So, when you want to figure out what's going on in your project, you look there first.

It follows that you would modify your documents and check them back into the CMS after any change has been approved too.

Every scope change goes through the Control Scope process.

Control Scope Process Magnets

Whenever you make a scope change, you need to go through all of the steps of change control. So what are those steps? Arrange the magnets to show the order that you handle changes to the scope.

1._____   5._____

2._____   6._____

3._____   7._____

4._____   8._____

Control Scope Process Magnets Solutions

Arrange all of the activities you do to control scope in the right order.

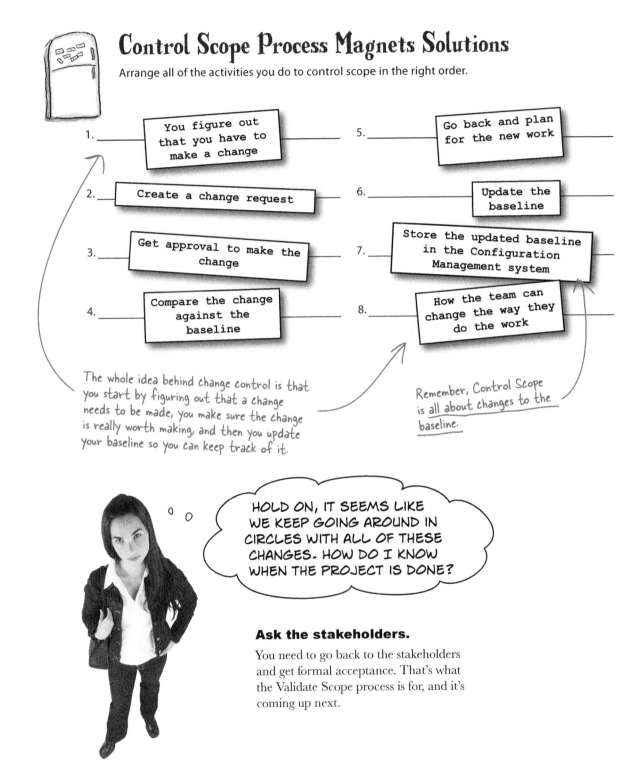

1. You figure out that you have to make a change

2. Create a change request

3. Get approval to make the change

4. Compare the change against the baseline

5. Go back and plan for the new work

6. Update the baseline

7. Store the updated baseline in the Configuration Management system

8. How the team can change the way they do the work

The whole idea behind change control is that you start by figuring out that a change needs to be made, you make sure the change is really worth making, and then you update your baseline so you can keep track of it.

Remember, Control Scope is all about changes to the baseline.

HOLD ON, IT SEEMS LIKE WE KEEP GOING AROUND IN CIRCLES WITH ALL OF THESE CHANGES. HOW DO I KNOW WHEN THE PROJECT IS DONE?

Ask the stakeholders.

You need to go back to the stakeholders and get formal acceptance. That's what the Validate Scope process is for, and it's coming up next.

Make sure the team delivered the right product

When the team is done, what happens? You still have one more thing you need to do before you can declare victory. You need to gather all the stakeholders together and have them make sure that all the work really was done. We call that the **Validate Scope** process.

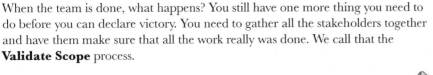

Monitoring & Controlling process group

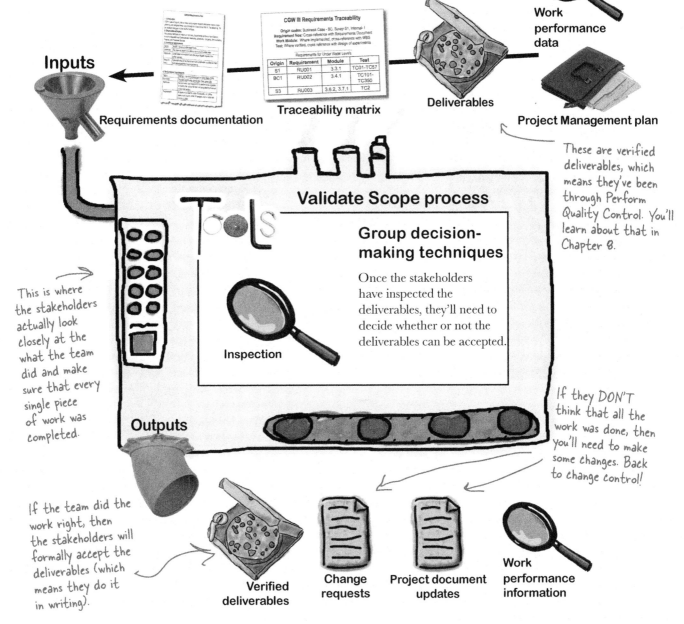

Inputs

Requirements documentation

CGW III Requirements Traceability

Origin codes: Business Case - BC, Survey-S1, Internal-1
Requirement Nos: Cross-reference with Requirements Document
Work Module: Where implemented, cross-reference with WBS
Test: Where verified, cross references with design of experiments

Requirements for Under Water Levels

Origin	Requirement	Module	Test
S1	RU001	3.3.1	TC01-TC57
BC1	RU002	3.4.1	TC101-TC350
S3	RU003	3.6.2, 3.7.1	TC2

Traceability matrix

Deliverables

Work performance data

Project Management plan

These are verified deliverables, which means they've been through Perform Quality Control. You'll learn about that in Chapter 8.

Validate Scope process

T**ools**

Inspection

Group decision-making techniques

Once the stakeholders have inspected the deliverables, they'll need to decide whether or not the deliverables can be accepted.

This is where the stakeholders actually look closely at the what the team did and make sure that every single piece of work was completed.

Outputs

If the team did the work right, then the stakeholders will formally accept the deliverables (which means they do it in writing).

If they DON'T think that all the work was done, then you'll need to make some changes. Back to change control!

Verified deliverables

Change requests

Project document updates

Work performance information

The stakeholders decide when the project is done

As you deliver the stuff in your scope statement, you need to make sure that each of the deliverables has everything in it that you listed in the scope statement. You inspect all of your deliverables versus the scope statement, the WBS, and the Scope Management plan. If your deliverables have everything in those documents, then they should be acceptable to stakeholders. When all of the deliverables in the scope are done to their satisfaction, *then* you're done.

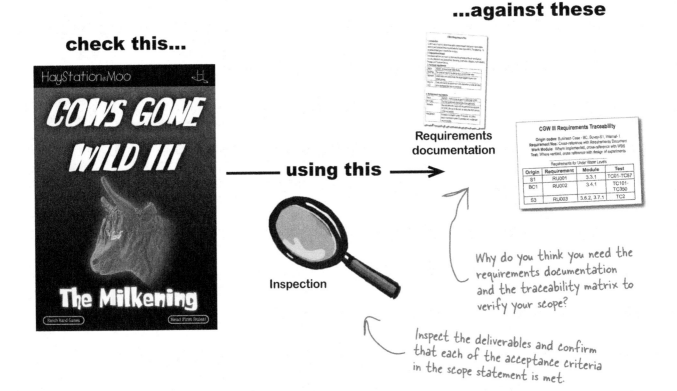

check this...

...against these

Requirements documentation

using this

Inspection

Why do you think you need the requirements documentation and the traceability matrix to verify your scope?

Inspect the deliverables and confirm that each of the acceptance criteria in the scope statement is met.

Formal acceptance means that you have <u>written confirmation</u> from all of the stakeholders that the <u>deliverables</u> match the <u>requirements</u> and the <u>Project Management</u> plan.

Is the project ready to go?

Once the deliverables are ready for prime time, you inspect them with the stakeholders to make sure that they meet acceptance criteria. The purpose of Validate Scope is to obtain formal, written acceptance of the work products. If they are found to be unsatisfactory, the specific changes requested by the stakeholders get sent to change control so that the right changes can be made.

Inspection just means sitting down with the stakeholders and looking at each deliverable to see if it's acceptable.

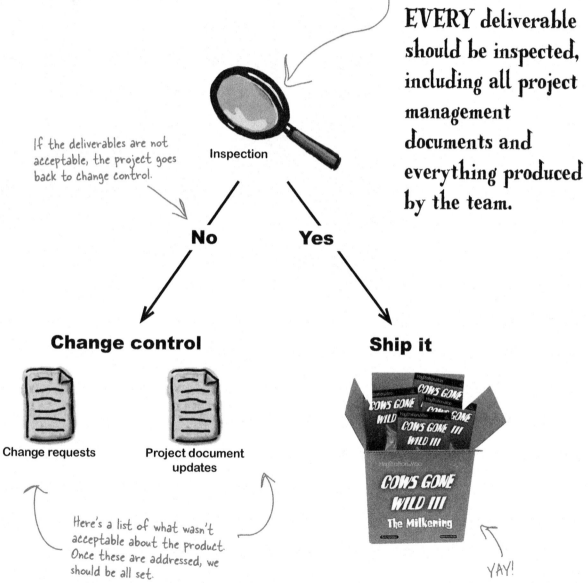

EVERY deliverable should be inspected, including all project management documents and everything produced by the team.

If the deliverables are not acceptable, the project goes back to change control.

Inspection

No   **Yes**

Change control   **Ship it**

Change requests   **Project document updates**

Here's a list of what wasn't acceptable about the product. Once these are addressed, we should be all set.

COWS GONE WILD III
COWS GONE WILD III
COWS GONE III
WILD III

COWS GONE WILD III
The Milkening

YAY!

Scopecross

Take some time to sit back and give your right brain something to do. It's your standard crossword; all of the solution words are from this chapter.

Across

2. Bob used an _____ diagram to get a handle on all of the ideas he collected from stakeholders.

4. The details of every work package in the WBS are stored in the WBS _____.

5. The process where you write the project scope statement is called _____ Scope.

7. Figuring out how big a change is by comparing it to the baseline is called _____ analysis.

8. When one change leads to another and another and another, it's called scope _____.

10. Getting work packages out of deliverables.

12. Exploring all of the ways that you can do the work so that you can find the best way to do the work in your project is called _____ identification.

13. A quantified and documented need or expectation of a sponsor, customer, or other stakeholder.

Down

1. A JAD session is an example of a _____ workshop.

3. Looking closely at the product to see if you completed all of the work is called _____.

4. When you're making a WBS, you can break the work down by phase or _____.

6. A version of the Scope Management plan, work breakdown structure, and product scope that you will compare your project to is called the scope _____.

9. _____ scope means the features or functions of the thing or service that you are building.

11. A good way to gather requirements is to _____ how the people who will use your deliverables perform their jobs.

Answers on page 230.

The project is ready to ship!

There were a few unexpected changes to the scope along the way. But, for the most part, everything went according to plan. The stakeholders and the CEO got together with the team and went through everything they did—and it's ready to go. Great job, guys!

The team finally finished Cows Gone Wild III. Pizza's on the CEO—and then it's time to get cracking on CGW IV!

Sharpen your pencil Solution

Here are some attributes of *Cows Gone Wild III*. Which are project scope and which are product scope?

1. Programming

☑ Project scope ☐ Product scope

2. 34 levels in the game

☐ Project scope ☑ Product scope

3. Graphic design

☑ Project scope ☐ Product scope

4. Four playable characters

☐ Project scope ☑ Product scope

5. Great graphics

☐ Project scope ☑ Product scope

6. Testing

☑ Project scope ☐ Product scope

7. Mac and PC compatible

☐ Project scope ☑ Product scope

8. A "boss battle" milk fight level at the end

☐ Project scope ☑ Product scope

WHAT'S MY PURPOSE

Here are a few things that Mike left out of the *CGWIII* project scope statement. Can you figure out where each of them should go?

1. The game must have fewer than 15 defects per 10,000 lines of code.

2. There will be four graphic designers reporting to the art director, and six programmers and four testers reporting to the development manager.

3. No more than 15 people can be allocated to work on the game at any time.

4. Scenery artwork.

5. The product will not include bug fixes for the previous version.

6. The game needs to run on a machine with 1 GB of memory or less.

A. Project exclusions

B. Project deliverables

C. Project constraints

D. Project assumptions

E. Project requirements

F. Acceptance criteria

Scopecross

Take some time to sit back and give your right brain something to do. It's your standard crossword; all of the solution words are from this chapter.

Exam Questions

1. Which of the following is TRUE about a work breakdown structure?

- A. It contains work packages that are described in a linear, unstructured list.
- B. Each item in the WBS represents a feature in the product scope.
- C. The WBS represents all of the work that must be done on the project.
- D. The WBS is created by the product sponsor and stakeholders.

2. Which is NOT an output of a Scope Management process?

- A. Business case
- B. WBS dictionary
- C. Change requests
- D. Accepted deliverables

3. Which of the following is NOT TRUE about a work breakdown structure?

- A. It describes procedures to define the scope, verify work, and manage scope changes.
- B. It contains a graphical, hierarchical list of all work to be performed.
- C. It can be broken down by project phase or deliverable.
- D. It is an important element of the baseline.

4. What is the correct order of the Scope Management processes?

- A. Plan Scope Management, Define Scope, Create WBS, Collect Requirements, Validate Scope, Control Scope
- B. Plan Scope Management, Collect Requirements, Control Scope, Create WBS, Validate Scope
- C. Plan Scope Management, Collect Requirements, Define Scope, Create WBS, Validate Scope, Control Scope
- D. Plan Scope Management, Collect Requirements, Baseline, Define Scope, Control Scope, Validate Scope

5. You are managing a software project. Your team has been working for eight weeks, and so far the project is on track. The lead programmer comes to you with a problem: there is a work package that is causing trouble. Nobody seems to know who is responsible for it, the accounting department does not know what cost center to bill it against, and it's not even clear exactly what work should be performed. Which of the following would BEST help this situation?

- A. Alternatives analysis
- B. WBS dictionary
- C. Scope Management plan
- D. Scope validation

Exam Questions

6. The goal of Validate Scope is:

    A.   To inspect the scope statement for defects so that it is correct

    B.   To gain formal acceptance of the project deliverables from the sponsor and stakeholders

    C.   To get everyone in the project working together toward a common goal

    D.   To verify that all *PMBOK Guide* processes are complied with

7. Historical information and lessons learned are part of:

    A.   Organizational process assets

    B.   Enterprise environmental factors

    C.   Project management information system (PMIS)

    D.   Work performance information

8. You've taken over as a project manager on a highway construction project, and the execution is already under way. Your sponsor tells you that moving forward, all asphalt should be laid down with a 12" thickness. The scope statement and the WBS call for 9" thick asphalt. What is the BEST course of action?

    A.   Look for a cheaper supplier so the cost impact is minimized.

    B.   Tell the sponsor that the work is already under way, so you can't accommodate his request.

    C.   Refuse to alter the plans until the change control system has been used.

    D.   Tell the team to accommodate the request immediately.

9. Which of the following BEST describes the purpose of a requirements traceability matrix?

    A.   It describes how WBS dictionary entries are traced to work packages, and how work packages are decomposed from deliverables.

    B.   It's used to make sure that all of the subplans of the Project Management plan have been created.

    C.   It helps you understand the source of each requirement, and how that requirement was verified in a later deliverable.

    D.   It's used to trace the source of every change, so that you can keep track of them through the entire Control Scope process and verify that the change was properly implemented.

10. It's the end of execution for a large highway construction project. The work has been done, and the workers are ready to pack up their equipment. The project manager and project sponsor have come by with specialists to check that each requirement has been met, and that all of the work in the WBS has been performed. What process is being done?

    A.   Control Scope

    B.   Validate Scope

    C.   Scope Testing

    D.   Define Scope

Exam Questions

11. You have just been put in charge of a project that is already executing. While reviewing the project documentation, you discover that there is no WBS. You check the Scope Management plan and discover that there should be one for this project. What is the BEST thing for you to do:

    A.   Immediately alert the sponsor and make sure the project work doesn't stop.

    B.   Stop project work and create the WBS, and don't let work continue until it's created.

    C.   Make sure you closely manage communications to ensure the team doesn't miss any undocumented work.

    D.   Mark it down in the lessons learned so it doesn't happen on future projects.

12. A project manager on an industrial design project finds that the sponsor wants to make a change to the scope after it has been added to the baseline, and needs to know the procedure for managing changes. What is the BEST place to look for this information?

    A.   WBS

    B.   Scope Management plan

    C.   Change request form template

    D.   Business case

13. You have just started work on the project scope statement. You are analyzing the expected deliverables when you discover that one of them could be delivered in three different ways. You select the best method for creating that deliverable. What is the BEST way to describe what you are doing?

    A.   Alternatives analysis

    B.   Decomposition

    C.   Define scope process

    D.   Stakeholder analysis

14. You're the project manager on a software project. Your team has only completed half of the work when the sponsor informs you that the project has been terminated. What is the BEST action for you to take?

    A.   Verify the deliverables produced by the team against the scope, and document any place they do not match.

    B.   Call a team meeting to figure out how to spend the rest of the budget.

    C.   Work with the sponsor to see if there is any way to bring the project back.

    D.   Tell the team to stop working immediately.

15. You are managing an industrial design project. One of your team members comes to you with a suggestion that will let you do more work while at the same time saving the project 15% of the budget. What is the BEST way for you to proceed?

    A.   Tell the team to make the change because it will deliver more work for less money.

    B.   Refuse to make the change until a change request is documented and change control is performed.

    C.   Refuse to consider the change because it will affect the baseline.

    D.   Do a cost-benefit analysis and then make sure to inform the sponsor that the project scope changed.

Exam Questions

16. You are the project manager for a telecommunications project. You are working on the project scope statement. Which of the following is NOT included in this document?

    A. Authorization for the project manager to work on the project

    B. Requirements that the deliverables must meet

    C. A description of the project objectives

    D. The list of deliverables that must be created

17. Which of the following is NOT an input to Control Scope?

    A. WBS dictionary

    B. Approved change requests

    C. Requested changes

    D. Project scope statement

18. Which of these processes is not a part of Scope Management?

    A. Scope Identification

    B. Collect Requirements

    C. Control Scope

    D. Validate Scope

19. You are the project manager for a new project, and you want to save time creating the WBS. Which is the BEST way to do this?

    A. Make decomposition go faster by cutting down the number of deliverables.

    B. Use a WBS from a previous project as a template.

    C. Don't create the WBS dictionary.

    D. Ask the sponsor to provide the work packages for each deliverable.

20. The project manager for a design project is using the Define Scope process. Which BEST describes this?

    A. Creating a document that lists all of the features of the product

    B. Creating a plan for managing changes to the baseline

    C. Creating a document that describes all of the work the team does to make the deliverables

    D. Creating a graphical representation of how the phases or deliverables decompose into work packages

21. You are the project manager for a construction project. You have completed project initiation activities, and you are now creating a document that describes processes to document the scope, decompose deliverables into work packages, verify that all work is complete, and manage changes to the baseline. What process are you performing?

    A. Develop Project Management plan

    B. Define Scope

    C. Create WBS

    D. Develop Project Charter

Exam Questions

22. You are a project manager working on a project. Your sponsor wants to know who a certain work package is assigned to, what control account to bill it against, and what work is involved. What document do you refer her to?

    A. Scope Management plan

    B. WBS

    C. WBS dictionary

    D. Scope statement

23. You are the project manager for a software project. One of the teams discovers that if they deviate from the plan, they can actually skip one of the deliverables because it's no longer necessary. They do the calculations, and realize they can save the customer 10% of the cost of the project without compromising the features in the product. They take this approach, and inform you the following week what they did during the status meeting. What is the BEST way to describe this situation?

    A. The project team has taken initiative and saved the customer money.

    B. A dispute is resolved in favor of the customer.

    C. The team informed the project manager of the change, but they should have informed the customer, too.

    D. The team did not follow the Control Scope process.

24. Which of the following BEST describes the purpose of the project scope statement?

    A. It describes the features of the product of the project.

    B. It is created before the Scope Management plan.

    C. It decomposes deliverables into work packages.

    D. It describes the objectives, requirements, and deliverables of the project, and the work needed to create them.

25. A project manager at a cable and networking company is gathering requirements for a project to build a new version of their telecommunications equipment. Which of the following is NOT something that she will use?

    A. Specific descriptions of work packages that will be developed

    B. One-on-one interviews with the senior executives who need the new equipment for their teams

    C. An early working model of the telecommunications equipment to help get feedback from stakeholders

    D. Notes that she took while being "embedded" with the team that will eventually use the equipment being developed

26. Which of the following is NOT an output of Collect Requirements?

    A. Requirements observations

    B. Requirements traceability matrix

    C. Requirements documentation

    D. Requirements Management plan

Answers

Exam ~~Questions~~

1. Answer: C

The work breakdown structure is all about breaking down the work that your team needs to do. The WBS is graphical and hierarchical, not linear and unstructured. Did you notice that answer B was about *product* scope, not *project* scope?

2. Answer: A

There are two ways you can get to the right answer for this question. You can recognize that the WBS dictionary, change requests, and accepted deliverables are all Scope Management process outputs. (You'll see change requests in every knowledge area!) But you can also recognize that the business case was created by the Develop Project Charter, which is part of the Initiating process group.

3. Answer: A

Did you recognize that answer A was describing the Scope Management plan? Once you know what the WBS is used for and how to make one, questions like this make sense.

4. Answer: C

You'll need to know what order processes come in, and one good way to do that is to think about how the outputs of some processes are used as inputs for another. For example, you can't create the WBS until the scope is defined, which is why A is wrong. And you can't do change control until you have a baseline WBS, which is why B is wrong.

Take a minute and think about how there's no "right" order for Validate Scope and Control Scope. You could have a scope change at the beginning of the project, so Control Scope would come first. But a change could happen late in the project, too! If there's a major change to the project after the scope's verified, you need to redo it.

5. Answer: B

An important tactic for a lot of exam questions is to be able to recognize a particular tool, technique, input, or output from a description. What have you learned about that tells you who is responsible for a work package, tells what control account to associate with it, and describes the work associated with it? That's a good description of the WBS dictionary.

Inspection isn't just done at the end of the project. You do Validate Scope on every single deliverable made by you and the team.

6. Answer: B

There are some questions where you'll just have to know what a process is all about, and this is one of them. That's why it's really helpful to know why Validate Scope is so helpful to you on a project. You use Validate Scope to check that all of the work packages were completed, and get the stakeholders and sponsor to formally accept the deliverables.

Exam ~~Questions~~ Answers

7. Answer: A

It's easy to forget that organizational process assets is more than just an input. It's a real thing that's part of your company. Take a second and think about what **assets** are in your **organization** that help you carry out each **process**. Get it? Good! So what is historical information, anyway? It's stuff like reports and data that you or another project manager wrote down on a previous project and stored in a file cabinet or a database. That's an asset you can use now! What are lessons learned? Those are lessons you wrote down at the end of a previous project and stuck in a file cabinet or a database. And now those lessons are another asset you can use.

The PMBOK Guide says this stuff is stored in a "corporate knowledge base," but that's just another word for a file cabinet or a folder on your network.

> THESE INPUTS AND OUTPUTS MAKE SENSE WHEN I THINK ABOUT HOW I'D USE THEM ON A PROJECT. ORGANIZATIONAL PROCESS ASSETS ARE JUST THINGS THAT MY ORGANIZATION KEEPS TRACK OF TO HELP ME DO MY JOB, LIKE INFORMATION FROM OLD PROJECTS AND PROCEDURES.

8. Answer: C

One thing to remember about change control is that if you want to make the sponsor and stakeholders happy with the project in the end, sometimes you have to tell them "no" right now. When you're doing Control Scope, the most important tool you use is the change control system. It tells you how to take an approved change and put it in place on a project, and there's no other way that you should ever make a change to any part of the baseline. That means that once everyone has approved the scope statement and WBS, if you want to make any change to them, then you need to get that change approved and put it through the change control system.

9. Answer: C

The requirements traceability matrix is a tool that you use to trace each requirement back to a specific business case, and then forward to the rest of the scope deliverables (like specific WBS work packages), as well as other parts of the project: the product design (like specific levels in *Cows Gone Wild*) or test strategy (like test plans that the Ranch Hand Games testers use to make sure that the game works).

The idea is that you're tracing a deliverable from its initial description all the way through the project to testing, so that you can make sure that every single deliverable meets all of its requirements.

Answers

~~Exam Questions~~

10. Answer: B

When you're getting the sponsor and stakeholders to formally accept the results of the project, you're doing Validate Scope. There's only one tool for it: inspection. That means carefully checking the deliverables (in this case, what the workers built on the highway) to make sure they match the WBS.

Answer D is a good idea, but it's not as important as creating a new WBS.

11. Answer: B

This question is a little tricky. The most important thing about a WBS is that if your Scope Management plan says it should be there, then your project absolutely cannot be done without it. And a general rule is that if you ever find that there is no WBS, you should always check the Scope Management plan to find out why.

12. Answer: B

This is another question that is testing you on the definition of a specific document, in this case the Scope Management plan, which is one of the subsidiary plans of the Project Management plan. Think about what you use a Scope Management plan for. It gives you specific procedures for defining the scope, breaking down the work, verifying the deliverables, and **managing scope changes**—which is what this question is asking. All of the other answers don't have anything to do with managing changes.

13. Answer: A

Here's another example of how there are two correct answers but only one BEST one. Answer C is true—you are doing scope definition. But is that really the best way to describe this situation? Alternatives analysis is part of scope definition, and it's a more accurate way to describe what's going on here.

When you look at a few ways to create a deliverable and then decide on the best one, that's alternatives analysis.

> THAT WAY, IF I NEED TO RESTART THE PROJECT LATER OR REUSE SOME OF ITS DELIVERABLES, I'LL KNOW EXACTLY WHERE MY TEAM LEFT OFF WHEN IT ENDED.

14. Answer: A

This question is an example of how you need to rely on more than just common sense to pass the PMP exam. All four of these answers could be good ways to handle a terminated project, but there's only one of those answers that corresponds to what the *PMBOK Guide* says. When a project is terminated, you still need to complete the Validate Scope process. That way, you can document all of the work that has been completed, and the work that has not been completed.

Exam ~~Questions~~ Answers

15. Answer: B

Are you starting to get the hang of how this change control stuff works? The baseline isn't etched in stone, and you need to be able to change it, but you can't just go ahead and make changes whenever you want. You need to document the change request and then put that request through change control. If it's approved, then you can update the baseline so that it incorporates the change.

You definitely can't just make the change and inform the sponsor later. All changes need to be approved.

16. Answer: A

When a question asks you about what a particular document, input, or output contains, be on the lookout for answers that talk about a different document. What document do you know about that gives the project manager authorization to do the work? That's what the project charter is for.

17. Answer: C

Sometimes Control Scope is easiest to think about as a kind of machine that turns approved changes into updates. It sucks in the approved changes and all of the other Scope Management stuff (the scope statement, WBS, and WBS dictionary), does all the stuff that it needs to do to update those things, and then spits out updates. And sometimes it spits out new requested changes because when you're making changes to the WBS or scope statement you realize that you need to make even more changes.

18. Answer: A

Scope Identification is a made-up process. It didn't appear in this chapter, and even though it sounds real, it's wrong.

I THINK OF PERFORM INTEGRATED CHANGE CONTROL AS A MACHINE THAT TURNS CHANGE REQUESTS INTO APPROVED CHANGES, AND CONTROL SCOPE AS THE MACHINE THAT TURNS APPROVED CHANGES INTO UPDATES TO THE BASELINE.

19. Answer: B

WBS templates are a great way to speed up creating the WBS, and the easiest way to create a template is to use one from a previous project. It is **not** a good idea to cut out deliverables, skip important outputs like the WBS dictionary, or make the sponsor do your job for you.

You can also use a template for the Scope Management plan.

Answers

Exam ~~Questions~~

20. Answer: C

This question asked you about the Define Scope process, but all of the answers describe various outputs. Which of these outputs matches Define Scope? Well, the main output of Define Scope is the scope statement, and answer C is a good description of the scope statement.

Did you guess "Create WBS" because it was a Scope Management process and the question mentioned decomposing deliverables into work packages?

21. Answer: A

This question asked you where you defined the procedures for doing all of the Scope Management processes. Where do you find those procedures? You find them in the Project Management plan—specifically, the Scope Management subplan. And you build that in the Develop Project Management Plan process.

22. Answer: C

There's only one document you've seen that shows you details of individual work packages and contains a control account, a statement of work, and a resource assignment. It's the WBS dictionary.

Did you notice how the question made it sound like the team did a good thing by ignoring Control Scope and making changes that were never approved?

23. Answer: D

When you read the question, it looks like the team really helped the project, right? But think about what happened: the team abandoned the plan, and then they made a change to the project without getting approval from the sponsor or stakeholders. Maybe they discovered a useful shortcut. But isn't it possible that the shortcut the team found was already considered and rejected by the sponsor? That's why change control is so important.

HEY, I'LL BET A GOOD WAY TO STUDY FOR THE EXAM IS TO LOOK AT ANSWERS A, B, AND C IN QUESTION 23 AND FIGURE OUT WHAT EACH OF THEM IS DESCRIBING. IT'LL BE GREAT PRACTICE IDENTIFYING AN OUTPUT FROM A DESCRIPTION!

Answers

Exam ~~Questions~~

24. Answer: D

Some questions are just definition questions. When that definition is a "which-is-BEST" question, there could be an answer that makes some sense, and it's tempting to stop with it. In this case, answer A sounds like it might be right. But if you read answer D, it's much more accurate.

25. Answer: A

The question asked about the tools and techniques for Collect Requirements, and answer A is the only answer that has to do with the Create WBS process. The rest of the answers were descriptions of Collect Requirements tools and techniques: interviews (answer B), prototypes (answer C), and observations (answer D).

26. Answer: A

The three outputs of Collect Requirements are requirements documentation and the requirements traceability matrix. "Requirements Observations" isn't really an output.

> The easiest way to make sure you get questions like this right is to think about how each of those outputs are actually used later in the project.

Keep an eye out for questions that describe an input or output and then ask you to name it. Look at each answer and think up your own descriptions for them—one of them will match the question.

6 Time management

Getting it done on time

TIME MANAGEMENT MAKES ALL OF OUR COCKTAIL PARTIES A SUCCESS! NOW THAT WE KNOW HOW TO SEQUENCE OUR ACTIVITIES AND BUILD OUR SCHEDULES, WE ALWAYS HAVE TIME TO LIMBO BEFORE WE'VE HAD TOO MANY MARTINIS.

Time management is what most people think of when they think of project managers. It's where the deadlines are set and met. It starts with **figuring out what work** you need to do, how you will do it, what **resources you'll use**, and how long it will take. From there, it's all about developing and controlling that **schedule**.

Reality sets in for the happy couple

Rob and Rebecca have decided to tie the knot, but they don't have much time to plan their wedding. They want the big day to be unforgettable. They want to invite a lot of people and show them all a great time.

But just thinking about all of the details involved is overwhelming. Somewhere around picking the paper for the invitations, the couple realize they need help...

They've always dreamed of a June wedding, but it's already January.

EVERYTHING HAS TO BE PERFECT! BUT IT SEEMS SO HUGE. I DON'T KNOW WHERE TO START.

Rebecca's been dreaming of the big day since she was 12, but it seems like there's so little time to do it all. She needs some help.

DON'T WORRY. MY SISTER'S WEDDING PLANNER WAS GREAT. LET ME GIVE HER A CALL.

Meet the wedding planner

TAKE IT EASY, GUYS. I'VE GOT IT UNDER CONTROL.

Kathleen, the wedding planner

Rob: We want everything to be perfect.

Rebecca: There is so much to do! Invitations, food, guests, music…

Rob: Oh no, we haven't even booked the place.

Rebecca: And it's all got to be done right. We can't print the invitations until we have the menu planned. We can't do the seating arrangements until we have the RSVPs. We aren't sure what kind of band to get for the reception, or should it be a DJ? We're just overwhelmed.

Rob: My sister said you really saved her wedding. I know she gave you over a year to plan.

Rebecca: But I've always dreamed of a June wedding, and I'm not willing to give that up. I know it's late, but can you help us?

⚛ BRAIN POWER

What should Kathleen do first to make sure they have time to get everything done?

Time management helps with aggressive timelines

Since there are so many different people involved in making the wedding go smoothly, it takes a lot of planning to make sure that all of the work happens in the right order, gets done by the right people, and doesn't take too long. That's what the **Time Management** knowledge area is all about.

Initially, Kathleen was worried that she didn't have enough time to make sure everything was done properly. But she knew that she had some powerful time management tools on her side when she took the job, and they'll help her make sure that everything will work out fine.

WE'VE GOT A LOT OF PEOPLE AND ACTIVITIES TO GET UNDER CONTROL. YOU GUYS REALLY SHOULD HAVE CALLED SIX MONTHS AGO, BUT WE'LL STILL MAKE THIS WEDDING HAPPEN ON TIME.

There's a lot to get done before June. Kathleen's going to need to figure out what work needs to be done before she does anything else.

To-Do List

- Invitations
- Flowers
- Wedding cake
- Dinner menu
- Band

Time Management Magnets

You need to know the order of the Time Management processes for the exam. Luckily, they are pretty intuitive. Can you figure out the order?

❶

❷

❸

❹

❺

❻

❼

Control Schedule

Estimate Activity Durations

Develop Schedule

Sequence Activities

Define Activities

Estimate Activity Resources

Plan Schedule Management

Time Management Magnets Solution

Here are the correct order and the main output for each of the Time Management processes.

 Plan Schedule Management

First you define the processes you'll use to plan and control your schedule.

Schedule Management plan

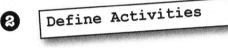 **Define Activities**

Next you come up with a list of all of the activities that will need to be completed.

Activity list

The activity list is the basis for the network diagram that you create in the next process.

 Sequence Activities

Project schedule network diagram

Next, you figure out which activities need to come before others, and put them in the right order. The main output here is **a project schedule network diagram**, a picture of how activities are related.

Knowing the stuff that needs to happen and the sequence is half the battle. Now you need to figure out who will do the work.

④ Estimate Activity Resources

First you estimate the resources you'll need to do the job, and create a list of them...

Activity resource requirements

Once you have the network diagram, you can start to figure out who and what are needed to get the project done.

⑤ Estimate Activity Durations

Activity duration estimates

…and then estimate the time it will take to do each activity.

You can use lots of different estimation techniques to determine how long the project will take.

The last process, Control Schedule, is in the Monitoring and Controlling process group.

Monitoring & Controlling process group

⑦ Control Schedule

Finally, you monitor and control changes to the schedule to make sure that it is kept up to date.

Keeping track of the issues that require schedule changes and dealing with them is as important in Time Management as it was in Scope Management.

⑥ Develop Schedule

Then you build a schedule from all of the estimates and the resource and activity information you've created.

The schedule pulls all of the information together to predict the project end date.

Planning process group

The first six Time Management processes are in the Planning process group because they're all about coming up with the schedule—and you need that before you can start executing your project.

⚛ BRAIN POWER

Time Management is all about breaking the work down into activities, so you can put them in order and come up with estimates for each of them.

What do you need to know before you can figure out what activities are needed for a project?

Plan your scheduling processes

The **Plan Schedule Management** process is just like all of the other planning processes you've seen so far. In fact, you've already seen all of the inputs and tools that are used to create it in previous processes. Just like with the Plan Scope Management process from Chapter 5, your goal is to build a Schedule Management plan from the other project management plans, your company's culture and existing documents, and the project charter.

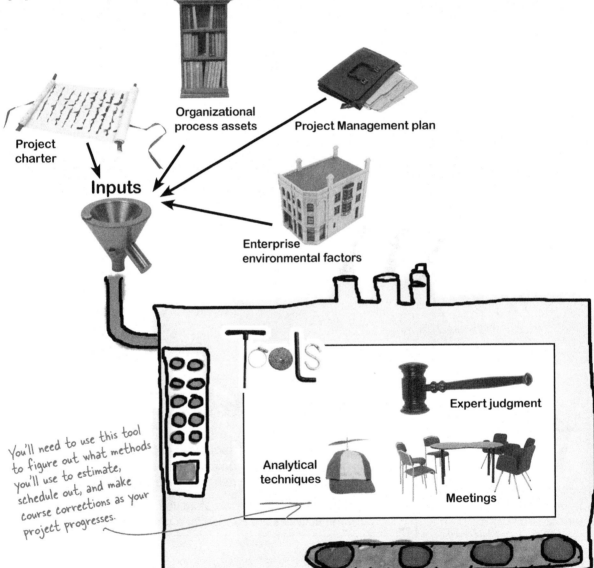

Project charter

Organizational process assets

Project Management plan

Inputs

Enterprise environmental factors

Tools

Expert judgment

Analytical techniques

Meetings

You'll need to use this tool to figure out what methods you'll use to estimate, schedule out, and make course corrections as your project progresses.

Now you know how you'll track your schedule

The only output of the Plan Schedule Management process is the Schedule Management plan. It describes the way you'll estimate your work, track your progress, and report on it.

Outputs

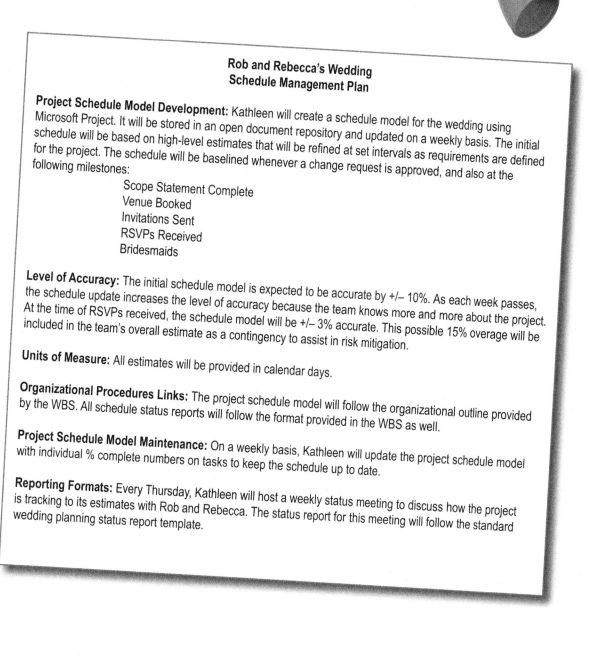

Rob and Rebecca's Wedding
Schedule Management Plan

Project Schedule Model Development: Kathleen will create a schedule model for the wedding using Microsoft Project. It will be stored in an open document repository and updated on a weekly basis. The initial schedule will be based on high-level estimates that will be refined at set intervals as requirements are defined for the project. The schedule will be baselined whenever a change request is approved, and also at the following milestones:

> Scope Statement Complete
> Venue Booked
> Invitations Sent
> RSVPs Received
> Bridesmaids

Level of Accuracy: The initial schedule model is expected to be accurate by +/– 10%. As each week passes, the schedule update increases the level of accuracy because the team knows more and more about the project. At the time of RSVPs received, the schedule model will be +/– 3% accurate. This possible 15% overage will be included in the team's overall estimate as a contingency to assist in risk mitigation.

Units of Measure: All estimates will be provided in calendar days.

Organizational Procedures Links: The project schedule model will follow the organizational outline provided by the WBS. All schedule status reports will follow the format provided in the WBS as well.

Project Schedule Model Maintenance: On a weekly basis, Kathleen will update the project schedule model with individual % complete numbers on tasks to keep the schedule up to date.

Reporting Formats: Every Thursday, Kathleen will host a weekly status meeting to discuss how the project is tracking to its estimates with Rob and Rebecca. The status report for this meeting will follow the standard wedding planning status report template.

Use the Define Activities process to break down the work

Define Activities uses everything we already know about the project to divide the work into activities that can be estimated. The inputs for this process all come from the processes in the Scope Management and Integration Management knowledge areas. The first step in Time Management is figuring out how the project work breaks down into activities—and that's what the **Define Activities process** is for.

Planning
process group

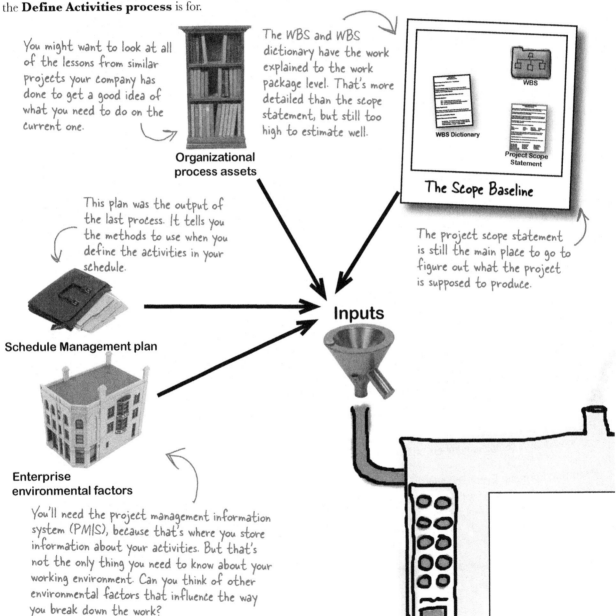

You might want to look at all of the lessons from similar projects your company has done to get a good idea of what you need to do on the current one.

Organizational process assets

The WBS and WBS dictionary have the work explained to the work package level. That's more detailed than the scope statement, but still too high to estimate well.

WBS

WBS Dictionary

Project Scope Statement

The Scope Baseline

This plan was the output of the last process. It tells you the methods to use when you define the activities in your schedule.

Schedule Management plan

The project scope statement is still the main place to go to figure out what the project is supposed to produce.

Inputs

Enterprise environmental factors

You'll need the project management information system (PMIS), because that's where you store information about your activities. But that's not the only thing you need to know about your working environment. Can you think of other environmental factors that influence the way you break down the work?

Tools and techniques for Define Activities

Kathleen wrote down everything she knew about the project. She used the activity list from her last wedding as a guide and then thought about the things that Rob and Rebecca wanted that were different from her past projects. She broke those things down into activities and pulled everything together into an activity list.

This "Tools" icon means we're showing you the tools and techniques for the process. Get the picture?

Decomposition

This means taking the work packages you defined in the Scope Management processes and breaking them down even further into activities that can be estimated.

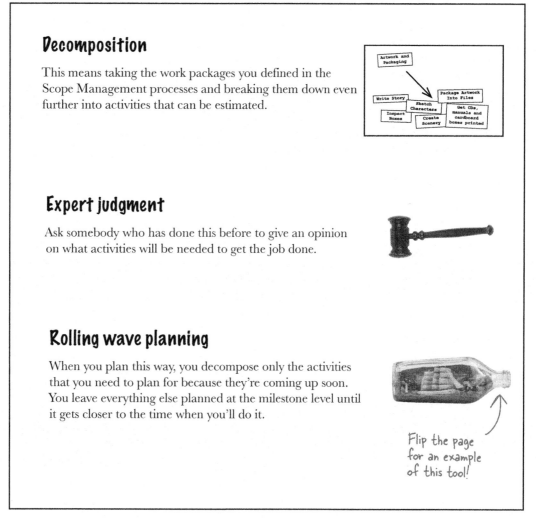

Expert judgment

Ask somebody who has done this before to give an opinion on what activities will be needed to get the job done.

Rolling wave planning

When you plan this way, you decompose only the activities that you need to plan for because they're coming up soon. You leave everything else planned at the milestone level until it gets closer to the time when you'll do it.

Flip the page for an example of this tool!

Rolling wave planning lets you plan as you go

Sometimes you start a project without knowing a lot about the work that you'll be doing later. **Rolling wave planning** lets you plan and schedule only the stuff that you know enough about to plan well.

If Kathleen were using rolling wave planning, she might write a schedule for only the tasks it takes to do the invitations, and leave the planning for the menu and the seating up in the air until she knows who will RSVP.

Rob and Rebecca probably wouldn't be happy hearing that Kathleen was only going to plan for the invitations to be sent, though. They want to know that their wedding is going to happen on time. That's why rolling wave planning should only be used in cases where it's not possible to plan any other way.

Think back to the definition of a project in Chapter 2. Remember how projects are **progressively elaborated**? Rolling wave planning takes advantage of the fact that you know more about the project as you go to make plans more accurate.

there are no Dumb Questions

Q: How would you use experts to help you define tasks?

A: A wedding is something that a lot of people have experience with, but some projects are not as easy to get a handle on. If you were asked to manage a project in a new domain, you might want to ask an expert in that field to help you understand what activities were going to be involved.

Even in Kathleen's case, access to a catering expert might help her think of some activities that she wouldn't have planned for on her own.

It could be that you create an activity list and then have the expert review it and suggest changes. Or, you could involve the expert from the very beginning and ask to have a Define Activities conversation with him before even making your first draft of the activity list.

Q: I still don't get rolling wave planning.

A: One way to develop a project is to divide it up into phases of work, and gather requirements for each phase as the previous one is completed. Sometimes projects are done iteratively, where you divide the work up into phases and then plan out each phase before you execute on it. Rolling wave planning is all about committing to planning out one portion of the work that you'll do, executing it, and then moving on to the next portion.

Software projects using **agile methodologies** use a form of rolling wave planning to make sure that everything they sign on to do gets done. They might do user stories for a release of the software up front, build it, and deliver it, and then gather more requirements based on the users' ideas after working with the released version.

Activity Magnets

Here is part of a WBS. Arrange the activities underneath the WBS to show how the work items decompose into activities.

This is part of the WBS that Kathleen made for the wedding project.

This is one work package from the wedding WBS. How does it decompose into activities?

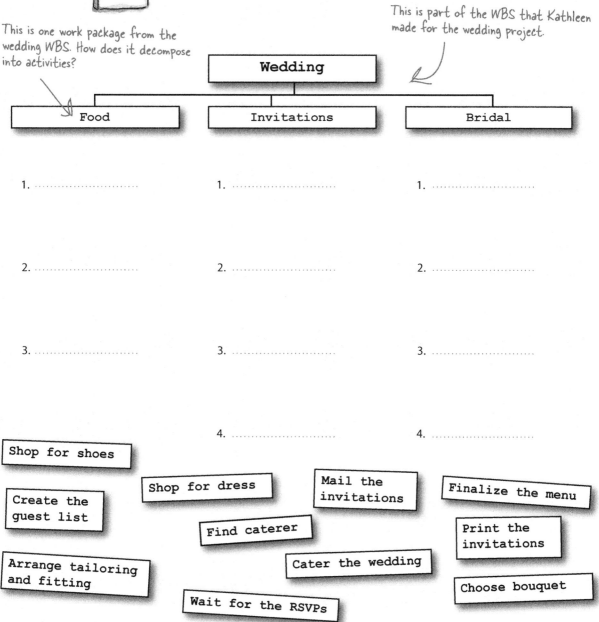

```
                    ┌─────────────────┐
                    │     Wedding     │
                    └─────────────────┘
        ┌───────────────────┼───────────────────┐
┌───────────────┐  ┌─────────────────┐  ┌─────────────────┐
│     Food      │  │   Invitations   │  │     Bridal      │
└───────────────┘  └─────────────────┘  └─────────────────┘
```

1. ....   1. ....   1. ....

2. ....   2. ....   2. ....

3. ....   3. ....   3. ....

                              4. ....   4. ....

Shop for shoes

Create the guest list

Shop for dress

Find caterer

Mail the invitations

Finalize the menu

Print the invitations

Arrange tailoring and fitting

Cater the wedding

Choose bouquet

Wait for the RSVPs

Activity Magnets Solution

Here is part of a WBS. Arrange the activities underneath the WBS to show how the work items decompose into activities.

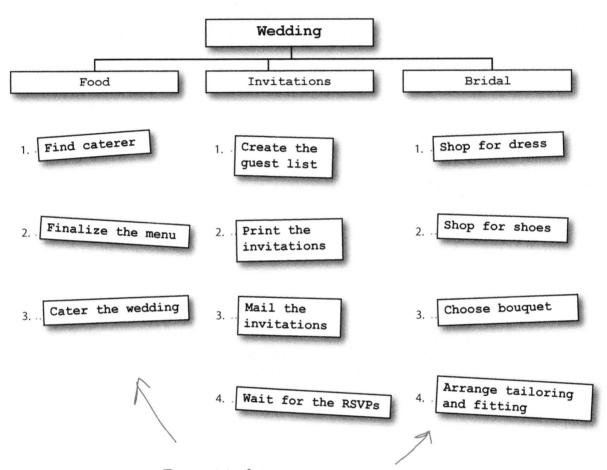

```
                          ┌──────────────────┐
                          │     Wedding      │
                          └──────────────────┘
          ┌───────────────────────┼───────────────────────┐
   ┌────────────┐          ┌────────────┐          ┌────────────┐
   │    Food    │          │ Invitations│          │   Bridal   │
   └────────────┘          └────────────┘          └────────────┘
```

1. Find caterer

2. Finalize the menu

3. Cater the wedding

1. Create the guest list

2. Print the invitations

3. Mail the invitations

4. Wait for the RSVPs

1. Shop for dress

2. Shop for shoes

3. Choose bouquet

4. Arrange tailoring and fitting

There are lots of other activities that could be defined for the three work packages in Kathleen's WBS.

The important thing to remember about activities, though, is that they are broken down to the level at which they can be estimated accurately.

Define activities outputs

The main output of this process is the **activity list**. It's the basis for all of the estimation and scheduling tasks you will do next. But there are a few other outputs that go along with it, and help to make the estimates more detailed and accurate.

Outputs

Activity list

This is a list of everything that needs to be done to complete your project. This list is lower-level than the WBS. It's all the activities that must be accomplished to deliver the work packages.

Activity list

Activity attributes

Here's where the description of each activity is kept. All of the information you need to figure out the order of the work should be here, too. So any predecessor activities, successor activities, or constraints should be listed in the attributes, along with descriptions and any other information about resources or time that you need for planning.

Activity attributes

Milestone list

All of the important checkpoints of your project are tracked as milestones. Some of them could be listed in your contract as requirements of successful completion; some could just be significant points in the project that you want to keep track of. The milestone list needs to let everybody know which are required and which are not.

Milestone list

Some milestones for the wedding:

* Invitations sent
* Menu finalized
* Church booked
* Bridesmaids' dresses fitted

WE JUST GOT THE PROGRAMS BACK FROM THE PRINTER, AND THEY'RE ALL WRONG!

Rob: The quartet cancelled. They had another wedding that day.

Rebecca: Aunt Laura is supposed to do the reading at the service, but after what happened at Uncle Stu's funeral, I think I want someone else to do it.

Rob: Should we really have a pan flute player? I'm beginning to think it might be overkill.

Rebecca: Maybe we should hold off printing the invitations until this stuff is worked out.

Kathleen: OK, let's think about exactly how we want to do this. We need to be sure about how we want the service to go before we do any more printing.

The Sequence Activities process puts everything in order

Now that we know what we have to do to make the wedding a success, we need to focus on the order of the work. Kathleen sat down with all of the activities she had defined for the wedding and decided to figure out exactly how they needed to happen. That's where she used the **Sequence Activities** process.

The **activity attributes** and the **activity list** she had created had most of the predecessors and successors written in them. Her **milestone list** had major pieces of work written down, and there were a couple of changes to the scope she had discovered along the way that were approved and ready to go.

Planning
process group

The plan will list any methodologies or tools you planned to use when sequencing.

This includes information about each activity, including known predecessors and successors.

Schedule Management plan

Activity list

Activity attributes

Knowing the full scope of the project helps Kathleen be sure she's got all of the activities needed to do the work.

Project Scope Statement

Milestone list

Organizational process assets

Kathleen looked through past project files to find one that might help her sequence the activities for Rob and Rebecca's wedding.

Rob and Rebecca had asked that the invitations be printed at least three months in advance to be sure that everyone had time to RSVP. That's a milestone on Kathleen's list.

Inputs

Enterprise environmental factors

Diagram the relationship between activities

One way to visualize the way activities relate is to create a network diagram. Kathleen created this one to show how the activities involved in producing the invitations depend on one another.

For example, the calligrapher is the person who's hired to write the addresses on the invitations, so Rob and Rebecca need to pick a calligrapher before the invitations can be addressed. But the invitations also need to be printed before they can be addressed, because otherwise the calligrapher won't have anything to write on! See how predecessors can get all complicated? Luckily, a diagram makes sense of them!

Showing the activities in rectangles and their relationships as arrows is called a **precedence diagramming method (PDM)**.

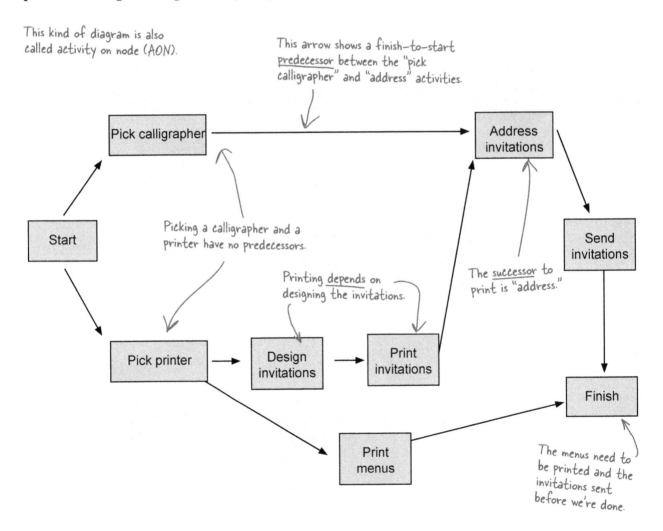

This kind of diagram is also called activity on node (AON).

This arrow shows a finish–to–start predecessor between the "pick calligrapher" and "address" activities.

Picking a calligrapher and a printer have no predecessors.

Printing depends on designing the invitations.

The successor to print is "address."

The menus need to be printed and the invitations sent before we're done.

Network diagrams put your tasks in perspective

Just looking at the way all of these tasks relate to one another can help you figure out what's important at any time in the project. Once Rob and Rebecca looked at the network diagram below, they realized they needed to get online and start looking for a venue for their wedding right away, even before they'd figured out their budget and guest list.

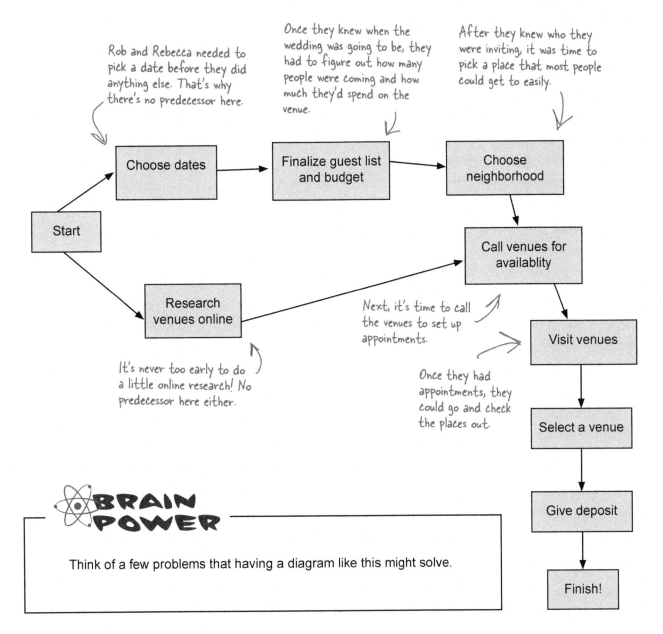

Rob and Rebecca needed to pick a date before they did anything else. That's why there's no predecessor here.

Once they knew when the wedding was going to be, they had to figure out how many people were coming and how much they'd spend on the venue.

After they knew who they were inviting, it was time to pick a place that most people could get to easily.

It's never too early to do a little online research! No predecessor here either.

Next, it's time to call the venues to set up appointments.

Once they had appointments, they could go and check the places out.

⚛ BRAIN POWER

Think of a few problems that having a diagram like this might solve.

Sharpen your pencil

You'll need to know how to turn a table of nodes into a network diagram, so here's your chance to get some practice! Here's a list of nodes for a PDM network diagram. Try drawing the diagram based on it:

Name	Predecessor
Start	—
A	Start
B	A
C	B
D	Start
E	D
F	B
G	C
H	D
I	E, H
Finish	F, G, I

Now try another one!

Name	Predecessor
Start	—
1	Start
2	1
3	2
4	Start
5	3
6	Start
7	6
Finish	7, 4, 5

Answers on page 322.

Dependencies help you sequence your activities

The most common kind of predecessor is the Finish to Start. It means that one task needs to be completed before another one can start. There are a few other kinds of predecessors, though. They can all be used in network diagrams to show the order of activities. The three main kinds of dependency are **Finish to Start (FS)**, **Start to Start (SS),** and **Finish to Finish (FF)**.

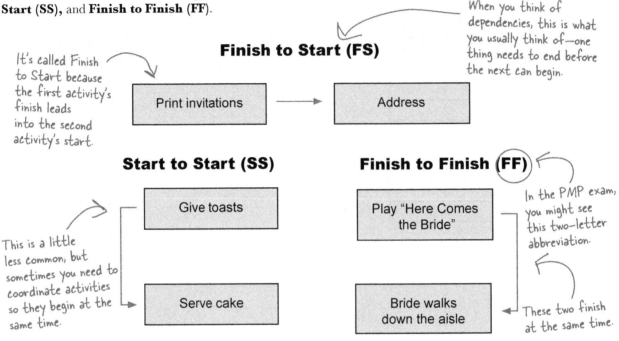

When you think of dependencies, this is what you usually think of—one thing needs to end before the next can begin.

Finish to Start (FS)

It's called Finish to Start because the first activity's finish leads into the second activity's start.

| Print invitations | → | Address |

Start to Start (SS)

This is a little less common, but sometimes you need to coordinate activities so they begin at the same time.

| Give toasts |

| Serve cake |

Finish to Finish (FF)

In the PMP exam, you might see this two-letter abbreviation.

| Play "Here Comes the Bride" |

| Bride walks down the aisle |

These two finish at the same time.

External dependencies

Sometimes your project will depend on things outside the work you are doing. For the wedding, we are depending on the wedding party before us to be out of the reception hall in time for us to decorate. The decoration of the reception hall then depends on that as an external dependency.

Discretionary dependencies

Rob and Rebecca really want the bridesmaids to arrive at the reception before the couple. There's no necessity there—it's just a matter of preference. For the exam, know that you should set discretionary dependencies based on your knowledge of the best practices for getting the job done.

Mandatory dependencies

You can't address an invitation that hasn't been printed yet. So, printing invitations is a mandatory predecessor for addressing them. Mandatory predecessors are the kind that have to exist just because of the nature of the work.

Internal dependencies

The rehearsal dinner can't begin until the happy couple leaves the church. Some dependencies are completely within the team's control.

Leads and lags add time between activities

Sometimes you need to give some extra time between activities. **Lag time** is when you purposefully put a delay between the predecessor task and the successor. For example, when the bride and her father dance, the guests wait awhile before they join them.

Lead time is when you give a successor task some time to get started before the predecessor finishes. So you might want the caterer preparing dessert an hour before everybody is eating dinner.

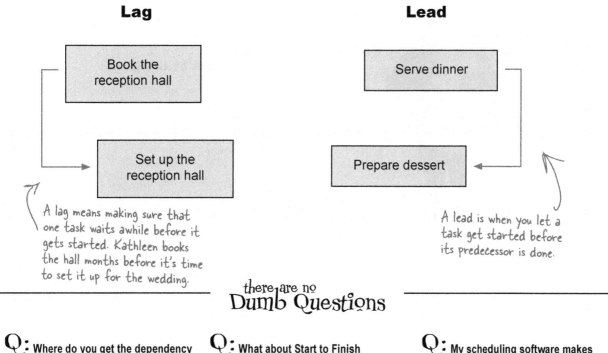

Lag

Book the reception hall

Set up the reception hall

A lag means making sure that one task waits awhile before it gets started. Kathleen books the hall months before it's time to set it up for the wedding.

Lead

Serve dinner

Prepare dessert

A lead is when you let a task get started before its predecessor is done.

there are no Dumb Questions

Q: Where do you get the dependency information to figure out your network diagram?

A: Your **activity attributes** should list the predecessors and successors for each activity. As you build the network diagram, you might discover new dependencies as well. Your project team will determine the dependencies necessary for each of the activities.

Q: What about Start to Finish dependencies?

A: It's possible for activities to require that a task has been started before it can finish. An example might be that singing couldn't start until after the music had started. But tasks like that are pretty rare and almost never show up in network diagrams.

Q: My scheduling software makes network diagrams for me. Why do I need to know this?

A: Most scheduling software does create one of these diagrams automatically. But spending the time to think through your dependencies and examine them visually can really help you find places where you might need to give some tasks more priority if you want to get your project done on time. So you should know how to make them too.

BUT I DO ALL THIS STUFF AT THE SAME TIME. WHY DO I NEED ALL THESE SEQUENCE DIAGRAMS?

You should still think of things in sequence.

For the test, it's important to know the order of these processes. And even though you might do it all at once, you probably spend some time thinking about each of these things.

What's the advantage of thinking about Define Activities and Sequence Activites separately?

Create the network diagram

As you sequence the activities, you will find new activities that need to be added and new attributes for activities that you didn't know about. So, while the main product of this process is the network diagram, you also produce updates to some of the Define Activities documents and outputs of other processes, too.

Outputs

Sometimes sequencing will show that two tasks rely on each other. If you find new predecessors or successors, their attributes will need to be changed.

For the test, you won't need to know exactly which documents change as an output to this process. All you need to know is that project document updates are an output.

Activity attributes updates

Risk Register Updates

When you sequence your activities, you can find activities that carry a lot of risk to the project. Any risks you find in the process need to be added to your risk register. You'll read more about that in Chapter 11.

Project schedule network diagram

Here's where you work out how all of the tasks fit together based on their predecessors and determine the critical path through the project.

Project documents updates

When you've sequenced your activities, you might find that some of the documents you've created as part of other processes need to be updated. The *PMBOK Guide* calls documents like these "project document updates." The pictures to the right are some examples, but there could be other documents that require updates as well.

Activity list updates

If you find a new activity while sequencing, updates need to be made to the activity list.

Rob and Rebecca have resource problems

Getting a handle on all of the tasks that have to be done is a great start. But it's not enough to know the tasks and the order they come in. Before you can put a schedule together, you need to know who is going to do each job, and the things they need available to them in order to do it! Those are **resources**, and getting a handle on them is a very important part of Time Management.

WE'VE GOT SO MUCH TO DO! INVITATIONS, CATERING, MUSIC... AND I'VE GOT NO IDEA WHO'S GOING TO DO IT ALL. I'M TOTALLY OVERWHELMED.

AND IT'S NOT JUST PEOPLE—WE NEED FOOD, FLOWERS, A CAKE, A SOUND SYSTEM, A VENUE! HOW DO WE GET A HANDLE ON THIS?

Rebecca's worried about human resources.

Rob realizes that not all resources are people!

Resources are people, equipment, locations, or anything else that you need in order to do all of the activities that you planned for. Every activity in your activity list needs to have resources assigned to it.

BRAIN POWER

What do you need to know about a project before you can assign resources?

What you need to estimate resources

Good news: you've already seen most of the inputs to the **Estimate Activity Resources process** already! Before you can assign resources to your project, you need to know which ones you're authorized to use on your project. That's an input, and it's called **resource calendars**. You'll also need the activity list that you created earlier, and you'll need to know about how your organization typically handles resources. Once you've got a handle on these things, you're set for resource estimation.

Planning
process group

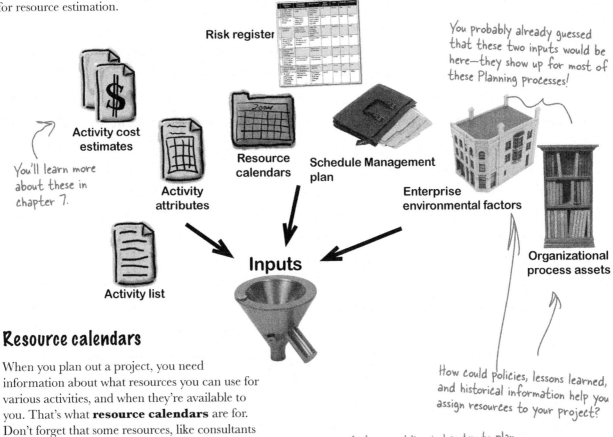

Risk register

You probably already guessed that these two inputs would be here—they show up for most of these Planning processes!

Activity cost estimates

You'll learn more about these in chapter 7.

Activity attributes

Resource calendars

Schedule Management plan

Enterprise environmental factors

Organizational process assets

Activity list

Inputs

How could policies, lessons learned, and historical information help you assign resources to your project?

Resource calendars

When you plan out a project, you need information about what resources you can use for various activities, and when they're available to you. That's what **resource calendars** are for. Don't forget that some resources, like consultants or training rooms, have to be scheduled in advance, and they might only be available at certain times. You'll need to know this before you can finish planning your project.

A June wedding is harder to plan than one in December, because the wedding halls are all booked up. That's a resource constraint!

Resource calendars are the only new input to the Estimate Activity Resources process. You've already seen the rest of the inputs.

Estimating the resources

The goal of **Estimate Activity Resources** is to assign resources to each activity in the activity list. There are **five tools and techniques** for the Estimate Activity Resources process. Some of them have technical-sounding names, but they're all actually pretty sensible when you think about it. They should all make sense to you when you think about what you have to do to figure out what resources your project needs.

Expert judgment means bringing in experts who have done this sort of work before and getting their opinions on what resources are needed.

Alternative analysis means considering several different options for how you assign resources. This includes varying the number of resources as well as the kind of resources you use.

Published estimating data is something that project managers in a lot of industries use to help them figure out how many resources they need. They rely on articles, books, journals, and periodicals that collect, analyze, and publish data from other people's projects.

Project management software like Microsoft Project will often have features designed to help project managers play around with resources and constraints and find the best combination of assignments for the project.

Bottom-up estimating is a technique that you may have used before without even knowing it! It means breaking down complex activities into pieces, and working out the resource assignments for each of those simpler pieces using the other four tools and techniques.

---

there are no Dumb Questions

Q: In my company, I'm given my resources—I don't get to assign them myself. How do these tools help me?

A: When you work in a functional organization or some matrix organizations, you don't have as much freedom in selecting resources as you do in a projectized organization. But that doesn't mean these tools aren't important! Whoever is doing the resource selection and assignment should be using them. And they'll be on the PMP exam, so you need to understand them all.

Q: Is choosing a consultant, contractor, or vendor to do project work part of resource estimation?

A: When you're working with a resource outside your company, like a contractor or consultant, you consider that resource the same way you consider any other resource. But actually negotiating the contract and selecting the vendor is not part of the Estimate Activity Resources process. There's a whole other knowledge area for that—Procurement Management.

Q: What if I need a resource that isn't available when my project needs it?

A: This is one of the reasons that project management is a tough job! When you need a resource that isn't available, you need to negotiate for it. Think about it...your project depends on getting this resource, and without it your project won't get done. You need it, or you'll face delays! You have to do whatever you can to get that resource for your project.

Sharpen your pencil

You'll need to understand the different Estimate Activity Resources tools and techniques for the exam. Look at each of these scenarios and write down which of the five activity resource estimation tools and techniques is being used.

1. Kathleen has to figure out what to do for the music at Rob and Rebecca's wedding. She considers using a DJ, a rock band, or a string quartet.

2. The latest issue of *Wedding Planner's Journal* has an article on working with caterers. It includes a table that shows how many waiters work with various guest-list sizes.

3. There's a national wedding consultant who specializes in Caribbean-themed weddings. Kathleen gets in touch with her to ask about menu options.

4. Kathleen downloads and fills out a specialized spreadsheet that a project manager developed to help with wedding planning.

5. There's so much work that has to be done to set up the reception hall that Kathleen has to break it down into five different activities in order to assign jobs.

6. Kathleen asks Rob and Rebecca to visit several different caterers and sample various potential items for the menu.

7. Kathleen calls up her friend who knows specifics of the various venues in their area for advice on which one would work best.

⟶ Answers on page 319.

Figuring out how long the project will take

Once you're done with Estimate Activity Resources, you've got everything you need to figure out how long each activity will take. That's done in a process called **Estimate Activity Durations**. This is where you look at each activity in the activity list, consider the scope and the resources, and estimate how long it will take to perform.

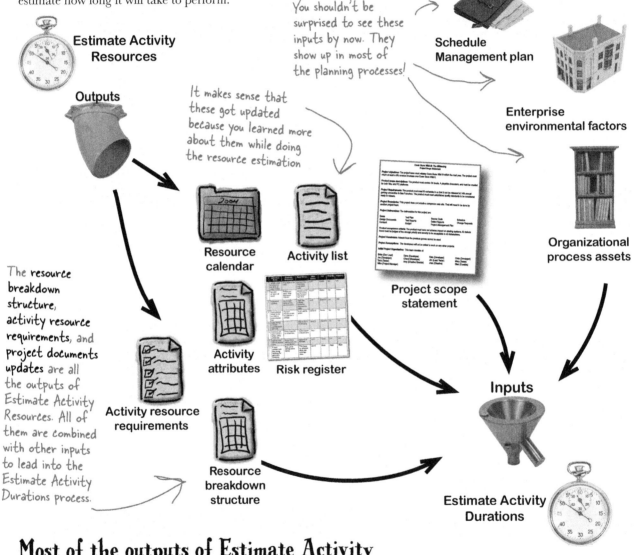

Planning process group

Estimate Activity Resources

You shouldn't be surprised to see these inputs by now. They show up in most of the planning processes!

Schedule Management plan

Outputs

It makes sense that these got updated because you learned more about them while doing the resource estimation

Enterprise environmental factors

Organizational process assets

Resource calendar

Activity list

Project scope statement

The resource breakdown structure, activity resource requirements, and project documents updates are all the outputs of Estimate Activity Resources. All of them are combined with other inputs to lead into the Estimate Activity Durations process.

Activity attributes

Risk register

Activity resource requirements

Inputs

Resource breakdown structure

Estimate Activity Durations

Most of the outputs of Estimate Activity Resources are immediately used as inputs for Estimate Activity Durations.

Exercise

You'll need to understand the various inputs and outputs for each process for the exam. Here's a list of some of the inputs for Estimate Activity Durations. Write down what you think each of them will be used for when you actually sit down and estimate how long each activity will take.

1. Activity list and activity attributes

2. Activity resource requirements

3. Resource calendar

4. Project scope statement

5. Enterprise environmental factors

6. Organizational process assets

Exercise

You'll need to understand the various inputs and outputs for each process for the exam. Here's a list of some of the inputs for Estimate Activity Durations. Write down what you think each of them will be used for when you actually sit down and estimate how long each activity will take.

1. Activity list and activity attributes

You need these because the goal of this process is to estimate the duration of each activity.

Contains information about the activities that are being estimated

2. Activity resource requirements

The more resources you add to an activity, the less time it takes.

But sometimes adding people won't get the job done any faster! Remember, nine women can't have a baby in one month.

Shows which resources are assigned to each activity

3. Resource calendar

You need to know when the resources are available, because that's going to impact the final estimate for the activity.

Shows the availability, capabilities, and skills of each human resource, or the quantity and availability of equipment and other resources

4. Project scope statement

Lists constraints and assumptions for each activity

You're probably not the first person in your company to do this sort of project. Information from people around you will be very valuable when you're creating estimates.

5. Enterprise environmental factors

Other people or databases in my company can help with estimation

This input is always about looking elsewhere in your organization for information.

6. Organizational process assets

Contains historical information and records from past projects

Any time you see this, think about historical information and project records!

The more you know about how past projects went, the more accurate your estimates will be.

Estimation tools and techniques

Estimating the duration of an activity means starting with the information you have about that activity and the resources that are assigned to it, and then working with the project team to come up with an estimate. Most of the time you'll start with a rough estimate and then refine it (maybe a few times!) to make it more accurate. You'll use these five tools and techniques to create the most accurate estimates.

Expert judgment will come from your project team members who are familiar with the work that has to be done. If you don't get their opinion, then there's a huge risk that your estimates will be wrong!

Parametric estimating means plugging data about your project into a formula, spreadsheet, database, or computer program that comes up with an estimate. The software or formula that you use for parametric estimating is built on a database of actual durations from past projects.

Reserve analysis means adding extra time to the schedule (called a *contingency reserve* or a *buffer*) to account for extra risk.

Analogous estimating is when you look at activities from previous projects that were similar to this one and look at how long it took to do similar work before. But this only works if the activities and the project team are similar!

Three-point estimates are when you come up with three numbers: a **most likely** estimate that probably will happen, an **optimistic** one that represents the best-case scenario, and a **pessimistic** one that represents the worst-case scenario. The final estimate is the average.

Group decision techniques help the team decide on the best estimates for the activities they've defined.

Exercise

Each of these scenarios describes a different tool or technique from Estimate Activity Durations. Write down which tool or technique is being described.

1. Kathleen comes up with three estimates (one where everything goes wrong, one where some things go wrong, and one where nothing goes wrong) for printing invitations, and averages them together to come up with a final number. ....

2. There will be two different catering companies at the wedding. Kathleen asks the head chef at each of them to give her an estimate of how long it will take to do the job. ....

3. There's a spreadsheet Kathleen always uses to figure out how long it takes guests to RSVP. She enters the number of guests and their ZIP codes, and it calculates an estimate for her. ....

4. Kathleen's done four weddings that are very similar to Rob and Rebecca's, and in all four of them it took exactly the same amount of time for the caterers to set up the reception hall. ....

Answers on page 319.

Three-Point Estimates Up Close

PERT (Program Evaluation Review Technique) is the most common form of three-point estimation. It's a technique that was developed in the 1960s by consulting firms working with the U.S. government as a way of getting more accurate project duration predictions up front. To do a PERT estimate, you start with three estimates—pessimistic, most likely, and optimistic estimates. Since the pessimistic and optimistic estimates are less likely to happen than the normal estimate, the normal estimate is weighted (multiplied by 4) and added to the optimistic and pessimistic estimates, and then the whole thing is divided by 6 to give an expected duration. The formula looks like this:

$$\left(\text{Optimistic duration} + 4\;\text{Most likely duration} + \text{Pessimistic duration} \right) \div 6 = \text{Expected duration}$$

Kathleen used a PERT estimate for the all of the wedding planning activities to make sure that she could get it all done in time for Rob and Rebecca's big day. They only have six months until the wedding, so all of the planning needs to be done within the month to leave enough time to actually get everything done. She wrote down the **assumptions** she made for each estimate, coming up with all of the reasons she could think of that she took into account when coming up with her estimates.

When Kathleen assumed the best-case scenario, these assumptions led her to her 9 day estimate.

$$(9 + 4(15) + 30) \div 6 = 16.5$$

An assumption is a decision that you make to account for things you don't know when you make an estimate.

Optimistic = 9 days	Most likely = 15 days	Pessimistic = 30 days
All guests RSVP early.	Half of the guests won't RSVP until the very last week; a few won't RSVP at all but will still show up.	Nobody RSVPs and many bring guests unannounced.
The couple settles on the first venue they visit.	They'll visit four or five and spend weeks negotiating with venue operators.	They'll comb the neighborhood and visit every possible place for weeks.
The printer can get the invitations done in two weeks.	They'll want to talk to a few printers and most of them will ask for at least a month.	All the printers will be booked, so we'll have to use somebody from out of town and it'll take six work weeks (30 business days).

Exercise

Here are some examples of three-point estimates. Use the formula to figure out the expected time for each of these.

1. A software team gathered estimates for all of the work they'd have to do to build the next major release of their flagship product. Last time it took them around 45 days, but they're hoping that the lessons learned from the past release could bring the time down to 30 days. However, the infrastructure team needs to upgrade their servers, and they are concerned that procurement delays could potentially extend the project out to 90 days.

Expected duration = _____   Optimistic duration = _____   Most Likely duration = _____
Pessimistic duration = _____

2. A construction team gathered estimates for all of the work they'd have to do to build a garage. In general, they can build a garage in 20 days, but rain or cooler temperatures could stretch the project out to 30 days. If, however, the forecast is correct, warm, sunny weather might bring the duration down to just 12 days.

Expected duration = _____   Optimistic duration = _____   Most Likely duration = _____
Pessimistic duration = _____

3. A project manager used data from past projects to come up with an estimate for an upcoming software system replacement project. She felt confident about a 25-day duration, but also noted that adding an extra resource could bring the schedule down to 10 days. The test team felt that the complexity of some completely new features would add additional test cases, adding a few weeks for a 40-day estimate.

Expected duration = _____   Optimistic duration = _____   Most Likely duration = _____
Pessimistic duration = _____

4. A project manager in charge of a big civil engineering project came up with an estimate for a highway re-paving project. The worst-case scenario was 82 days, but the team felt more certain based on past experience that they could get it done in 49 days. If all went well with their equipment and materials, it might be done in 33 days instead.

Expected duration = _____   Optimistic duration = _____   Most Likely duration = _____
Pessimistic duration = _____

Answers on page 324.

Create the duration estimate

You've got a list of activities, you know what resources are needed to actually do each activity, and you've got your estimation tools and techniques…now you have enough to create the estimates! That's the whole point of the **Estimate Activity Durations** process, and it's also the main output.

You don't always know exactly how long an activity will take, so you might end up using a range (like 3 weeks +/− 2 days).

Outputs

The **activity duration estimates** are estimates of how long each activity in the activity list will take. The estimate can be in hours, days, weeks…any work period is fine, and you'll use different work periods for different jobs. A small job (like booking a DJ) may just take a few hours; a bigger job (like catering—including deciding on a menu, ordering ingredients, cooking food, and serving guests on the big day) could take days.

You'll also learn more about the specific activities while you're estimating them. That's something that always happens—you have to really think through all of the aspects of a task in order to estimate it. So the other output of Estimate Activity Durations is **updates to the project documents**.

You may have guessed from the name that the activity duration estimates are always **duration estimates**, not effort estimates, so they show you calendar time and not just person-hours.

Activity duration estimates

Project document updates

The activity duration estimate consists of estimates for each activity. It's the main output of the Estimate Activity Durations process.

---

there are no
Dumb Questions

Q: When you use parametric estimation, how does the program or formula know how much to estimate?

A: When people design a system for parametric estimation, they collect a lot of data from past projects and condense it into a table or database. And then they come up with a **heuristic** (like a rule of thumb) that lets you boil your estimation down into just a few parameters that you need to enter. Most successful parametric estimation systems need a lot of time to develop.

Q: Since reserve analysis lets me use buffers, why can't I just put everything I don't know about into the reserve?

A: The idea behind reserve analysis is that there are always unknowns on any project, but you can account for these unknowns by taking your best guess at what's going to go wrong and inserting a buffer. But you can't just make an enormous reserve, because then there's no reason to ever do any estimation! The entire project becomes one big unknown, and that's not particularly useful to anyone.

Q: Wait a minute! I don't quite get the difference between a duration estimate and an effort estimate. Can you explain?

A: Duration is the amount of time that an activity takes, while effort is the total number of person-hours expended. If it takes two people six hours to carve the ice sculpture for the centerpiece of a wedding, the duration is six hours. But since 2 people worked on it for the whole time, it took 12 person-hours of effort to create!

Back to the wedding

Kathleen's really got a handle on how long things are going to take, but that's not enough to get the job done. She's still got some work to do before she's got the whole project under control.

Rob and Rebecca know where they want to get married, and they've got the place booked now.

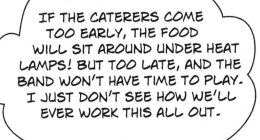

IF THE CATERERS COME TOO EARLY, THE FOOD WILL SIT AROUND UNDER HEAT LAMPS! BUT TOO LATE, AND THE BAND WON'T HAVE TIME TO PLAY. I JUST DON'T SEE HOW WE'LL EVER WORK THIS ALL OUT.

But what about the caterer? They have no idea who's going to be providing the food.

And what about the band that they want? Will the timing with their schedule work out?

It's not easy to plan for a lot of resources when they have tight time restrictions and overlapping constraints. How would you figure out a schedule that makes everything fit together?

Bringing it all together

The **Develop Schedule process** is the core of Time Management. It's the process where you put it all together—where you take everything you've done so far and combine it into one final schedule for the whole project. A lot of project managers consider this the most important part of their job. The schedule is your most important tool for managing a project.

Planning process group

There are some assets that you'll need for your schedule, like a calendar of shifts or holidays.

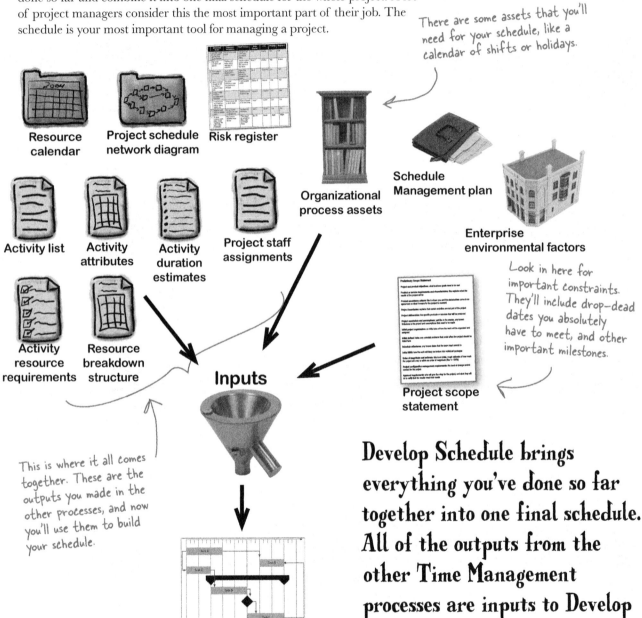

Resource calendar

Project schedule network diagram

Risk register

Organizational process assets

Schedule Management plan

Enterprise environmental factors

Activity list

Activity attributes

Activity duration estimates

Project staff assignments

Activity resource requirements

Resource breakdown structure

Inputs

Look in here for important constraints. They'll include drop-dead dates you absolutely have to meet, and other important milestones.

Project scope statement

This is where it all comes together. These are the outputs you made in the other processes, and now you'll use them to build your schedule.

Develop Schedule

Develop Schedule brings everything you've done so far together into one final schedule. All of the outputs from the other Time Management processes are inputs to Develop Schedule.

HOLD ON! I DON'T REALLY WORK LIKE THAT ALL THE TIME WHEN I'M MANAGING PROJECTS! I FIGURE SOME THINGS OUT, THEN GO BACK AND MAKE CHANGES. LIKE WHAT IF I'M WORKING ON THE SCHEDULE AND I REALIZE I NEED TO CHANGE MY RESOURCES? THIS SAYS I SHOULD HAVE FIGURED THAT ALL OUT BY NOW, RIGHT?

Each of the processes allows updates to an output from a previous one, so when you discover changes, you can include them in the schedule.

Don't worry, even though you're done with the Estimate Activity Resources process, you're not done with the resources.

You're never going to have the complete resource picture until you're done building the schedule. And the same goes for your activity list and duration estimates, too! It's only when you lay out the schedule that you'll figure out that some of your activities and durations didn't quite work.

That's why the processes have the word "Estimating" in their names! Because you're taking an educated guess, but you won't know for sure until you've actually developed the schedule.

You're not done with activity attributes yet. When you estimate resources, you'll learn more about some activities and update their attributes.

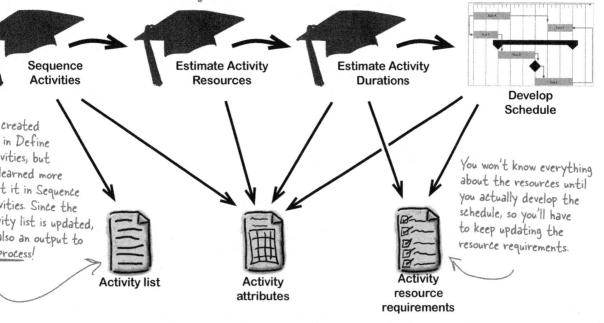

Sequence Activities

Estimate Activity Resources

Estimate Activity Durations

Develop Schedule

You created this in Define Activities, but you learned more about it in Sequence Activities. Since the activity list is updated, it's also an output to the process!

You won't know everything about the resources until you actually develop the schedule, so you'll have to keep updating the resource requirements.

Activity list

Activity attributes

Activity resource requirements

Question Clinic: The "which-comes-next" question

IF YOU WANT TO PASS THE PMP EXAM, YOU'LL NEED TO HAVE A GOOD FEEL FOR THE ORDER THAT THE PROCESSES ARE PERFORMED IN, BECAUSE YOU'LL BE ASKED A LOT OF "WHICH-COMES-NEXT" QUESTIONS! THESE ARE QUESTIONS THAT QUIZ YOU ON HOW THE PROCESSES FIT TOGETHER INTO ONE BIG FRAMEWORK. THESE QUESTIONS AREN'T HARD, BUT THEY CAN BE A LITTLE MISLEADING.

Hold on—this question doesn't look like it's asking about the order of the processes! But a lot of which-comes-next questions describe a situation and ask you what you'd do.

Don't be thrown if the question asks about an industry you don't know much about. All projects follow the same processes.

In other words, you've used decomposition and created an activity list. These are part of the Define Activities process.

27. You're the project manager for a highway construction project. You've analyzed the work that has to be done and come up with a list of activities. You consulted with the project sponsor in order to find out any important milestones that you need to meet. What's the next thing that you do?

The milestone list is an input that you've seen before.

The Develop Schedule process needs more than an activity list and resource availability.

A. Create the project schedule.

B. Perform the Define Activities process.

C. Consult your Project Management plan to figure out how to handle any schedule changes.

D. Figure out the dependencies between activities and create a diagram of the activity network.

The question described the Define Activities process, so you've already performed it.

You only do this during Control Schedule, but since there's no schedule yet, there's nothing to control.

Did you notice the question said "diagram of the activity network" and not "project network diagram"? The exam might not use the exact same phrasing as the PMBOK Guide. That's why you're learning how these things are used, not just memorizing their names.

This answer describes Sequence Activities, which happens after Define Activities and takes the activity list and milestone list as inputs. That's the right answer.

THE WHICH-COMES-NEXT QUESTION DOESN'T ALWAYS LOOK LIKE IT'S ASKING ABOUT THE ORDER OF THE PROCESSES! KEEP AN EYE OUT FOR QUESTIONS THAT DESCRIBE INPUTS, OUTPUTS, TOOLS, OR TECHNIQUES AND ASK YOU WHAT YOU'RE SUPPOSED TO DO NEXT.

HEAD LIBS

Fill in the blanks to come up with your own "which-comes-next" question! Start by thinking of a process to be the correct answer, and then figure out which process came right before it—that's the one you'll describe in the question!

You are managing a _____ . You've finished creating the
(an industry or the name of a project)

_____ , you've come up with _____
(an output from the previous process)   (another output from the previous process)

and you've just finished _____ . What's the next thing you do?
(a tool or technique from the previous process)

A. _____
(the correct answer—a brief description of what happens during the process)

B. _____
(a description of a different process)

C. _____
(the name of a tool or technique that's part of a totally different process)

D. _____
(the name of an irrelevant process)

One thing leads to another

AUNT LAURA IS A VEGETARIAN. THAT WON'T BE A PROBLEM, RIGHT?

Rob thought this was just a little problem...

Rebecca: Well, let's see. What menu did we give to the caterers?

Rob: We didn't give it to them yet, because we won't have the final menu until everyone RSVPs and lets us know which entrée they want.

Rebecca: But they can't RSVP because we haven't sent out the invitations! What's holding that up?

Rob: We're still waiting to get them back from the printer. We can't send them out if we don't have them yet!

Rebecca: Oh no! I still have to tell the printer what to print on the invitations, and what paper to use.

Rob: But you were waiting on that until we finished the guest list.

Rebecca: What a mess!

...but it turns out to be a lot bigger than either Rob or Rebecca realized at first! How'd a question about one guest's meal lead to such a huge mess?

**BRAIN POWER**

Can you think of a situation where a delay in an activity early on in a project can lead to a problem in a later activity, which leads to another problem in another activity, leading to a cascade of problems that makes the project late?

Use the <u>critical path method</u> to avoid big problems

The **critical path method** is an important tool for keeping your projects on track. Every network diagram has something called the **critical path**. It's the string of activities that, if you add up all of the durations, is longer than any other path through the network. It usually starts with the first activity in the network and usually ends with the last one.

The reason that the critical path is, well, *critical*, is that every single activity on the path must finish on time in order for the project to come in on time. A ***delay in any one of the critical path activities*** will cause the **entire project to be delayed**.

The critical path is the string of activities that will delay the whole project if any one of them is delayed.

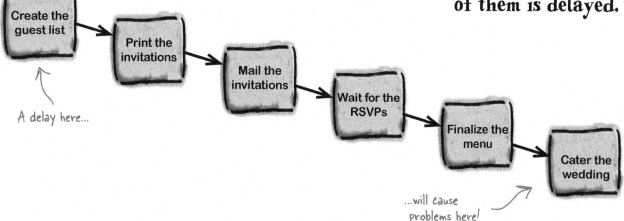

A delay here...

...will cause problems here!

How does knowing your critical path help?

Knowing where your critical path is can give you a lot of freedom. If you know an activity is *not* on the critical path, then you know a delay in that activity may not *necessarily* delay the project.

This can really help you handle emergency situations. Even better, it means that if you need to bring your project in earlier, you know that adding resources to the critical path will be much more effective than adding them elsewhere.

How to find the critical path

It's easy to find the critical path in any project! With a little practice, you'll get the hang of it. Of course, on a large project with dozens or hundreds of tasks, you'll probably use software like Microsoft Project to find the critical path for you. But when it does, it's following the same exact steps that you'll follow here.

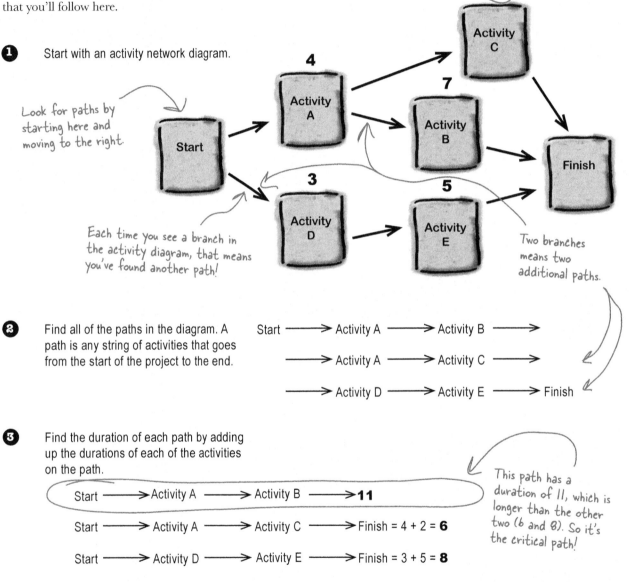

① Start with an activity network diagram.

You'll usually write the duration above each node in the diagram.

Look for paths by starting here and moving to the right.

Each time you see a branch in the activity diagram, that means you've found another path!

Two branches means two additional paths.

② Find all of the paths in the diagram. A path is any string of activities that goes from the start of the project to the end.

Start ⟶ Activity A ⟶ Activity B ⟶

⟶ Activity A ⟶ Activity C ⟶

⟶ Activity D ⟶ Activity E ⟶ Finish

③ Find the duration of each path by adding up the durations of each of the activities on the path.

Start ⟶ Activity A ⟶ Activity B ⟶ **11**

Start ⟶ Activity A ⟶ Activity C ⟶ Finish = 4 + 2 = **6**

Start ⟶ Activity D ⟶ Activity E ⟶ Finish = 3 + 5 = **8**

This path has a duration of 11, which is longer than the other two (6 and 8). So it's the critical path!

The critical path is the one with the longest duration!

You may get questions on the exam asking you to identify the critical path in a network diagram. Here's some practice for doing that! Find the critical path and duration for this PDM.

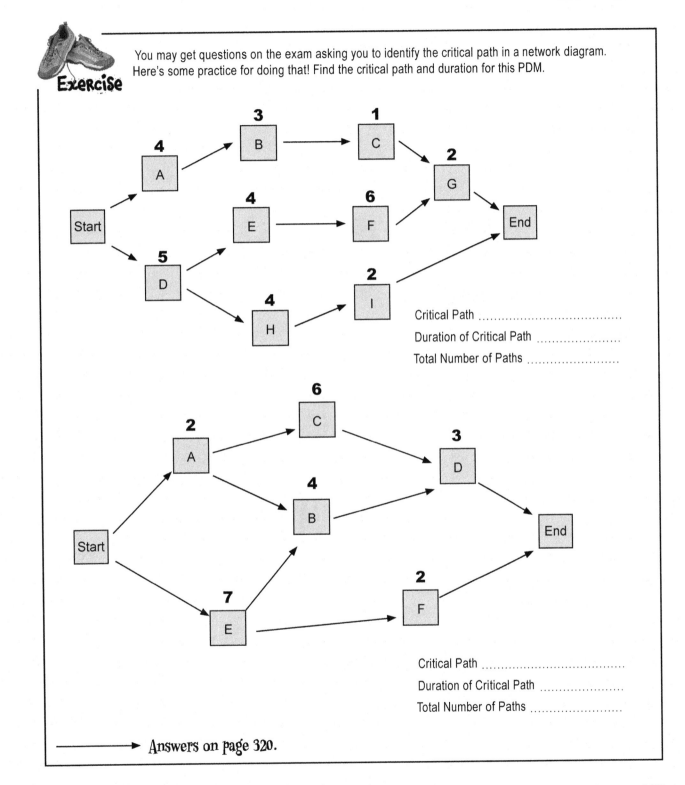

Critical Path ....

Duration of Critical Path ....

Total Number of Paths ....

Critical Path ....

Duration of Critical Path ....

Total Number of Paths ....

Answers on page 320.

Finding the float for any activity

Once you've figured out the critical path, there's all sorts of useful stuff you can do with it. One of the most useful things you can do is calculate the **float**. The float for any activity is the amount of time that it can slip before it causes your project to be delayed. You might also see the word *slack*—it's the same thing.

Luckily, it's not hard to figure out the float for any activity in a network diagram. First you write down the list of all of the paths in the diagram, and you identify the critical path. The float for every activity in the critical path is zero.

The goal is to find the float for each activity. We're not really concerned with finding a total float for each path—we're looking at the activities independently.

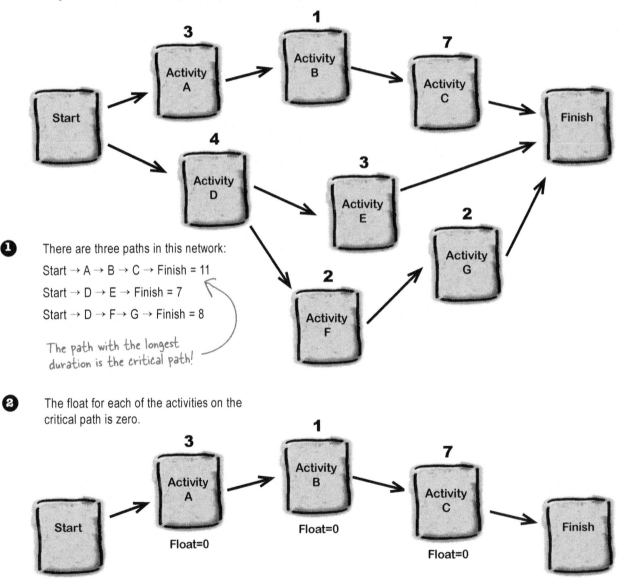

❶ There are three paths in this network:

Start → A → B → C → Finish = 11

Start → D → E → Finish = 7

Start → D → F→ G → Finish = 8

The path with the longest duration is the critical path!

❷ The float for each of the activities on the critical path is zero.

Float=0   Float=0   Float=0

❸ Find the **next longest path.** Subtract its duration from the duration of the critical path, and that's the float for each activity on it.

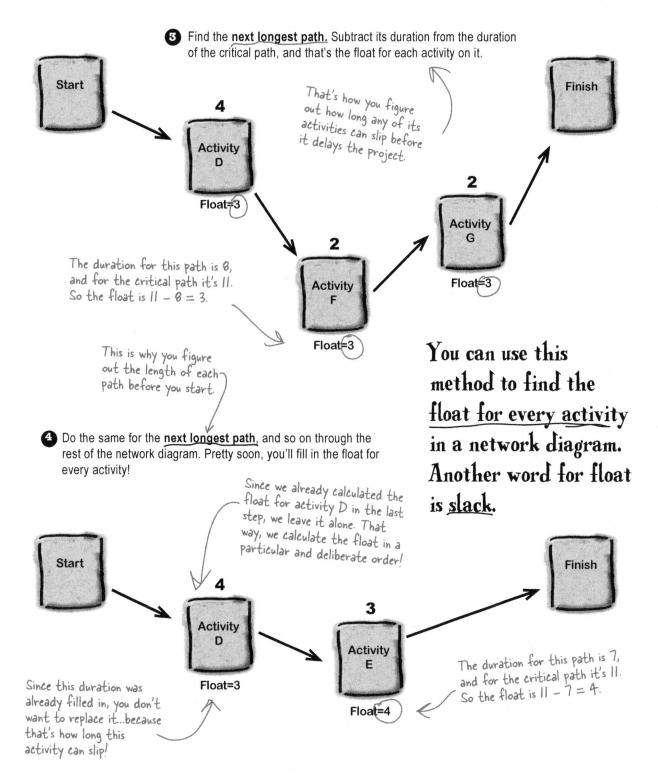

That's how you figure out how long any of its activities can slip before it delays the project.

4

Activity D

Float=3

The duration for this path is 8, and for the critical path it's 11. So the float is 11 − 8 = 3.

2

Activity F

Float=3

This is why you figure out the length of each path before you start.

2

Activity G

Float=3

Finish

Start

You can use this method to find the float for every activity in a network diagram. Another word for float is slack.

❹ Do the same for the **next longest path,** and so on through the rest of the network diagram. Pretty soon, you'll fill in the float for every activity!

Since we already calculated the float for activity D in the last step, we leave it alone. That way, we calculate the float in a particular and deliberate order!

Start

4

Activity D

Float=3

Since this duration was already filled in, you don't want to replace it...because that's how long this activity can slip!

3

Activity E

Float=4

The duration for this path is 7, and for the critical path it's 11. So the float is 11 − 7 = 4.

Finish

Float tells you how much extra time you have

Once you know the float, you know how much play you have in your schedule. If an activity has a float of 2 days, it can slip by that much without affecting the end date.

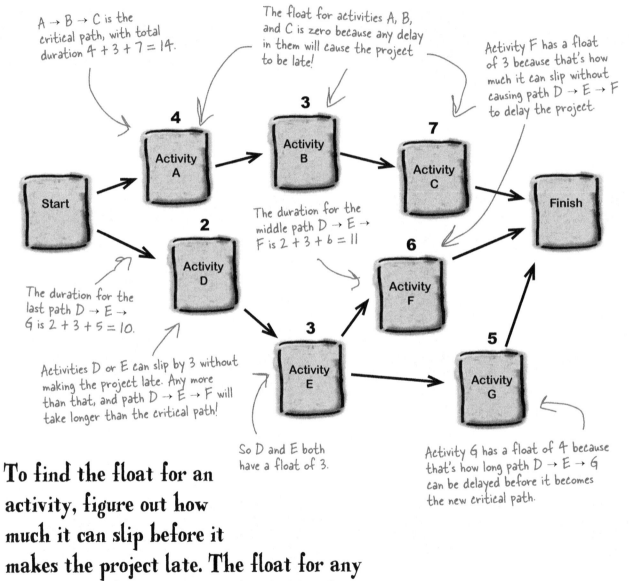

A → B → C is the critical path, with total duration 4 + 3 + 7 = 14.

The float for activities A, B, and C is zero because any delay in them will cause the project to be late!

Activity F has a float of 3 because that's how much it can slip without causing path D → E → F to delay the project.

The duration for the middle path D → E → F is 2 + 3 + 6 = 11

The duration for the last path D → E → G is 2 + 3 + 5 = 10.

Activities D or E can slip by 3 without making the project late. Any more than that, and path D → E → F will take longer than the critical path!

So D and E both have a float of 3.

Activity G has a float of 4 because that's how long path D → E → G can be delayed before it becomes the new critical path.

To find the float for an activity, figure out how much it can slip before it makes the project late. The float for any activity on the critical path is <u>ZERO!</u>

Exercise

You'll need to be able to calculate the float of an activity in a network diagram for the exam. Take another look at this PDM from the last exercise. Can you calculate the float for each activity?

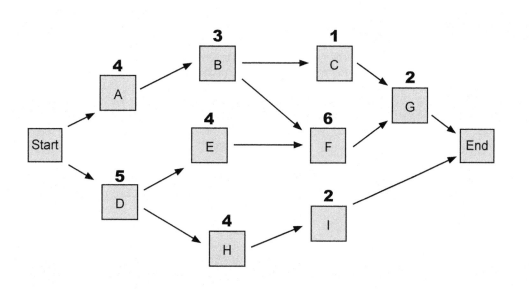

1. What is the float for each activity on the critical path? ....

2. What is the total duration for path A → B → C → G? ....

3. What is the total duration for path A → B → F → G? ....

4. What is the total duration for path D → E → F → G? ....

5. What is the total duration for path D → H → I? ....

6. Which path is the critical path? .... → .... → .... → ....

7. Write down the float for each activity:

A .... B .... C .... D .... E ....

F .... G .... H .... I ....

Hint: First fill in the float for the critical path activities. Then move on to the next-longest path, and then the next-longest one, filling in any float that hasn't been filled in yet.

⟶ Answers on page 321.

there are no
Dumb Questions

Q: Where do the duration numbers come from on each activity?

A: A lot of people ask that question. It's easy to forget that everything you do in Sequence Activities builds on the stuff you did in the other Time Management processes. Remember the estimates that you came up with Estimate Activity Durations? You used techniques like three-point estimates, analogous estimating, and parametric estimating to come up with an estimate for each activity. Those are the estimates that you use on your network diagrams!

Q: What if there's a path that's not critical, but where even a small slip in one activity would delay the project?

A: This is exactly why it's important to know the float for each of your activities. When you're managing your project, it's not enough to just pay attention to the activities on the critical path. You need to look for any activity with a low float. And don't forget that there may be some activities that aren't on the critical path but still have a float of zero! These are the ones where you really want to pay attention and watch out for potential resource problems.

o O

> I SEE—SO WHEN I CREATED THE NETWORK DIAGRAM IN SEQUENCE ACTIVITIES, I WAS BUILDING ON WHAT I DID IN DEFINE ACTIVITIES. IT ALL TIES TOGETHER!

All of the processes in Time Management tie together! When you develop your schedule, you're using the durations for your activities that you came up with in Estimate Activity Durations.

Figure out the early start and early finish

Coming up with the float for each activity is useful, but you can actually do better! When you have a long critical path, but the other paths in your network diagram are short, then you have a lot of freedom in when you can start and finish each of the activities that are not on the critical path. You can use **early start** and **early finish** to get a handle on exactly how much freedom you have in your schedule.

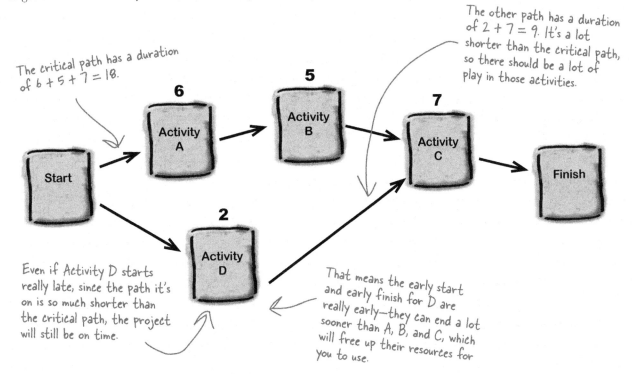

The critical path has a duration of 6 + 5 + 7 = 18.

The other path has a duration of 2 + 7 = 9. It's a lot shorter than the critical path, so there should be a lot of play in those activities.

Even if Activity D starts really late, since the path it's on is so much shorter than the critical path, the project will still be on time.

That means the early start and early finish for D are really early—they can end a lot sooner than A, B, and C, which will free up their resources for you to use.

Early start

Is the earliest time that an activity can start. An activity near the end of the path will only start early if all of the previous activities in the path also started early. If one of the previous activities in the path slips, that will push it out.

Early finish

Is the earliest time that an activity can finish. It's the date that an activity will finish if all of the previous activities started early and none of them slipped.

When you find the early start and early finish for each task, you know exactly how much freedom you have to move the start dates for those activities around without causing problems.

Figure out the latest possible start and finish

It's also important to know how late any activity can run before it delays the project. That's what **late start** and **late finish** are for! They let you figure out how late you can start a certain task and how much it can slip before it delays your project.

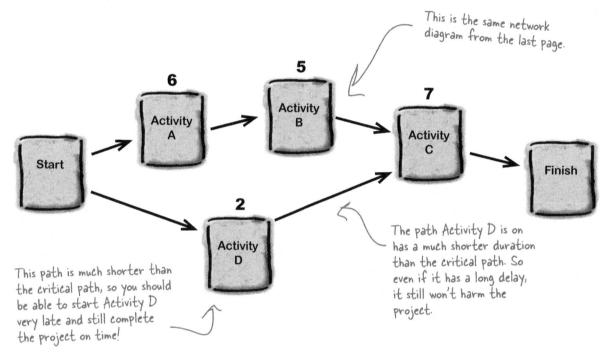

This is the same network diagram from the last page.

This path is much shorter than the critical path, so you should be able to start Activity D very late and still complete the project on time!

The path Activity D is on has a much shorter duration than the critical path. So even if it has a long delay, it still won't harm the project.

Late start

Is the latest time that an activity can start. If an activity is on a path that's much shorter than the critical path, then it can start very late without delaying the project—but those delays will add up quickly if other activities on its path also slip!

Late finish

Is the latest time that an activity can finish. If an activity is on a short path and all of the other activities on that path start and finish early, then it can finish very late without causing the project to be late.

Figuring out the late start and late finish will help you see how much "play" you have in your schedule. An activity with a large late start or late finish means you have more options.

Add early and late durations to your diagrams

You can use a method called **forward pass** to add the early start and finish to each path in your network diagram. Once you've done that, you can use **backward pass** to add the late start and finish. It makes your network diagrams look a little more complicated, but it gives you a lot of valuable information.

Early start and finish go in the upper corners. Write the name of the activity above it, and the duration and float inside the box.

You can use this special node in your network diagram to write down the early and late start and finish.

The early start for this activity is 4.

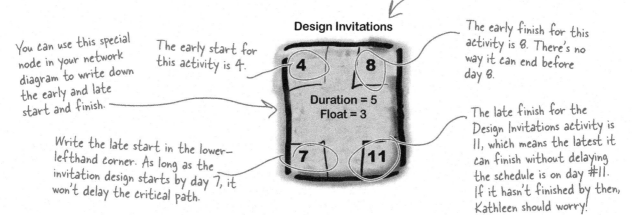

Design Invitations

The early finish for this activity is 8. There's no way it can end before day 8.

The late finish for the Design Invitations activity is 11, which means the latest it can finish without delaying the schedule is on day #11. If it hasn't finished by then, Kathleen should worry!

Write the late start in the lower-lefthand corner. As long as the invitation design starts by day 7, it won't delay the critical path.

Take a forward pass through the network diagram.
Start at the beginning of the critical path and move forward through each activity. Follow these three steps to figure out the early start and early finish!

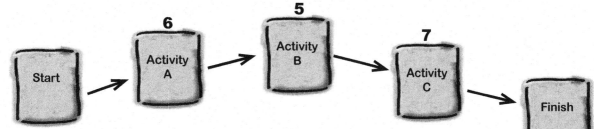

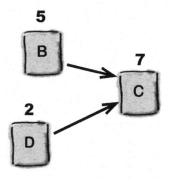

1. The ES (early start) of the first activity in the path is 1. The EF (early finish) of any task is its ES plus its duration minus one. So start with Activity A. It's the first in the path, so ES = 1, and EF = 1 + 6 – 1 = 6.

2. Now move forward to the next activity in the path, which is Activity B in this diagram. To figure out ES, take the EF of the previous task and add one. So for Activity B, you can calculate ES = 6 + 1 = 7, and EF = 7 + 5 – 1 = 11.

3. **Uh-oh! Activity C has two predecessors.** Which one do you use to calculate EF? Since C can't start until both B and D are done, use **the one with the latest EF**. That means you need to figure out the EF of Activity D (its ES is 1, so its EF is 1 + 2 – 1 = 2). Now you can move forward to Activity C and calculate its EF. The EF of Activity D is 2, which is smaller than B's EF of 11, so for Activity C the ES = 11 + 1 = 12, and EF = 12 + 7 – 1 = 18.

Take a backward pass to find late start and finish

You can use a **backward pass** through the same network diagram to figure out the late finish and start for each activity.

The backward pass is just as easy as the forward pass. Start at the end of the path you just took a pass through and work your way backward to figure out the late start and finish.

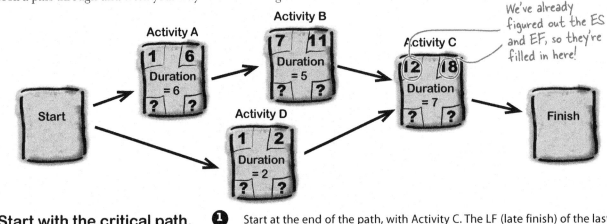

Start with the critical path.

You're calculating the latest any activity can start and finish, so it makes sense that you need to start at the end of the project and work backward—and the last activity on the critical path is always the last one in the project. Then do these three steps, working backward to the next-longest path, then the next-longest, and so on, until you've filled in the LS and LF for all of the activities. Fill in the LF and LS for the activities on each path, but **don't replace** any LF or LS you've already calculated.

1 Start at the end of the path, with Activity C. The LF (late finish) of the last activity is the same as the EF. Calculate its LS (late start) by subtracting its duration from the LF and adding one. LS = 18 – 7 + 1 = 12.

2 Now move backward to the previous activity in the path—in this case, Activity B. Its LF is the LS of Activity C minus one, so LF = 12 – 1 = 11. Calculate its LS in the same way as step 1: LS = 11 – 5 + 1 = 7.

3 Now do the same for Activity A. LF is the LS for Activity B minus one, so LF = 7 – 1 = 6. And LS is LF minus duration plus one, so LF = 6 – 6 + 1 = 1.

4 Now you can move onto the next-longest path, Start-D-C-Finish. If there were more paths, you'd then move on to the next-longest one, and so on, filling in LF and LS for any nodes that **haven't already been filled in**.

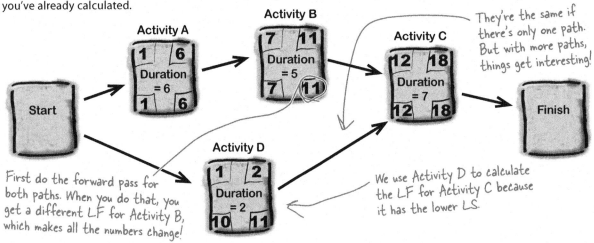

Let's take some time out to walk through this!

All of this critical path stuff seems pretty serious, right? It's one of the toughest concepts on the exam. But don't sweat it, because it's actually not hard! It just takes a little practice. Once you do it yourself, you'll see that there's really nothing to worry about.

Calculating the ES, EF, LS, and LF may seem complicated, but it only takes a little practice to get the hang of it. Once you walk through it step by step, you'll see that it's actually pretty easy!

Sharpen your pencil

There are four paths in this network diagram. Fill in each of the activity names and durations for each of the paths.

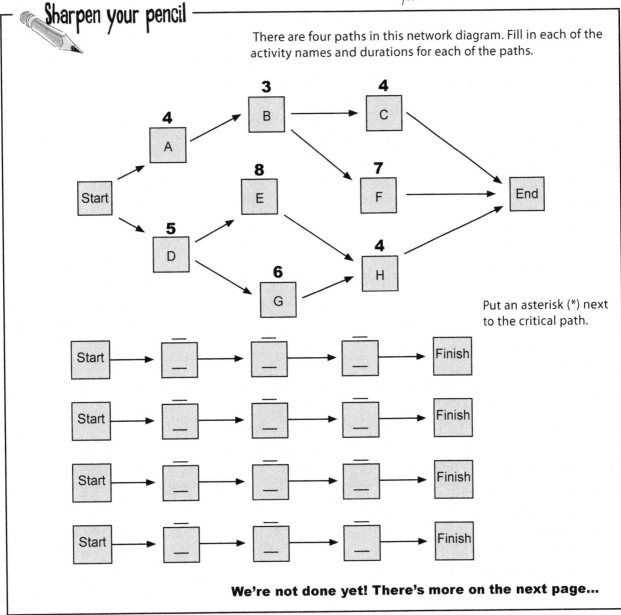

Put an asterisk (*) next to the critical path.

We're not done yet! There's more on the next page...

Sharpen your pencil

Take a forward pass through each of the four paths in the diagram and fill in the early starts and early finishes for each activity. Start with the first one.

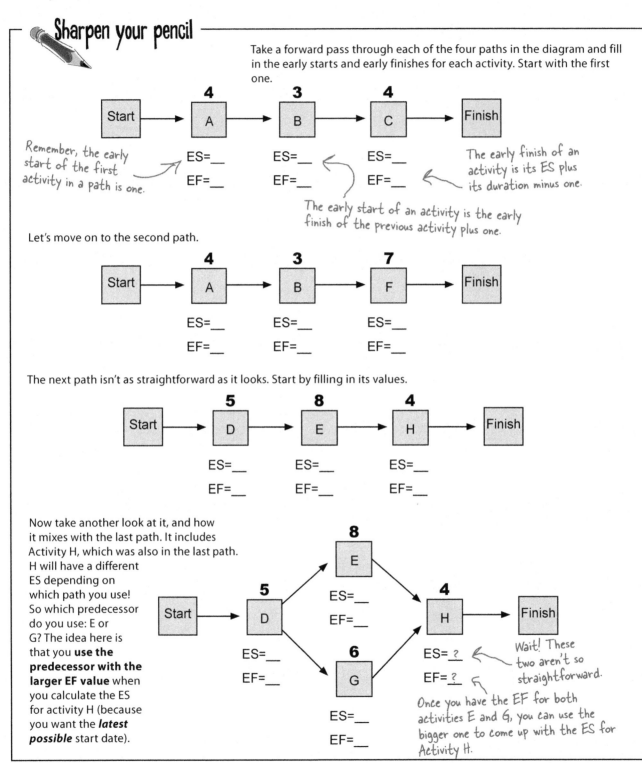

Remember, the early start of the first activity in a path is one.

The early start of an activity is the early finish of the previous activity plus one.

The early finish of an activity is its ES plus its duration minus one.

Let's move on to the second path.

The next path isn't as straightforward as it looks. Start by filling in its values.

Now take another look at it, and how it mixes with the last path. It includes Activity H, which was also in the last path. H will have a different ES depending on which path you use! So which predecessor do you use: E or G? The idea here is that you **use the predecessor with the larger EF value** when you calculate the ES for activity H (because you want the *latest possible* start date).

Wait! These two aren't so straightforward.

Once you have the EF for both activities E and G, you can use the bigger one to come up with the ES for Activity H.

You've calculated the ES for each activity. Use that information and take a backward pass through the paths, starting with the first two paths.

First start with the critical path. Take the EF of the last activity in the critical path and use it as the LF for the last activity in *every* path. If you take a minute to think about it, it makes sense to do that. The point of LF is to figure out the absolute latest that the activity can end without making the project late. And as long as every non-critical-path activity ends before the last activity in the critical path, then it won't be late.

We'll start by giving you the LF of critical path, Start-D-E-H-Finish, which is 17.

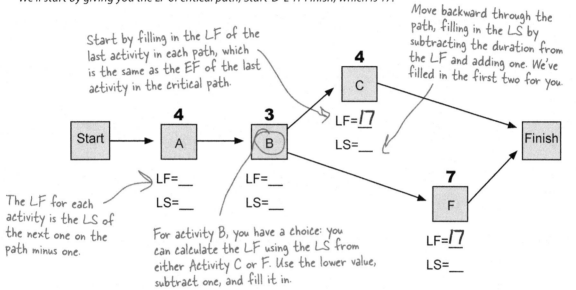

Start by filling in the LF of the last activity in each path, which is the same as the EF of the last activity in the critical path.

Move backward through the path, filling in the LS by subtracting the duration from the LF and adding one. We've filled in the first two for you.

The LF for each activity is the LS of the next one on the path minus one.

For activity B, you have a choice: you can calculate the LF using the LS from either Activity C or F. Use the lower value, subtract one, and fill it in.

Finish up by calculating the LS and LF for the last two paths!

Activities B and D have two possible choices for which LS to use for the calculation. For Activity B, do you use the LS of C or the LS of F? And for Activity D, do you use Activity E or G? The answer is that you always **use the lowest value of LS to calculate the LF**. The reason is that you're trying to find the latest possible start date that *won't make the project late*. If you use an activity with a later LS, and the activity really is delayed by that much, then it'll cause a delay in both following activities. And that will make the one with the lower LS start too late.

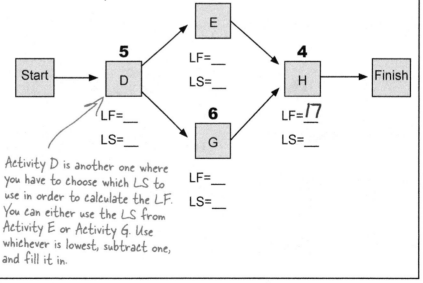

Activity D is another one where you have to choose which LS to use in order to calculate the LF. You can either use the LS from Activity E or Activity G. Use whichever is lowest, subtract one, and fill it in.

Sharpen your pencil Solution

If you got a few of these wrong, don't worry. It's easy to miss one calculation, and that leads to a problem on the whole path.

For the exam, you'll only have to do one or two of these calculations, not a whole string of them like this. You'll definitely be able to handle the exam questions now!

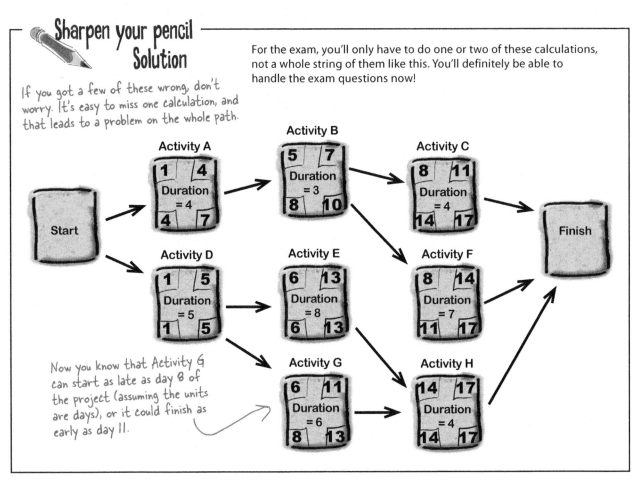

Now you know that Activity G can start as late as day 8 of the project (assuming the units are days), or it could finish as early as day 11.

WAIT A MINUTE...I'VE NEVER HAD TO DO THIS FOR MY PROJECTS AT WORK! I'VE GOT PROJECTS WITH DOZENS OF ACTIVITIES, AND THIS WOULD TAKE ALL DAY!

You won't have to do this kind of thing on the job...that's what computers are for!

Project management software like Microsoft Project will do these calculations for you. But you need to know how to do it yourself, because when the computer is doing critical path analysis, this is exactly how it figures it out!

there are no
Dumb Questions

Q: Would I really use this critical path stuff in real life, or is it just something I need to memorize for the PMP exam?

A: Yes, critical path analysis really is important in real life! Sure, for a small project with a dozen or so activities, it's pretty easy to figure out which activities are critical and which can slip by a little bit. But what happens if you've got a project with dozens of team members and hundreds of activities? That's where critical path analysis can come in very handy. For a project like that, you'd probably be using project management software rather than calculating the critical path yourself, and the software will be able to highlight that path for you. Pay special attention to all of the activities that are on the critical path—those are the ones that could potentially delay the project.

Q: What about the other numbers? How do I use float?

A: Float is a very powerful planning tool that you can use to figure out how well your project is going, and to predict where your trouble spots might be. Any activity with a low or zero float absolutely must come in on time, while the people performing an activity with a larger float have more freedom to slip without delaying the project. So you might want to assign your "superstar" resources to the low-float activities, and those people who need a little more mentoring to the ones with higher float.

Q: OK, but what about late start, early finish, and those other numbers? Do those do me any good?

A: Early and late start and finish numbers are also very useful. How many times have you been in a situation where you've been asked, "If we absolutely had to have this in two months, can we do it?" Or, "How late can this project realistically be?" Now you can use these numbers to give you real answers, with actual evidence to back them up.

Here's an example. Let's say you've got an activity in the middle of your project, and one of your team members wants to plan a vacation right at the time that the activity will start. Do you need to find someone to fill in for him? If he'll be back before the late start date, then your project won't be late! But that comes at a cost—you'll have used up the extra slack in the schedule.

Q: I can see how the critical path is useful on its own, but what does it have to do with the rest of Time Management?

A: If you start putting together your schedule but the activities are in the wrong order, that's really going to cause serious problems...and sometimes doing critical path analysis is the only way you'll really figure out that you've made that particular mistake. That's why you need to pay a lot of attention to the Sequence Activities tools and techniques. If you've come up with an inefficient or inaccurate sequence, with too many or incorrect predecessors and dependencies, then your entire critical path analysis will be useless.

BULLET POINTS: AIMING FOR THE EXAM

- The **critical path** is the path that has the longest duration.

- You should be able to figure out the number of paths in a **project network diagram**, and the duration of each path.

- The **float** for an activity is the amount that its duration can slip without causing the project to be delayed. The float for any activity on the critical path is zero.

- You'll need to know how to calculate the **early start**, **late start**, **early finish**, and **late finish** for an activity in a network diagram using the forward pass and backward pass. This is the core of critical path analysis.

- You may see a **PDM** (or **activity-on-node**) diagram with special nodes that have extra boxes in the corners for the ES, EF, LF, and LS.

- Don't forget that when two paths intersect, you have to decide which ES or LF value to take for the calculation in the next node. For the **forward pass**, use the larger value; for the **backward pass**, use the smaller one.

Crash the schedule

There are two important **schedule compression** techniques that you can use to bring in your project's milestone dates…but each has its own cost. When you absolutely have to meet the date and you are running behind, you can sometimes find ways to do activities more quickly by adding more resources to critical path tasks. That's called **crashing**.

> A LOT MORE PEOPLE RSVP'D THAN WE EXPECTED!

> NO PROBLEM! WE'LL JUST ADD MORE COOKS AND WAITERS TO SERVE MORE PEOPLE.

Hmm. That'll cost more...

Crashing the schedule means adding resources or moving them around to shorten it. Crashing ALWAYS costs more and doesn't always work!

> YEAH, BUT WHAT IF YOU'RE AT THE BUDGET LIMIT AND CAN'T AFFORD EXTRA RESOURCES?

Then you can't crash the schedule.

There's no way to crash a schedule without raising the overall cost of the project. So, if the budget is fixed and you don't have any extra money to spend, you can't use this technique.

Fast-tracking the project

Another schedule compression technique is called **fast-tracking**. Sometimes you've got two activities planned to occur in sequence, but you can actually do them at the same time. On a software project, you might do both your UAT testing and your functional testing at the same time, for example. This is pretty risky, though. There's a good chance you will need to redo some of the work you have done concurrently.

On the exam, if you see something about "overlapping activities," it's talking about fast-tracking.

WE CAN SAVE TIME BY HAVING THE FLORIST WORK ON THE RECEPTION HALL FLOWERS WHILE WE FIGURE OUT THE REST OF THE DECORATIONS.

If the decorations don't match the flowers well enough, we'll have to do some rework.

Crashing and fast-tracking are SCHEDULE COMPRESSION tools.

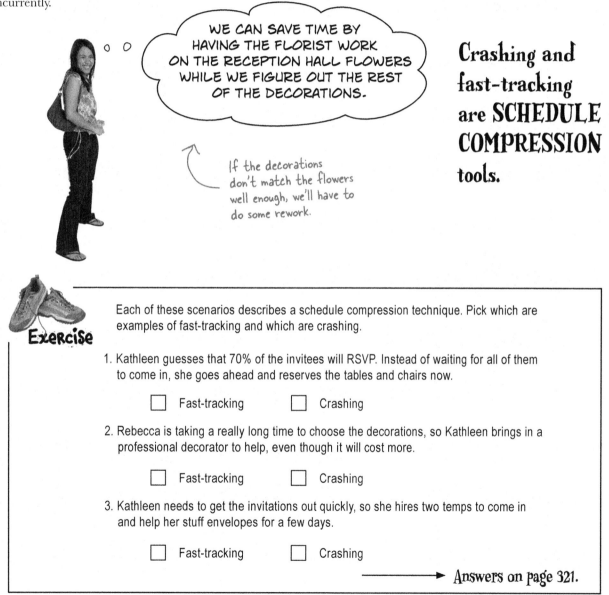

Exercise

Each of these scenarios describes a schedule compression technique. Pick which are examples of fast-tracking and which are crashing.

1. Kathleen guesses that 70% of the invitees will RSVP. Instead of waiting for all of them to come in, she goes ahead and reserves the tables and chairs now.

    ☐ Fast-tracking   ☐ Crashing

2. Rebecca is taking a really long time to choose the decorations, so Kathleen brings in a professional decorator to help, even though it will cost more.

    ☐ Fast-tracking   ☐ Crashing

3. Kathleen needs to get the invitations out quickly, so she hires two temps to come in and help her stuff envelopes for a few days.

    ☐ Fast-tracking   ☐ Crashing

⟶ Answers on page 321.

Modeling techniques

It's always a good idea to think about all of the things that could go wrong on your project in advance. Trying to think through all of the possible problems your project could run into is called **what-if analysis.**

- What if the limo breaks down?

- What if the florist cancels at the last minute?

- What if the dress doesn't fit?

- What if the band gets sick?

- What if the guests get food poisoning?

- What if there's a typo in the church address on the invitation?

- What if the bridesmaids don't show up?

- What if the cake tastes horrible?

- What if we lose the rings?

That way, you can figure out how to deal with any problems that might come your way. Sometimes there's no way to still meet your dates and deal with these scenarios. But it always makes sense to try to understand the impact they will have on your schedule.

Simulation

This is a specific kind of what-if analysis where you model uncertainty using a computer. There are some packages that will help to calculate risk using random numbers and **Monte Carlo analysis** algorithms. While this is not a commonly used technique, there might be a question or two about it on the PMP exam, and you should know what it is.

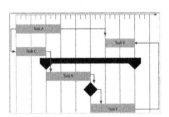

Scheduling tool

Using a project management software package to create a model of the schedule and adjusting various elements to see what might happen is another technique for analyzing network diagrams.

Other Develop Schedule tools and techniques

There are just a few more tools and techniques in the
Develop Schedule process that you should know.

Critical chain method

In this method, resource dependencies are used to determine
the critical path. Then, you add buffers, working backward
from the delivery date into the schedule at strategic points, and
managing the project so that each milestone is hit on time.

Resource optimization techniques

Sometimes only one resource can do a given activity. If that
resource is busy doing another activity on the critical path, the
path itself needs to change to include that dependency. That's
the point of resource leveling. It evaluates all of the resources to
see if the critical path needs to change to accommodate resource
assignments.

Adjusting leads and lags

If you made any mistakes in your leads and your
lags, you might be able to adjust them to change
the projected end date.

And don't forget...

Schedule compression and schedule network analysis

The last two tools and techniques in the Develop Schedule process are the
ones you just learned over the last few pages: **schedule compression**
and **schedule network analysis** using critical path, float, and the other
schedule analysis techniques you just learned.

Outputs of Develop Schedule

Of course, the main product of Develop Schedule is the schedule. But there are a few other supporting documents that help you understand how the work will get done as well.

Outputs

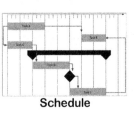

Schedule

Project schedule

All of that analysis and modeling should produce a schedule that everyone can get behind. After thinking your way through everything that can go wrong and assigning resources, you should have a pretty accurate prediction of the work required to complete the project.

The reason you go through all of that what—if analysis is to make sure everybody agrees that this schedule is achievable!

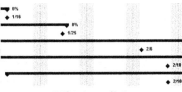

Milestone list

Milestone chart

Technically, the milestone chart is a part of the project schedule (so they're really all part of the same output), but it's very important. All of the major chunks of work can be marked with milestones to track their completion. Usually this list is published for other teams or stakeholders who are depending on parts of the work to be accomplished by a certain time.

The main thing to remember here is that milestones are usually used to track stuff that will be used by people outside the project.

Schedule data

Schedule data

The schedule data is a collection of information about your schedule. It will include things that you'll need to analyze your schedule later on in the project: alternative schedules, specific requirements for resources, milestone charts, bar charts, project schedule network diagrams, and other data and metrics about your schedule.

Before you can do change control, you need requested changes. Once the change is approved, you can update the baseline!

The Schedule Baseline

This is just like the scope baseline in the last chapter.

Schedule baseline

When the Develop Schedule process is complete, a baseline is created so that actual progress can be compared to the plan.

Project calendars

Calendars will help you keep track of the time when team members will be away on vacation or unavailable to work on your project.

Project Management plan updates

Since the schedule baseline and Schedule Management plan are both part of the Project Management plan, it makes sense that it would have to be updated.

Project documents updates

While you're creating the schedule, you might find that you need to update calendars, your resource requirements, the attributes of the activities themselves, or even your risk register, to name a few possibilities.

For the exam, you need to know Develop Schedule outputs. Several outputs from the wedding's Develop Schedule process are on the left. Match them to the correct description on the right.

A. Project schedule

1. Kathleen gives a list of dates to the caterer telling him when he will need to have his menu plans, and when the shopping for the ingredients will need to be complete for the reception and rehearsal dinner.

B. Schedule data

2. Kathleen realizes that she needs to make a change to how she keeps track of the waiters' time, so she makes a change to the document that describes it.

C. Schedule baseline

3. While making the schedule, Kathleen realizes that the catering company can't work from 3-4 on the day of the event because they'll be travelling from another event.

D. Project calendar update

4. Kathleen makes a copy of the schedule when it's done so that she can compare how she is doing to the original plan.

E. Project document updates

5. There's a big poster on the wall where Kathleen keeps track of who does what, and when.

Answers: A–5; B –1; C– 4; D–3; E–2

there are no
Dumb Questions

Q: Don't we need to go through change control before we update the resource requirements or the activity attributes?

A: No. You need to go through change control if you are requesting changes to, say, your Cost Management plan. But while you are working on creating your schedule, everything you have created as part of the Time Management knowledge area is fair game.

As you work your way through your network diagram and figure out new dependencies, you will find that you need more resources for some items or that the activity itself has changed. That's why this process gives you the freedom to refine your earlier idea and make all of the Time Management documents sync with your new understanding.

The Develop Schedule process is about taking all of the information you can think of up front and putting it into a schedule that is realistic. When you are done with this process, you should have a really good idea of what you are going to do, who will do it, and how long it will take.

Q: We always want to do our projects as quickly as we can. Why don't we always fast-track and crash our schedules?

A: Because crashing is expensive and fast-tracking is risky. While it may look good on paper to add a lot of resources to a project that is running late, it often adds so much management overhead and training issues that the project just comes in later.

Even though it might seem like some predecessors are really unnecessary, you usually planned them for a reason. So when you break your dependencies to fast-track your project, you can significantly compromise the quality of the work that gets done. That means you might have to redo it altogether—which would probably take a lot of time.

While fast-tracking and crashing might work sometimes, they always add both risk and cost to your project.

Q: Do people really do Monte Carlo analysis to figure out their schedules? I have never heard of that before.

A: It's true that most people don't use this technique to figure out what might go wrong on their projects, so don't feel bad if you've never heard of it. Some people think that this is just one of those things that is on the PMP exam, so you have to know what it is. But there really are some project managers who use it and get great results!

Q: The critical chain method sounds complicated. Do I need to know how to do it?

A: Not really. You need to know that it is a technique for developing schedules that takes resource assignment into account early on.

When project managers use the critical chain method, they identify strategic points to put buffers in their schedule and then manage the size of the buffers so that each milestone in the schedule is met.

Don't worry, you won't be asked to create a schedule using this technique. You just need to know the definition.

Updates refine the outputs of previous processes so you don't have to go back and redo them.

Influence the factors that cause change

You might get a question on the PMP exam that asks you about this.

Kathleen doesn't just sit around and wait for schedule changes to happen...

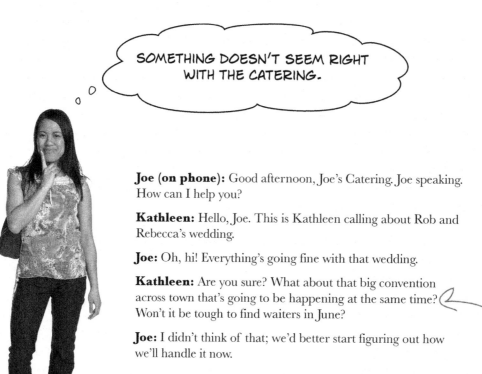

> SOMETHING DOESN'T SEEM RIGHT WITH THE CATERING.

Joe (on phone): Good afternoon, Joe's Catering. Joe speaking. How can I help you?

Kathleen: Hello, Joe. This is Kathleen calling about Rob and Rebecca's wedding.

Joe: Oh, hi! Everything's going fine with that wedding.

Kathleen: Are you sure? What about that big convention across town that's going to be happening at the same time? Won't it be tough to find waiters in June?

Joe: I didn't think of that; we'd better start figuring out how we'll handle it now.

By realizing that the convention across town will need waiters, too, Kathleen prevents a lot of changes <u>before</u> they cause schedule problems!

The project manager doesn't just wait for change to happen! She finds the things that cause change and influences them.

Control Schedule inputs and outputs

As the project work is happening, you can always discover new information that makes you re-evaluate your plan, and use the **Control Schedule process** to make the changes. The inputs to Control Schedule cover the various ways you can discover that information. The outputs are the changes themselves.

Monitoring & Controlling process group

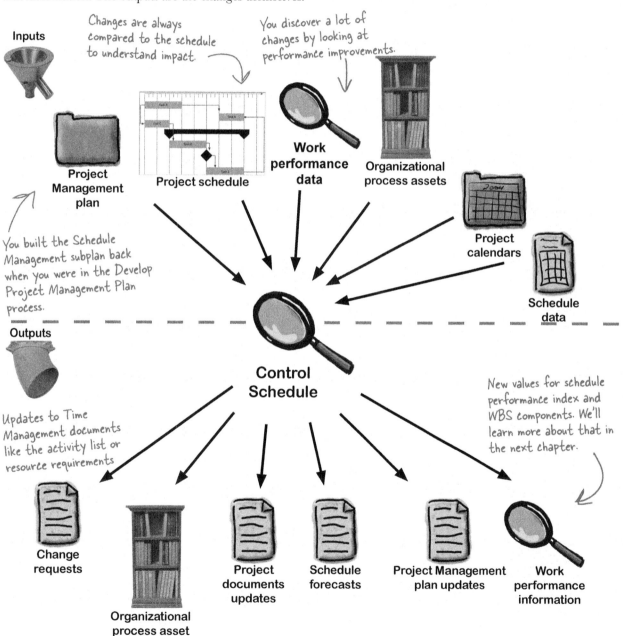

Inputs

Changes are always compared to the schedule to understand impact.

You discover a lot of changes by looking at performance improvements.

Project Management plan

You built the Schedule Management subplan back when you were in the Develop Project Management Plan process.

Project schedule

Work performance data

Organizational process assets

Project calendars

Schedule data

Outputs

Updates to Time Management documents like the activity list or resource requirements

Control Schedule

New values for schedule performance index and WBS components. We'll learn more about that in the next chapter.

Change requests

Organizational process asset updates

Project documents updates

Schedule forecasts

Project Management plan updates

Work performance information

What Control Schedule updates

All of the stuff you made during the Develop Schedule process gets updated using the Control Schedule process. Here's a closer look at what those updates mean.

The Schedule Baseline

Schedule baseline

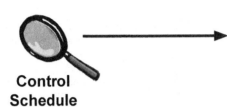

Control Schedule

Updates

Whenever a change is approved to your schedule, the baseline needs to be updated. That way, you will always be comparing your results to the right plan.

Organizational process assets

As you make changes to your project schedule, you should be tracking your lessons learned so that other projects can benefit from your experience. Sometimes you might find changes to templates that will help future projects, too.

Activity list   **Activity attributes**

If the work you need to do changes, then you need to update your activity list and attributes to match the new information.

Schedule data

Some scenarios for what might go wrong on your project might show up when you are already doing the work. You need to update your schedule accordingly.

Project Management plan

It could happen that the way you manage Control Schedule needs to change, and those changes would need to be updated in the **Project Management plan**.

Managing schedule change means keeping all of your schedule documents up to date.

Measuring and reporting performance

Most often, you identify changes by looking at performance data. It's just as important once you make a change to gather performance data as it was when you found the change in the first place. Here's how performance data feeds into the Control Schedule process.

The Control Schedule process turns Work Performance Data into Work Performance Information.

Work Performance Data

Control Schedule

Work Performance Information

Routine performance measurements might show that you are lagging in your schedule. We'll learn more about how to measure this stuff in the next chapter.

Putting the recommended changes through Control Schedule will help you to evaluate the impact and update all the necessary documents.

During Control Schedule, you'll use the routine measurements you've taken on your schedule to come up with better understand of how your project is doing in relation to it's baseline schedule. You might come up with new forecasts, or other information that will help your stakeholders know how you're doing. Those new forecasts and status indicators are called work performance information.

Now that you've been through the Control Schedule process, the Work Performance Data has been evaluated and turned into Work Performance Information.

Control Schedule tools and techniques

The tools and techniques for Control Schedule are all about figuring out where your project schedule stands. By comparing your actual project to the schedule you laid out in the baseline and looking at how people are performing, you can figure out how to handle every schedule change.

Performance reviews

There are two important calculations called schedule variance (SV) and schedule performance index (SPI) that give you valuable information about how your project is doing. You'll learn all about them in the next chapter.

Project management software

This is software like Microsoft Project that helps you organize and analyze all of the information you need to evaluate the schedule of any project.

Leads and lags, modeling techniques, schedule compression, scheduling tool

Most of the tools from the last process apply to this one too. As you find variances in the schedule, you need to figure out the impact of those issues and change your schedule to account for the new information.

Resource optimization techniques

As things change in your project, you need to make sure that resources are covering all of the activities in your plan. That means you need to distribute resources so that the work that needs to get done always has a resource available to do it.

A lot of scheduling software contains logic to do this automatically.

> **HOLD IT! ALMOST ALL THE SCHEDULING IN MY JOB IS DONE WITH GANTT CHARTS, BUT I BARELY SEE THEM HERE. WHAT GIVES?**

Remember, Gantt charts—the bar charts you make with MS Project—are just one tool for scheduling. You may use them a lot in your day-to-day work, but they're only one piece of Time Management. And remember, on the exam they're called **bar charts**, not Gantt charts!

there are no
Dumb Questions

Q: When I create work performance information, who uses them?

A: The work performance information that you create are used by a lot of people. The team uses them to keep an eye on the project. If there's a schedule problem coming up, it alerts the team so that they can help you figure out how to avoid it.

Performance information is also used by your project's sponsor and stakeholders, who are very interested in whether or not your project is on track. That information gives them a good picture of how the project is doing...and that's especially important in Control Schedule, because most change control systems require that every change is approved by a change control board that includes sponsor and stakeholders.

Q: What's schedule data used for?

A: You use the schedule data to build the schedule, and you'll usually generate and analyze it using a schedule tool (like Microsoft Project). It includes detailed information about things like resource requirements, alternate best-case and worst-case schedules, and contingency reserves.

When you put together your schedule, you should look at all of these things in order to create an accurate plan. The more information you have when you're building your schedule, the more likely it is that you'll catch those little problems that add up to big schedule slips.

Q: One of the tools is project management software. Do I need to know how to use software in order to pass the exam?

A: No. The PMP exam does not require that you know how to use software like Microsoft Project. However, if you spend a lot of time using project management software, then you probably have become very familiar with a lot of the Time Management concepts. It's a good way to learn the basics of time management.

Q: How often am I supposed to update the project calendar?

A: The project calendar shows you the working days for your team, holidays, nonworking days, planned training, and the dates that could affect your project. Luckily, in most companies these dates don't change very often. You probably won't need to update it—and most project managers just use their company's existing project calendar.

When you're doing Develop Schedule, you may discover that you need to make a change to the project calendar. That's why updates to the project calendar are an output of Develop Schedule.

Q: What do I do with work performance data and work performance information once I've collected it?

A: When you're planning your project, you'll often look to your company's past projects to see what went well and what could have been planned better. And where do you look? That information is in the organizational process assets. So where do you think that information comes from? It comes from project managers like you who added their work performance data and information to the company's Organizational Process Asset library.

Any time you generate data about your project, you should add it to your organizational process assets so you can use it for future projects.

Control Schedule Magnets

You'll see change control over and over again—every single knowledge area has its own change control process! Luckily, you'll start to see how similar they all are. But Control Schedule has its own quirks, and they're important for understanding Time Management.

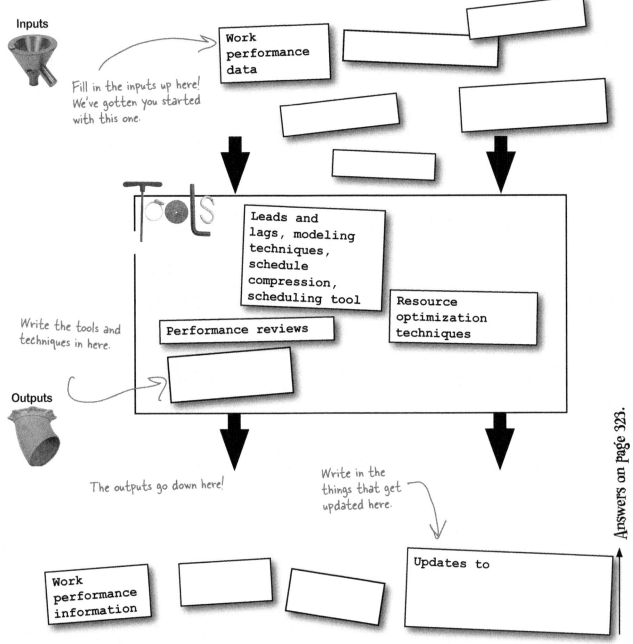

Inputs

Work performance data

Fill in the inputs up here!
We've gotten you started
with this one.

Tools

Leads and lags, modeling techniques, schedule compression, scheduling tool

Resource optimization techniques

Performance reviews

Write the tools and techniques in here.

Outputs

The outputs go down here!

Write in the things that get updated here.

Updates to

Work performance information

Answers on page 323.

Timecross

Take some time to sit back and give your right brain something to do. It's your standard crossword; all of the solution words are from this chapter.

Across

1. Taking work packages from the WBS and breaking them down into activities.

5. The "P" in PDM.

7. The kind of analysis where you ask a lot of questions about possibilities.

8. A snapshot of the schedule that you can use for later comparison.

10. _____ estimation means plugging data about your project into a database of historical information to get an estimate of how long it will take to do the work.

12. Giving a successor task some time to start before a predecessor finishes.

13. Adding more resources to a project so you can get it done faster is called _____ the schedule.

14. A PERT three-point estimate is optimistic time + 4 x most likely time + _____ time.

Down

2. Any delay in an activity on the _____ path will delay the entire project.

3. You do this sort of planning when you get more information as the project progresses.

4. What you're doing to resources when you evaluate all of them to see if the critical path needs to change to accommodate their restrictions.

6. An activity with a dependency on something outside the project has an _____ predecessor.

9. _____ Activities is the process where you put the activities in order.

11. How long an activity can slip before the whole project is delayed.

———————▶ Answers on page 324

Another satisfied customer!

Rob and Rebecca had a beautiful wedding! Everything was perfect. The guests were served their meals, the band was just right, and everyone had a blast…

…and Kathleen got lots of referrals!

OH, THAT ROCKS. FOUR MORE WEDDINGS TO PLAN RIGHT AWAY!

Sharpen your pencil
Solution

You'll need to understand the different Estimate Activity Resources tools and techniques for the exam. Look at each of these scenarios and write down which of the five activity resource estimation tools and techniques is being used.

1. Kathleen has to figure out what to do for the music at Rob and Rebecca's wedding. She considers using a DJ, a rock band, or a string quartet. **Alternatives analysis**

2. The latest issue of *Wedding Planner's Journal* has an article on working with caterers. It includes a table that shows how many waiters work with various guest-list sizes. **Published estimating data**

3. There's a national wedding consultant who specializes in Caribbean-themed weddings. Kathleen gets in touch with her to ask about menu options. **Expert judgment**

4. Kathleen downloads and fills out a specialized spreadsheet that a project manager developed to help with wedding planning. **Project management software**

5. There's so much work that has to be done to set up the reception hall that Kathleen has to break it down into five different activities in order to assign jobs. **Bottom-up estimating**

6. Kathleen asks Rob and Rebecca to visit several different caterers and sample various potential items for the menu. **Alternatives analysis**

7. Kathleen calls up her friend who knows specifics of the various venues in their area for advice on which one would work best. **Expert judgment**

Exercise Solution

Each of these scenarios describes a different tool or technique from Estimate Activity Durations. Write down which tool or technique is being described.

1. Kathleen comes up with three estimates (one where everything goes wrong, one where some things go wrong, and one where nothing goes wrong) for printing invitations, and averages them together to come up with a final number. **Three-point estimate**

2. There will be two different catering companies at the wedding. Kathleen asks the head chef at each of them to give her an estimate of how long it will take to do the job. **Expert judgment**

3. There's a spreadsheet Kathleen always uses to figure out how long it takes guests to RSVP. She enters the number of guests and their ZIP codes, and it calculates an estimate for her. **Parametric estimating**

4. Kathleen's done four weddings that are very similar to Rob and Rebecca's, and in all four of them it took exactly the same amount of time for the caterers to set up the reception hall. **Analogous estimating**

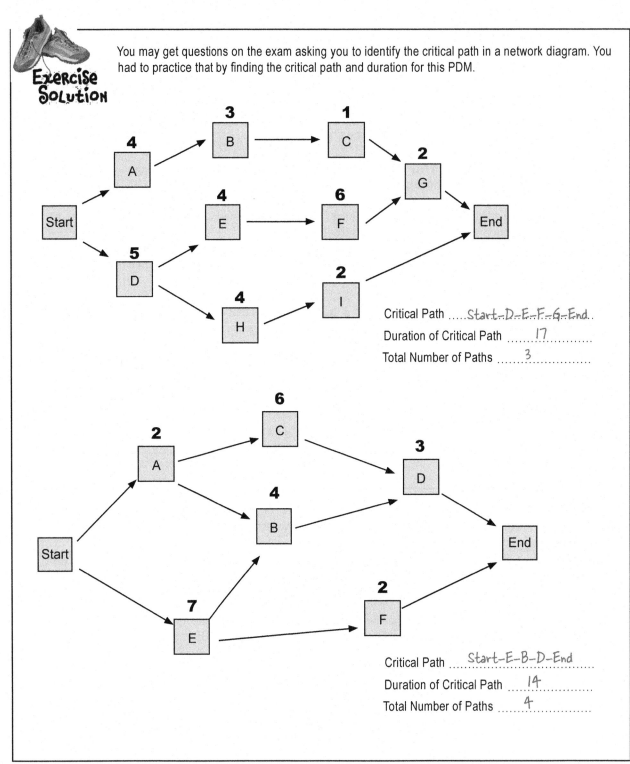

You may get questions on the exam asking you to identify the critical path in a network diagram. You had to practice that by finding the critical path and duration for this PDM.

Exercise Solution

Critical Path ....._Start–D–E–F–G–End_.....

Duration of Critical Path ....17........

Total Number of Paths ....3........

Critical Path ....._Start–E–B–D–End_.....

Duration of Critical Path ....14........

Total Number of Paths ....4........

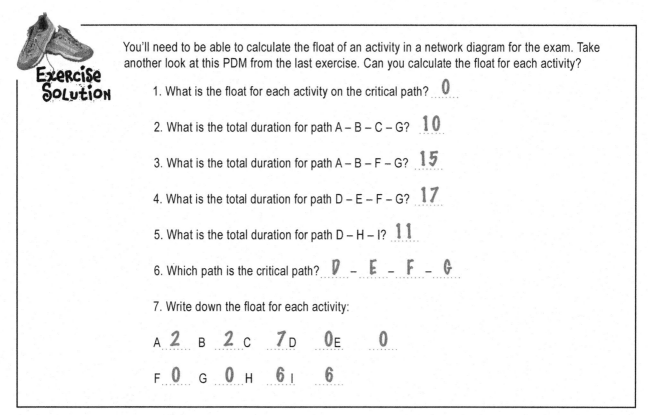

Exercise Solution

You'll need to be able to calculate the float of an activity in a network diagram for the exam. Take another look at this PDM from the last exercise. Can you calculate the float for each activity?

1. What is the float for each activity on the critical path? **0**

2. What is the total duration for path A – B – C – G? **10**

3. What is the total duration for path A – B – F – G? **15**

4. What is the total duration for path D – E – F – G? **17**

5. What is the total duration for path D – H – I? **11**

6. Which path is the critical path? **D – E – F – G**

7. Write down the float for each activity:

A **2**   B **2**   C **7** D **0**E   **0**

F **0**   G **0** H **6** I **6**

Exercise Solution

Each of these scenarios describes a schedule compression technique. Pick which are examples of fast-tracking and which are crashing.

1. Kathleen guesses that 70% of the invitees will RSVP. Instead of waiting for all of them to come in, she goes ahead and reserves the tables and chairs now.

   ☑ Fast-tracking   ☐ Crashing

2. Rebecca is taking a really long time to choose the decorations, so Kathleen brings in a professional decorator to help, even though it will cost more.

   ☐ Fast-tracking   ☑ Crashing

3. Kathleen needs to get the invitations out quickly, so she hires two temps to come in and help her stuff envelopes for a few days.

   ☐ Fast-tracking   ☑ Crashing

Sharpen your pencil
Solution

You'll need to know how to turn a table of nodes into a network diagram, so here's your chance to get some practice! Here's a list of nodes for a PDM network diagram. Try drawing the diagram based on it:

Name	Predecessor
Start	—
A	Start
B	A
C	B
D	Start
E	D
F	B
G	C
H	D
I	E, H
Finish	F, G, I

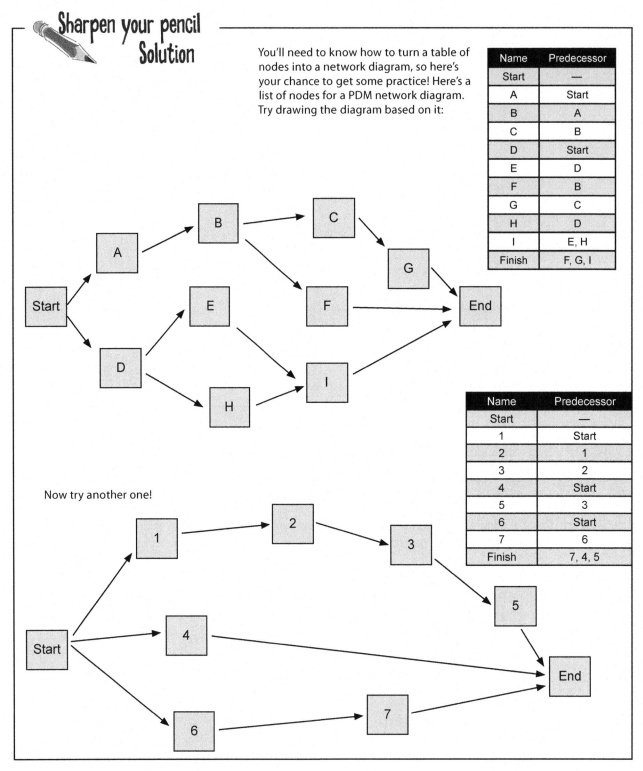

Now try another one!

Name	Predecessor
Start	—
1	Start
2	1
3	2
4	Start
5	3
6	Start
7	6
Finish	7, 4, 5

Control Schedule Magnets Answers

You'll see change control over and over again—every single knowledge area has its own change control process! Luckily, you'll start to see how similar they all are. But Control Schedule has its own quirks, and they're important for understanding Time Management.

This is just like Scope Management! You start with a plan, a baseline, and change requests.

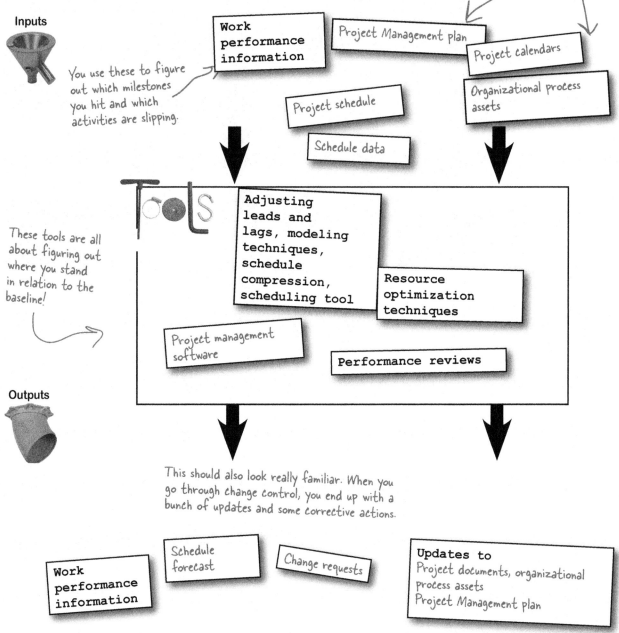

Inputs

You use these to figure out which milestones you hit and which activities are slipping.

Work performance information

Project Management plan

Project calendars

Organizational process assets

Project schedule

Schedule data

These tools are all about figuring out where you stand in relation to the baseline!

T⚬⚬ls

Adjusting leads and lags, modeling techniques, schedule compression, scheduling tool

Resource optimization techniques

Project management software

Performance reviews

Outputs

This should also look really familiar. When you go through change control, you end up with a bunch of updates and some corrective actions.

Work performance information

Schedule forecast

Change requests

Updates to
Project documents, organizational process assets
Project Management plan

Timecross solution

Did you get thrown because you thought using a historical database meant that you were doing analogous estimation? If you're plugging values into a database or spreadsheet, you're doing parametric estimation. A lot of people consider this a special type of analogous estimation, but describing it as parametric is more accurate.

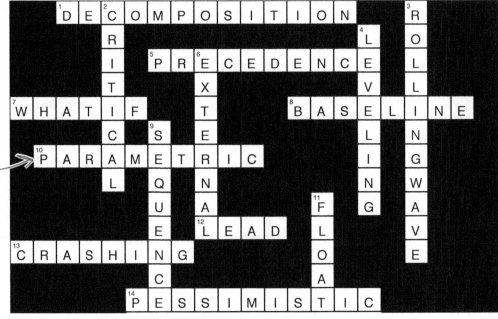

Exercise

Here are some examples of three-point estimates. Use the formula to figure out the expected time for each of these.

1. Expected duration = 50d; Optimistic duration = 30d; Most likely duration = 45d; Pessimistic duration = 90d

2. Expected duration = 20.3d; Optimistic duration = 12d; Most likely duration = 20d; Pessimistic duration = 30d

3. Expected duration = 25d; Optimistic duration = 10d; Most likely duration = 25d; Pessimistic duration = 40d

4. Expected duration = 51.8d; Optimistic duration = 33d; Most likely duration = 49d; Pessimistic duration = 82d

Exam Questions

1. You're managing a project when your client tells you that an external problem happened, and now you have to meet an earlier deadline. Your supervisor heard that in a situation like this, you can use schedule compression by either crashing or fast-tracking the schedule, but he's not sure which is which. What do you tell him?

    A.   Crashing the project adds risk, while fast-tracking adds cost.

    B.   When you crash a project, it always shortens the total duration of the project.

    C.   Crashing the project adds cost, while fast-tracking adds risk.

    D.   When you fast-track a project, it always shortens the total duration of the project.

2. Given this portion of the network diagram to the right, what's the ES of activity F?

    A.   9

    B.   10

    C.   12

    D.   13

3. Given this portion of the network diagram to the right, what's the LF of activity F?

    A.   10

    B.   11

    C.   16

    D.   17

D: ES = 7, EF = 9, LS = 3, LF = 4

E: ES = 10, EF = 12, LS = 3, LF = 4

F

G: ES = 17, EF = 26, LS = 17, LF = 26

H: ES = 17, EF = 19, LS = 11, LF = 13

4. You are managing a software project. Your QA manager tells you that you need to plan to have her team start their test planning activity so that it finishes just before testing begins. But other than that, she says it can start as late in the project as necessary. What's the relationship between the test planning activity and the testing activity?

    A.   Start-to-Start (SS)

    B.   Start-to-Finish (SF)

    C.   Finish-to-Start (FS)

    D.   Finish-to-Finish (FF)

5. You're managing an industrial design project. You've come up with the complete activity list, created network diagrams, assigned resources to each activity, and estimated their durations. What's the next thing that you do?

    A.   Use rolling wave planning to compensate for the fact that you don't have complete information.

    B.   Create the schedule.

    C.   Consult the project scope statement and perform Sequence Activities.

    D.   Use fast-tracking to reduce the total duration.

Exam Questions

6. Which of the following is NOT an input to Develop Schedule?

   A.   Activity list

   B.   Project schedule network diagrams

   C.   Resource calendars

   D.   Schedule baseline

7. Three members of your project team want to pad their estimates because they believe there are certain risks that might materialize. What is the BEST way to handle this situation?

   A.   Estimate the activities honestly, and then use a contingency reserve to cover any unexpected costs.

   B.   Allow more time for the work by adding a buffer to every activity in the schedule.

   C.   Tell the team members not to worry about it, and if the schedule is wrong it's OK for the project to be late.

   D.   Crash the schedule.

8. Which of the following tools is used for adding buffers to a schedule?

   A.   Three-point estimates

   B.   Critical chain method

   C.   Expert judgment

   D.   Critical path analysis

9. What is the critical path in the activity list to the right?

   A.   Start-A-B-C-Finish

   B.   Start-A-D-E-F-Finish

   C.   Start-G-H-I-J-Finish

   D.   Start-A-B-J-Finish

10. What is the float for activity F in the activity list to the right?

   A.   0

   B.   7

   C.   8

   D.   10

Name	Predecessor	Duration
Start	—	—
A	Start	6
B	A	4
C	B	8
D	A	1
E	D	1
F	E	2
G	Start	3
H	G	3
I	H	2
J	B, I	3
Finish	F, J, C	—

11. You're managing an interior decoration project when you find out that you need to get it done earlier than originally planned. You decide to fast-track the project. This means:

   A.   Starting the project sooner and working overtime

   B.   Assigning more people to the tasks at a greater total cost, especially for activities on the critical path

   C.   Starting activities earlier and overlapping them more, which will cost more and could add risks

   D.   Shortening the durations of the activities and asking people to work overtime to accommodate that

Exam Questions

12. Slack is a synonym for:

   A. Float

   B. Lag

   C. Buffer

   D. Reserve

13. You're managing a construction project. You've decomposed work packages into activities, and your client needs a duration estimate for each activity that you come up with. Which of the following will you use for this?

   A. Milestone list

   B. Activity list

   C. Critical path analysis

   D. Project schedule network diagram

14. What's the correct order of the Time Management planning processes?

   A. Sequence Activities, Define Activities, Estimate Activity Resources, Estimate Activity Durations, Develop Schedule

   B. Plan Schedule Management, Define Activities, Sequence Activities, Develop Schedule, Estimate Activity Resources, Estimate Activity Durations

   C. Plan Schedule Management, Define Activities, Sequence Activities, Estimate Activity Resources, Estimate Activity Durations, Develop Schedule

   D. Plan Schedule Management, Develop Schedule, Define Activities, Sequence Activities, Estimate Activity Resources, Estimate Activity Durations

15. Which of the following is NOT a tool or technique used in Estimate Activity Durations?

   A. SWAG estimation

   B. Parametric estimation

   C. Analogous estimation

   D. Three-point estimation

16. You're managing a project to build a new project management information system. You work with the team to come up with an estimate of 27 weeks. In the best case, this could be shortened by two weeks because you can reuse a previous component. But there's a risk that a vendor delay could cause the project to be delayed by five weeks. Use PERT to calculate a three-point estimate for this project.

   A. 25.83 weeks

   B. 26 weeks

   C. 27.5 weeks

   D. 28.3 weeks

Exam Questions

17. Given the network diagram below, what's the critical path?

    A.   Start-A-B-C-End

    B.   Start-A-D-G-End

    C.   Start-E-D-C-End

    D.   Start-E-F-G-End

18. For that same network diagram below, what's the float for activity A?

    A.   0 weeks

    B.   1 week

    C.   2 weeks

    D.   4 weeks

19. For that same network diagram below, what's the float for activity E?

    A.   0 weeks

    B.   1 week

    C.   2 weeks

    D.   4 weeks

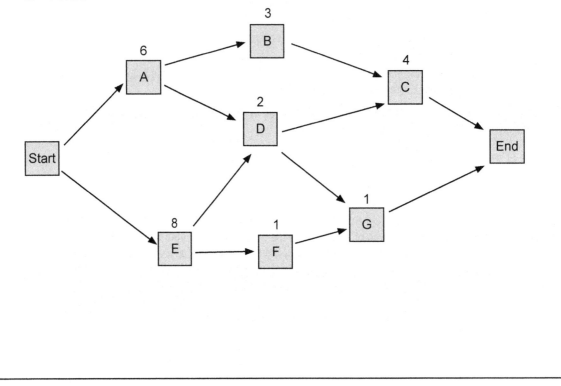

Exam Questions

20. You're managing a software project when your customer informs you that a schedule change is necessary. Which is the BEST thing to do?

    A.   Consult the schedule management plan.

    B.   Notify the team and the sponsor that there's going to be a schedule change.

    C.   Influence the factors that cause change.

    D.   Refuse to make the change because there's already a schedule baseline.

21. Your company has previously run other projects similar to the one you're currently managing. What is the BEST way to use that information?

    A.   Check the organizational process assets for lessons learned and other information about the past projects.

    B.   Use parametric estimation to estimate your project based on past projects' performance.

    C.   Start from scratch because you don't want mistakes from past projects to influence you.

    D.   Reuse the Project Management plan from a past project.

22. You're planning the schedule for a highway construction project, but the final date you came up with will run into the next budget year. The state comes up with capital from a reserve fund, and now you can increase the budget for your resources. What's the BEST way to compress the schedule?

    A.   Go back to your three-point estimates and use the most optimistic ones.

    B.   Use the extra budget to increase your contingency reserve.

    C.   Hire more experts to use expert judgment so your estimates are more accurate.

    D.   Crash the schedule.

23. You're managing a software project. You've created the schedule, and you need to figure out which activities absolutely cannot slip. You've done critical path analysis, identifying the critical path and calculating the early start and early finish for each activity. Which activities cannot slip without making the project late?

    A.   The ones with the biggest difference between ES and LF

    B.   The activities on the critical path

    C.   The activity with the most lag

    D.   The last activity in the project, because it has no float

24. You're managing a construction project. You've decomposed work packages into activities, and your client needs a duration estimate for each activity that you came up with. Which of the following BEST describes what you are doing?

    A.   Evaluating each activity to figure out how much effort it will take

    B.   Estimating the number of person-hours that will be required for each activity

    C.   Understanding, in calendar time, how long each activity will take

    D.   Estimating how many people it will take to perform each activity

Answers

Exam ~~Questions~~

1. Answer: C

You're likely to get some questions that ask you about crashing and fast-tracking, and it's important to know the difference between them. When you crash the project, it means that you add resources to it, especially to the critical path. There's no real risk in doing that—in the worst-case scenario, the extra people just sit around!—but it does cost more. Fast-tracking means adjusting the schedule so that activities overlap. The same resources are doing the work, so it's not going to cost more, but it's definitely riskier, because now you've eliminated buffers and possibly broken some dependencies! And remember that crashing or fast-tracking won't always work to make the project go faster!

2. Answer: D

Calculating the early start (ES) of an activity isn't hard. All you need to do is look at the early finish (EF) of the previous activity and add one. If there's more than one predecessor, then you take the largest EF and add one. In this case, the predecessors to activity F are D, with an EF of 9, and E, with an EF of 12. So the ES of F is 12 + 1 = 13.

3. Answer: A

It's just as easy to calculate the late finish (LF). Look at the following activity, take its LS (late start), and subtract one. If there's more than one following activity, use the one with the lowest LS. So, for activity F in the question, the following activities are G, with an LS of 17, and H, with an LS of 11. So the LF of F is 11 – 1 = 10.

4. Answer: C

Don't let the jargon fool you! You don't need to know anything about software testing to answer this question. When you have two activities, and the first activity has to be timed so that it finishes before the second one starts, then you've got a Finish-to-Start relationship, or FS.

Did answer A trick you? No need for rolling wave planning when you've got enough info to define all the activities!

5. Answer: B ⟵

This is a which-is-next question that describes a project that's completed the Define Activities, Sequence Activities, Estimate Activity Resources, and Estimate Activity Durations processes. The next process in Time Management is Develop Schedule, which means that the next thing you do is create the schedule!

Exam ~~Questions~~ Answers

6. Answer: D

The schedule baseline is an output of the Develop Schedule process, not an input. You should definitely know what goes into the schedule baseline: it's a specific version of the schedule that you set aside and use for comparison later on, when you want to know if the project is running late.

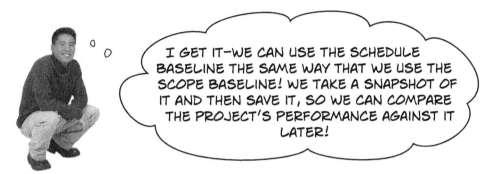

I GET IT—WE CAN USE THE SCHEDULE BASELINE THE SAME WAY THAT WE USE THE SCOPE BASELINE! WE TAKE A SNAPSHOT OF IT AND THEN SAVE IT, SO WE CAN COMPARE THE PROJECT'S PERFORMANCE AGAINST IT LATER!

7. Answer: A

You always want to be honest with your estimates. Every project has unknowns and risks, and there's no way to estimate any activity exactly! Luckily, we have tools to deal with this. You can use reserve analysis, a tool of Estimate Activity Durations, to come up with a contingency reserve that you can use to plan for these risks.

8. Answer: B

The critical chain method is a technique that's part of Develop Schedule that lets you figure out how to handle the problems that come with having limited resources. You use it to shuffle both activities and resources on your critical path. One important aspect of the critical chain method is that you can use it to add buffers to the schedule to reduce the risk of certain activities.

9. Answer: A

When you draw out a network diagram for the activities in the table, you end up with four paths. And you definitely should draw out the activity diagram for a question like this! You're allowed to use scratch paper on the exam, and this is one place where you should definitely do it. Of the four paths, only one has the longest duration: Start-A-B-C-Finish, which has a duration of 6 + 4 + 8 = 18. That's the critical path.

Answers
Exam ~~Questions~~ ——————————————

10. Answer: C

Activity F is in the path Start-A-D-E-F-Finish. This path has a duration of 6 + 1 + 1 + 2 = 10. The float of an activity is the longest time it can slip before it affects the critical path. In this case, activity F can slip by 8 without causing the path that it's on to go beyond the critical path. But any more than that, and its path becomes the new critical path!

Did you notice answer A? Don't forget that the float of any activity in the critical path is zero!

11. Answer: C

This is the definition of fast-tracking, and you're probably getting the hang of this one by now. You may get a question like this, but you'll almost certainly see fast-tracking as an incorrect answer to several questions!

12. Answer: A

Remember that when you see *slack*, it's the same thing as float. Either term could appear on the exam!

When a question asks what you'd use for a process, it's asking you to pick an input, tool, or technique that's part of the process.

13. Answer: B

This question is asking about the Estimate Activity Durations process. Take a look at the answers—there's only one answer that's used in that process: you need to start with the activity list in order to do the estimates for the activities! The other answers are things that are inputs, tools, or techniques for other processes.

14. Answer: C

It's not hard to remember the order in which the Time Management processes are performed. If you use a little common sense, you can reason your way through a question like this. You need to define your activities before you can sequence them, you need to know who's going to be doing an activity before you can estimate how long it's going to take, and you need to do all of that before you can build a schedule!

Control Schedule isn't included in the list of processes because if a schedule change happens, you'll have to go back and revisit the other Time Management processes. So it doesn't have a specific order!

15. Answer: A

You'll have to know the different kinds of estimating techniques for the exam. You don't necessarily have to be good at doing them, but you should recognize which are which. Parametric estimating is when you plug values into a formula, program, or spreadsheet and get an estimate. Analogous estimating uses similar activities from past projects to calculate new estimates. Three-point estimating uses an optimistic, pessimistic, and realistic estimate.

Exam Questions ~~Answers~~

16. Answer: C

This question is asking you to apply the PERT three-point estimation formula: (optimistic time + 4 × most likely time + pessimistic time) ÷ 6. When a question gives you these values directly, it's easy. But in this case, to answer the question you had to figure out the values for the optimistic time and pessimistic time, which meant that you needed to look at the assumptions that the team was making. The most likely time was given: 27 weeks. The best-case scenario would come in two weeks earlier, at 25 weeks, and the worst case would come in five weeks late, at 32 weeks. So the estimate is (25 weeks + 4 x (27 weeks) + 32 weeks) ÷ 6 = 27.5 weeks.

Sometimes you'll get a question about applying a formula, but you'll need to read the text in the question to figure out all of the variables.

17. Answer: C

The path Start-E-D-C-End has a duration of 8 + 2 + 4 = 14, which is the longest total duration in the entire network.

18. Answer: B

Activity A is on three different paths: Start-A-B-C-End (13), Start-A-D-C-End (12), and Start-A-D-G-End (9). To calculate its float, you take the longest path's length and subtract it from the length of the critical path: 14 – 13 = 1.

Can you think of how a question might quiz you on this information without actually asking you to look at a network diagram?

IT LOOKS LIKE THERE WILL BE A BUNCH OF QUESTIONS ON THE CRITICAL PATH METHOD! IT'S A GOOD THING I'VE GOT SO MUCH PRACTICE WITH IT.

19. Answer: A

Since activity E is on the critical path, its float is zero, because the float of any activity on the critical path is zero.

20. Answer: A

The Schedule Management plan tells you how changes to the schedule are to be handled. Any time there's a change, the first thing you should do is consult the plan to see how it should be handled.

Exam ~~Questions~~ Answers

21. Answer: A

The organizational process assets contain historical information about past projects. When you write up your lessons learned, or create work performance information, you store it in your company's organizational process asset library! Also, did you notice that answer B was the wrong definition of parametric estimation?

22. Answer: D

Crashing the schedule is the form of schedule compression that increases cost. This is a difficult question because all of the answers sound good, and one or two are a little misleading! Don't fall into the trap of choosing an answer because you recognize a valid tool or technique in it. Reserve analysis and three-point estimates are very useful techniques, but they're not the answer to this question.

23. Answer: B

The critical path is the path in the network diagram where any delay will cause a delay in the schedule. These are the activities that cannot slip without making the project late!

24. Answer: C

This question was really about the definition of *duration*, and the key to answering it is to understand how duration is different from effort. The correct answer talks about "calendar time," which is what a duration is: it's a measurement (or estimate) of how long the activity will take in real life, taking into account the number of people who will be doing the work, the availability of the people and other resources, everyone's vacation time, time taken away from the schedule because people are pulled off of the activity to work on higher-priority activities, and other real-world factors. That's different from effort (which is often measured in person-hours), and it's different from resource estimating (which involves estimating how many people and what other resources will be used for the activity).

7 Cost management

Watching the bottom line

THE POPULAR BOYS NEVER USED TO ASK ME OUT. BUT NOW THAT I USE COST MANAGEMENT, I'M SUCH A CHEAP DATE THAT I NEVER HAVE TO SPEND FRIDAY NIGHT ALONE.

Every project boils down to money. If you had a bigger **budget**, you could probably get more people to do your project more quickly and deliver more. That's why no project plan is complete until you come up with a budget. But no matter whether your project is big or small, and no matter how many **resources** and **activities** are in it, the process for figuring out the bottom line is *always the same!*

Time to expand the Head First Lounge

The Head First Lounge is doing so well that the guys are going to go ahead and open another Lounge near you! They're renting a basement bar, and now all they need to do is renovate it.

WE HAVE TO MAKE SURE THE NEW LOUNGE LOOKS ROCKIN', TOO!

SO WHERE DO YOU WANT TO START? HEY, WITH THE FURNITURE, RIGHT?!

Jeff

Charles

These guys own the Head First Lounge, a local hangout with good tunes, refreshing elixirs, and wireless access... Oh, and it also has a cool website!

The guys go overboard

When they start planning out what to buy, they want really expensive original retro stuff—the biggest bar they can find, and seventies textiles for the walls, floor, and upholstery, plus accessories—this is going to cost a lot of money…

Lounge conversation

GUYS, YOU DON'T HAVE THAT KIND OF MONEY TO SPEND. I'M THE PMP IN THIS SETUP, AND I KNOW A LITTLE SOMETHING ABOUT CONTROLLING COSTS.

Alice, the Head First Lounge's elixir mixer— and project manager, since she got her PMP certification.

Jeff: That bar is soooo cool! I can just imagine mixing up some crazy elixirs at parties!

Alice: Look, I know you want the new Lounge to look as good as the original, but you only have a little spare cash to spend on this. That means you have a limit of $10,000.

Charles: We should be able to get the new place looking so sweet with that!

Alice: Costs can creep up on you if you don't watch what you're doing. The best way to handle this is to create a budget and check your progress against it as you go.

Jeff: You always turn everything into a project, even mixing elixirs! Can't we just have fun with this?

Alice: Not if you don't want to go into debt.

Introducing the Cost Management processes

To make sure that they don't go over budget, Jeff, Charles, and Alice sit down and come up with detailed estimates of their costs. Once they have that, they add up the cost estimates into a budget, and then they track the project according to that budget while the work is happening.

Plan Cost Management

Just like all of the other knowledge areas, you need to plan out all of the processes and methodologies you'll use for Cost Management up front.

Estimate Costs process

This means figuring out exactly how much you expect each work activity you are doing to cost. So each activity is estimated for its time and materials cost, and any other known factors that can be figured in.

You need to have a good idea of the work you're going to do and how long it will take to do that work.

Determine Budget process

Here's where all of the estimates are added up and baselined. Once you have figured out the baseline, that's what all future expenditures are compared to.

This is just like the scope baseline from Chapter 5 or the schedule baseline from Chapter 6.

Control Costs process

This just means tracking the actual work according to the budget to see if any adjustments need to be made.

Controlling costs means always knowing how you are doing compared to how you thought you would do.

Plan how you'll estimate, track, and control your costs

Planning
process group

When you've got your project charter written and you're starting to put together your Project Management plan, you need to think about all of the processes and standards you'll follow when you estimate your budget and track to that estimate. By now, you're pretty familiar with the inputs, outputs, tools, and techniques you'll use in the **Plan Cost Management** process.

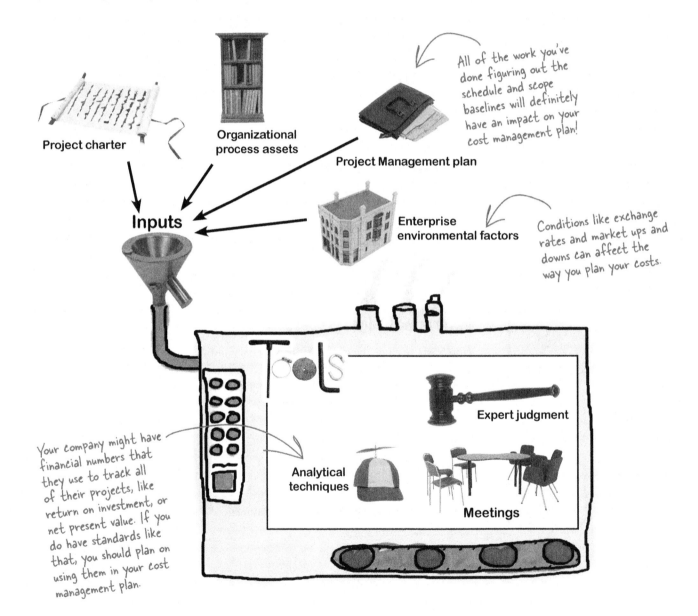

Project charter

Organizational process assets

Project Management plan

All of the work you've done figuring out the schedule and scope baselines will definitely have an impact on your cost management plan!

Inputs

Enterprise environmental factors

Conditions like exchange rates and market ups and downs can affect the way you plan your costs.

Expert judgment

Your company might have financial numbers that they use to track all of their projects, like return on investment, or net present value. If you do have standards like that, you should plan on using them in your cost management plan.

Analytical techniques

Meetings

Now you've got a consistent way to manage costs

There's only one output of the Plan Cost Management process, and that's the Cost Management plan. You'll use this document to specify the accuracy of your cost estimates, the rules you'll use to determine whether or not your cost processes are working, and the way you'll track your budget as the project progresses. When you've planned out your Cost Management processes, you should be able to estimate how much your project will cost using a format consistent with all the rest of your company's projects. You should also be able to tell your management how you'll know if your project starts costing more than you estimated.

Cost Management plan

Here's where you write down the subsidiary plan inside the Project Management plan that deals with costs. You plan out all of the work you'll do to figure out your budget and make sure your project stays within it.

> You'll want to define the units you'll use to manage your budget. For some projects, that's total person-hours; for others, it's an actual value in money. However you plan to track your costs, you need to let everybody on the project know up front.

The Plan Cost Management process is where you plan out all the work you'll do to make sure your project doesn't cost more than you've budgeted.

What Alice needs before she can estimate costs

Alice wants to keep the Lounge project's costs under control, and that starts with the **Estimate Costs** process. Before Alice can estimate costs, she needs the scope baseline. Once she knows who's doing what work, and how long it'll take, she can figure out how much it will cost.

Planning
process group

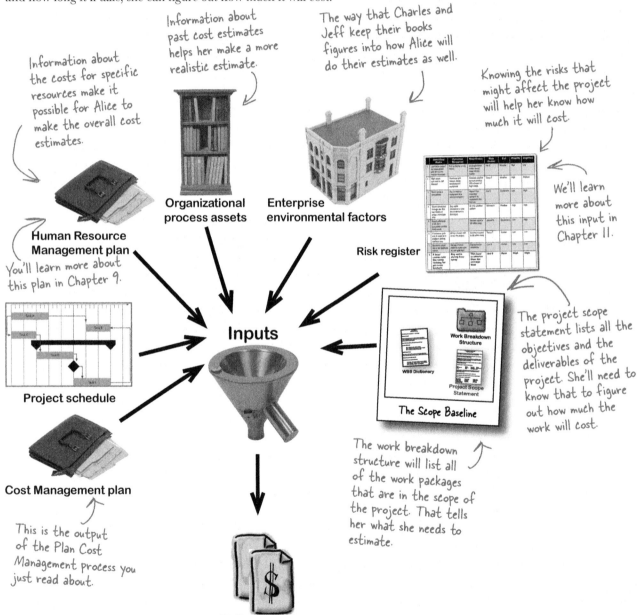

Information about the costs for specific resources make it possible for Alice to make the overall cost estimates.

Information about past cost estimates helps her make a more realistic estimate.

The way that Charles and Jeff keep their books figures into how Alice will do their estimates as well.

Knowing the risks that might affect the project will help her know how much it will cost.

Organizational process assets

Enterprise environmental factors

Human Resource Management plan

You'll learn more about this plan in Chapter 9.

Risk register

We'll learn more about this input in Chapter 11.

Inputs

Project schedule

Work Breakdown Structure

WBS Dictionary

Project Scope Statement

The Scope Baseline

The project scope statement lists all the objectives and the deliverables of the project. She'll need to know that to figure out how much the work will cost.

Cost Management plan

This is the output of the Plan Cost Management process you just read about.

The work breakdown structure will list all of the work packages that are in the scope of the project. That tells her what she needs to estimate.

Estimate costs

Sharpen your pencil

You've actually seen five of the tools and techniques in the Estimate Costs process before. Take a look at the list of tools below, and try to figure out which one of them Alice is using when she estimates costs. Can you write down which tool is being used in each scenario?

A. Bottom-up estimating

B. Analogous estimating

C. Expert judgment

D. Parametric estimating

E. Three-point estimates

1. The Starbuzz across the street opened just a few months ago. Alice sits down with the contractor who did the work there and asks him to help her figure out how much it will cost. He takes a look at the equipment Charles and Jeff want to buy and the specs for the cabinets and seating and tells her what she can afford to do with the budget she has.

Tool: ....

2. Alice creates a spreadsheet with all of the historical information from similar remodeling projects that have happened on her block. She sits down and types in the guys' desired furnishings and the square footage of the room to generate an estimated cost.

Tool: ....

3. Before Alice finishes her schedule, she gathers all of the information she has about previous projects' costs (like how much labor and materials cost). She also talks to a contractor, who gives valuable input.

Tool: ....

4. Alice sits down and estimates each and every activity and resource that she is going to need. Then she adds up all of the estimates into "rolled-up" categories. From there she adds up the categories into an overall budget number.

Tool: ....

5. Jeff sets up an appointment with the same contractor his friend used for some remodelling work. The contractor comes to the house, takes a look at the room, and then gives an estimate for the work.

Tool: ....

6. Alice figures out a best-case scenario, a most likely scenario, and a worst-case scenario. Then she used a formula to come up with an expected cost for the project.

Tool: ....

Sharpen your pencil
Solution

Did you work out which estimating tool from Time Management Alice is using when she estimates costs in each of these scenarios?

1: B. Analogous estimating
Since Alice is using the contractor's experience with a similar project to figure out how long her project will take, she is assuming that her project will go like the Starbuzz one did.

2: D. Parametric estimating
In this one Alice is just applying some numbers particular to her project to some historical information she has gathered from other projects and generating an estimate from that.

3: C. Expert judgment
Expert judgment often involves going back to historical information about past projects as well as consulting with experts or using your own expertise.

4: A. Bottom-up estimating
Starting at the lowest level and rolling up estimates is bottom-up estimating. Alice started with the activities on her schedule and rolled them up to categories and finally to a budget number.

5: C. Expert judgment
This is another example of asking somebody who has direct experience with this kind of work to give an estimate.

6: E. Three-point estimate
Alice came up with the three estimates and then performed the PERT calculation on them.

Watch it!

Analogous estimating is sometimes called "top-down estimating"

Take a minute and think about why it would be called "top-down." When you're doing bottom-up estimating, first you break it down into pieces, estimate each piece, and add them up. Analogous estimation is the opposite: you start with the whole project (without breaking it up at all), find other projects that were like it, and use those projects to come up with a new estimate.

> HOLD ON! HOW CAN YOU USE THE SAME TOOLS TO ESTIMATE BOTH TIME AND COST?

Good question.

Not all of the estimation techniques for cost are the same as the ones we used for time. Often, people only have a certain amount of time to devote to a project, and a fixed amount of money too. So, it makes sense that some of the tools for estimating both would overlap. We'll learn a few new ones next.

Other tools and techniques used in Estimate Costs

A lot of times you come into a project and there is already an expectation of how much it will cost or how much time it will take. When you make an estimate really early in the project and you don't know much about it, that estimate is called a **rough order of magnitude** estimate. (You'll also see it called a **ROM** or a **ballpark estimate**.) It's expected that it will get more refined as time goes on and you learn more about the project. Here are some more tools and techniques used to estimate cost:

This estimate is REALLY rough! It's got a range of −25% to +75%, which means it can be anywhere from half to one and a half times the actual cost! So you only use it at the very beginning of the project.

Project management estimating software

Project managers will often use specialized estimating software to help come up with cost estimates (like a spreadsheet that takes resource estimates, labor costs, and materials costs and performs calculations).

Vendor bid analysis

Sometimes you will need to work with an external contractor to get your project done. You might even have more than one contractor bid on the job. This tool is all about evaluating those bids and choosing the one you will go with.

Reserve analysis

You need to set aside some money for cost overruns. If you know that your project has a risk of something expensive happening, better to have some cash laying around to deal with it. Reserve analysis means putting some cash away just in case.

You'll see this in action when we look at Risk Management in Chapter 11.

Cost of quality

Since the next chapter is all about quality, you'll be learning a lot about this in Chapter 8.

You will need to figure the cost of all of your quality-related activities into the overall budget, too. Since it's cheaper to find bugs earlier in the project than later, there are always quality costs associated with everything your project produces. Cost of quality is just a way of tracking the cost of those activities.

Group decision-making techniques

You'll need to work with groups of people to figure out your costs. It's important that your team feels like they can commit to the overall budget and schedule.

Cost of quality is how much money it takes to do the project right.

Let's talk numbers

There are a few numbers that can appear on the test as definitions. You won't need to calculate these, but you should know what each term means.

> You'll get exam questions asking you to use BCR or NPV to compare two projects. The higher these numbers are, the better!

Benefit cost ratio (BCR)

This is the amount of money a project is going to make versus how much it will cost to build it. Generally, if the benefit is higher than the cost, the project is a good investment.

Net present value (NPV)

> Money you'll get in three years isn't worth as much to you as money you're getting today. NPV takes the "time value" of money into consideration, so you can pick the project with the best value in <u>today's</u> dollars.

This is the actual value at a given time of the project minus all of the costs associated with it. This includes the time it takes to build it and labor as well as materials. People calculate this number to see if it's worth doing a project.

Opportunity cost

When an organization has to choose between two projects, it's always giving up the money it would have made on the one it doesn't do. That's called opportunity cost. It's the money you don't get because you chose not to do a project.

> If a project will make your company $150,000, then the opportunity cost of selecting another project instead is $150,000 because that's how much your company's missing out on by not doing the project.

Internal rate of return

This is the amount of money the project will return to the company that is funding it. It's how much money a project is making the company. It's usually expressed as a percentage of the funding that has been allocated to it.

Depreciation

This is the rate at which your project loses value over time. So, if you are building a project that will only be marketable at a high price for a short period of time, the product loses value as time goes on.

Lifecycle costing

Before you get started on a project, it's really useful to figure out how much you expect it to cost—not just to develop, but to support the product once it's in place and being used by the customer.

WHAT'S MY PURPOSE

Match each scenario to the cost numbers that Alice is using in each one.

1. Alice does such a good job planning out her entertainment center remodeling that the Smiths down the street ask if they can have her help with their home theater upgrade. Since she is too busy doing the work on the lounge, she has to say no. Rob Smith says, "That's a shame; we were willing to pay $1,000 to someone to help us out with this."

A. Opportunity cost

2. The minute the TV gets installed, Alice starts inviting all of her friends over to the lounge to watch the games on the weekend. She charges a $2 cover charge for her football Saturdays and has been clearing about $20 per week even though the room isn't finished.

B. Benefit cost ratio

3. Even though the system she is currently installing is state of the art, Alice knows that within a year or so it will be on sale for half as much as she is paying now.

C. Internal rate of return

4. Alice wants to figure out how much the project is worth so far. So she adds up the value of all of the materials she has used, and subtracts the labor and any depreciation that needs to be accounted for. The number she ends up with gives the value of the overall project right now.

D. Depreciation

5. Before Jeff and Charles decided to do the remodeling, they compared how much the project was going to cost to how much good they thought it would do for them.

E. Net present value

Answers: 1-A, 2-C, 3-D, 4-E, 5-B

Now Alice knows how much the Lounge will cost

Once you've applied all of the tools in the Estimate Costs process, you'll get an estimate for how much your project will cost. It's always important to keep all of your supporting estimate information, too. That way, you know the assumptions you made when you were coming up with your numbers.

Outputs

Activity cost estimates

This is the cost estimate for all of the activities in your activity list. It takes into account resource rates and estimated duration of the activities.

Activity cost estimates

Basis of estimates

Just like the WBS has a WBS dictionary, and the activity list has activity attributes, the cost estimate has a supporting detail called the **basis of estimates**. Here is where you list out all of the rates and reasoning you have used to come to the numbers you are presenting in your estimates.

Basis of estimates

Updates to project documents

Along the way, you might find that you need to change the way you measure and manage cost. These updates allow you to make changes to the Project Management plan to deal with those improvements.

Project document updates

Once Alice has an estimate of the project's cost, what should she do with that information?

Lounge conversation

LET'S BUILD A BUDGET!

Jeff: OK, how do we start? There are a lot of things to buy here.

Alice: We already have your savings, and the rest will come in July at the end of the quarter. The Lounge is having another great year, so the profits are pretty good. Your savings are around $4,000 and the profits will probably be closer to $6,000. That's definitely enough money to work with.

Charles: Well, the furniture I want isn't back in stock until June.

Alice: OK, so we have to time our costs so that they're in line with our cash flow.

Jeff: Oh! I see. So we can start building now, but we'll still have money in June and July when the furniture comes in. Perfect.

The Determine Budget process

Once Alice has cost estimates for each activity, she's ready to put a budget together. She does that using the **Determine Budget** process. Here's where you take the estimates that you came up with and build a budget out of them. You'll build on the activity cost estimates and basis of cost estimate that you came up with in Estimate Costs.

Planning
process group

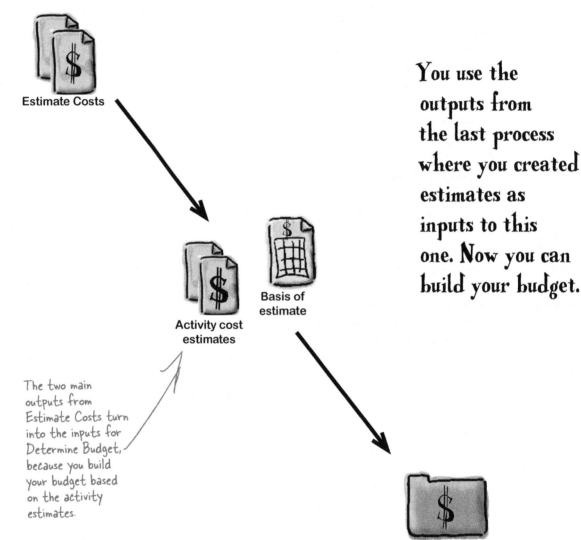

Estimate Costs

Activity cost estimates

Basis of estimate

The two main outputs from Estimate Costs turn into the inputs for Determine Budget, because you build your budget based on the activity estimates.

Determine Budget

You use the outputs from the last process where you created estimates as inputs to this one. Now you can build your budget.

What you need to build your budget

The **inputs** to Determine Budget are largely the same ones that you saw in Estimate Costs, with the notable additions of activity cost estimates and basis of cost estimate.

If you're doing work that's been contracted, then your agreement will have information (like fees or rates) that you'll need to take into account. You'll learn all about contracts in Chapter 12.

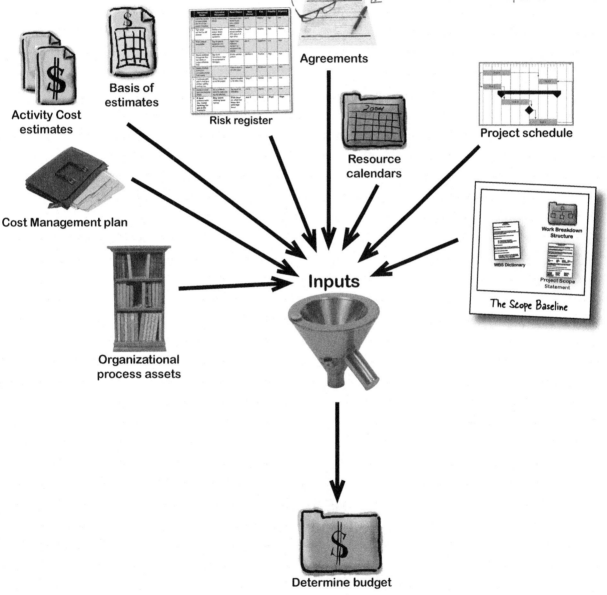

Activity Cost estimates

Basis of estimates

Risk register

Agreements

Resource calendars

Project schedule

Cost Management plan

Organizational process assets

Inputs

The Scope Baseline

Determine budget

Determine budget: how to build a budget

① **Roll up your estimates into control accounts.**

This tool is called **cost aggregation**. You take your activity estimates and roll them up into control accounts on your work breakdown structure. That makes it easy for you to know what each work package in your project is going to cost.

Outputs

② **Come up with your reserves.**

When you evaluate the risks to your project, you will set aside some cash reserves to deal with any issues that might come your way. This tool is called **reserve analysis**.

⑤ **Build a baseline.**

Just like your scope and schedule baselines, a **cost baseline** is a snapshot of the planned budget. You compare your actual performance against the baseline so you always know how you are doing versus what you planned.

Cost baseline

Your company's management plans for project overruns!

Just because you plan out a budget in your cost baseline, that doesn't mean your project is 100% guaranteed to fall inside that budget. It's common for a company to have a standard policy for keeping a **management reserve** to cover unexpected, unplanned costs. When you need to get your project funded, that funding has to cover both the budget in your cost baseline *and* the management reserve.

 Use your expert judgment.

Here's where you compare your project to historical data that has been collected on other projects to give your budget some grounding in real-world **historical relationships**, and you use your own expertise and the expertise of others to come up with a realistic budget to cover your project's costs.

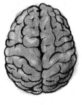

It's true that not everybody has access to historical data to do a check like this. But, for the purposes of the test, you need to know that it's a tool for making your budget accurate.

 Make sure you haven't blown your limits.

This tool is **funding limit reconciliation**. Since most people work in companies that aren't willing to throw unlimited money at a project, you need to be sure that you can do the project within the amount that your company is willing to spend.

If you blow your limit, you need to replan or go to your sponsor to figure out what to do. It could be that a scope change is necessary, or the funding limit can be increased.

 Figure out funding requirements.

It's not enough to have an overall number that everyone can agree to. You need to plan out how and when you will spend it, and document those plans in the **project funding requirements**. This output is about figuring out how you will make sure your project has money when it's needed, and that you have enough to cover unexpected risks as well as known cost increases that change with time.

So these requirements need to cover both the budget and the management reserve.

 Update your project documents.

Once you have estimated and produced your baseline and funding requirements, you need to update your Cost Management plan with anything you learned along the way.

Exercise

What tool or technique is Alice using to build the budget?

1. Alice reads a newspaper article that says that there has been a sharp increase in lumber costs recently. She knows this wasn't in her contractor's original plan and decides to put a few hundred dollars aside to deal with the price hike if it should happen.

☐ Parametric estimating   ☐ Reserve analysis   ☐ Cost aggregation   ☐ Funding limit reconciliation

2. Jeff helps Alice add up all of the estimates they have done into control accounts so that they can figure out how much the stereo installation is going to cost versus building the entertainment center.

☐ Parametric estimating   ☐ Reserve analysis   ☐ Cost aggregation   ☐ Funding limit reconciliation

3. Once the budget is close to done, Alice looks over their financial plans for the year to be sure that they can afford everything at the time that it is needed.

☐ Parametric estimating   ☐ Reserve analysis   ☐ Cost aggregation   ☐ Funding limit reconciliation

Answers: 1. Reserve analysis 2. Cost aggregation 3. Funding limit reconciliation

BULLET POINTS: AIMING FOR THE EXAM

- **Parametric estimation** is used in Estimate Costs and Determine Budget.

- **Cost aggregation** is rolling up costs from the work package level to the control account level so that the numbers can be followed down through the WBS hierarchy.

- **Control accounts** are high-level WBS items that are used to track cost estimates. They do not represent activities or work packages. They represent the cost of the work packages and activities that appear under them in the WBS.

- The main outputs of Estimate Costs are the activity cost estimate and the basis of estimates. The main outputs of Determine Budget are the cost baseline and project funding requirements.

- You will get questions on the exam asking you to select between projects using **net present value** (NPV) or **benefit cost ratio** (BCR). Always choose the project with the biggest NPV or BCR!

- **Lifecycle costing** means estimating the money it will take to support your product or service when it has been released.

- **Rough order of magnitude Estimation** is estimating with very little accuracy at the beginning of a project and then refining the estimate over time. It's got a range of –25% to +75%.

- A **management reserve** is money set aside to cover unplanned, unexpected costs. Your project's funding requirements need to cover both the budget in the cost baseline and the management reserve.

there are no
Dumb Questions

Q: Isn't it enough to know my project's scope and schedule, and then trust the budget to come out all right?

A: Even if you don't have a strict budget to work within, it makes sense to estimate your costs. Knowing your costs means that you have a good idea of the value of your project all the time. That means you will always know the impact (in dollars) of the decisions you make along the way. Sometimes understanding the value of your project will help you to make decisions that will keep your project healthier.

Many of us do have to work within a set of cost expectations from our project sponsors. The only way to know if you are meeting those expectations is to track your project against the original estimates.

It might seem like fluff. But knowing how much you are spending will help you relate to your sponsor's expectations much better as well.

Q: In my job I am just handed a budget. How does estimating help me?

A: In the course of estimating, you might find that the budget you have been given is not realistic. Better to know that while you're planning, before you get too far into the project work, rather than later.

You can present your findings to the sponsor and take corrective action right away if your estimate comes in pretty far off target. Your sponsor and your project team will thank you for it.

Take a minute to think about what "value" really means. How does the sponsor know if he's getting his money's worth halfway through the project? Is there an easy way you can give the sponsor that information?

Q: What if I don't have all of this information and I am supposed to give a ballpark estimate?

A: This is where those rough order of magnitude estimates come in. That's just a fancy way of saying you take your best guess, knowing that it's probably inaccurate, and you let everybody know that you will be revising your estimates as you know more and more about the project.

Q: My company needs to handle maintenance of projects after we release them. How do you estimate for that?

A: That's called lifecycle costing. The way you handle it is just like you handle every other estimate. You sit down and try to think of all of the activities and resources involved in maintenance, and project the cost. Once you have an estimate, you present it along with the estimate for initially building the product or service.

Q: I still don't get net present value. What do I use it for?

A: The whole idea behind net present value is that you can figure out which of two projects is more valuable to you. Every project has a value—if your sponsor's spending money on it, then you'd better deliver something worth at least that much to him! That's why you figure out NPV by coming up with how much a project will be worth, and then subtracting how much it will cost. But for the exam, all you really need to remember are two things: net present value has the cost of the project built into it, and if you need to use NPV to select one of several projects, always choose the one with the biggest NPV. That's not hard to remember, because you're just choosing the one with the most value!

Q: Hold on just a minute. Can we go back to the rough order of magnitude estimate? I remember from my math classes that an order of magnitude has something to do with a fixed ratio. Wouldn't –50% to +100% make more sense as an order of magnitude?

A: Yes, it's true that in science, math, statistics, or engineering, an order of magnitude typically involves a series of magnitudes increasing by a fixed ratio. So if an order of magnitude down is 50%, then you'd typically maintain that same 2:1 ratio between orders of magnitude, so the next order of magnitude higher would be 100%.

However, if you check the *PMBOK Guide*, it defines it as follows: " a project in the initiation phase could have a rough order of magnitude (ROM) estimate in the range of –25% to +75%." [5th Edition, p201] Since that's the definition in the *PMBOK Guide*, that's what to remember for the exam.

Estimate Costs is just like Estimate Activity Durations. You get the cost estimate and the basis of the cost estimate, updates to the plan, and requested changes when you are done.

Question Clinic: The red herring

SOMETIMES A QUESTION WILL GIVE YOU A LOT OF EXTRA INFORMATION THAT YOU DON'T NEED. IT'LL INCLUDE A RAMBLING STORY OR A BUNCH OF EXTRA NUMBERS THAT ARE IRRELEVANT.

104. You are managing a highway construction project. You have to build a three-mile interchange at a cost of $75,000 per quarter-mile. Your project team consists of a road planner, an architect, an engineer, a foreman, and 16 highway workers. The workers will not be available until week 10 of the project. Your business case document is complete, and you have met with your stakeholders and sponsor. Your senior managers are now asking you to come up with an estimate. Your company has done four other highway projects very similar to this one, and you have decided to make your estimate by looking at the costs of those previous projects.

What kind of estimate involves comparing your project to a previous one?

A. Parametric

B. Analogous

C. Bottom-up

D. Rough order of magnitude

Did you read that whole paragraph, only to find out the question had nothing to do with it?

You only needed to read this sentence to get the answer right.

WHEN YOU SEE A RED HERRING QUESTION, YOUR JOB IS TO FIGURE OUT WHAT PART OF IT IS RELEVANT AND WHAT'S INCLUDED JUST TO DISTRACT YOU. IT SEEMS TRICKY, BUT IT'S ACTUALLY PRETTY EASY ONCE YOU GET THE HANG OF IT.

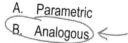

Red Herring

HEAD LIBS

Fill in the blanks to come up with your own red herring question!

You are managing a _____ project.
(kind of project)

You have _____ at your disposal, with _____ . Your
(describe a resource)   (how that resource is restricted)

_____ contains _____ . The _____
(a project document)   (something that document would contain)   (a team member)

alerts you that _____ , and suggests _____ .
(a problem that affected your project)   (a suggested solution)

_____ ?
(a question vaguely related to one of the things in the paragraph above)

A. _____
(wrong answer)

B. _____
(trickily wrong answer)

C. _____
(correct answer)

D. _____
(ridiculously wrong answer)

The Control Costs process is a lot like schedule control

When something unexpected comes up, you need to understand its impact on your budget and make sure that you react in the best way for your project. Just like changes can cause delays in the schedule, they can also cause cost overruns. The **Control Costs process** is all about knowing how you are doing compared to your plan and making adjustments when necessary.

Monitoring & Controlling process group

I JUST TALKED TO OUR ACCOUNTANT. SHE SAYS WE NEED TO PUT A THIRD OF OUR PROFITS AWAY FOR TAXES.

TIME TO CONTROL COSTS!

⚛ BRAIN POWER

Given what you already know about controlling your scope and schedule, how would you handle this problem?

Sharpen your pencil

Using what you already know about the Control Scope and Control Schedule processes, can you take a guess at what each of these inputs will be used for?

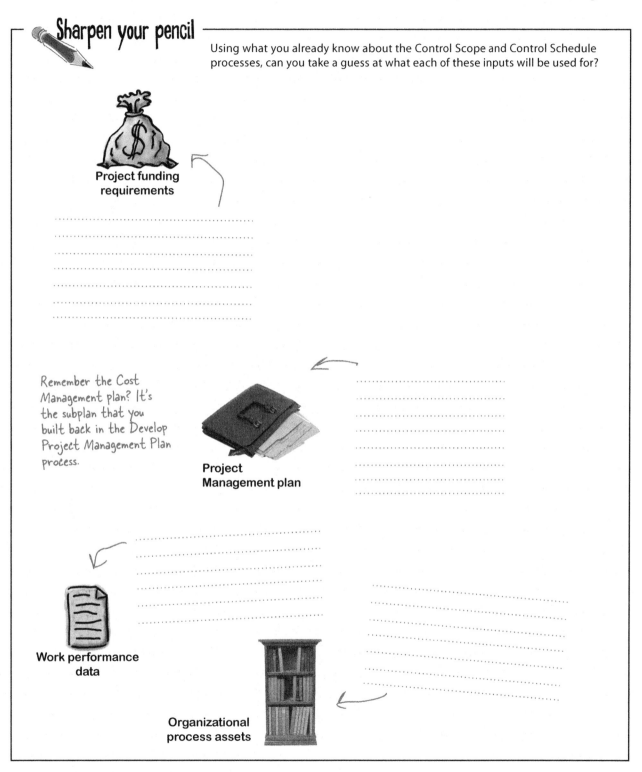

Project funding requirements

....
....
....
....
....
....
....

Remember the Cost Management plan? It's the subplan that you built back in the Develop Project Management Plan process.

Project Management plan

....
....
....
....
....
....
....
....

Work performance data

Organizational process assets

Sharpen your pencil
Solution

Using what you already know about the Control Scope and Control Schedule processes, can you take a guess at what each of these inputs will be used for?

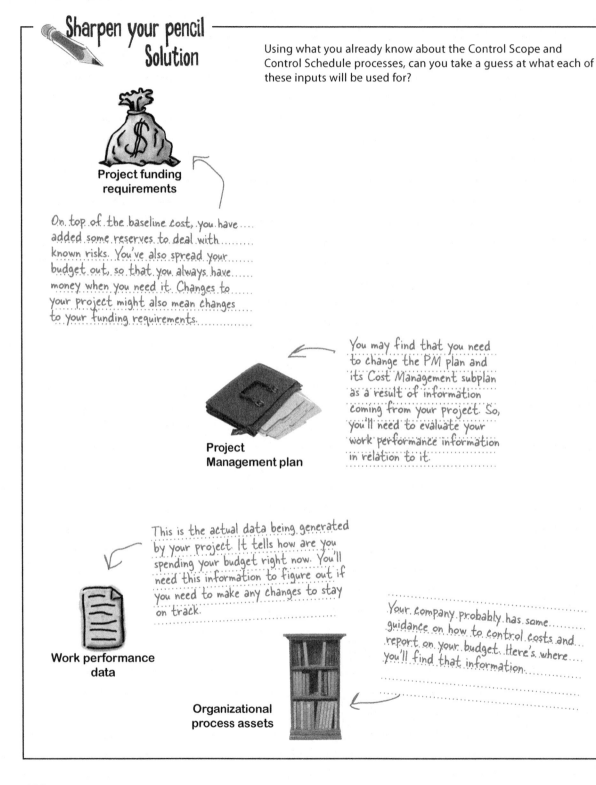

Project funding requirements

On top of the baseline cost, you have added some reserves to deal with known risks. You've also spread your budget out, so that you always have money when you need it. Changes to your project might also mean changes to your funding requirements.

You may find that you need to change the PM plan and its Cost Management subplan as a result of information coming from your project. So, you'll need to evaluate your work performance information in relation to it.

Project Management plan

This is the actual data being generated by your project. It tells how are you spending your budget right now. You'll need this information to figure out if you need to make any changes to stay on track.

Work performance data

Your company probably has some guidance on how to control costs and report on your budget. Here's where you'll find that information.

Organizational process assets

A few new tools and techniques

The tools in Control Costs are all about helping you figure out where to make changes so you don't overrun your budget.

Earned value management

Here's where you measure how your project is doing compared to the plan. This involves using the earned value formulas to assess your project.

You'll learn more about the formulas in just a few pages!

To-complete performance index

The to-complete performance index (TCPI) is a calculation that you can use to help you figure out how well your project needs to perform in the future in order to stay on budget.

You'll learn more about TCPI, too!

Performance reviews

Reviews are meetings where the project team reviews performance data to examine the variance between actual performance and the baseline. Earned value management is used to calculate and track the variance. Over time, these meetings are a good place to look into trends in the data.

Forecasting

Use the information you have about the project right now to predict how close it will come to its goals if it keeps going the way it has been. Forecasting uses some earned value numbers to help you come up with preventive and corrective actions that can keep your project on the right track.

Project management software

You can use software packages to track your budget and make it easier to know where you might run into trouble.

Reserve analysis

Throughout your project, you are looking at how you are spending versus the amount of reserve you've budgeted. You might find that you are using reserved money at a faster rate than you expected or that you need to reserve more as new risks are uncovered.

Forecasting and performance measurement are very important! You use them to find the changes you need to make in your project.

Look at the schedule to figure out your budget

The tools in Control Costs are all about helping you figure out
where to make changes so you don't overrun your budget.

$10,000

Budget at completion (BAC)

How much money are you planning on spending on your project?
Once you add up all of the costs for every activity and resource,
you'll get a final number…and that's the total project budget. If
you only have a certain amount of money to spend, you'd better
make sure that you haven't gone over!

Once you figure this out, you can figure out your project's planned value.

How to calculate planned value

If you look at your schedule and see that you're **supposed to have done** a certain percentage of the work, then that's the percent of the total budget that you've "earned" so far. This value is known as planned value. Here's how you calculate it.

 First, write down your

BAC—Budget at completion

This is the ***first number you think of*** when you work on your project costs. It's the **total budget** that you have for your project—how much you plan to spend on your project.

BAC x

The name "BAC" should make sense—it's the budget of your project when it's complete!

 Then multiply that by your

Planned % complete

If the schedule says that your team should have done 300 hours of work so far, and they will work a total of 1,000 hours on the project, then your planned % complete is 30%.

BAC x **Planned % complete**

Planned % complete is easy to work out, as it's just the calculation Given amount ÷ Total amount.

 The resulting number is your

PV—Planned value

This is how much of your budget you planned on using so far. If the BAC is $200,000, and the schedule says your planned % complete is 30%, then the planned value is $200,000 × 30% = $60,000.

BAC x **Planned % complete** = PV

BAC x **Planned % complete** = PV   PV = BAC x **Planned % complete**

 You may also see the planned value formula flipped around and written with the PV out front, but it's exactly the same formula.

Sharpen your pencil

Now it's your turn! See if you can figure out BAC and PV for a typical project.

1. You're managing a project to install 200 windows in a new skyscraper and need to figure out your budget. Each week of the project costs the same: your team members are paid a total of $4,000 every week, and you need $1,000 worth of parts each week to do the work. If the project is scheduled to last 16 weeks, what's the BAC for the project?

   BAC = ...

   Even though we are at the beginning of the project now, we can still figure out what the PV will be in four weeks.

2. What will the planned % complete be four weeks into the project?

   Planned % complete = ...

   This is the part that takes some thinking. How do you know what % you are through the project?

3. What should the PV be four weeks into the project?

   PV = ... x .... = ...

   ⟶ Answers on page 388.

> HMM. OK, BUT THAT DOESN'T TELL ME MUCH ABOUT MY BUDGET, DOES IT?

Not yet, it doesn't.

But wouldn't be nice if, when your schedule said you were supposed to be 37.5% complete with the work, then you knew that you'd actually spent 37.5% of your budget?

Well, in the real world things don't always work like that, but there are ways to work out—approximately—how far on (or off) track your budget actually is.

Earned value tells you how you're doing

When Alice wants to track how her project is doing versus the budget, she uses **earned value**. This is a technique where you figure out how much of your project's value has been delivered to the customer so far. You can do this by comparing the value of **what your schedule says** you should have delivered against the value of what you **actually** delivered.

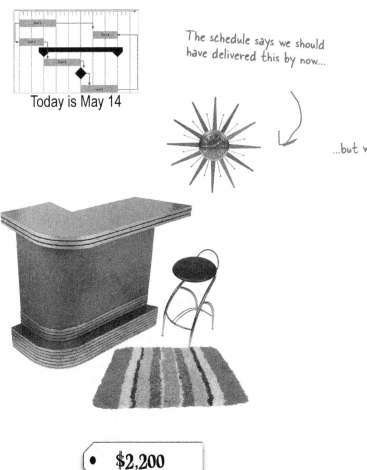

Today is May 14

The schedule says we should have delivered this by now...

...but we only delivered this.

Your schedule tells you a lot about where you are supposed to be right now.

• **$2,200**

• **$1,650**

The <u>actual cost</u> of this project on May 14th is **$1,650**.
The <u>planned value</u> was **$2,200**.

How to calculate earned value

If you could estimate each activity exactly, every single time, you wouldn't need earned value. Your schedule would always be perfectly accurate, and you would always be exactly on budget.

But you know that real projects don't really work that way! That's why earned value is so useful—it helps you put a number on how far off track your project actually is. And that can be a really powerful tool for evaluating your progress and reporting your results. Here's how you calculate it.

> When you do work, you convert the money your sponsor invests in your project into value. So, **earned value** is about **how much work you have been able to accomplish with the money you've been given**. When you calculate earned value, you're showing your sponsor how much value that investment has earned.

 1 *First, write down your*

BAC—Budget at completion

Remember, this is the **total budget** that you have for your project.

BAC   x

 2 *Then multiply that by your*

Actual % complete

Say the schedule says that your team should have done 300 hours of work so far, out of a total of 1,000. But you talk to your team and find out they actually completed 35% of the work. That means the actual % complete is 35%.

BAC   x   Actual % complete

If your team actually got 35% of the work done when the schedule says they should only have gotten 30% done, that means they're more efficient than you planned!

3 *The resulting number is your*

EV—Earned value

This figure tells you how much your project *actually* earned. Every hour that each team member works adds value to the project. You can figure it out by taking the percentage of the hours that the team has actually worked and multiplying it by the BAC. If the total cost of the project is $200,000, then the earned value is $200,000 × 35% = $70,000.

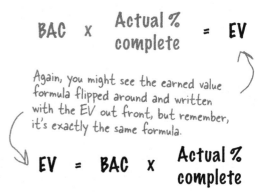

BAC   x   Actual % complete   =   EV

Again, you might see the earned value formula flipped around and written with the EV out front, but remember, it's exactly the same formula.

EV   =   BAC   x   Actual % complete

BRAIN POWER

What's the difference between actual cost and planned value? What does it mean if your AC is bigger than your PV? What if it's smaller?

This is a harder problem to solve than it seems! Take a minute and really think about it before you turn the page.

You'll be doing a lot of calculations in a minute. The best way to approach any calculation is to understand what it's for and why you use it. So before you go on, grab a cup of coffee and think for a few minutes about the difference between BAC and PV. Also, how do you compute actual % complete on a real project? Doing that will really help you get these ideas firmly embedded into your brain!

Put yourself in someone else's shoes

Earned value is one of the most difficult concepts that you need to understand for the PMP exam. The reason it's so confusing for so many people is that these calculations seem a little weird and arbitrary to a lot of project managers.

But *they make a lot more sense* if you **think about your project the way your sponsor thinks about it**. If you put yourself into the sponsor's shoes, you'll see that this stuff actually makes sense!

> Think about earned value from the sponsor's perspective. It all makes a lot more sense then.

Let's say you're an executive:
You're making a decision to spend $300,000 of your company's money on a project. To a project manager, that's a project's budget. But to you, the sponsor, that's $300,000 of value you expect to get!

That's the total budget, or the BAC.

So how much value is the project delivering?
If you're the sponsor, you're thinking about the bottom line. And that bottom line is whether or not you're getting your money's worth from the project. If the team's done 50% of the work, then you've gotten $150,000 of value so far.

But if the schedule says that they should have done 60% of the work by now, then *you're getting less value than you were promised!*

If you put the value in dollar terms, your sponsor knows what return he's getting for his investment.

That's earned value—it's based on how much work the team actually did.

Look at the schedule to figure out how much value you planned to deliver to your sponsor.

The sponsor doesn't care as much about how you spend the budget. He just wants to get the most value for his money!

Sharpen your pencil

Let's get back to that 16-week project from page 354. In the last exercise you figured out what the project should look like by using planned value. Now you can use earned value to figure out if your project is really going the way you planned.

1. Fast-forward four weeks into the project installing those 200 skyscraper windows. Fill in the BAC and PV you figured out before. (Check your answer at the top of page 378 to make sure you got it right!)

BAC = ... PV = ...

Figure out the actual % complete by dividing the actual work done into the total amount you're planning on.

2. You've checked with your team, but they have bad news. The schedule says they were supposed to have installed 50 windows by now, but they've only installed 40. Can you figure out the actual % complete?

Actual % complete = $\dfrac{..............}{..............}$ = ...

← Fill in the number of windows the team's actually installed.

Fill in the total number of windows that will be installed over the course of the project.

3. What should the earned value be right now?

EV = ... x .... = ...

Fill in the BAC   Fill in the actual % complete

4. Look at the planned value, and then look at the earned value. Are you delivering all the value you planned on delivering?

☐ Yes   ☐ No   ⟶ **Answers on page 388.**

> **NEAT. I THINK I CAN USE THESE FORMULAS TO TRACK MY SCHEDULE AND MY BUDGET!**

You can definitely use them to track the schedule and budget on smaller projects.

But once your projects start getting more complex, your formulas are going to need to take into account that you've got several people all doing different activities, and that could make it harder to track whether you're ahead of schedule or over budget.

So now that you know how to calculate PV and EV, they're all you need to stay on top of everything. What are you waiting for? Flip the page to find out how!

Is your project behind or ahead of schedule?

Figuring out if you're on track in a small project with just a few people is easy. But what if you have dozens or hundreds of people doing lots of different activities? And what if some of them are on track, some are ahead of schedule, and some of them are behind? It starts to get hard to even figure out whether you're meeting your goals.

Wouldn't it be great if there were an easy way to figure out if you're ahead or behind schedule? Well, good news: that's exactly what earned value is used for!

Schedule performance index (SPI)

If you want to know whether you're ahead of or behind schedule, use SPIs. The key to using this is that when you're **ahead of schedule**, you've **earned more value** than planned! So **EV will be bigger than PV**.

To work out your SPI, you just divide your EV by your PV.

$$SPI = \frac{EV}{PV}$$

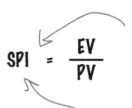

If SPI is greater than one, that means EV is bigger than PV, so you're ahead of schedule!

If SPI is less than one, then you're behind schedule because the amount you've actually worked (EV) is less than what you'd planned (PV).

Schedule variance (SV)

It's easy to see how variance works. The **bigger the difference** between *what you planned* and *what you actually earned*, the **bigger the variance**.

So, if you want to know how much ahead or behind schedule you are, just subtract PV from EV.

Remember, for the sponsor's benefit, we measure this in dollars...

$$SV = EV - PV$$

... so if the variance is positive, it tells you exactly how many dollars you're ahead. If it's negative, it tells you how many dollars you're behind.

> **Don't get freaked out by the thought of all these formulas.**
>
> They're really not very complex. All you need to remember is that they all use EV and PV in different ways. Once you've learned how EV and PV interact in each one, you're golden!

Relax

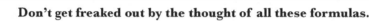

Sharpen your pencil

Meanwhile, back in the Lounge, Alice is working out if the project's coming in on schedule and on budget. Here are the steps she's taking and her notes. She was called away, so it's up to you to work out whether the guys need to push the schedule.

1 **Start with the schedule and budget.** Figure out how much work you planned, how much the team has done, and the total budget (BAC).

BAC = _____

Jeff and Charles have a <u>total budget</u> of <u>$10,000,</u> and they're currently <u>halfway through</u> the schedule.

Planned % complete = _____

2 **Figure out PV.** Multiply the BAC by the percentage of the work that your schedule says the team should have worked so far to get the planned value.

So their planned value is?

PV = $ ____ × % = $ ____

PV = BAC × Planned % complete

3 **Figure out EV.** This is the part that actually takes some thinking! You need to figure out what percentage of work the team has actually done. Once you have that, multiply it with the BAC to find the earned value.

Uh-oh! On a closer look, it seems they've really only gotten 40% of the work done.

EV = $ ____ × ____ = $ ____

EV = BAC × Actual % complete

4 **Now you can calculate SPI and SV.** Once you've figured out EV and PV, you can do the calculations.

Now that you have the EV and PV, you can tell Jeff and Charles if they're getting their money's worth!

SPI = $ ____ ÷ $ ____ = 0.8

SV = $ ____ − $ ____ = $ ____

5 **How's the schedule looking?** What do all these figures tell us?

So are we ahead of schedule or behind it?

⟶ Answers on page 389.

Are you over budget?

You can do the same thing for your budget that you can do for your schedule. The calculations are almost exactly the same, except instead of using planned value—which tells you how much work the schedule says you should have done so far—you use **actual cost** (AC). That's the amount of money that you've spent so far on the project.

Remember, EV measures the work that's been done, while AC tells you how much you've spent so far.

Measuring your cost difference in dollars is easy, but what if your schedule variance is negative?

A lot of people worry about that, but it's actually not bad. Planned value just means that you planned on delivering a certain amount of value to your sponsor at a certain time. An SV of, say, −$5,000 tells you that you haven't delivered all the value you promised.

Cost performance index (CPI)

$$CPI = \frac{EV}{AC}$$

If you want to know whether you're over or under budget, use CPI.

Cost variance (CV)

This tells you the difference between what you planned on spending and what you actually spent.

So, if you want to know how much under or over budget you are, just take AC away from EV.

$$CV = EV - AC$$

Remember what CV means to the sponsor: EV says how much of the total value of the project has been earned back so far. If CV is negative, then he's not getting good value for his money.

To-complete performance index (TCPI)

This tells you how well your project will need to perform to stay on budget.

$$TCPI = \frac{(BAC-EV)}{(BAC-AC)}$$

We'll talk about this in just a few pages...

You're within your budget if...

CPI is greater than or equal to 1 and CV is positive. When this happens, your actual costs are less than earned value, which means the project is delivering more value than it costs.

Now Alice can take a look at the Lounge's checkbook. She figures out that she spent $5,750 on the project so far.

CPI = $4,000 ÷ $5,750 = 0.696

Since CPI is less than 1, it means that Jeff and Charles have blown their budget.

You've blown your budget if...

CPI is less than 1 and CV is negative. When your actual costs are more than earned value, that means that your sponsor is not getting his money's worth of value from the project.

CV = $4,000 − $5,750 = −$1,750

And that's how much they've gone over! Jeff, Charles, and Alice had better figure out how to contain those runaway costs, or they'll have a nasty surprise later.

The earned value management formulas

Earned value management (EVM) is just one of the tools and techniques in the Control Costs process, but it's a big part of PMP exam preparation. When you use these formulas, you're measuring and analyzing how far off your project is from your plan. Remember, think of everything in terms of how much value you're delivering to your sponsor! Take a look at the formulas one more time:

> Remember, your sponsor always cares most about what the project is worth to him. BAC says how much value he's getting for the whole project, and EV tells him how much of that value he's gotten so far.

Name	Formula	What it says	Why you use it
BAC—Budget at completion	No formula—it's the project budget	How much money you'll spend on the project	To tell the sponsor the total amount of value that he's getting for the project
PV—Planned value	$PV = BAC \times \text{Planned \% complete}$	What your schedule says you should have spent	To figure out what value your plan says you should have delivered so far
EV—Earned value	$EV = BAC \times \text{Actual \% complete}$	How much of the project's value you've really earned	To translate how much work the team's finished into a dollar value
AC—Actual cost	What you've actually spent on the project	How much you've actually spent so far	The amount of money you spend doesn't always match the value you get!
SPI—Schedule performance index	$SPI = \dfrac{EV}{PV}$	Whether you're behind or ahead of schedule	To figure out whether you've delivered the value your schedule said you would
SV—Schedule variance	$SV = EV - PV$	How much behind or ahead of schedule you are	To put a dollar value on exactly how far ahead or behind schedule you are
CPI—Cost performance index	$CPI = \dfrac{EV}{AC}$	Whether you're within your budget or not	Your sponsor is always most interested in the bottom line!
TCPI—To-complete performance index	$TCPI = \dfrac{BAC-EV}{BAC-AC}$	How well your project must perform to stay on budget	To forecast whether or not you can stick to your budget
CV—Cost variance	$CV = EV - AC$	How much above or below your budget you are	Your sponsor needs to know how much it costs to get him the value you deliver

Interpret CPI and SPI numbers to gauge your project

The whole idea behind earned value management is that you can use it to easily put a number on how your project is doing. That's why there will be exam questions that test you on your ability to interpret these numbers! Luckily, it's pretty easy to evaluate a project based on the EVM formulas.

If the SPI is below 1, then your project is behind schedule. But if you see a CPI under 1, your project is over budget!

If your project is on track, that means you're delivering the value you promised.

You can tell that your project is on track because the two index numbers—CPI and SPI—are both very close to 1, and the variance numbers—CV and SV—are very close to zero dollars. It's very rare that you'll get exactly to a CPI of 1 or a SV of $0, but a SPI of 1.02 means you're very close to on time, and a CV of –$26 means you're very close to on budget.

A lot of PMOs have a rule where a CPI or SPI between 0.95 and 1.10 is absolutely fine!

Sometimes you'll see negative values written in parentheses— in this case, ($26).

Ahead of schedule or under budget

You can tell if your project is ahead of schedule or under budget by looking for larger numbers.

If the **CPI** is much **bigger than 1**, it means you're ***under budget***. And you can tell how much under by looking at the CV—that's what variance is for! It helps you see just how much the actual cost **varies** from the value you were supposed to earn by now.

Being a long way under budget isn't always a good thing. It means you asked for and were given resources that you didn't need—and which your company could have used elsewhere.

Behind schedule or over budget

A project that's behind schedule or over budget will have lower numbers.

When you see a **SPI** that's **between 0 and 1**, that tells you that the project is ***behind schedule***…and that means you're not delivering enough value to the sponsor! That's when you check the SV to see how much less value you're delivering. And the same goes for cost—a **low CPI** means that your project is ***over budget***, and CV will tell you how much more value you promised to deliver to the sponsor.

CPI and SPI can't be below zero, because they're ratios!

OH, I GET IT: SPI AND CPI ARE JUST RATIOS! IF SPI IS REALLY CLOSE TO 1, THEN SV WILL BE REALLY CLOSE TO ZERO—AND IT MEANS THAT MY PROJECT IS GOING AS PLANNED!

Exactly! And when your CPI is really close to 1, it means that every dollar your sponsor's spending on the project is earning just about a dollar in value.

The biggest thing to remember about all of these numbers is that **the lower they are, the worse your project is doing**. If you've got a SPI of 1.1 and a CPI of 1.15, then you're within your budget and ahead of schedule. But if you calculate a SPI of 0.6 and a CPI of 0.45, then you're behind schedule and you've blown your budget. And when these ratios are below 1, then you'll see a negative variance!

Make it Stick

EV = BAC × Actual % Complete

LOWER = LOSER

wash the dog

Remember:

Lower = Loser

If CPI or SPI is below 1, or if CV or SV is negative, then you've got trouble!

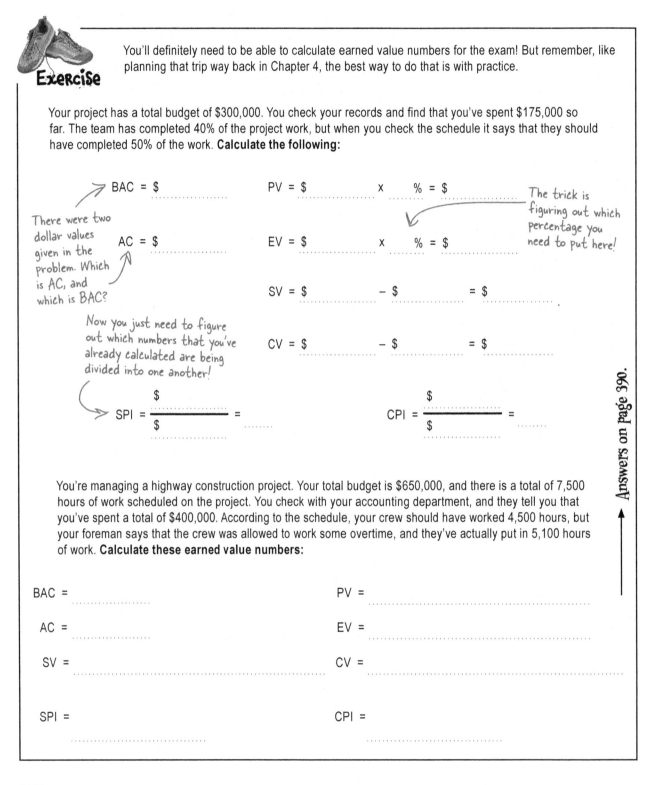

Exercise

You'll definitely need to be able to calculate earned value numbers for the exam! But remember, like planning that trip way back in Chapter 4, the best way to do that is with practice.

Your project has a total budget of $300,000. You check your records and find that you've spent $175,000 so far. The team has completed 40% of the project work, but when you check the schedule it says that they should have completed 50% of the work. **Calculate the following:**

BAC = $ ....

There were two dollar values given in the problem. Which is AC, and which is BAC?

AC = $ ....

Now you just need to figure out which numbers that you've already calculated are being divided into one another!

PV = $ .... x ...... % = $ ....

The trick is figuring out which percentage you need to put here!

EV = $ .... x ...... % = $ ....

SV = $ .... – $ .... = $ ....

CV = $ .... – $ .... = $ ....

$$SPI = \frac{\$ ....}{\$ ....} = ...$$

$$CPI = \frac{\$ ....}{\$ ....} = ...$$

You're managing a highway construction project. Your total budget is $650,000, and there is a total of 7,500 hours of work scheduled on the project. You check with your accounting department, and they tell you that you've spent a total of $400,000. According to the schedule, your crew should have worked 4,500 hours, but your foreman says that the crew was allowed to work some overtime, and they've actually put in 5,100 hours of work. **Calculate these earned value numbers:**

BAC = ...

AC = ...

SV = ...

SPI = ...

PV = ...

EV = ...

CV = ...

CPI = ...

Answers on page 390.

Exercise

You are the project manager at an industrial design firm. You expect to spend a total of $55,000 on your current project. Your plan calls for six people working on the project eight hours a day, five days a week, for four weeks. According to the schedule, your team should have just finished the third week of the project. When you review what the team has done so far, you find that they have completed 50% of the work, at a cost of $25,000. **Based on this information, calculate the earned value numbers:**

BAC = ...

PV = ...

AC = ...

EV = ...

SV = ...

CV = ...

SPI = ...

CPI = ...

Check all of the following that apply:

.... The project is ahead of schedule

.... The project is over budget

.... The project is behind schedule

.... The project is under budget

.... You should consider crashing the schedule

.... You should find a way to cut costs

Your current project is an $800,000 software development effort, with two teams of programmers that will work for six months, at a total of 10,000 hours. According to the project schedule, your team should be done with 38% of the work. You find that the project is currently 40% complete. You've spent 50% of the budget so far. **Calculate these numbers:**

BAC = ...

PV = ...

AC = ...

EV = ...

SV = ...

CV = ...

SPI = ...

CPI = ...

Check all of the following that apply:

.... The project is ahead of schedule

.... The project is over budget

.... The project is behind schedule

.... The project is under budget

.... You should consider crashing the schedule

.... You should find a way to cut costs

Answers on page 391.

Forecast what your project will look like when it's done

There's another piece of earned value management, and it's part of the last tool and technique in Cost Management: **forecasting**. The idea behind forecasting is that you can use earned value to come up with a pretty accurate prediction of what your project will look like at completion.

If you know your CPI now, you can use it to predict what your project will actually cost when it's complete. Let's say that you're managing a project with a CPI of 0.8 today. If you assume that the CPI will be 0.8 for the rest of the project—and that's not an unreasonable assumption when you're far along in the project work—then you can predict your total costs when the project is complete. We call that **estimate at completion** (EAC).

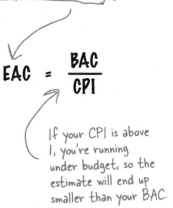

If your CPI is _below 1_, that means you're running _over budget_—which will give you an EAC that's _larger_ than your current budget.

$$EAC = \frac{BAC}{CPI}$$

If your CPI is above 1, you're running under budget, so the estimate will end up smaller than your BAC.

> There are a bunch of different ways to calculate EAC, but this one is sufficient for the PMP exam.

Meanwhile, back in the Lounge

Alice is forecasting how the new Lounge project will look when it's done.

IF JEFF AND CHARLES HAVE A CPI OF 0.869 AND A TOTAL BUDGET OF $10,000, THEN THEY CAN FORECAST THEIR FINAL COSTS: EAC = BAC ÷ CPI...

Here's what Alice wrote down first...

$$\frac{\$10,000}{0.869} = \$11,507$$

...now Alice can take a look at the Lounge's checkbook. She figures out that she spent $5,750 on the project so far...

Once you've got an estimate, you can calculate a variance!

There are two useful numbers that you can compute with the EAC. One of them is called **estimate to complete** (ETC), which tells you how much more money you'll probably spend on your project. And the other one, **variance at completion** (VAC), predicts what your variance will be when the project is done.

You can use **EAC, ETC, and VAC** to predict what your earned value numbers will look like when your project is complete.

$$ETC = EAC - AC$$

Since EAC predicts how much money you'll spend, if you subtract the AC, you'll find out how much money the rest of the project will end up costing.

$$VAC = BAC - EAC$$

If you end up spending more than your budget, the VAC will be negative...just like CV and SV!

> IF THE EAC IS $11,507, AND THE AC IS $5,750, THEN I CAN FIGURE OUT WHAT JEFF AND CHARLES HAVE LEFT TO SPEND. ETC = EAC − AC.

> SO, WE'RE OVER BUDGET. BUT WHAT WILL THE DAMAGE BE? AT LEAST NOW I CAN FIGURE OUT THE FINAL VARIANCE. VAC = BAC − EAC.

$$\$11,507 - \$5,750 = \$5,757$$

...and now she knows how much money the rest of the project is likely to cost...

$$\$10,000 - \$11,507 = -\$1,507$$

...but will the guys be able to come up with the extra money?

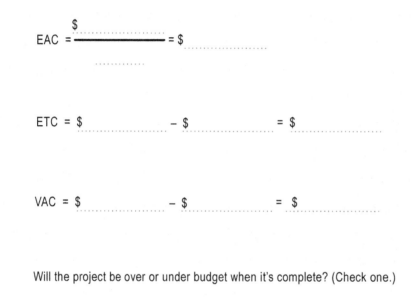

Exercise You're a project manager working on a large project scheduled to last for two years. You've got six different teams working on five major functional areas. Some teams are ahead of schedule, and others are falling behind. That means that you have cost overruns in some areas, but you've saved costs in others—and that's making it very hard to get an intuitive grasp on whether your project is over or under budget!

It's nine months into your project. The total budget for your project is $4,200,000. You've spent $1,650,000 so far, and you've got a CPI of .875. Use the earned value management formulas from forecasting to figure out where things stand.

$$\text{EAC} = \frac{\$\,\rule{2cm}{0.15mm}}{\rule{1.5cm}{0.15mm}} = \$\,\rule{2.5cm}{0.15mm}$$

$$\text{ETC} = \$\,\rule{2cm}{0.15mm} - \$\,\rule{2cm}{0.15mm} = \$\,\rule{2.5cm}{0.15mm}$$

$$\text{VAC} = \$\,\rule{2.5cm}{0.15mm} - \$\,\rule{2.5cm}{0.15mm} = \$\,\rule{2cm}{0.15mm}$$

Will the project be over or under budget when it's complete? (Check one.)

_____ The project will be over budget   _____ The project will be within its budget

How much will the project be over or under budget? $ _____

Now it's six months later, and your project looks very different. You need to work out a new forecast for what your budget situation will be like at project completion. You've now spent a total of $2,625,000. You look at all of the activities done by the team, and you find that the project is 70% complete. Can you come up with a new forecast for your project?

BAC = $ ....

AC = $ ....

EV = ...

CPI =

....

EAC =

....

ETC =

....

VAC =

....

Your project will be **over/under** budget at completion.
(Circle one.)

How much will the project be over or under budget?

$ ....

———————➤ Answers on page 392.

Finding missing information

Most of the earned value questions on the exam will be pretty straightforward: you'll be given the numbers that you need to plug into a formula, and when you do it you'll get the answer. But once in a while, you'll get a question that isn't quite so straightforward.

Let's say you're given...

...the CPI and earned value, and you want to figure out the actual costs. Why would you ever see this? Sometimes it's hard to figure out how important a project is unless you know how much it's really spending—if a project is more expensive, people in your company probably care more about it. If you're told that **a project's CPI is 1.14 and its EV is $350,000**, how do you figure out the actual costs?

...the earned value and actual percent complete, and you want to figure out the project's budget. This can be really helpful when you need to "read between the lines" to make a decision about a project when someone doesn't want to give you all the information you need. When you have **a project's EV of $438,750 and its actual % complete of 32.5%**, how do you figure out the total budget (BAC)?

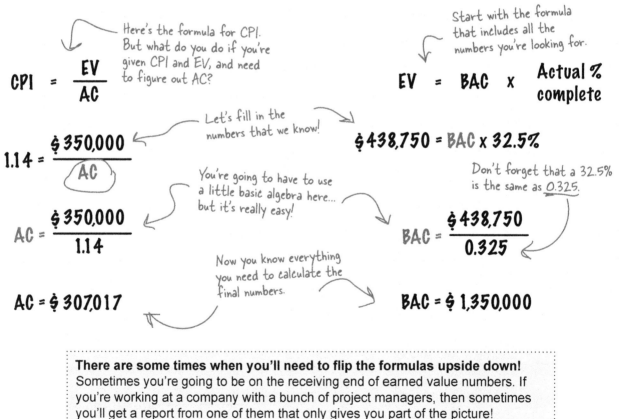

Here's the formula for CPI. But what do you do if you're given CPI and EV, and need to figure out AC?

$$CPI = \frac{EV}{AC}$$

Let's fill in the numbers that we know!

$$1.14 = \frac{\$350,000}{AC}$$

You're going to have to use a little basic algebra here... but it's really easy!

$$AC = \frac{\$350,000}{1.14}$$

Now you know everything you need to calculate the final numbers.

$$AC = \$307,017$$

Start with the formula that includes all the numbers you're looking for.

$$EV = BAC \times Actual\ \%\ complete$$

$$\$438,750 = BAC \times 32.5\%$$

Don't forget that a 32.5% is the same as 0.325.

$$BAC = \frac{\$438,750}{0.325}$$

$$BAC = \$1,350,000$$

> **There are some times when you'll need to flip the formulas upside down!** Sometimes you're going to be on the receiving end of earned value numbers. If you're working at a company with a bunch of project managers, then sometimes you'll get a report from one of them that only gives you part of the picture!

Sharpen your pencil

You'll probably get a question or two where you'll need to flip your formulas over to figure out one of the values you'd normally be given. **Don't worry if you're math-phobic!** This is really easy—you'll definitely get it with a little practice.

If EV is $93,406 and SPI is 0.91, what is the planned value?

Write down the formula for SPI.

$$SPI = \frac{\text{............}}{\text{............}} = \text{............} = \$ \text{_____}$$

Fill in the numbers that you have.

Now flip around the formula so PV is on the left.

$$PV = \frac{\text{............}}{\text{............}} \qquad PV = \text{............}$$

And now you can solve for PV!

If PV is $252,000 and BAC is $350,000, what is the planned percent complete?

Fill in the numbers that you have.

Start with the formula for PV.

$$PV = \text{............} \quad X \qquad \$ \text{............} = \$ \text{............} \quad X$$

Now flip around the formula so % complete is on the left.

$$\% \text{ complete} = \frac{\$ \text{............}}{\$ \text{............}} \qquad \% \text{ complete} = \text{............}$$

And now you can solve it!

Now try one on your own. If BAC is $126,500 and EAC is $115,000, what is the CPI?

❶ First write out the formula that has EAC, CPI, and BAC.

❷ Next fill in the numbers that you know.

❸ Now flip around the formula so the number you're looking for is on the left.

❹ Now you can solve the problem!

Answers on page 393.

there are no
Dumb Questions

Q: What does CPI really mean, and why can it predict your final budget?

A: Doesn't it seem a little weird that you can come up with a pretty accurate forecast of what you'll actually spend on your project just by dividing CPI into your BAC, or the total amount that you're planning to spend on the project? How can there be one "magic" number that does that for you?

But when you think about it, it actually makes sense. Let's say that you're running 15% over budget today. If your budget is $100,000, then your CPI will be $100.000 ÷ $115,000 = .87. One good way to predict what your final budget will look like is to assume that you'll keep running 15% over budget. Let's say your total budget is $250,000. If you're still 15% over at the end of the budget, your final CPI will still be $250,000 ÷ $287,500 = .87! Your CPI will always be .87 if you're 15% over budget.

That's why we call that forecast EAC—it's an *estimate* of what your budget will look like *at completion*. By dividing CPI into BAC, all you're doing is calculating what your final budget will be if your final budget overrun or underrun is exactly the same as it is today.

Q: Is that really the best way to estimate costs? What if things change between now and the end of the project?

A: EAC is a good way to estimate costs, because it's easy to calculate and relatively accurate—assuming that nothing on the project changes too much. But you're right, if a whole lot of unexpected costs happen or your team members figure out a cheaper and better way to get the job done, then an EAC forecast could be way off!

It turns out that there are over 25 different ways to calculate EAC, and the one in this chapter is just one of them. Some of those other formulas take risks and predictions into account. But for the PMP exam, you just need to know EAC = BAC ÷ CPI.

Q: Wow, there are a lot of earned value formulas! Is there an easy way to remember them?

A: Yes, there are a few ways that help you remember the earned value formulas. One way is to notice that the performance Reporting formulas all have something either being divided into or subtracted from EV. This should make sense—the whole point of earned value management is that you're trying to figure out how much of the value you're delivering to your sponsor has been earned so far. Also, remember that a variance is always subtraction, and an index is always division. The schedule formulas SV and SPI both involve PV numbers you got from your schedule, while the cost formulas CV and CPI both involve AC numbers from your budget.

And remember, the lower the index or variance, the worse your project is doing! A negative variance or an index that's below 1 is bad, while a positive variance or an index that's above 1 is good!

The earned value formulas have numbers divided into or subtracted from EV. SV and SPI use PV, while CV and CPI use AC.

Keep your project on track with TCPI

You can use earned value to gauge where you need to be to get your project in under budget. TCPI can help you find out not just whether or not you're on target, but exactly where you need to be to make sure you get things done with the money you have.

Have you ever wondered halfway through a project just how much you'd have to cut costs in order to get it within your budget? This is how you figure that out!

To-complete performance index (TCPI)

This number represents a **target that your CPI would have to hit** in order to hit your forecasted completion cost. If you're performing within your budgeted cost, it'll be based on your BAC. If you're running over your budget, you'll have to estimate a new EAC and base your TCPI on that.

There are two different formulas for TCPI. One is for when you're trying to get your project within your original budget, and the other is for when you are trying to get your project done within the EAC you've determined from earned value calculations.

BAC based:

$$TCPI = \frac{(BAC-EV)}{(BAC-AC)}$$

How much budgeted work is left divided by how much budgeted money is left

EAC based:

$$TCPI = \frac{(BAC-EV)}{(EAC-AC)}$$

How much budgeted work is left divided by how much estimated money is left

TCPI for the Head First Lounge renovation project

Alice figured out the BAC and EAC for the bar project and realized that the lounge was over budget, so she did a TCPI calculation to figure out exactly where needed to keep her CPI if she wanted to get the project in without blowing the budget. Alice's earned value calculations have put the lounge renovation project's numbers here:

EAC= $11,507   **AC= $5,750**   **BAC=$10,000**   **EV=$5000**

The project is over budget! So Alice uses the BAC-based formula to figure out where she needs to keep the CPI for the project if she wants to complete it within the original budget. Here's the calculation:

$$TCPI = \frac{(BAC-EV)}{(BAC-AC)} \qquad TCPI = \frac{(\$10,000-\$5,000)}{(\$10,000-\$5,750)} = 1.17$$

So, if the project were going to get back on budget, it would have to run at a 1.17 CPI for the rest of the project to make up for the initial overage. Alice doesn't think that's going to happen. Jeff and Charles pushed for stucco in the lounge that cost an extra $750 in the beginning, and, the way things are going, it's probably a safe bet that there will be a few more cost overruns like that as the project goes on. She prepared a second TCPI to see what the numbers would be to complete the project based on the current EAC.

$$TCPI = \frac{(BAC-EV)}{(EAC-AC)} \qquad TCPI = \frac{(\$10,000-\$5,000)}{(\$11,507-\$5,750)} = .87$$

A high TCPI means a tight budget

When you're looking at the TCPI for a project, a higher number means it's time to take a stricter cost management approach. The higher the number, the more you're going to have to rein in spending on your project and cut costs. When the number is lower than one, you know you're well within your budget and you can relax a bit.

Remember "lower = loser"? Well, with TCPI, it's the opposite. A higher number means that your budget is too tight. You want it lower to give you more room to spend money!

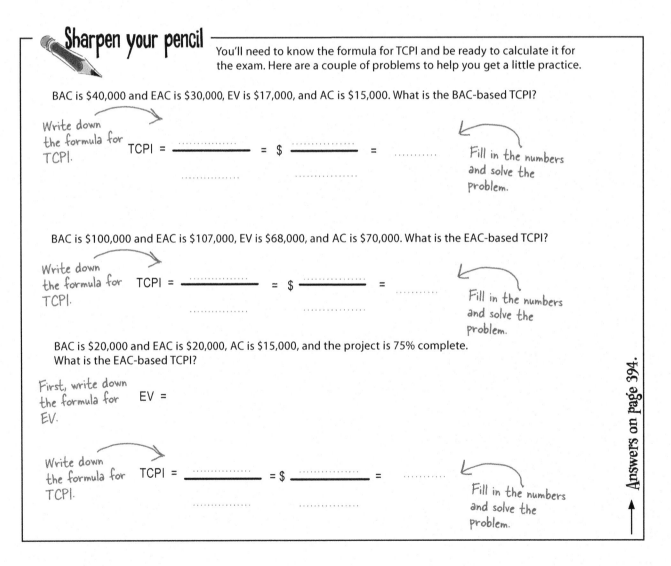

Sharpen your pencil

You'll need to know the formula for TCPI and be ready to calculate it for the exam. Here are a couple of problems to help you get a little practice.

BAC is $40,000 and EAC is $30,000, EV is $17,000, and AC is $15,000. What is the BAC-based TCPI?

Write down the formula for TCPI.

TCPI = ———— = $ ———— = ...

Fill in the numbers and solve the problem.

BAC is $100,000 and EAC is $107,000, EV is $68,000, and AC is $70,000. What is the EAC-based TCPI?

Write down the formula for TCPI.

TCPI = ———— = $ ———— = ...

Fill in the numbers and solve the problem.

BAC is $20,000 and EAC is $20,000, AC is $15,000, and the project is 75% complete. What is the EAC-based TCPI?

First, write down the formula for EV.

EV =

Write down the formula for TCPI.

TCPI = ———— = $ ———— = ...

Fill in the numbers and solve the problem.

Answers on page 394.

Party time!

Jeff and Charles finished the new Lounge! It looks great, and they're really happy about it…because Alice managed their costs well. She used earned value to correct their budget problems, and they managed to cut a few costs while they still had time. And they had just enough money left over at the end to throw a great party for her!

Since she used earned value to stay on top of the budget throughout the project, Jeff and Charles were able to renovate the Lounge in style and still manage to stay within their spending limits.

Sharpen your pencil Solution

Now it's your turn! See if you can figure out BAC and PV for a typical project.

1. You're managing a project to install 200 windows in a new skyscraper and need to figure out your budget. Each week of the project costs the same: your team members are paid a total of $4,000 every week, and you need $1,000 worth of parts each week to do the work. If the project is scheduled to last 16 weeks, what's the BAC for the project?

BAC = **$ 5,000 × 16 = $ 80,000**

The project's 16 weeks long. Multiply that by the costs per week to get the total budget for the project.

Each week costs $4,000 for labor and $1,000 for parts.

2. What will the planned % complete be four weeks into the project?

Planned % complete = **25%**

You're 4 weeks into a 16-week project. That means you're 25% of the way through.

3. What should the PV be four weeks into the project?

Fill in the BAC from question 1.

PV = **$ 80,000 × 25% = $ 20,000**

Fill in the planned % complete from question 2. Now multiply them to get the PV.

Sharpen your pencil Solution

Let's get back to that 16-week project from page 364. Can you figure out how to use EV?

1. Fast-forward four weeks into the project installing those 200 skyscraper windows. Fill in the BAC and PV you figured out above. (Check your answer above to make sure you got it right!)

BAC = **$ 80,000**   PV = **$ 20,000**

2. You've checked with your team, but they have bad news. The schedule says they were supposed to have installed 50 windows by now, but they've only installed 40. Can you figure out the actual % complete?

Actual % complete $= \dfrac{40}{200} =$ **20%**

The team installed 40 windows out of a total of 200. That means they're 20% of the way done with the work.

3. What should the earned value be right now?

EV = **$ 80,000 × 20% = $ 16,000**

4. Look at the planned value, and then look at the earned value. Are you delivering all the value you planned on delivering?

☐ Yes   ☑ No

You planned on delivering $20,000 worth of value, but you've only delivered $16,000 worth. That means the customer isn't getting all the value he's paying for!

Sharpen your pencil
Solution

Meanwhile, back in the Lounge, Alice is working out if the project's coming in on schedule and on budget. Here are the steps she's taking and her notes. She was called away, so it's up to you to work out whether the guys need to push the schedule.

1 **Start with the schedule and budget.** Figure out how much work you planned, how much the team has done, and the total budget (BAC).

Jeff and Charles have a <u>total budget</u> of <u>$10,000</u>, and they're currently <u>halfway through</u> the schedule.

BAC = <u>$10,000</u>

Planned % complete = <u>50%</u>

2 **Figure out PV.** Multiply the BAC by the percentage of the work that your schedule says the team should have worked so far to get the planned value.

So their planned value is?

PV = <u>$10,000</u> × <u>50%</u> = <u>$5,000</u>

$$PV = BAC \times Planned\ \%\ complete$$

3 **Figure out EV.** This is the part that actually takes some thinking! You need to figure out what percentage of work the team has actually done. Once you have that, multiply it with the BAC to find the earned value.

Uh-oh! On a closer look, it seems they've really only gotten 40% of the work done.

EV = <u>$10,000</u> × <u>40%</u> = <u>$4,000</u>

$$EV = BAC \times Actual\ \%\ complete$$

4 **Now you can calculate SPI and SV.** Once you've figured out EV and PV, you can do the calculations.

Now that you have the EV and PV, you can tell Jeff and Charles if they're getting their money's worth!

SPI = <u>$4,000</u> ÷ <u>$5,000</u> = <u>0.8</u>

SV = <u>$4,000</u> - <u>$5,000</u> = -<u>$1,000</u>

5 **How's the schedule looking?** What do all these figures tell us?

So are we ahead of schedule or behind it?

The Lounge project is behind schedule.

Exercise

You'll definitely need to be able to calculate earned value numbers for the exam! But remember, like planning that trip way back in Chapter 4, the best way to do that is with practice.

Your project has a total budget of $300,000. You check your records and find that you've spent $175,000 so far. The team has completed 40% of the project work, but when you check the schedule it says that they should have completed 50% of the work. **Calculate the following:**

BAC = $ **300,000**

PV = $ **300,000** × **50** % = $ **150,000**

> *Planned value uses what's on the schedule; earned value uses what actually happened.*

AC = $ **175,000**

EV = $ **300,000** × **40** % = $ **120,000**

Did you notice how the formulas for SV and SPI use the same numbers? You subtract for one, and divide for the other!

SV = $ **120,000** − $ **150,000** = $ **-30,000**

> *You may have to round the CPI and SPI numbers. Don't worry; since the PMP exam is multiple choice, you'll see a match!*

CV = $ **120,000** − $ **175,000** = $ **-55,000**

$$SPI = \frac{\$120,000}{\$150,000} = 0.8$$

The formulas for CV and CPI use the same numbers, too.

$$CPI = \frac{\$120,000}{\$175,000} = 0.68$$

You're managing a highway construction project. Your total budget is $650,000, and there is a total of 7,500 hours of work scheduled on the project. You check with your accounting department, and they tell you that you've spent a total of $400,000. According to the schedule, your crew should have worked 4,500 hours, but your foreman says that the crew was allowed to work some overtime, and they've actually put in 5,100 hours of work. **Calculate these earned value numbers:**

BAC = $ **650,000**

4,500 out of a total of 7,500 hours you planned to work:
4,500 ÷ 7,500 = 60%

PV = $ **650,000** × **60%** = $ **390,000**

AC = $ **400,000**

Do the same for actual hours:
5,100 ÷ 7,500 = 68%

EV = $ **650,000** × **68%** = $ **442,000**

SV = $ **442,000** − $ **390,000** = $ **52,000**

CV = $ **442,000** − $ **400,000** = $ **42,000**

$$SPI = \frac{\$442,000}{\$390,000} = 1.13$$

$$CPI = \frac{\$442,000}{\$400,000} = 1.11$$

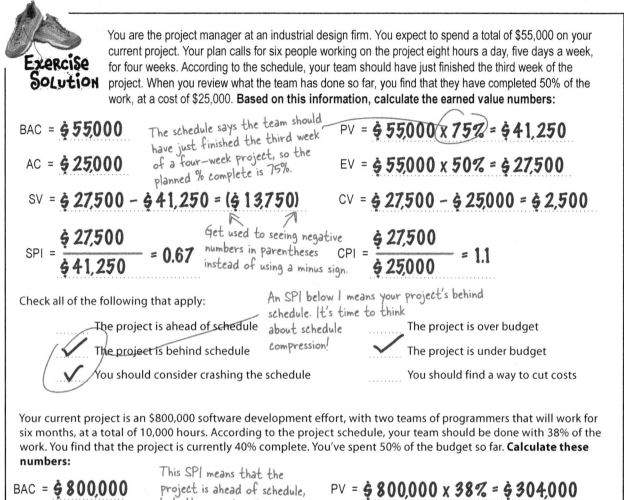

Exercise Solution

You are the project manager at an industrial design firm. You expect to spend a total of $55,000 on your current project. Your plan calls for six people working on the project eight hours a day, five days a week, for four weeks. According to the schedule, your team should have just finished the third week of the project. When you review what the team has done so far, you find that they have completed 50% of the work, at a cost of $25,000. **Based on this information, calculate the earned value numbers:**

BAC = $55,000

The schedule says the team should have just finished the third week of a four-week project, so the planned % complete is 75%.

PV = $55,000 x 75% = $41,250

AC = $25,000

EV = $55,000 x 50% = $27,500

SV = $27,500 − $41,250 = ($13,750)

CV = $27,500 − $25,000 = $2,500

Get used to seeing negative numbers in parentheses instead of using a minus sign.

SPI = $\dfrac{\$27,500}{\$41,250}$ = 0.67

CPI = $\dfrac{\$27,500}{\$25,000}$ = 1.1

Check all of the following that apply:

An SPI below 1 means your project's behind schedule. It's time to think about schedule compression!

☐ The project is ahead of schedule

✓ The project is behind schedule

✓ You should consider crashing the schedule

☐ The project is over budget

✓ The project is under budget

☐ You should find a way to cut costs

Your current project is an $800,000 software development effort, with two teams of programmers that will work for six months, at a total of 10,000 hours. According to the project schedule, your team should be done with 38% of the work. You find that the project is currently 40% complete. You've spent 50% of the budget so far. **Calculate these numbers:**

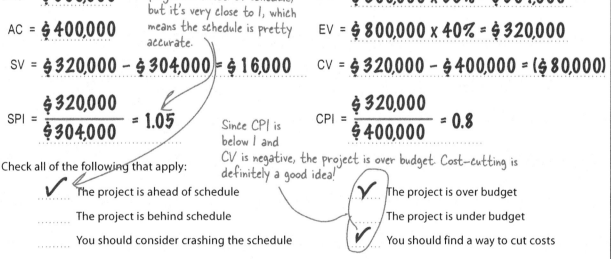

BAC = $800,000

This SPI means that the project is ahead of schedule, but it's very close to 1, which means the schedule is pretty accurate.

PV = $800,000 x 38% = $304,000

AC = $400,000

EV = $800,000 x 40% = $320,000

SV = $320,000 − $304,000 = $16,000

CV = $320,000 − $400,000 = ($80,000)

SPI = $\dfrac{\$320,000}{\$304,000}$ = 1.05

CPI = $\dfrac{\$320,000}{\$400,000}$ = 0.8

Since CPI is below 1 and CV is negative, the project is over budget. Cost-cutting is definitely a good idea!

Check all of the following that apply:

✓ The project is ahead of schedule

☐ The project is behind schedule

☐ You should consider crashing the schedule

✓ The project is over budget

☐ The project is under budget

✓ You should find a way to cut costs

Exercise Solution

You're a project manager working on a large project scheduled to last for two years. You've got six different teams working on five major functional areas. Some teams are ahead of schedule, and others are falling behind. That means that you have cost overruns in some areas, but you've saved costs in others—and that's making it very hard to get an intuitive grasp on whether your project is over or under budget!

It's nine months into your project. The total budget for your project is $4,200,000. You've spent $1,650,000 so far, and you've got a CPI of .875. Use the earned value management formulas from forecasting to figure out where things stand.

$$EAC = \frac{\$4,200,000}{0.875} = \$4,800,000$$

You're starting to get the hang of this stuff! These formulas look a little intimidating at first, but they're really not that bad once you get used to them.

$$ETC = \$4,800,000 - \$1,650,000 = \$3,150,000$$

$$VAC = \$4,200,000 - \$4,800,000 = (\$600,000)$$

Since VAC is negative, it means that you'll be $600,000 over budget at the end of the project.

Will the project be over or under budget when it's complete? (Check one.)

✓ The project will be over budget .... The project will be within its budget

How much will the project be over or under budget? **$600,000**

Now it's six months later, and your project looks very different. You need to work out a new forecast for what your budget situation will be like at project completion. You've now spent a total of $2,625,000. You look at all of the activities done by the team, and you find that the project is 70% complete. Can you come up with a new forecast for your project?

$$BAC = \$4,200,000$$

$$EV = \$4,200,000 \times 70\% = \$2,940,000$$

$$EAC = \frac{\$4,200,000}{1.12} = \$3,750,000$$

$$VAC = \$4,200,000 - \$3,750,000$$
$$= \$450,000$$

$$AC = \$2,625,000$$

$$CPI = \frac{\$2,940,000}{\$2,625,000} = 1.12$$

$$ETC = \$3,750,000 - \$2,625,000$$
$$= \$1,125,000$$

Your project will be **over** / **under** budget at completion. (Circle one.)

How much will the project be over or under budget? **$450,000**

Take a second to think about what these numbers really mean. <u>Are you delivering good value to the sponsor?</u>

This VAC means your project is $450,000 <u>under</u> budget.

Sharpen your pencil Solution

You'll probably get a question or two where you'll need to flip your formulas over to figure out one of the values you'd normally be given. **Don't worry if you're math-phobic!** This is really easy—you'll definitely get it with a little practice.

If EV is $93,406 and SPI is 0.91, what is the planned value?

$$SPI = \frac{EV}{PV} = 0.91 = \$\frac{\$93{,}406}{PV}$$

When you're dividing, you just need to swap these two numbers.

$$PV = \frac{\$93{,}406}{0.91}$$

$$PV = \$102{,}644$$

Sometimes your answers aren't nice, round numbers. That doesn't mean that they're wrong!

If PV is $252,000 and BAC is $350,000, what is the planned percent complete?

$$PV = \underset{BAC}{\ } \times \text{Scheduled \% complete}$$

$$\$252{,}000 = \$350{,}000 \times \text{Scheduled \% complete}$$

$$\% \text{ complete} = \frac{\$252{,}000}{\$350{,}000}$$

$$PV = 72\%$$

Don't forget that when you're calculating a percentage, 72% is the same as 0.72.

Now try one on your own. If BAC is $126,500 and EAC is $115,000, what is the CPI?

1 First write out the formula that has EAC, CPI, and BAC.

$$EAC = \frac{BAC}{CPI}$$

If you're still stumped here, don't worry! You'll only see one or two questions like this on the exam.

2 Next fill in the numbers that you know.

$$\$115{,}000 = \frac{\$126{,}500}{CPI}$$

3 Now flip around the formula so the number you're looking for is on the left.

$$CPI = \frac{\$126{,}500}{\$115{,}000}$$

4 Now you can solve the problem!

$$CPI = 1.1$$

Sharpen your pencil
Solution

You'll need to know the formula for TCPI and be ready to calculate it for the exam. Here are a couple of problems to help you get a little practice.

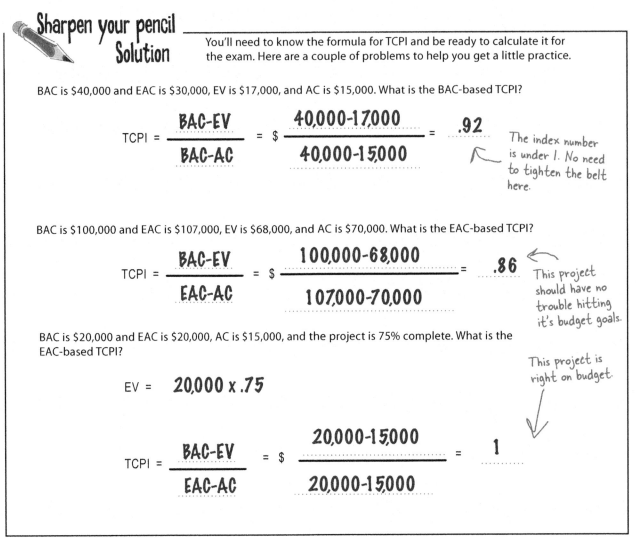

BAC is $40,000 and EAC is $30,000, EV is $17,000, and AC is $15,000. What is the BAC-based TCPI?

$$TCPI = \frac{BAC-EV}{BAC-AC} = \$\frac{40,000-17,000}{40,000-15,000} = .92$$

The index number is under 1. No need to tighten the belt here.

BAC is $100,000 and EAC is $107,000, EV is $68,000, and AC is $70,000. What is the EAC-based TCPI?

$$TCPI = \frac{BAC-EV}{EAC-AC} = \$\frac{100,000-68,000}{107,000-70,000} = .86$$

This project should have no trouble hitting it's budget goals.

BAC is $20,000 and EAC is $20,000, AC is $15,000, and the project is 75% complete. What is the EAC-based TCPI?

This project is right on budget.

$$EV = 20,000 \times .75$$

$$TCPI = \frac{BAC-EV}{EAC-AC} = \$\frac{20,000-15,000}{20,000-15,000} = 1$$

Exam Questions

1. You are creating your cost baseline. What process are you in?

   A.   Determine Budget

   B.   Control Costs

   C.   Estimate Costs

   D.   Cost Baselining

> Some of the earned value numbers have alternate four-letter abbreviations. This one stands for "budgeted cost of work scheduled." Don't worry—you don't need to memorize them!

2. You're working on a project that has an EV of $7,362 and a PV (BCWS) of $8,232. What's your SV?

   A.   −$870

   B. $870

   C.   0.89

   D.   Not enough information to tell

3. You are managing a project for a company that has previously done three projects that were similar to it. You consult with the cost baselines, lessons learned, and project managers from those projects, and use that information to come up with your cost estimate. What technique are you using?

   A.   Parametric estimating

   B.   Net present value

   C.   Rough order of magnitude estimation

   D.   Analogous estimating

4. You are working on a project with a PV of $56,733 and an SPI of 1.2. What's the earned value of your project?

   A. $68,079.60

   B. $47,277.50

   C. $68,733

   D.   .72

5. Your company has two projects to choose from. Project A is a billing software project for the Accounts Payable department; in the end it will make the company around $400,000 when it has been rolled out to all of the employees in that department. Project B is a payroll application that will make the company around $388,000 when it has been put to use throughout the company. After a long deliberation, your board chooses to go ahead with Project B. What is the opportunity cost for choosing Project B over Project A?

   A. $388,000

   B. $400,000

   C. $12,000

   D.   1.2

Exam Questions

6. Your company has asked you to provide a cost estimate that includes maintenance, installation, support, and upkeep costs for as long as the product will be used. What is that kind of estimate called?

    A.   Benefit cost ratio

    B.   Depreciation

    C.   Net present value

    D.   Lifecycle costing

7. You are working on a project with an SPI of .72 and a CPI of 1.1. Which of the following BEST describes your project?

    A.   Your project is ahead of schedule and under budget.

    B.   Your project is behind schedule and over budget.

    C.   Your project is behind schedule and under budget.

    D.   Your project is ahead of schedule and over budget.

8. Your project has a BAC of $4,522 and is 13% complete. What is the earned value (EV)?

    A. $3,934.14

    B.   There is not enough information to answer.

    C. $587.86

    D. $4,522

9. A project manager is working on a large construction project. His plan says that the project should end up costing $1.5 million, but he's concerned that he's not going to come in under budget. He's spent $950,000 of the budget so far, and he calculates that he's 57% done with the work, and he doesn't think he can improve his CPI above 1.05. Which of the following BEST describes the current state of the project?

    A.   The project is likely to come in under budget.

    B.   The project is likely to exceed its budget.

    C.   The project right on target.

    D.   There is no way to determine this information.

10. You are managing a project laying underwater fiber optic cable. The total cost of the project is $52/meter to lay 4 km of cable across a lake. It's scheduled to take 8 weeks to complete, with an equal amount of cable laid in each week. It's currently the end of week 5, and your team has laid 1,800 meters of cable so far. What is the SPI of your project?

    A.   1.16

    B.   1.08

    C.   .92

    D.   .72

Exam Questions

11. During the execution of a software project, one of your programmers informs you that she discovered a design flaw that will require the team to go back and make a large change. What is the BEST way to handle this situation?

    A.   Ask the programmer to consult with the rest of the team and get back to you with a recommendation.

    B.   Determine how the change will impact the project constraints.

    C.   Stop all work and call a meeting with the sponsor.

    D.   Update the cost baseline to reflect the change.

12. If AC (ACWP) is greater than your EV (BCWP), what does this mean?

    A.   The project is under budget.

    B.   The project is over budget.

    C.   The project is ahead of schedule.

    D.   The project is behind schedule.

13. A junior project manager is studying for her PMP exam, and asks you for advice. She's learning about earned value management, and wants to know which of the variables represents the difference between what you expect to spend on the project and what you've actually spent so far. What should you tell her?

    A.   Actual cost (AC)

    B.   Cost performance index (CPI)

    C.   Earned value (EV)

    D.   Cost variance (CV)

14. You are managing an industrial architecture project. You've spent $26,410 so far to survey the site, draw up preliminary plans, and run engineering simulations. You are preparing to meet with your sponsor when you discover that there is a new local zoning law will cause you to have to spend an additional estimated $15,000 to revise your plans. You contact the sponsor and initiate a change request to update the cost baseline.

What variable would you use to represent the $26,410 in an earned value calculation?

    A.   PV

    B.   BAC

    C.   AC

    D.   EV

Exam Questions

15. You are working on the project plan for a software project. Your company has a standard spreadsheet that you use to generate estimates. To use the spreadsheet, you meet with the team to estimate the number of functional requirements, use cases, and design wireframes for the project. Then you categorize them into high, medium, or low complexity. You enter all of those numbers into the spreadsheet, which uses a data table derived from past projects' actual costs and durations, performs a set of calculations, and generates a final estimate. What kind of estimation is being done?

    A.   Parametric

    B.   Rough order of magnitude

    C.   Bottom-up

    D.   Analogous

16. Project A has a NPV of $75,000, with an internal rate of return of 1.5% and an initial investment of $15,000. Project B has a NPV of $60,000 with a BCR of 2:1. Project C has a NPV of $80,000, which includes an opportunity cost of $35,000. Based on these projects, which is the BEST one to select:

    A.   Project A

    B.   Project B

    C.   Project C

    D.   There is not enough information to select a project.

17. What is the range of a rough order of magnitude estimate?

    A.   −5% to +10%

    B.   −25% to +75%

    C.   −50% to +50%

    D.   −100% to +200%

18. You are managing a software project when one of your stakeholders needs to make a change that will affect the budget. What defines the processes that you must follow in order to implement the change?

    A.   Perform Integrated Change Control

    B.   Monitoring and Controlling process group

    C.   Change control board

    D.   Cost baseline

Exam Questions

19. You are managing a software project when one of your stakeholders needs to make a change that will affect the budget. You follow the procedures to implement the change. Which of the following must get updated to reflect the change?

    A.   Project Management plan

    B.   Project cost baseline

    C.   Cost change control system

    D.   Project performance reviews

20. You are managing a project with a BAC of $93,000, EV (BCWP) of $51,840, PV (BCWS) of $64,800, and AC (ACWP) of $43,200. What is the CPI?

    A.   1.5

    B.   0.8

    C.   1.2

    D. $9,000

> *Again, don't panic if you see these four-letter abbreviations. You'll always be given the ones you're used to on the exam!*

21. You are managing a project that has a TCPI of 1.19. What is the BEST course of action?

    A.   You're under budget, so you can manage costs with lenience.

    B.   Manage costs aggressively.

    C.   Create a new schedule.

    D.   Create a new budget.

22. You are starting to write your project charter with your project sponsor when the senior managers ask for a time and cost estimate for the project. You have not yet gathered many of the project details. What kind of estimate can you give?

    A.   Analogous estimate

    B.   Rough order of magnitude estimate

    C.   Parametric estimate

    D.   Bottom-up estimate

23. You are managing a project for a defense contractor. You know that you're over budget, and you need to tell your project sponsor how much more money it's going to cost. You've already given him a forecast that represents your estimate of total cost at the end of the project, so you need to take that into account. You now need to figure out what your CPI needs to be for the rest of the project. Which of the following BEST meets your needs?

    A.   BAC

    B.   ETC

    C.   TCPI (BAC calculation)

    D.   TCPI (EAC calculation)

Answers

Exam ~~Questions~~

1. Answer: A

This is really a question about the order of the processes. Determine Budget and Control Costs both use the cost baseline, so it has to be created before you get to them. Cost Baselining isn't a process at all, so you should exclude that from the choices right away. The main output of Determine Budget is the cost baseline and supporting detail, so that's the right choice here.

    D.   Cost Baselining

Watch out for fake processes! This isn't a real process name.

2. Answer: A

This one is just testing whether or not you know the formula for schedule variance. Just plug the values into the SV formula: SV = EV − PV and you get answer A. Watch out for negative numbers, though! Answer B is a trap because it's a positive value. Also, the test will have answers like C that check if you're using the right formula. If you use the SPI formula, that's the answer you'll get! You can throw out D right away—you don't need to do any calculation to know that you have enough information to figure out SV!

    2. You're working on a project that has an EV of $7,362 and a PV (BCWS) of $8,232. What's your SV?

Don't get thrown off by four-letter abbreviations like BCWS—some people have different abbreviations for PV, EV, and AC. The PMP exam will always give you the abbreviations you're familiar with.

3. Answer: D

When you're using the past performance of previous projects to help come up with an estimate, that's called analogous estimation. This is the second time you saw this particular technique—it was also in Chapter 6. So there's a good chance that you'll get an exam question on it.

4. Answer: A

The formula for SPI is: SPI = EV ÷ PV. So you just have to fill in the numbers that you know, which gets you 1.2 = EV ÷ $56,733. Now flip it around. You end up with EV = 1.2 x $56,733, which multiplies out to $68,079.60.

Did you notice the red herring in the question? It didn't matter what the projects were about, only how much they cost!

5. Answer: B

If you see a question asking the opportunity cost of selecting one project over another, the answer is the value of the project that was not selected! So even though the answers were all numbers, there's no math at all in this question.

Exam ~~Questions~~ Answers

Answers

6. Answer: D

This is one of those questions that gives you a definition and asks you to pick the term that's being defined. So which one is it?

Try using the process of elimination to find the right answer! It can't be benefit cost ratio, because you aren't being asked to compare the overall cost of the project to anything to figure out what its benefit will be. Depreciation isn't right—that's about how your project loses value over time, not about its costs. And it's not net present value, because the question didn't ask you about how much value your project is delivering today. That leaves lifecycle costing.

If you don't know the answer to a question, try to eliminate all the answers you know are wrong.

Don't forget: Lower = Loser!

7. Answer: C

When you see an SPI that's lower than one, that means your project is behind schedule. But your CPI is above one, which means that you're ahead on your budget!

8. Answer: C

Use the formula: EV = BAC × actual % complete. When you plug the numbers into the formula, the right answer pops out!

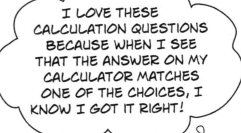

I LOVE THESE CALCULATION QUESTIONS BECAUSE WHEN I SEE THAT THE ANSWER ON MY CALCULATOR MATCHES ONE OF THE CHOICES, I KNOW I GOT IT RIGHT!

9. Answer: B

You might not have recognized this as a TCPI problem immediately, but take another look at the question. It's asking you whether or not a project is going to come in under budget, and that's what TCPI is for. Good thing you were given all of the values you need to calculate it! The actual % complete is 57%, the BAC is $1,500,000, and the AC is $950,000. You can calculate the EV = BAC x actual % complete = $1,500,000 x 57% = $855,000. So now you have everything you need to calculate TCPI: this means he needs a TCPI of 1.17 in order to come in under budget. Since he knows that he can't get better than 1.05, he's likely to blow the budget.

$$TCPI = \frac{BAC - EV}{BAC - AC} \quad \frac{(\$1,500,000 - \$855,000)}{(\$1,500,000 - \$950,000)} = 1.17$$

Answers

Exam ~~Questions~~

10. Answer: D

Some of these calculation questions can get a little complicated—but that doesn't mean they're difficult! Just relax—you can do them!

The formula you need to use is: SPI = EV ÷ PV. But what do you use for EV and PV? If you look at the question again, you'll find everything you need to calculate them. First, figure out earned value: EV = BAC × actual % complete. But wait! You weren't given these in the question!

OK, no problem—you just need to think your way through it. The project will cost $52/meter to lay 4 km (or 4,000 meters) of cable, which means the total cost of the project will be $52 x 4,000 = $208,000. And you can figure out actual % complete too! You've laid 1,800 meters so far out of the 4,000 meters you'll lay in total…so that's 1,800 ÷ 4,000 = 45% complete. All right! Now you know your earned value: EV = $208,000 × 45% = $93,600.

So what's next? You've got half of what you need for SPI—now you have to figure out PV. The formula for it is: PV = BAC × scheduled % complete. So how much of the project were you supposed to complete by now? You're 5 weeks into an 8-week project, so 5 ÷ 8 = 62.5%. Your PV is $208,000 × 62.5% = $130,000. Now you've got everything you need to calculate SPI! EV ÷ PV = $93,600 ÷ $130,000 = .72

> SO THAT QUESTION WAS REALLY ABOUT WHETHER I COULD FIGURE OUT HOW TO CALCULATE EV AND PV FROM WHAT I WAS GIVEN.

Did you think that this was a red herring? It wasn't—you needed all the numbers you were given.

11. Answer: B

You'll run into a lot of questions like this where a problem happens, a person has an issue, or the project runs into trouble. When this happens, the first thing you do is stop and gather information. And that should make sense to you, since you don't know if this change will really impact cost or not. It may seem like a huge change to the programmer, but may not actually cost the project anything. Or it may really be huge. So the first thing to do is figure out the impact of the change on the project constraints, and that's what answer B says!

Answers

Exam Questions

12. Answer: B

What formula do you know that has AC and EV? Right, the CPI formula does! Take a look at it: CPI = EV ÷ AC. So what happens if AC is bigger than EV? Make up two numbers and plug them in! You get CPI that's below 1, and you know what that means…it means that you've blown your budget!

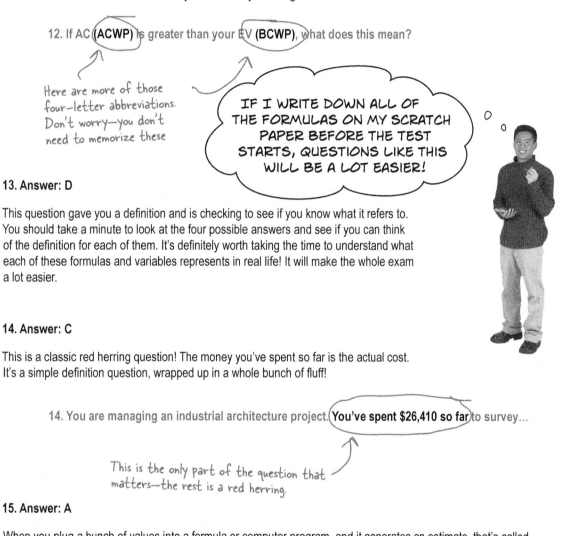

12. If AC (**ACWP**) is greater than your EV (**BCWP**), what does this mean?

Here are more of those four-letter abbreviations. Don't worry—you don't need to memorize these

IF I WRITE DOWN ALL OF THE FORMULAS ON MY SCRATCH PAPER BEFORE THE TEST STARTS, QUESTIONS LIKE THIS WILL BE A LOT EASIER!

13. Answer: D

This question gave you a definition and is checking to see if you know what it refers to. You should take a minute to look at the four possible answers and see if you can think of the definition for each of them. It's definitely worth taking the time to understand what each of these formulas and variables represents in real life! It will make the whole exam a lot easier.

14. Answer: C

This is a classic red herring question! The money you've spent so far is the actual cost. It's a simple definition question, wrapped up in a whole bunch of fluff!

14. You are managing an industrial architecture project. You've spent $26,410 so far to survey…

This is the only part of the question that matters—the rest is a red herring.

15. Answer: A

When you plug a bunch of values into a formula or computer program, and it generates an estimate, that's called parametric estimation. Parametric estimation often uses some historical data, but that doesn't mean it's the same as analogous estimation!

Answers

Exam ~~Questions~~

16. Answer: C

You've been given a net present value (NPV) for each project. NPV means the total value that this project is worth to your company! It's got the costs—including opportunity costs—built in already. So all you need to do is select the project with the biggest NPV.

17. Answer: B

The rough order of magnitude estimate is a very preliminary estimate that everyone knows is only within an order of magnitude of the actual cost (or –25 to +75%).

18. Answer: A

You should definitely have a pretty good idea of how change control works by now! The change control system defines the procedures that you use to carry out the changes. And Control Costs has its own set of procedures, which are part of the Perform Integrated Change Control process you learned about in Chapter 4.

19. Answer: B

You use the project cost baseline to measure and monitor your project's cost performance. The idea behind a baseline is that when a change is approved and implemented, the baseline gets updated.

> I RECOGNIZE THIS—A CHANGE IS REQUESTED, APPROVED, AND IMPLEMENTED, AND THEN THE BASELINE IS UPDATED. SO I'M USING THE COST BASELINE JUST LIKE I USED THE SCOPE BASELINE BACK IN CHAPTER 5!

20. Answer: C

You should have the hang of this by now! Plug the numbers into the formula ($CPI = EV \div AC$), and it spits out the answer. Sometimes the question will give you more numbers than you actually need to use—just ignore them like any other red herring and only use the ones you need!

21. Answer: B

If your TCPI is above 1, you need to manage costs aggressively. It means that you need to meet your goals without spending as much money as you have been for the rest of the project.

Answers

~~Exam Questions~~

22. Answer: B

If you are just starting to work on your project charter, it means you're just starting the project and you don't have enough information yet to do analogous, parametric, or bottom-up estimates.

The only estimation technique that you can use that early in the project is the rough order of magnitude estimate. That kind of estimate is not nearly as accurate as the other kinds of estimate and is used just to give a rough idea of how much time and cost will be involved in doing a project.

23. Answer: D

This question may have seemed a little wordy, but it's really just a question about the definition of TCPI. You're being asked to figure out where you need to keep your project's CPI in order to meet your budget. And you know it's the EAC-based TCPI number, because the question specified that you already gave him a forecast, which means you gave him an EAC value already. So now you can calculate the EAC-based TCPI number to figure out where you need to keep your CPI for the rest of the project.

By calculating this based on the EAC, you show your sponsor just how much money he needs to kick in (or less, if you've got good news!) in order to come in under budget.

8 Quality management

Getting it right

I CAN'T REMEMBER IF I USED CHOCOLATE CHIPS OR MARBLES, BUT I GUESS IT'S OK. SOMEONE'S SURE TO FIGURE IT OUT BEFORE THE BIG BAKE SALE.

It's not enough to make sure you get it done on time and under budget. You need to be sure you make the right product to suit your stakeholders' needs. Quality means making sure that you build what you said you would, and that you do it as efficiently as you can. That means trying not to make too many mistakes and always keeping your project working toward the goal of creating the right product!

What is quality?

Everybody "knows" what **quality** is. But the way the word is used in everyday life is a little different than how it is used in project management. You manage quality on your project by setting goals and taking measurements. That's why you need to understand the quality levels your stakeholders believe are acceptable, and that your project meets those targets…just like it needs to meet their budget and schedule goals.

How do you know if this is a high-quality product?

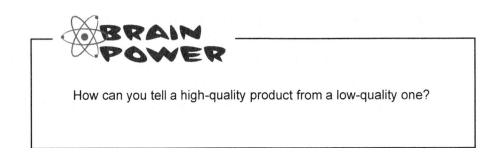

How can you tell a high-quality product from a low-quality one?

You need more than just tests to figure out quality

A lot of people confuse quality with testing. When projects run into quality problems, some project managers will respond by adding more testers to the project to try to find more bugs. But testing is only one part of the story. To know your product's quality, you need to do more than test it:

The Black Box 3000™

Scenario 1

Lisa presses the button, but nothing happens.

> HMM. I HAVE NO IDEA WHAT THESE TESTS PROVE!

Lisa, our tester, is testing the Black Box 3000™, but she isn't sure what she's supposed to be testing for.

Scenario 2

Lisa presses the button and a voice comes out of the box that says, "You pressed the button incorrectly."

Scenario 3

Lisa presses the button and the box heats up to 628°F. Lisa drops the box and it shatters into hundreds of pieces.

How does Lisa know which of these boxes is working, and which failed her test?

Once you know what the product is supposed to do, it's easy to tell which tests pass and which fail

Testing is all about checking to be sure that the product does what it is supposed to do. That means that you need to have a good idea of what it is supposed to do to judge its quality. That's why the most important concept in defining quality for the PMP exam is **conformance to requirements**. That just means that your product is only as good as the requirements you have written for it. To say that something is a high-quality product means that it fulfills the requirements your team agreed to when you started the work.

That's why getting the Collect Requirements process right is so important!

Quality is the measurement of how closely your product meets its requirements.

BLACK BOX 3000™
Specification Manual

The BB3K™ is a heating element for an industrial oven.

BB3K™ must heat up to exactly 628°F in 0.8 seconds.

BB3K™ must have a large, easy-to-press button.

The spec lists all of the requirements that must be met by the product.

> SCENARIO #3 WAS THE TEST THAT PASSED! THE PRODUCT LOOKS LIKE IT'S CONFORMING TO THAT REQUIREMENT. BUT SCENARIOS 1 AND 2 COULD BE DEFECTS. I DON'T SEE ANYTHING ABOUT THEM IN THE SPEC.

Now that she knows what she is supposed to be testing for, Lisa can report on what behavior was correct and what wasn't.

Quality up close

There are a few general ideas about quality that will help you understand a little better where the PMP exam is coming from. A lot of work has been done on quality engineering in the past 50 years or so that was originally focused on manufacturing. Those ideas have been applied to product quality over lots of different industries. Here are a few concepts that are important for the exam.

Customer needs should be written down as requirements before you start to build your product. That way, you can always plan on building the right thing.

Customer satisfaction is about making sure that the people who are paying for the end product are happy with what they get. When the team gathers requirements for the specification, they try to write down all of the things that the customers want in the product so that you know how to make them happy.

Some requirements can be left **unstated**, too. Those are the ones that are implied by the customer's explicit needs. In the end, if you fulfill all of your requirements, your customers should be really satisfied.

Some requirements are just common sense—like a product that people hold can't be made from toxic stuff that kills you. It <u>might not be stated,</u> but it's definitely a requirement.

Fitness for use is about making sure that the product you build has the best design possible to fit the customer's needs. Which would you choose: a product that's beautifully designed, well constructed, solidly built, and all-around pleasant to look at but does not do what you need, or a product that does what you want despite being really ugly to look at and a pain in the butt to work with?

You'll always choose the product that fits your needs, even if it's seriously limited. That's why it's important that the product both does what it is supposed to do and does it well.

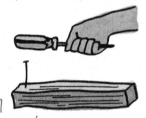

You could pound in a nail with a screwdriver, but a hammer is more fit for the job.

This idea came from a quality theorist named <u>Joseph Juran</u>.

Conformance to requirements is the core of both customer satisfaction and fitness for use. Above all, your product needs to do what you wrote down in your requirements specification. Your requirements should take into account both what will satisfy your customer <u>and</u> the best design possible for the job.

Philip Crosby made this idea popular in the early 1980s. It's been really important to quality engineering ever since.

In the end, your product's quality is judged by whether you built what you said you would build.

Quality is a measure of how well your product does what you intend.

That means conforming to both stated and implied requirements.

> I KNOW QUALITY WHEN I SEE IT. CAN'T I JUST LOOK AT THE FINAL PRODUCT AND REJECT IT IF IT'S LOW QUALITY?

It's easy to mistake a low-grade product for a low-quality one.

When people talk about the quality of their car or their meal, they are often talking about its **grade**. You can judge something's *grade* without knowing too much about its requirements. But that's a lot different than knowing its *quality*.

Quality vs. grade

You can eat a lobster platter for dinner, or you can eat a hot dog. They are both types of food, right? But they have very different tastes, looks, feels, and most importantly, cost. If you order the lobster in a restaurant, you'll be charged a lot more than if you order a hot dog. But that doesn't mean the lobster is a higher-quality meal. If you'd ordered a salad and got lobster or a hot dog instead, you wouldn't be satisfied.

Quality means that something does what you <u>needed</u> it to do. Grade describes how much people <u>value</u> it.

Higher-grade stuff typically <u>costs</u> more, but just because you pay more for something doesn't mean it does what you need it to do.

The lobster is a high-grade meal; the hot dog is a low-grade one. But they're both low quality if you actually wanted a salad.

Sharpen your pencil

Take a look at each of these situations and figure out if they're talking about quality or grade.

1. You ordered mushrooms on your pizza, but you got onions.

☐ Quality   ☐ Grade

2. You called the pizza parlor to complain and the guy yelled at you.

☐ Quality   ☐ Grade

3. The pizza arrived, but it had canned mushrooms.

☐ Quality   ☐ Grade

4. The pizza was cold.

☐ Quality   ☐ Grade

---

5. You just got a brand new luxury car that cost a whole lot of money.

☐ Quality   ☐ Grade

6. But it's in the shop every two weeks.

☐ Quality   ☐ Grade

You probably didn't tell the salesman you needed the car to work, but you expected it to. That's an unstated requirement.

7. Your neighbors make fun of you because your chrome hubcaps aren't very classy…

☐ Quality   ☐ Grade

8. …even though they do a great job of protecting the wheels from dirt, which is why you bought them in the first place.

☐ Quality   ☐ Grade

⟶ Answers on page 446.

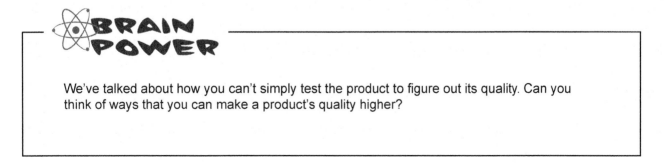

BRAIN POWER

We've talked about how you can't simply test the product to figure out its quality. Can you think of ways that you can make a product's quality higher?

"An ounce of prevention..."

It's not enough to go to the dentist to get your cavities filled. You need to brush your teeth every day. The same goes with product quality. If you focus on preventing mistakes on your project before they happen, you are more likely to get the product done on time and without spending too much money.

10% of the Black Boxes have buttons that stick when you press them.

When it comes to defects, prevention is always better than inspection!

We could hire a lot more inspectors to check to see if each of the products has a sticky button and send it for repair…

OR WE COULD CHANGE THE DESIGN TO MAKE THE BUTTON A MILLIMETER SMALLER AND ELIMINATE THE PROBLEM ALTOGETHER.

And that's why you need the three Quality Management processes!

There are **three processes** in the Quality Management knowledge area, and they're all designed to make sure that you and your team deliver the highest quality product that you can.

Plan Quality Management is like the other planning processes you've learned about— you create a Quality Management plan to help guide you and your team through quality activities.

Control Quality is the Monitoring and Controlling process where you look at each deliverable and inspect it for defects.

Perform Quality Assurance is where you take a step back and look at how well your project fits in with your company's overall quality standards and guidelines.

Exercise

Which of these activities are prevention, and which are inspection?

1. You find that 40% of the sneakers your factory makes have the left foot insole put into the right shoe and the right insole put into the left shoe. So, you print an L on the underside of the left insole so that factory workers can tell them apart more easily.

☐ Prevention   ☐ Inspection

2. The applications being built by your programming team have lots of bugs. So you add extra test cycles and make them longer and more intensive to try to find more problems before you ship.

☐ Prevention   ☐ Inspection

3. The applications being built by your programming team have lots of bugs. So you write up coding standards that will guide everyone in building the product with more attention to quality.

☐ Prevention   ☐ Inspection

4. Some of the Black Boxes being built at the factory are only heating up to 500 degrees when the button is pushed. So you set up an automated button presser to press each one and measure its temperature as it comes off of the assembly line.

☐ Prevention   ☐ Inspection

5. You set up code reviews at important milestones in your project to catch defects as early as you can.

☐ Prevention   ☐ Inspection

6. The programmers on your team write unit tests before they write the code for the application they're writing. That helps them to think of ways that the application's design might go wrong and avoid major pitfalls.

☐ Prevention   ☐ Inspection

Exercise
Solution

Which of these activities are prevention, and which are inspection?

1. You find that 40% of the sneakers your factory makes have the left foot insole put into the right shoe and the right insole put into the left shoe. So, you print an L on the underside of the left insole so that factory workers can tell them apart more easily.

☑ Prevention　　　☐ Inspection

The focus here is on making sure that no more defects happen, rather than on finding them.

2. The applications being built by your programming team have lots of bugs. So you add extra test cycles and make them longer and more intensive to try to find more problems before you ship.

☐ Prevention　　　☑ Inspection

Catching the bugs after they've been put in the product is not the most efficient way to deal with this problem. It will cost more money and take longer.

3. The applications being built by your programming team have lots of bugs. So, you write up coding standards that will guide everyone in building the product with more attention to quality.

☑ Prevention　　　☐ Inspection

This is a much better way of dealing with the same problem. It focuses on making sure the bugs never make it into the software rather than finding them and fixing them.

4. Some of the Black Boxes being built at the factory are only heating up to 500 degrees when the button is pushed. So you set up an automated button presser to press each one and measure its temperature as it comes off of the assembly line.

☐ Prevention　　　☑ Inspection

This one is also focused on finding the problems once they're in the product.

5. You set up code reviews at important milestones in your project to catch defects as early as you can.

☐ Prevention　　　☑ Inspection

6. The programmers on your team write unit tests before they write the code for the application they're writing. That helps them to think of ways that the application's design might go wrong and avoid major pitfalls.

☑ Prevention　　　☐ Inspection

Plan Quality is how you prevent defects

Since prevention is the best way to deal with defects, you need to do a lot of
planning to make sure that your product is made with as few defects as possible.
The **Plan Quality Management process** focuses on taking all of the
information available to you at the beginning of your project and figuring out
how you will measure your quality and prevent defects.

Planning
process group

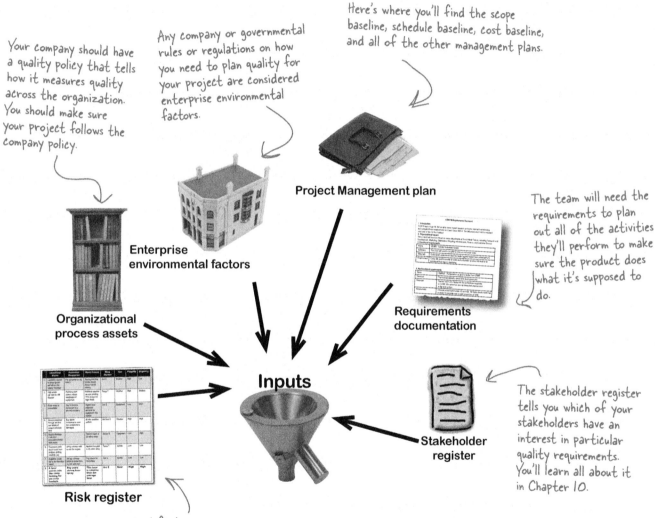

Your company should have
a quality policy that tells
how it measures quality
across the organization.
You should make sure
your project follows the
company policy.

Any company or governmental
rules or regulations on how
you need to plan quality for
your project are considered
enterprise environmental
factors.

Here's where you'll find the scope
baseline, schedule baseline, cost baseline,
and all of the other management plans.

Project Management plan

The team will need the
requirements to plan
out all of the activities
they'll perform to make
sure the product does
what it's supposed to
do.

**Enterprise
environmental factors**

**Organizational
process assets**

**Requirements
documentation**

Inputs

Risk register

**Stakeholder
register**

The stakeholder register
tells you which of your
stakeholders have an
interest in particular
quality requirements.
You'll learn all about it
in Chapter 10.

Risks that have been identified
will help you find the places
where quality might suffer.

How to plan for quality

You need to plan out which activities you're going to use to measure the quality of the product of your project. And you need to be sure that the activities you plan are going to pay off in the end. So you'll need to think about the cost of all of the quality-related activities you want to do. Then you'll need to set some guidelines for what you're going to measure against. Finally, you'll need to design the tests you're going to run when the product is ready to be tested.

Cost-benefit analysis is looking at how much your quality activities will cost versus how much you will gain from doing them. The costs are easy to measure; the effort and resources it takes to do them are just like any other task on your schedule. Since quality activities don't actually produce a product, though, it is harder for people to measure the benefits sometimes. The main benefits are less rework, higher productivity and efficiency, and more satisfaction from both the team and the customer.

That makes sense. A team that is making a high-quality product will be really proud of their work.

Benchmarking means using the results of Plan Quality on other projects to set goals for your own. You might find that the last project your company did had 20% fewer defects than the one before it. You would want to learn from a project like that, and put in practice any of the ideas the company used to make such a great improvement. Benchmarks can give you some reference points for judging your own project before you even get started with the work.

Design of experiments is where you apply the scientific method to create a set of tests for your project's deliverables. It's a *statistical* method, which means you use statistics to analyze the results of your experiments to determine how your deliverables best meet the requirements. A lot of quality managers use this technique to produce a list of tests that they'll run on the deliverables, so they have data to analyze later.

In the software world, this is usually called test planning.

Seven basic quality tools are the main methods used for measuring quality across your project. You'll learn more about them later in the chapter.

Meetings are used to figure out how your team will do all of the quality-related activities your project requires. The whole team might collaborate on the Quality Management plan in these meetings.

Cost of quality is what you get when you add up the cost of all of the prevention and inspection activities you are going to do on your project. It doesn't just include the testing. It includes any time spent writing standards, reviewing documents, meeting to analyze the root causes of defects, doing rework to fix the defects once the team finds them—absolutely everything you do to ensure quality on the project.

Statistical sampling is when you look at a representative sample of something to make decisions. For example, you might take a look at a selection of widgets produced in a factory to figure out which quality activities would help you prevent defects in them.

There are **additional quality planning tools** that project managers might use:

- **Brainstorming** (which you'll learn all about in Chapter 11).
- **Affinity diagrams** (which you learned about in Chapter 5).
- **Force field analysis** is how engineers analyze structures to see what forces affect their use.
- **Nominal group techniques** mean brainstorming with small groups, and then working with larger groups to review and expand the results.
- **Matrix diagrams** are tables, spreadsheets or pivot tables that help you analyze complex relationships.
- **Prioritization matrices** let you analyze multiple issues and prioritize, so you can attack the important ones first.

Cost of quality can be a good number to check whether your project is doing well or having trouble. If your company tracks cost of quality on all of its projects, you could tell if you were spending more or less than the others are, so you can get your project up to snuff.

Don't worry, you don't need to know how to use these techniques to pass the PMP exam!

Exercise

Read each of these scenarios and identify which tool or technique is being used.

1. You look through your company's asset library and find that a recent project was able to reduce defects by 20% by inserting defect prevention meetings early in the construction phase. You put the same process in your quality plan and set the target for shipped defects to be 20% lower than the company average for your project.

   Tool/technique: ....

2. You add up all of the costs projected for quality activities and track that number in your Quality Management plan. You use this number to gauge the health of your project compared to other projects in your company.

   Tool/technique: ....

3. You write up a list of all of the tests you are going to run on the Black Box 3000™ when it rolls off the assembly line. You determine what kinds of failures might cause you to stop testing, what it would take for you to resume test activities, and requirements that the product would need to fulfill to be considered accepted into test.

   Tool/technique: ....

Answers on page 447.

The Quality Management plan gives you what you need to manage quality

Once you have your Quality Management plan, you know your guidelines for managing quality on your project. Your strategies for monitoring your project quality should be included in the plan, as well as the reasons for all of the steps you are taking. It's important that everyone on the team understands the rationale behind the metrics being used to judge success or failure of the project.

The Quality Management plan is the main tool for preventing defects on your project.

Outputs

> The Quality Management plan is the main output of Plan Quality Management. It's a subplan of the Project Management plan.

> A metric is just a number you use to measure your product's quality.

> Even though this number is part of Time Management, you'll often measure it in your Quality Management plan because it's part of customer satisfaction on the project.

BLACK BOX 3000™
Quality Management Plan

Project Background:

The project goal is to create as many industrial heating elements as possible with no defects. Past problems included sticky buttons and difficulty testing the product. This was corrected when a specification was given to the test team.

Goals for Project Metrics:

Metric	Goal	Rationale	How we'll do it
Schedule variance	<5%	Because shipments of Black Boxes are planned with clients in advance, very few delays are acceptable.	Track any activities that might cause delays. Use extra resources if necessary to meet the deadline.
Defect density	0 High priority 2 Medium priority 5 Low priority (defects per thousand Black Boxes)	Defect repair is extremely costly. We need to get as many products shipped as possible on the first try.	Set up defect prevention activities early in the process. Monitor the results of inspections and adjust if necessary.

Defect Prevention Plan:

Outputs

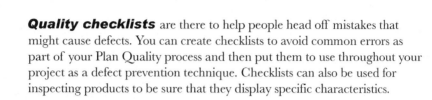

Quality checklists are there to help people head off mistakes that might cause defects. You can create checklists to avoid common errors as part of your Plan Quality process and then put them to use throughout your project as a defect prevention technique. Checklists can also be used for inspecting products to be sure that they display specific characteristics.

> This means you need to think about more than just building the product of the project. You also need to think about how your company does all of its projects.

The Process Improvement plan is a plan for improving the process you are using to do the work. In it, you come up with strategies for finding inefficiencies and places where the way you work might be slowing you down or creating a low-quality product. You set goals for how you can monitor the process during your project, and make recommendations to make it better.

> The Process Improvement plan is another subplan of the Project Management plan.

Quality metrics are the kinds of measurements you'll take throughout your project to figure out its quality. If you're running a project to install windows in a skyscraper, and 15% of them have to be reinstalled because they were broken, that's an important metric. You'll probably want to work with your company to bring that number down.

Here's where you document how you'll be figuring out the product's quality. You need to write down the formulas you'll use, when you will do the measurements, why you are taking them, and how you will interpret them.

> stakeholder register
> ht need to be
> .ated if you find
> stakeholders in
> course of planning
> lity activities.

Stakeholder register

Project document updates might need to be made because you have found new information in the course of planning your quality activities that affects one of the other plans you've already made. That's why this process includes an output for making those kinds of changes.

Match each Plan Quality output to its description.

Quality Management plan

Helps you to make sure that each deliverable is up to the project's standards.

Process Improvement plan

Helps you to plan out all of your quality activities.

Quality checklists

Describes how you'll measure a particular attribute of a deliverable during testing.

Quality metrics

Helps you change the way you work for the better.

Answers on page 447.

there are no Dumb Questions

Q: Why do you need to track the cost of testing?

A: You mean **cost of quality**, right? Cost of quality isn't just the cost of testing. It's the cost of all of your quality activities. Even preventive activities like spending time writing checklists and standards are part of it. The reason you track cost of quality is that it can tell you a lot about the health of your project as a whole.

Say you find you're spending twice as much on quality activities as you are on building your product. You need to use that number to start asking some questions about the way the work is being done.

Are people not doing enough up front to prevent defects, and adding a lot of expensive test activities at the end of the project to compensate? Is the design not clear, so your team needs to do a lot of rework trying to get what the customer needs? There are many reasons that could be causing a high cost-of-quality number, but you wouldn't even know to ask about them if *you didn't track it.*

Q: How do you know your benchmarks before you start building?

A: That's what your organizational process assets are for. Since your company keeps a record of all of the projects that have been done over the years, those projects' quality measurements can help you get a gauge on how your project will perform too. If your company knows that all of the projects in your division had a cost of quality that was 40% of the cost of the overall cost of development, you might set 40% cost of quality as a benchmark for your project as well. Your company might have stated a goal of having a schedule variance of plus or minus 10% on all projects for this calendar year. In that case, the schedule variance is a benchmark for your project.

Q: I don't really have good requirements for my projects because everyone on the team starts out with just a good idea of what we're building. How do I handle quality?

A: You should never do that. Remember how you spent all that time collecting requirements in the Collect Requirements process? Well, this is why you needed them. And it's why it's **your** responsibility to make sure that the project starts out with good, well-defined, and correct requirements. If you don't have them, you can't measure quality—and quality is an important part of project management.

Without requirements, you have no idea what the product is supposed to do, and that means you can't judge its quality. You can learn a lot about a product by testing it, but without knowing its requirements, a product could pass all of its tests and still not do what the customer expects it to do. So having good requirements really is the only way to know whether or not your product is high quality.

Monitoring
& Controlling
process group

Inspect your deliverables

It's not enough to inspect the final product. You must look at all of the things that you make throughout a project to find bugs. In fact, the earlier you find them, the easier they are to fix. The **Control Quality process** is all about inspecting work products to find defects.

LAST WEEK | **TODAY**

Lisa looked for defects in the parts as they were being made.

BB3K™ PARTS

She also inspected the blueprints for the Black Box when they were designed.

Lisa takes a good look at a sample of all of the products that are about to be shipped to Black Box 3000™ customers.

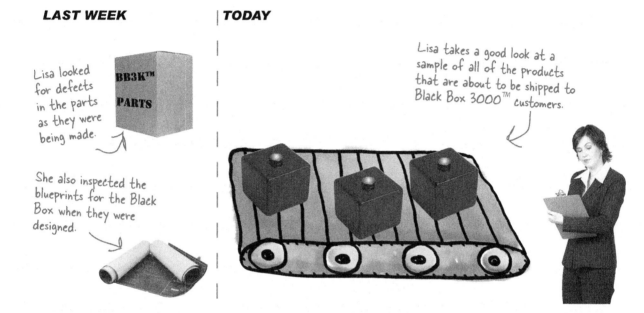

Control Quality is in the Monitoring and Controlling process group. Like Control Scope and Control Costs, you look at the work performance information that is coming from your project and compare it to your plan. *If there are problems, you recommend a change. That way, you can either fix the problem or make sure that it doesn't happen again.*

⚛BRAIN POWER

How would you use your checklists and metrics to inspect all of the deliverables and find defects?

Use the planning outputs for Control Quality

You've come up with a plan to make sure each deliverable is right. Now it's time to monitor the work that's being done to the requirements—and that's just a matter of following your plan! You'll need to look at everything that is being produced and make sure that it stands up to all of the requirements that have been gathered. And you'll need nearly everything you produced in Plan Quality Management in order to get a handle on your product's quality.

Metrics make it easy for you to check how well your product meets expectations.

Metrics tell what and how you are going to measure your product's quality. They give you some objective measures to help you make better judgments about it.

You'll use checklists to help you remember all of the things you need to include in your deliverables.

Deliverables are the things you inspect. Stuff like Black Boxes, specifications, or buttons.

Here's where you'll find your company's quality policy, company-wide metrics, and company-wide project goals.

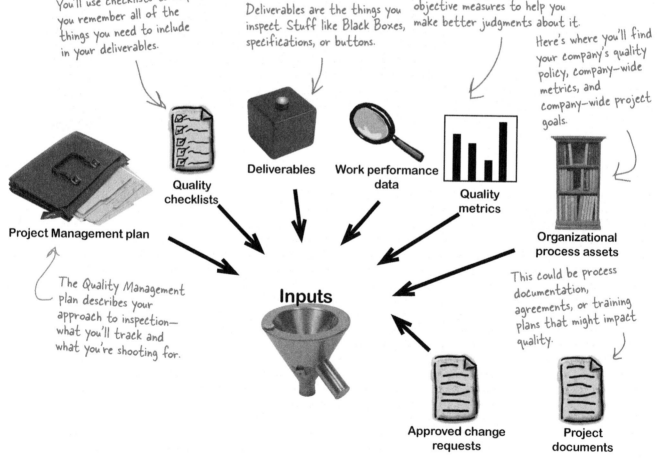

Project Management plan

Quality checklists

Deliverables

Work performance data

Quality metrics

Organizational process assets

Inputs

Approved change requests

Project documents

The Quality Management plan describes your approach to inspection—what you'll track and what you're shooting for.

This could be process documentation, agreements, or training plans that might impact quality.

The seven basic tools of quality

These charts and tools are so common in quality control that they have a name. They're called the **seven basic tools of quality**. Expect a bunch of questions on these in the exam!

These points are showing the Rule of Seven AND that this process is out of control.

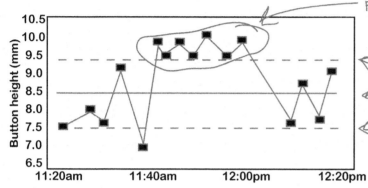

There are three lines on a control chart. The first one is the upper control limit.

Mean—the average height in your sample of buttons.

The lower control limit is the last line. This one represents the shortest that you want the buttons to be.

Control charts are a way of visualizing how processes are doing over time. Let's say that the button on each Black Box needs to be between 7.5 and 9.5 millimeters tall, and the chart above represents sample height measurements of boxes being made. We want the boxes to all be between 7.5mm and 9.5mm. The **lower control limit** of the chart is 7.5mm, and the **upper control limit** is 9.5mm. The chart above shows control limits as dashed lines. The **mean** is the solid line in the middle, and it shows the average height of all of the buttons in the sample. By looking at the chart above, you can see that there are a lot of buttons that were taller than 9.5mm manufactured and only one that was shorter than 7.5mm. When a data point falls outside of the control limits, we say that data point is **out of control**, and when this happens we say that the **entire process is out of control**.

It's pretty normal to have your data fluctuate from sample to sample. But when seven data points in a row fall **on one side of the mean**, that's an uncommon enough occurrence that it means your process might have a problem. So when you see this, you need to look into it and try to figure out what's going on. That's called the **rule of seven**, and *you'll definitely see questions about it on the PMP exam.*

When you're looking at the whole process, that's called Perform Quality Assurance—and it's coming up next.

The vertical "fishbone" lines are categories to help you find and organize the root causes of defects.

Horizontal lines show the root causes you've found for each category.

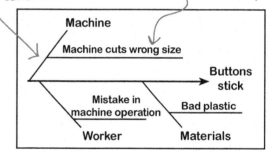

Fishbone or Ishikawa diagram

Cause and effect diagrams are also called **fishbone** and **Ishikawa** diagrams. They are used to figure out what caused a defect. You list all of the categories of the defects that you have identified and then write the possible causes of the defect you are analyzing from each category.

Fishbone diagrams help you **see all of the possible causes** in one place so you can think of how you might prevent the defect in the future.

Pareto charts, flowcharts, and histograms

Tools

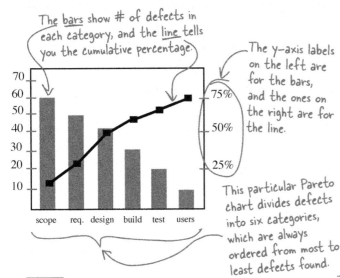

The <u>bars</u> show # of defects in each category, and the <u>line</u> tells you the cumulative percentage.

The y-axis labels on the left are for the bars, and the ones on the right are for the line.

This particular Pareto chart divides defects into six categories, which are always ordered from most to least defects found.

Pareto charts help you to figure out which problems need your attention right away. They're based on the idea that a large number of problems are caused by a small number of causes. In fact, that's called the **80/20 rule**—80% of the defects are usually caused by 20% of the causes. Pareto charts plot out the frequency of defects and sort them in descending order. The right axis on the chart shows the cumulative percentage. For the example here, the most defects are caused by scope issues. So improving the way projects are scoped would be the best way to prevent defects in upcoming projects.

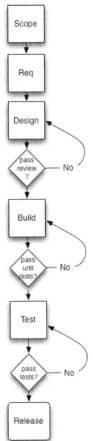

Flowcharts let you show how processes work visually. You can use a flowchart to show how the tasks in your project interrelate and what they depend on. They are also good for showing decision-making processes.

The example on the left shows a high-level view of a software development process. First, the high-level scope is decided, then the requirements, and then the design. After design there is a decision to be made: does the design pass a review? If yes, then move on to the build phase; if no, there's still some design work to do. After the build process, the product needs to pass its unit tests to make it into the test phase.

The flowchart helps you to see how all of the phases relate to each other. Sometimes the way you are working is responsible for defects in your product. Flowcharts help you get a handle on the way you are working by showing you a picture of the whole process.

This product probably isn't ready to ship—it still has a lot of bugs. But at least you know that the bugs aren't all critical!

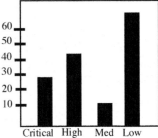

Don't call this a "bar chart"! In the PMP world, a bar chart is another name for a Gantt chart, which is a kind of project schedule.

The PMBOK Guide does refer to it as a "<u>vertical bar chart</u>," so you might see that term where you might normally see the term "bar chart."

Histograms give you a good idea of how your data breaks down. If you heard that your product had 158 defects, you might think that they were all critical. So looking at a chart like the one above would help you to get some perspective on the data. A lot of the bugs are low priority. It looks like only 28 or so are critical. Histograms are great for helping you to compare characteristics of data and make more informed decisions.

Checksheets and scatter diagrams

Checksheets allow you to collect data on the product under test. Checksheets are sometimes called *checklists* or *tally sheets*. You can use them to organize the test activities you'll be performing and track whether the product passes or fails tests. Checksheets are often used as a means of gathering the data that's displayed in Pareto diagrams or other trending and charting tools in the seven basic tools of quality.

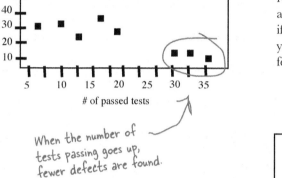

Scatter diagrams show how two different types of data relate to each other. If you worked with your test team to create a bunch of new tests, you might use a scatter diagram to see if the new test cases had any impact on the number of defects you found. The chart here shows that as more test cases pass, fewer defects are found.

When the number of tests passing goes up, fewer defects are found.

BRAIN POWER

The seven basic tools are all about charting defects. Why do you think that would be useful in Control Quality?

More quality control tools

Inspection is what you're doing when you look at the deliverables and see if they conform to requirements. It's important to remember that you don't just inspect the final product. You also look at all of the deliverables that are made along the way.

Approved change request review is when you inspect a repaired defect to be sure that it is actually fixed.

It's not enough to fix defects. You need to be sure that they don't cause more damage once they're fixed.

Statistical sampling helps you make decisions about your product without looking at each and every thing you make. Lisa is responsible for the quality of the Black Box 3000™, but there's no way she can inspect each one as it comes off the assembly line. It makes sense for her to take a sample of the products and inspect those. From that sample she can learn enough about the project to make good judgments.

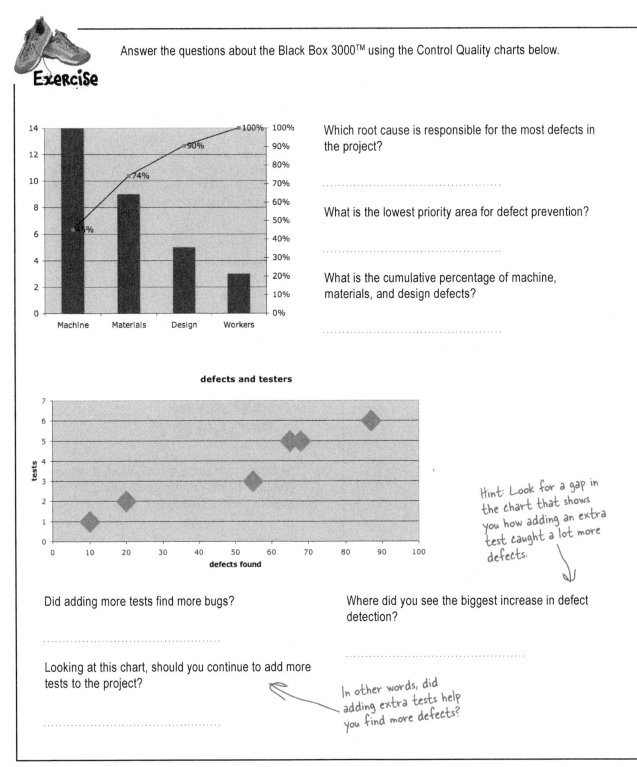

Exercise

Answer the questions about the Black Box 3000™ using the Control Quality charts below.

Which root cause is responsible for the most defects in the project?

....

What is the lowest priority area for defect prevention?

....

What is the cumulative percentage of machine, materials, and design defects?

....

defects and testers

Hint: Look for a gap in the chart that shows you how adding an extra test caught a lot more defects.

Did adding more tests find more bugs?

....

Where did you see the biggest increase in defect detection?

....

Looking at this chart, should you continue to add more tests to the project?

....

In other words, did adding extra tests help you find more defects?

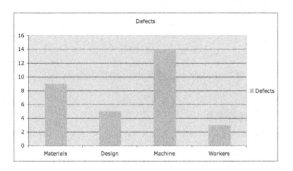

How many machine defects were found?

....

How many defects were caused by workers?

....

How many total defects are shown on this chart?

....

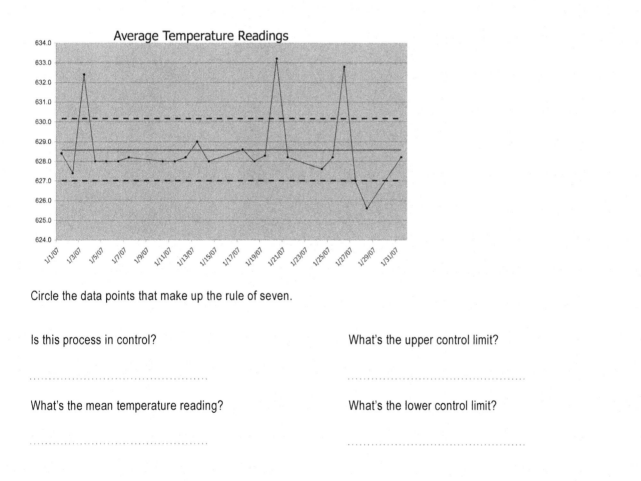

Circle the data points that make up the rule of seven.

Is this process in control?

....

What's the mean temperature reading?

....

What's the upper control limit?

....

What's the lower control limit?

....

Answer the questions about the Black Box 3000™ using the Control Quality charts below.

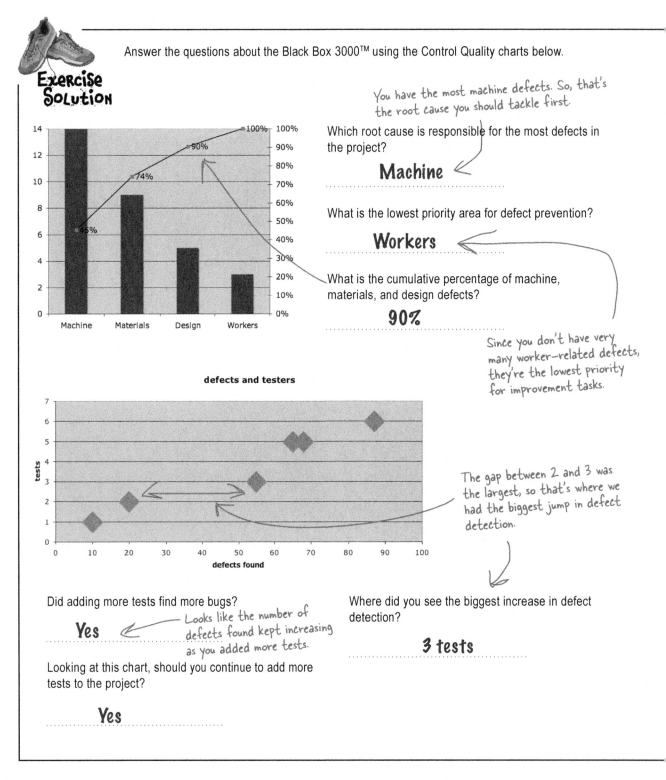

You have the most machine defects. So, that's the root cause you should tackle first.

Which root cause is responsible for the most defects in the project?

Machine

What is the lowest priority area for defect prevention?

Workers

What is the cumulative percentage of machine, materials, and design defects?

90%

Since you don't have very many worker-related defects, they're the lowest priority for improvement tasks.

The gap between 2 and 3 was the largest, so that's where we had the biggest jump in defect detection.

Did adding more tests find more bugs?

Yes

Looks like the number of defects found kept increasing as you added more tests.

Looking at this chart, should you continue to add more tests to the project?

Yes

Where did you see the biggest increase in defect detection?

3 tests

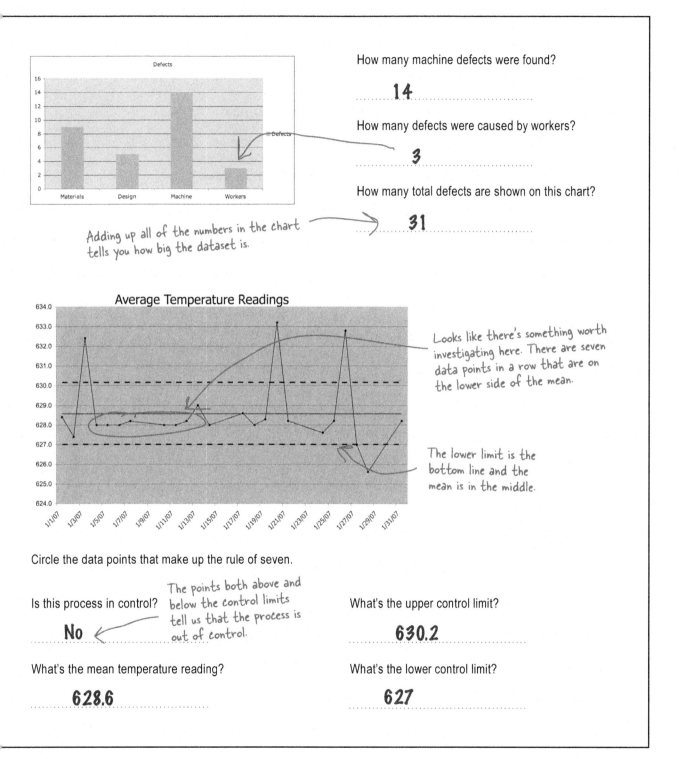

How many machine defects were found?

14

How many defects were caused by workers?

3

How many total defects are shown on this chart?

31

Adding up all of the numbers in the chart tells you how big the dataset is.

Average Temperature Readings

Looks like there's something worth investigating here. There are seven data points in a row that are on the lower side of the mean.

The lower limit is the bottom line and the mean is in the middle.

Circle the data points that make up the rule of seven.

Is this process in control?

The points both above and below the control limits tell us that the process is out of control.

No

What's the mean temperature reading?

628.6

What's the upper control limit?

630.2

What's the lower control limit?

627

Question Clinic: The "which-one" question

YOU'LL SEE A LOT OF QUESTIONS ON THE EXAM THAT DESCRIBE A SITUATION AND ASK YOU TO IDENTIFY THE TOOL, TECHNIQUE, OR PROCESS THAT'S BEING USED OR IS MOST APPROPRIATE. LUCKILY, PROCESS OF ELIMINATION IS REALLY USEFUL WHEN YOU SEE A "WHICH-ONE" QUESTION.

83. You're managing a project to install 13,000 light switches in a new strip mall. You hire a team of inspectors to help your lead electrician find any defective light switches. They check a sample of 650 light switches, and find that 15% of them are defective. You ask your lead electrician to produce a chart that shows you these defects, broken down by category, and shown in order of importance. Which quality control tool will show this information?

- A. Rule of seven
- B. Run chart
- C. Histogram
- D. Pareto chart

This can't be right—it's not even a tool! It's just a rule.

The run chart just tells you trends, and that's not what you're looking for.

Getting closer...the histogram will show you categories, but not importance.

Aha! This is what a Pareto chart is for. It shows you categories of defects, and which category is the most important because it has the most defects.

WHEN YOU THINK ABOUT IT, ALL QUESTIONS ARE "WHICH-ONE" QUESTIONS...BUT WHEN THE QUESTION ASKS TO CHOOSE ONE ITEM FROM A LIST OF FOUR REALLY SIMILAR OR RELATED THINGS, THEN THAT'S WHEN YOU REALLY GET TO WORK YOUR WAY BACKWARD AND START ELIMINATING THEM ONE AT A TIME.

HEAD LIBS

Fill in the blanks to come up with your own "which-one" question! Start by thinking of the correct tool and then figure out three really similar answers that sound right, but can't be because the question gives more specific details, allowing you to eliminate the wrong ones.

You're working on a _____ project, and you want to
(kind of project)

measure _____ . Which of the seven basic tools of
(something you'd measure on a project)

quality is best for doing that?

A. _____
(an obviously wrong tool)

B. _____
(something that isn't a tool at all)

C. _____
(another incorrect tool)

D. _____
(the right answer)

Quality control means finding and correcting defects

When you look for bugs in your deliverables, you produce two kinds of things: outputs from the inspections and outputs from the repairs you've made. All of the **outputs of the Control Quality process** fall into those two categories.

Outputs

Quality control measurements are all of the results of your inspections: the numbers of defects you've found, numbers of tests that passed or failed—stuff like that. You'll use them when you look at the overall process you are using in your company to see if there are trends across projects.

That's coming up in the Quality Assurance process. That's next!

You might need to update templates for quality metrics or checklists.

Organizational process assets updates

Lessons learned updates are where you keep a record of all of the major problems that you solve in the course of your project so that you can use them later.

Completed checklists are records of quality activities that are performed through the course of the project and their results. It's a good idea to keep records of the results of reviews and quality tests.

There's just one output here, organizational process assets updates. You'll store your lessons learned and records of your completed checklists in your organizational process asset library.

Project Management plan updates You may need to update the Quality Management plan and the Process Improvement plan, which are both subplans of the Project Management plan.

You might need to update the PM Plan because of what you find in Control Quality.

Project Management plan updates

Verified deliverables and validated changes are two of the most important outputs of Control Quality. Every single deliverable on the project needs to be inspected to make sure it meets your quality standards. If you find defects, the team needs to fix them—and then those repairs need to be checked, to make sure the defects are now gone.

First the team inspects every deliverable to find defects that need to be fixed.

Deliverables

When you've finished inspecting your product, you know whether or not your fixes worked.

Validated defect repair

Change requests are recommended or preventive actions that also require changes to the way you are doing your project. Those kinds of changes will need to be **put through change control**, and the appropriate baselines and plans will need to be updated if they are approved.

Project documents updates might need to be made. You might discover that your company's quality standards need to be updated, because something you thought was a defect might not be a defect after all!

Work performance information

Work performance information might include all of the data your quality processes are producing. Once you've looked at the results of your quality tools, you might find places where the processes you're using to build your product need to be changed. The data you collect in the Control Quality process can help you make those kinds of changes.

there are no
Dumb Questions

Q: What exactly are Pareto charts for?

A: Pareto charts go together with the **80/20 rule**. It says that 80 percent of the problems you'll encounter in your project are caused by 20 percent of the root causes you can find. So if you find that most of your problems come from misunderstanding requirements, changing the way you gather requirements and making sure that everybody understands them earlier in the process will have a big impact on your project's quality.

To get the data for your Pareto chart, first you have to categorize all of the defects that have been found in your project by their root causes. Then you can graph them in a Pareto chart to show the frequency of bugs found with each root cause and the percentage of the cumulative defects that are caused by each root cause. The one with the highest frequency is the root cause that you should work on first.

Q: If I am trying to prevent quality problems, why can't I just test more?

A: You can find a lot of problems by testing. If you find them during testing, then you have to go back and fix them. The later you find them, the more expensive they are to fix. It's much better for everybody if you never put the bugs in the product in the first place. It's much easier to fix a problem in a specification document than it is to fix it in a finished product. That's why most of the Plan Quality Management process group is centered on setting standards and doing reviews to be sure that bugs are never put into your product and, if they are, they're caught as early as possible.

Q: I still don't get that thing where a control chart can show you defects that are out of control, but also show you that your process is out of control.

A: The reason that's a little confusing to some people is that you use the same tool to look at defects that you do when you're looking at the whole process.

A lot of the time, you'll use charts to measure processes, not just projects. They're used to look at sample data from processes and make sure that they operate within limits over time. But they are considered quality control tools because those data samples come from inspecting deliverables as they are produced. Yes, it's a little confusing, but if you think of control charts as the product of inspection, you'll remember that they are Control Quality tools for the test.

BULLET POINTS: AIMING FOR THE EXAM

- **Inspection** means **checking each deliverable** for defects. That means checking your specs and your documentation, as well as your product, for bugs.

- The better you plan the quality activities for your project, the **less inspection** you need.

- Ishikawa diagrams help you to **pinpoint the causes** of defects.

- The **rule of seven means** that any time you have seven data points in a row that fall on the same side of the mean on a control chart, you need to figure out why.

- When data points fall above the upper limit or below the lower limit on a control chart, the process is out of control.

- For the test, using any of the seven basic quality tools is **usually** a good indication that you are in the Control Quality process.

- **Ishikawa**, **fishbone**, and **cause-and-effect** diagrams are all the same thing.

- **Scatter charts** help you look at the **relationship** between two different kinds of data.

- **Flowcharts** help you get a handle on how processes work by showing all of the decision points graphically.

- **Grade** refers to the **value** of a product, but not its quality. So, a product can be low-grade by design, and that's fine. But if it's a low-quality product, that's a big problem.

Trouble at the Black Box 3000™ factory

It's not enough to inspect your deliverables. Sometimes it's the way you work
that's causing your problems. That's why you need to spend some time thinking
about how you will make sure you are doing the work efficiently and with as
few defects as possible. The **Perform Quality Assurance process** is about
tracking the way you work and improving it all of the time.

Executing process group

THE PRODUCTS ARE LOOKING
GREAT, AND CUSTOMERS ARE
REALLY HAPPY! BUT WAIT—WE HAVE
BOXES FULL OF PARTS COLLECTING DUST
IN OUR WAREHOUSE, AND WE'VE GOT SO
MANY INSPECTORS THAT OUR BUDGET'S
THROUGH THE ROOF. WE NEED TO DO
SOMETHING!

These boxes are full of
parts that the company
uses to build the Black
Box 3000™. They were
ordered from suppliers
who delivered them to the
BB3K™ warehouse, but
they've been sitting around
for weeks taking up valuable
space.

⚛ BRAIN POWER

What do you do if the quality is good but you aren't
satisfied with the speed or efficiency of the work?

Introducing Quality Assurance

In the **Perform Quality Assurance process**, you take all of the outputs from Plan Quality Management and Control Quality and look at them to see if you can find ways to improve your process. If you find improvements, you recommend changes to your process and your individual project plan to implement them.

This is in the Executing group because you need to make sure your project is done in a way that complies with your company's quality standards.

Executing process group

Quality metrics

Quality control measurements

Project documents

Quality Management plan

Process Improvement plan

Inputs

You'll use the outputs of the other quality processes and information about how the work is being performed to find ways to improve the way your company operates.

Tools

Plan Quality Management tools and techniques are all of the tools you used in Plan Quality. They come in handy when you're reviewing your process, too.

Control Quality tools and techniques are all of the tools from Control Quality. You can use histograms, control charts, and flowcharts—all of them can be used to help you figure out how your process is working.

Quality audits are reviews of your project by your company. They figure out whether or not you are following the company's process.

Process analysis is when you look at your process to find out how to make it better. You use your Process Improvement plan to do this one.

The project documents that get updated are quality audits, training plans, and process documentation.

Outputs

Project Management plan updates

Change requests

Project documents updates

Organizational process assets updates

A closer look at some tools and techniques

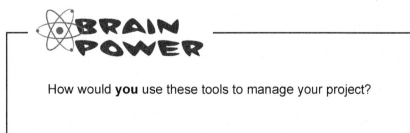

Fixing the bugs in your project solves the problems that give you trouble. But fixing bugs in your **process** means that other projects can learn from the problems you've faced and avoid your project's bugs altogether. The tools that are used in quality assurance are the same as the ones in quality control, but they're used to examine the process rather than the project.

Even if your company has the best process in the world, it doesn't do your project any good if you don't follow it!.

Quality audits are when your company reviews your project to see if you are following its processes. The point is to figure out if there are ways to help you be more effective by finding the stuff you are doing on your project that is inefficient or that causes defects. When you find those problem areas, you recommend corrective actions to fix them.

A lot of companies have Quality Assurance departments whose job is to perform these audits and report findings from projects to a process group.

Process analysis means following your Process Improvement plan to compare your project's process data to goals that have been set for your company. If you find that the process itself needs to change, you recommend those changes to the company and sometimes update the organizational process assets as well as your own Project Management plan to include your recommendations.

When Lisa noticed that the warehouse was full of Black Box parts that weren't needed yet, she was really noticing a problem with the process. Why spend money on overstocked inventory?

Quality management and control tools are the same ones you already know about from earlier in this chapter. But instead of using them to look for problems with specific defects, you'll use them to look at your overall process. A good example of this is using a control chart to see if your whole process is in control. If it's not, then you'll want to make a change to the whole way you do your work in order to bring it under control.

Here's another example. If you created a Pareto chart that showed all of the defects in all of your projects, you could find the one or two categories of defects that caused problems for the whole company. Then you could get all of the PMs together to figure out an improvement that they could all make that would help the whole company.

BRAIN POWER

How would **you** use these tools to manage your project?

More ideas behind quality assurance

There are a couple more things you need to know about quality assurance. These are some of the most important ideas behind modern quality and process improvement.

Kaizen means continuous improvement. It's all about constantly looking at the way you do your work and trying to make it better. *Kaizen* is a Japanese word that means **improvement**. It focuses on making small improvements and measuring their impact. Kaizen is a philosophy that guides management, rather than a particular way of doing quality assurance.

Just-in-Time means *keeping only the inventory you need on hand when you need it*. So, instead of keeping a big inventory of parts sitting around, the Black Box company might have only the parts it needs for that day. Some companies have done away with warehouses all together and have production lines take the parts directly off the trucks to do the work. If you're working in a Just-in-Time shop, quality is really important because **there isn't any extra inventory to deal with mistakes**.

Plan-Do-Check-Act was created by Walter Shewhart, who also created the control chart while he was working at Bell Labs in the 1920s.

Plan-Do-Check-Act is one way to go about improving your process, and it's used by a lot of Kaizen practitioners. It was popularized by a well-known quality theorist named W. Edwards Deming and is also known as the Deming Cycle. Plan-Do-Check-Act is about *making small improvements*, and *measuring* how much *benefit* they make **before** *you change your process* to include them. Here's how it works:

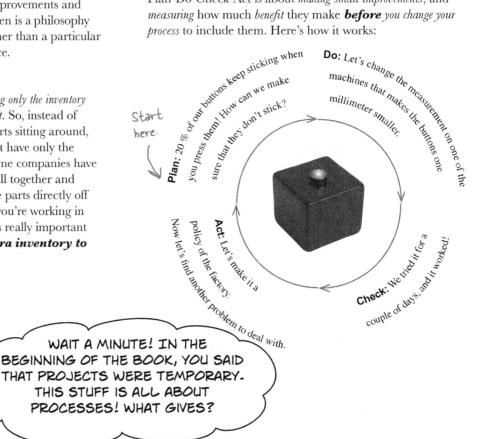

Start here.

Plan: 20% of our buttons keep sticking when you press them! How can we make sure that they don't stick?

Do: Let's change the measurement on one of the machines that makes the buttons one millimeter smaller.

Check: We tried it for a couple of days, and it worked!

Act: Let's make it a policy of the factory. Now let's find another problem to deal with.

> WAIT A MINUTE! IN THE BEGINNING OF THE BOOK, YOU SAID THAT PROJECTS WERE TEMPORARY. THIS STUFF IS ALL ABOUT PROCESSES! WHAT GIVES?

You're right. The Perform Quality Assurance process is all about improving the process, and that isn't what most of project management is about. But your project is really affected by the process you are working in, so you should fully understand it and help to make it better wherever you can. The bottom line is that your project has a better chance of succeeding if you stay involved with process improvement and keep your eye on how your project stacks up to your company's expectations of quality and process.

Qualitycross

Take some time to sit back and give your right brain something to do. It's your standard crossword; all of the solution words are from this chapter.

Across

3. When a process has data points above the upper limit or below the lower limit, those data points are out of _____.

5. The middle line on a control chart.

7. The quality theorist who popularized Plan-Do-Check-Act.

9. _____ is more important than inspection in Quality Management.

11. An important definition of quality is _____ to requirements.

12. Tool used to make sure your project is following the company's process.

13. What you compare your work performance information to.

14. Tool for finding the 20% of root causes responsible for 80% of defects.

15. Tool for comparing two kinds of data to see if they are related.

16. Tool used in Plan Quality Management to set numeric goals for your project.

Down

1. Quality theorist who came up with the idea of fitness for use.

2. Tool for finding the root cause of a defect.

4. Synonym for continuous improvement.

6. Process where you inspect deliverables to look for defects.

8. Tools that help you visualize processes and all of their decision points.

10. Heuristic that says that seven data points on one side of the mean requires investigation.

——————➤ Answers on page 448.

Fireside Chats

Tonight's talk: **Two quality processes discuss the best ways to correct problems on your project.**

Control Quality:

I'd like to go first, because I'm what most people think of as quality. Whenever you see one of those "Inspected by #8" stickers on the inside of your sneaker, that's me!

Perform Quality Assurance:

You're right—most people do think that quality begins and ends with inspection. Which is funny, because we wouldn't even need you if people paid attention to me.

Whoa, there, buddy. That's a strong statement!

Now don't get me wrong. Nobody's ever felt comfortable enough with me that they've eliminated inspection entirely. You always need someone at the end of the line to look at what's been produced and make sure that we delivered what we meant to.

That's right. And don't forget, I'm everywhere. Any time you call for customer service, I'm there to tell you that your call will be recorded for quality purposes. I'm always warning you to make sure package contents haven't shifted, and to check your car's emissions once a year.

Right, but don't you get tired of doing all of that tedious work? An ounce of prevention is worth a pound of cure, after all.

I guess I don't really understand exactly how you do your job, then, because I'm having a hard time figuring out how I would ever be able to take a long weekend.

Let's take a look at those sneakers you mentioned. What's the most common reason you throw a pair back to the factory floor to be restitched?

Well, last week it was because the company logo came out upside down on a bunch of the shoes. It turned out that the logo was being stitched into the leather and then put on another assembly line, and once in a while it was placed on the belt upside-down.

Control Quality:

We had to throw out about 10% of our sneakers last week. Let's just say that the boss wasn't happy. You could see the little veins in his forehead throbbing. It was kind of gross.

The boss yelled at everyone, and we'll check even more carefully to make sure we don't ship it.

Wow, I never thought of that.

We'd have to pay someone else to paint that on. This is no time to be *increasing* our costs!

Huh. Um. No.

Perform Quality Assurance:

That seems like an honest mistake. How much did it cost?

Wow, that sounds expensive. What's keeping it from happening again?

So next week your inspection costs will be even higher, and you'll probably still have to throw out just as many shoes, or more!

What if you painted a little arrow on the inside of the leather showing which direction the logo should be placed on the belt?

But a small increase in the cost of painting the leather will cause you to throw out a whole lot fewer sneakers.

I call that ***cost of quality***. You have to pay more to put quality in at the beginning, but you can reduce the number of inspectors and scrap a lot less product. In the end, I save you far more money that I cost. Can you say the same about yourself?

Choose whether the tools are being used for Control Quality or Perform Quality Assurance.

Exercise

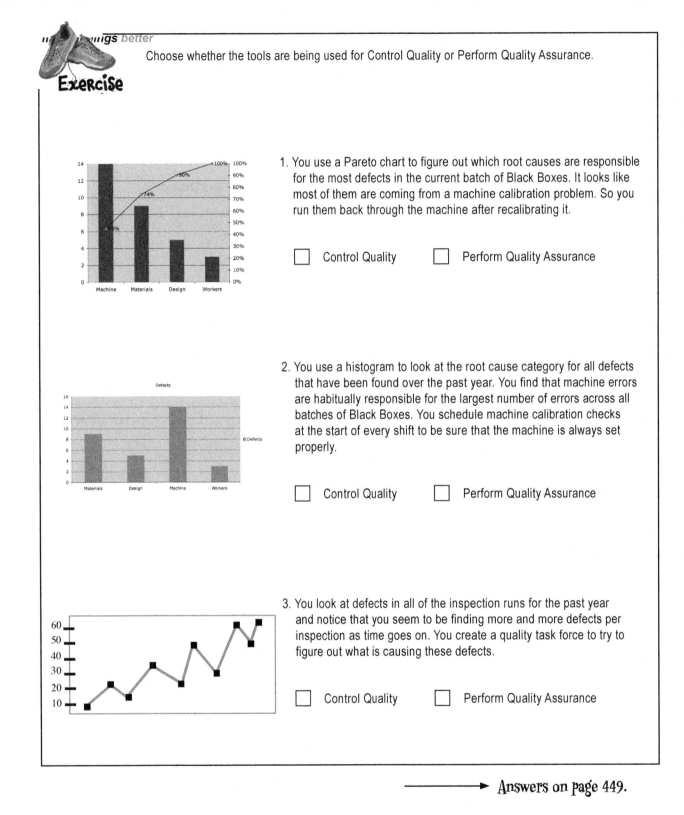

1. You use a Pareto chart to figure out which root causes are responsible for the most defects in the current batch of Black Boxes. It looks like most of them are coming from a machine calibration problem. So you run them back through the machine after recalibrating it.

   ☐ Control Quality   ☐ Perform Quality Assurance

2. You use a histogram to look at the root cause category for all defects that have been found over the past year. You find that machine errors are habitually responsible for the largest number of errors across all batches of Black Boxes. You schedule machine calibration checks at the start of every shift to be sure that the machine is always set properly.

   ☐ Control Quality   ☐ Perform Quality Assurance

3. You look at defects in all of the inspection runs for the past year and notice that you seem to be finding more and more defects per inspection as time goes on. You create a quality task force to try to figure out what is causing these defects.

   ☐ Control Quality   ☐ Perform Quality Assurance

Answers on page 449.

The Black Box 3000™ makes record profits!

People who bought the product were thrilled with it. They were happy that the Black Box company always kept its promises and the products were always high quality. The company managed to save a lot of money by implementing process improvement measures that caught defects before they cost too much money to fix. And Lisa got a big promotion—now she's in charge of quality assurance for the whole company. Great job, Lisa!

The number of inspectors the company needed went down as the quality got better and better.

Thanks to Quality Management, the Black Box 3000™ is always high quality.

Since the company focused on preventing defects, it was able to find and fix problems efficiently and didn't need to hire lots of testers.

BB3K™ PARTS

The company cut down on extra inventory by using Just-in-Time processes to track and forecast how much it would need. Just in Time means no parts sitting around anymore!

Sharpen your pencil
Solution

Take a look at each of these situations and figure out if they're talking about quality or grade.

1. You ordered mushrooms on your pizza, but you got onions.

☑ Quality   ☐ Grade

2. You called the pizza parlor to complain and the guy yelled at you.

☑ Quality   ☐ Grade

3. The pizza arrived, but it had canned mushrooms.

☐ Quality   ☑ Grade

4. The pizza was cold.

☑ Quality   ☐ Grade

---

5. You just got a brand new luxury car that cost a whole lot of money.

☐ Quality   ☑ Grade

6. But it's in the shop every two weeks.

☑ Quality   ☐ Grade

7. Your neighbors make fun of you because your chrome hubcaps aren't very classy...

☐ Quality   ☑ Grade

8. ...even though they do a great job of protecting the wheels from dirt, which is why you bought them in the first place.

☑ Quality   ☐ Grade

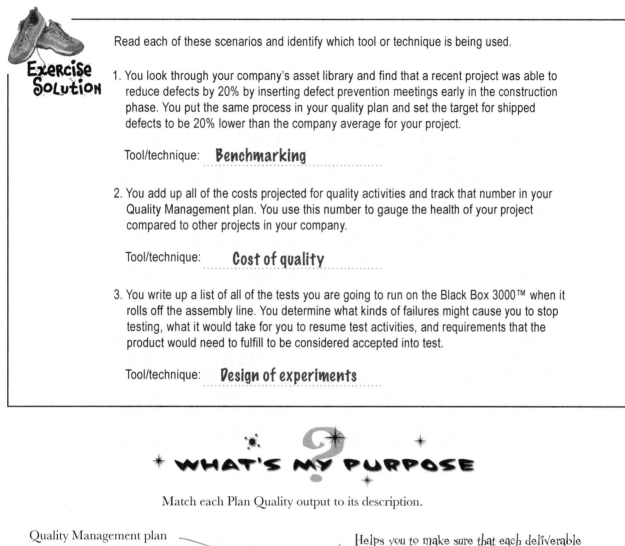

Read each of these scenarios and identify which tool or technique is being used.

Exercise Solution

1. You look through your company's asset library and find that a recent project was able to reduce defects by 20% by inserting defect prevention meetings early in the construction phase. You put the same process in your quality plan and set the target for shipped defects to be 20% lower than the company average for your project.

   Tool/technique: **Benchmarking**

2. You add up all of the costs projected for quality activities and track that number in your Quality Management plan. You use this number to gauge the health of your project compared to other projects in your company.

   Tool/technique: **Cost of quality**

3. You write up a list of all of the tests you are going to run on the Black Box 3000™ when it rolls off the assembly line. You determine what kinds of failures might cause you to stop testing, what it would take for you to resume test activities, and requirements that the product would need to fulfill to be considered accepted into test.

   Tool/technique: **Design of experiments**

☀ WHAT'S MY PURPOSE

Match each Plan Quality output to its description.

Quality Management plan

Process Improvement plan

Quality checklists

Quality metrics

Helps you to make sure that each deliverable is up to the project's standards.

Helps you to plan out all of your quality activities.

Describes how you'll measure a particular attribute of a deliverable during testing.

Helps you change the way you work for the better.

Qualitycross

Take some time to sit back and give your right brain something to do. It's your standard crossword; all of the solution words are from this chapter.

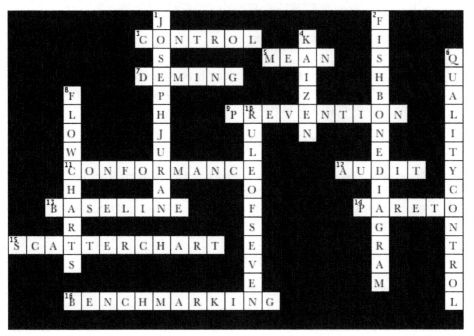

Across

3. When a process has data points above the upper limit or below the lower limit, those data points are out of _____.

5. The middle line on a control chart.

7. The theorist who came up with Plan-Do-Check-Act.

9. _____ is more important than inspection in Quality Management.

11. An important definition of quality is _____ to requirements.

12. Tool used to make sure your project is following the company's process.

13. What you compare your work performance information to.

14. Tool for finding the 20% of root causes responsible for 80% of defects.

15. Tool for comparing two kinds of data to see if they are related.

16. Tool used in Plan Quality to set numeric goals for your project.

Down

1. Quality theorist who came up with the idea of fitness for use.

2. Tool for finding the root cause of a defect.

4. Synonym for continuous improvement.

6. Process where you inspect deliverables to look for defects.

8. Tools that help you visualize processes and all of their decision points.

10. Heuristic that says that seven data points on one side of the mean requires investigation.

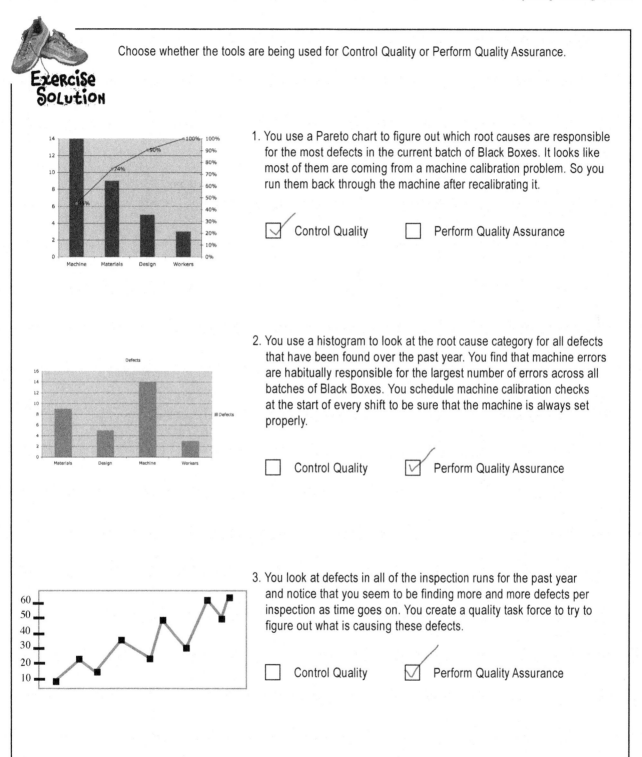

Exercise Solution

Choose whether the tools are being used for Control Quality or Perform Quality Assurance.

1. You use a Pareto chart to figure out which root causes are responsible for the most defects in the current batch of Black Boxes. It looks like most of them are coming from a machine calibration problem. So you run them back through the machine after recalibrating it.

☑ Control Quality   ☐ Perform Quality Assurance

2. You use a histogram to look at the root cause category for all defects that have been found over the past year. You find that machine errors are habitually responsible for the largest number of errors across all batches of Black Boxes. You schedule machine calibration checks at the start of every shift to be sure that the machine is always set properly.

☐ Control Quality   ☑ Perform Quality Assurance

3. You look at defects in all of the inspection runs for the past year and notice that you seem to be finding more and more defects per inspection as time goes on. You create a quality task force to try to figure out what is causing these defects.

☐ Control Quality   ☑ Perform Quality Assurance

Exam Questions

1. Which of the following is NOT a part of quality?

    A.   Fitness for use

    B.   Conformance to requirements

    C.   Value to the sponsor

    D.   Customer satisfaction

2. A project manager is using a histogram to analyze defects found by the team during inspection activities. What process is being performed?

    A.   Plan Quality Management

    B.   Control Quality

    C.   Perform Quality Assurance

    D.   Verify Scope

3. Which of the following is NOT an example of cost of quality?

    A.   Having team members spend extra time reviewing requirements with the stakeholders

    B.   Paying extra programmers to help meet a deadline

    C.   Hiring extra inspectors to look for defects

    D.   Sending a crew to repair a defective product that was delivered to the client

4. You're working with an audit team to check that your company's projects all meet the same quality standards. What process is being performed?

    A.   Plan Quality Management

    B.   Control Quality

    C.   Perform Quality Assurance

    D.   Perform Quality Management

5. You're managing a project to deliver 10,000 units of custom parts to a manufacturer that uses Just-in-Time management. Which of the following constraints is most important to your client?

    A.   The parts must be delivered on time.

    B.   The parts must be delivered in a specific order.

    C.   The parts must conform to ISO specifications.

    D.   The parts must be packaged separately.

Exam Questions

6. Which of the following is NOT part of the Quality Management plan?

   A.   Strategies for handling defects and other quality problems

   B.   Guidance on how the project team will implement the company's quality policy

   C.   Metrics for measuring your project's quality

   D.   A description of which deliverables don't have to be inspected

7. Which of the following tools and techniques is used to show which categories of defects are most common?

   A.   Control charts

   B.   Pareto charts

   C.   Checksheets

   D.   Flowcharts

8. You're managing a highway construction project. The foreman of your building team alerts you to a problem that the inspection team found with one of the pylons, so you use an Ishikawa diagram to try to figure out the root cause of the defect. What process is being performed?

   A.   Quality Management

   B.   Plan Quality Management

   C.   Control Quality

   D.   Perform Quality Assurance

9. Which tool or technique is used to analyze trends?

   A.   Scatter chart

   B.   Run chart

   C.   Checklist

   D.   Flowchart

10. When is inspection performed?

   A.   At the beginning of the project

   B.   Any time a project deliverable is produced

   C.   Just before the final product is delivered

   D.   At the end of the project

Exam Questions

11. What's the difference between Control Quality and Verify Scope?

A. Control Quality is done at the end of the project, while Verify Scope is done throughout the project.

B. Control Quality is performed by the project manager, while Verify Scope is done by the sponsor.

C. Control Quality is performed by the sponsor, while Verify Scope is done by the project manager.

D. Control Quality means looking for defects in deliverables, while Verify Scope means verifying that the product is acceptable to the stakeholders.

12. You're a project manager at a wedding planning company. You're working on a large wedding for a wealthy client, and your company has done several weddings in the past that were very similar to the one you're working on. You want to use the results of those weddings as a guideline to make sure that your current project's quality is up to your company's standards. Which tool or technique are you using?

A. Checklists

B. Benchmarking

C. Design of experiments

D. Cost-benefit analysis

13. You are using a control chart to analyze defects when something on the chart causes you to realize that you have a serious quality problem. What is the MOST likely reason for this?

A. The rule of seven

B. Upper control limits

C. Lower control limits

D. Plan-Do-Check-Act

14. Which of the following BEST describes defect repair review?

A. Reviewing the repaired defect with the stakeholder to make sure it's acceptable

B. Reviewing the repaired defect with the team to make sure they document lessons learned

C. Reviewing the repaired defect to make sure it was fixed properly

D. Reviewing the repaired defect to make sure it's within the control limits

15. The project team working on a project printing 3,500 technical manuals for a hardware manufacturer can't inspect every single manual, so they take a random sample and verify that the manuals have been printed correctly. This is an example of:

A. Root cause analysis

B. Cost-benefit analysis

C. Benchmarking

D. Statistical sampling

Exam Questions

16. What's the difference between Control Quality and Perform Quality Assurance?

    A. Control Quality involves charts like histograms and control charts, while Perform Quality Assurance doesn't use those charts.

    B. Control Quality and Perform Quality Assurance mean the same thing.

    C. Control Quality means inspecting for defects in deliverables, while Perform Quality Assurance means auditing a project to check the overall process.

    D. Perform Quality Assurance means looking for defects in deliverables, while Control Quality means auditing a project to check the overall process.

17. Which Control Quality tool is used to analyze processes by visualizing them graphically?

    A. Checklists

    B. Flowcharts

    C. Pareto charts

    D. Histograms

18. You are looking at a control chart to figure out if the way you are doing your project fits into your company's standards. Which process are you using?

    A. Plan Quality Management

    B. Perform Quality Assurance

    C. Control Quality

    D. Quality Management

19. Which of the following is associated with the 80/20 rule?

    A. Scatter chart

    B. Histogram

    C. Control chart

    D. Pareto chart

20. Validated defect repair is an output of which process?

    A. Integrated Change Control

    B. Plan Quality Management

    C. Control Quality

    D. Perform Quality Assurance

Answers

Exam ~~Questions~~

1. Answer: C

It's important for projects to produce a valuable product, but value isn't really a part of quality. That's why earned value is part of Cost Management, not Quality Management.

2. Answer: B

In the Control Quality process, the team inspects the product for defects and uses the seven basic tools to analyze them. Since the defects came from inspection, you know it's Control Quality.

3. Answer: B

Cost of quality is the time and money that you spend to prevent, find, or repair defects.

4. Answer: C

The Perform Quality Assurance process is all about how well your company meets its overall quality goals.

Keep an eye out for fake process names like Perform Quality Management.

5. Answer: A

A manufacturer that uses Just-in-Time management is relying on its suppliers to deliver parts exactly when they're needed. This saves costs, because it doesn't have to warehouse a lot of spare parts.

But those parts had better not have a lot of defects, because there aren't a lot of spare parts lying around to do repairs!

6. Answer: D

Your project team needs to inspect ALL of the deliverables! That means every single thing that gets produced needs to be reviewed by team members, so they can find and repair defects.

7. Answer: B

A Pareto chart divides your defects into categories, and shows you the percentage of the total defects each of those categories represents. It's really useful when you have a limited budget for Plan Quality Management and want to spend it where it's most effective!

Don't forget that ALL deliverables need to be inspected, including the stuff you create—like the schedule, WBS, and Project Management plan. So you'll get defects for them, too!

Exam ~~Questions~~ Answers

8. Answer: C

Keep your eye out for questions asking you about Ishikawa or fishbone diagrams. When you use those tools to analyze defects, you're in the Control Quality process.

Watch it!

Don't assume that just because you're using a fishbone diagram, you're always doing quality control!
It's also used in Risk Management; you'll see that in Chapter 11. The key thing to watch for here is that the fishbone diagram is being used to find the root cause of a DEFECT, not a risk or something else.

9. Answer: B

A run chart is one of the seven basic tools of quality. It's a long line graph that shows you the total number of defects that were found over time.

> OK, SO I CAN USE A RUN CHART TO FIGURE OUT WHETHER QUALITY IS GETTING BETTER OR WORSE OVER THE COURSE OF MY PROJECT.

10. Answer: B

Inspection is when your team examines something that they produced for defects…and every single deliverable needs to be inspected! That's what "prevention over inspection" means: if you produce a deliverable that's needed later in the project today, it's a lot cheaper to fix defects in it now than it will be when that deliverable is used later on in the project.

11. Answer: D

A lot of people get Control Quality and Verify Scope confused because they seem really similar. Both of them involve looking closely at deliverables to make sure that they meet requirements. But they serve really different purposes! You use Control Quality to find defects that you're going to repair. Verify Scope happens at the very end of the Executing phase; it's when you work with the stakeholder to get formal acceptance for the deliverables.

You'd better have found all the defects before you take the product to the customer!

12. Answer: B

Benchmarking is when you use previous projects to set quality guidelines for your current project. You can always find the results of the past projects in the organizational process assets.

Answers

~~Exam Questions~~

13. Answer: A

The rule of seven tells you that when seven consecutive data points on your control chart come out on the same side of the mean, you've got a process problem. That sounds a little complicated, but it's actually pretty straightforward. Defects tend to be scattered around pretty randomly; in any project that makes a lot of parts, even if they're all within the specification, you'll get a couple of parts that are a little bigger, and a couple that are a little smaller. But if you have a bunch of them in a row that all run a little big, that's a good indication that something's gone wrong on your assembly line!

14. Answer: C

Going back and repairing defects can be a pretty risky activity, because it's really easy to introduce new defects or not fully understand why the defect happened in the first place. Answer C says exactly that: you go back and review the defects to make sure they're fixed.

15. Answer: D

A lot of times it's impractical to check every single product that rolls off of your assembly line. Statistical sampling is a great tool for that; that's when you pull out a small, random sample of the products and inspect each of them. If they're all correct, then there's a very good chance that your whole product is acceptable!

16. Answer: C

A lot of people get confused about the difference between Control Quality and Perform Quality Assurance. Control Quality is where you inspect deliverables for defects, while Quality Assurance is where you audit the project to make sure the quality activities were performed properly.

You inspect products for defects, and you can remember that because you'll find an "inspected by #8" tag in a product. You audit processes, and you can remember that because when you get audited, they're making sure you did your taxes correctly—they're auditing your actions, not a product.

Inspected by
#8

Exam ~~Questions~~ Answers

17. Answer: B

A flowchart is one of the seven basic tools of quality. You use it to analyze processes that are part of your project in order to look for quality problems and inefficiencies.

18. Answer: B

You're analyzing the process, so you are using Perform Quality Assurance.

Just because you see a Perform Quality Control tool, that doesn't mean you're in the Perform Quality Control process... because they're also tools used in Perform Quality Assurance! You always need to figure out what you're using them for.

19. Answer: D

Pareto charts are based on the 80/20 rule. They sort your defects in descending order by root cause. So you always know which 20% of root causes are responsible for 80% of defects on your project.

20. Answer: C

Control Quality is where you inspect your work, including your repairs!

9 Human resource management

Getting the team together

OK, ON THREE... TWO...ONE... BE AN EFFECTIVE TEAM! RIGHT NOW! SERIOUSLY GUYS, PLEASE?

Behind every successful project is a great team. So how do you make sure that you get—and keep—the best possible team for your project? You need to **plan carefully**, set up a good **working environment**, and negotiate for the **best people** you can find. But it's not enough to put a good team together...if you want your project to go well, you've got to keep the team motivated and deal with any conflicts that happen along the way. **Human resource management** gives you the tools you need to get the best team for the job and lead them through a successful project.

Mike needs a new team

Cows Gone Wild III was a huge success! But now the Ranch Hand Games team is gearing up for their next big hit. How are things shaping up?

The box is done, but Brian, Mike, and Amy are just getting started on the game.

OK! LET'S GET THE TEAM BACK TOGETHER! HEY BRIAN, WHEN CAN YOUR TEAM GET STARTED?

NOT SO FAST, MIKE. REMEMBER THE ONLINE PORTION OF CGW III? WELL, THE TEAM'S SWAMPED WITH SERVER MAINTENANCE.

Mike, the project manager

Uh-oh! Looks like Mike's going to need to put together a new team.

Brian, the development team manager

Cubicle conversation

> HEY, WE'RE ALL ON THE SAME TEAM HERE. LOOK, WHY DON'T WE HAVE PEOPLE FROM OUR TEAMS JUST HELP OUT WHENEVER YOU NEED SOMEONE?

Amy, the creative director

Brian: Yeah, there's no reason you need our resources dedicated to your project. We can get **multiple** projects done that way.

Mike: Come on, guys. You don't really think that's gonna work, do you?

Amy: Sure, why not?

Mike: We can't just staff up as we go; that's going to cause huge problems.

Brian: You're overreacting, Mike. Look, I'm a team player, and I want to get the project done. You just tell me when you need someone off my team, and I'll make sure you've got the developers and testers you need. What's wrong with that?

Mike: OK, so what if I need three developers starting tomorrow for the next two weeks? Can you do that?

Brian: Well, no, I've got a deadline on Friday. It'll have to wait until next Monday. But that's just a couple of days.

Mike: See, that's what I'm talking about! A few days here, a few days there…if we have to wait a few days every time the team needs someone, we'll totally blow the schedule.

⚛ BRAIN POWER

How can Mike solve his problem? What can he do to make sure that he gets the team members he needs when he needs them?

Get your team together and keep them moving

You want to stay in control of your project team, right? But when you work in a matrix organization, your team members don't directly report to you. So how do you make sure you get the best people, and keep them motivated and productive? That's what the four processes in **Human Resource Management** are for: guiding you through all the things you need to do to make sure you get everyone for your project when you need them.

A lot of the stuff in this chapter applies mostly to matrix organizations...but you'll still find it really useful, even if you don't work in a matrix company!

This shouldn't be a surprise—every knowledge area has its planning process, and Plan Human Resource Management is no exception.

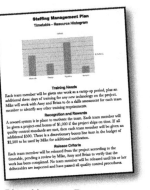

Planning process group

Staffing Management Plan
Timetable – Resource Histogram

Training Needs
Each team member will be given one week as a ramp-up period, plus an additional three days of training for any new technology on the project. Mike will work with Amy and Brian to do a skills assessment for each team member to identify any other training requirements.

Recognition and Rewards
A reward system is in place to motivate the team. Each team member will be given a project-end bonus of $1,000 if the project ships on time. If all quality control standards are met, then each team member will be given an additional $500. There is a discretionary bonus line item in the budget of $2,500 to be used by Mike for additional motivation.

Release Criteria
Each team member will be released from the project according to the timetable, pending a review by Mike, Amy and Brian to verify that the work has been completed. No team member will be released until his or her deliverables are inspected and have passed all quality control procedures.

Plan Human Resource Management

In the Plan Human Resource Management process, you plan out exactly which resources you'll need, what their roles and responsibilities are, and how you'll train your team and make sure they stay motivated.

This is where you plan out the staffing needs for your project, and how you'll manage and reward the team.

It makes sense that Acquire Project Team and Develop Project Team are in the Executing group—you only put the team together AFTER the project has started.

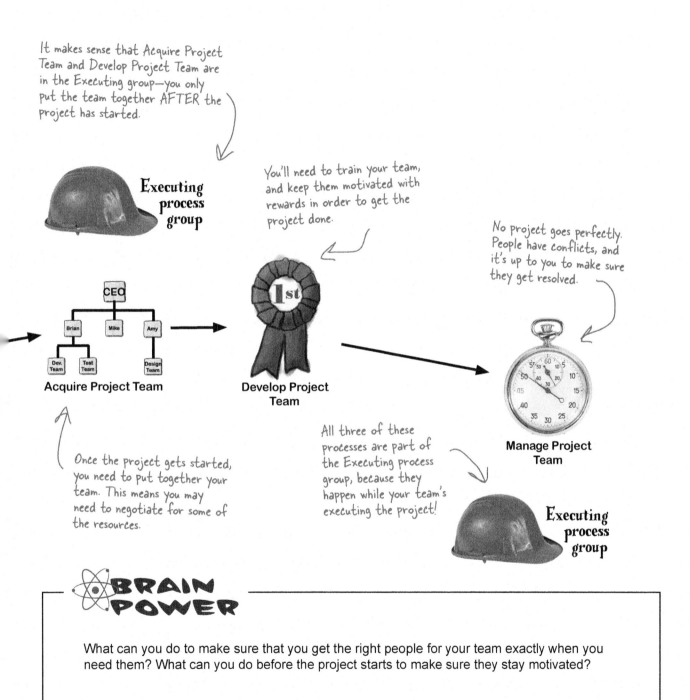

Executing process group

You'll need to train your team, and keep them motivated with rewards in order to get the project done.

No project goes perfectly. People have conflicts, and it's up to you to make sure they get resolved.

CEO
Brian   Mike   Amy
Dev. Team   Test Team   Design Team

Acquire Project Team

Develop Project Team

Manage Project Team

Once the project gets started, you need to put together your team. This means you may need to negotiate for some of the resources.

All three of these processes are part of the Executing process group, because they happen while your team's executing the project!

Executing process group

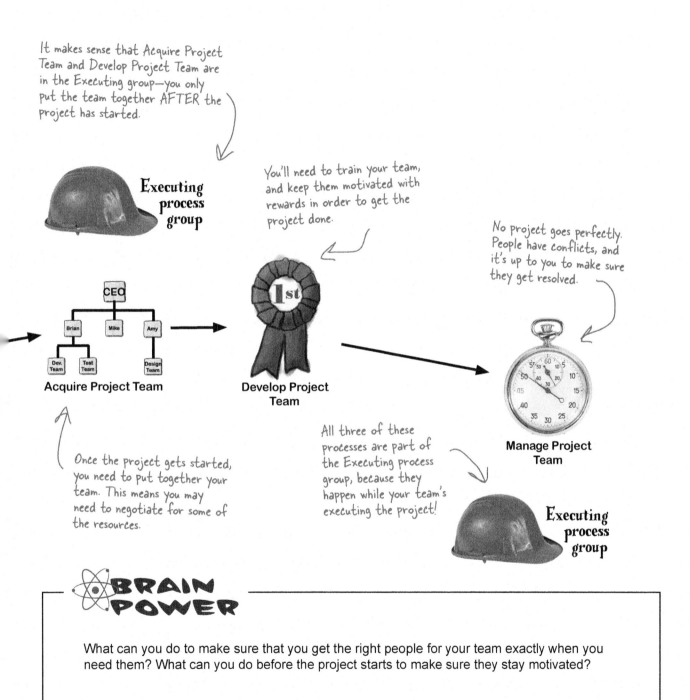

⚛ BRAIN POWER

What can you do to make sure that you get the right people for your team exactly when you need them? What can you do before the project starts to make sure they stay motivated?

Figure out who you need on your team

Project teams don't just assemble themselves spontaneously! It takes a lot of planning and guidance to get a team together, and that's the idea behind the **Plan Human Resource Management** process. Remember, in a matrix organization your team doesn't report directly to the project manager. You need to work with the functional managers to get the team members that you need for your project…which means there's a lot of information that you need to give to everyone so they know exactly who you need for your team.

Planning process group

You've seen these two inputs a whole bunch of times now!

Inputs

Activity resource requirements

Enterprise environmental factors

Organizational process assets

Project Management plan

Organization charts and position descriptions tell everyone how your team is structured.

Networking means both formally and informally interacting with other people in your company and industry to stay on top of everything.

Organizational theory is where you use proven principles to guide your decisions.

Expert judgment is used to figure out resource requirements and position descriptions.

Meetings help the team come together and agree on what's needed for the project.

This is a lot like the other planning processes you've seen already! You start with your project plan and what you know about your company, and you come up with a plan.

The Human Resource Management plan tells everyone on the project who you'll need on your team, when you need them, and what skills they'll need.

The Human Resource Management plan is divided into three major sections that help you manage the resources on your project.

Outputs

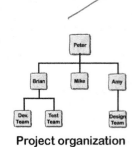

Project organization chart

Staffing Management plan

RACI Matrix		Role			
		Mike	Amy	Brian	Peter
Work Package	Project Management	R	I	I	I
	Design	C	R	C	I
	Construction	C	C	R	I
	Testing	C	C	R	I

R = Responsible A = Accountable C = Consult I = Inform

Roles and responsibilities

Sharpen your pencil

The inputs, tools and techniques, and outputs of **Plan Human Resource Management** should seem pretty familiar! Write down what you think you'd use each of them for. Notice that there are a few that you haven't seen before—take an educated guess at those.

INPUTS

Enterprise environmental factors

Organizational process assets

TOOLS AND TECHNIQUES

Organization charts and position descriptions   ←

There are a couple of other tools & techniques—this isn't the only one!

OUTPUTS

Human Resource Management plan:

• Staffing Management plan

• Roles and responsibilities

• Organization charts   ←

Think about how this is different than the organization charts listed under Tools and Techniques.

Sharpen your pencil
Solution

The inputs, tools and techniques, and outputs of **Plan Human Resource Management** should seem pretty familiar! Write down what you think you'd use each of them for. Notice that there are a few that you haven't seen before—take an educated guess at those.

INPUTS

Enterprise environmental factors

This is information about the company's culture and structure

Your company's culture is really important—stuff like common languages, technical disciplines, and how people normally relate to one another.

Organizational process assets

Templates and lessons learned from past projects

You've already seen lots of ways we use templates and checklists. They're just as important in Plan Human Resource Management.

It's easy to lose track of who reports to whom, and what different people do in your company. You need to know that stuff if you want to staff your project!

TOOLS AND TECHNIQUES

Organization charts and position descriptions

Shows the relationships between managers, team members, and other people inside and outside the company who will work on the project

OUTPUTS

Human Resource Management plan:

Your Staffing Management plan describes who will be on your project, when they'll do the work and for how long, and the reward system you'll use to keep the team motivated.

* Staffing Management plan

 Describes how you'll manage and control your resources

Every role on the project needs to be defined—it has a title, has authority to do certain things, and is responsible for specific deliverables.

* Roles and responsibilities

 Lists each role on the project that needs to be filled

This is a lot like the org chart for your whole company, except that it lists only the specific people on the project.

* Organization charts

 Shows the reporting structure of the resources assigned to the project team

HOLD ON—HOW CAN ORGANIZATION CHARTS BE BOTH TOOLS AND TECHNIQUES AND OUTPUTS?

You need two charts because you have two "organizations."

One organization is the whole company—along with any subcontractors or consultants you've got access to for your project team. The other organization is just the people who are on the team. The team might have people from different groups in your company, and they need to know how they interact. That's why you create a chart just for the team.

Some people will bring in a consultant or expert to manage or lead part of the team. Make sure the project's org chart shows that relationship, even if it's not part of the company!

The project organization chart shows how your team members relate with one another.

This might include people or relationships that may not necessarily show up on a company organization chart. If you've got a team built from multiple consultants and subcontractors, this chart will be the only place where everyone is listed at once.

Roles and responsibilities show who's responsible for what.

It's really common to see the roles and responsibilities for a project written out as a **RACI matrix**, which is just a table that lists the role or people on the top; the specific activities, work, or responsibilities down the side; and the level of responsibility that each person or role has for each of the activities or responsibilities. (RACI stands for "Responsible, Accountable, Consulted, and Informed.")

This could also list roles, like Project Manager, Creative Director, or Development Manager.

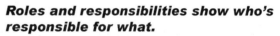

RACI Matrix		People			
		Mike	Amy	Brian	CEO
Work Package	Project Management	R	I	I	I
	Design	C	R	C	I
	Construction	C	C	R	I
	Testing	C	C	R	I
R = Responsible A = Accountable C = Consulted I = Informed					

The Staffing Management plan

An important component of the Human Resource Management plan is the **Staffing Management plan**. It tells you everything that you need in order to build your team, keep them motivated, and manage them to resolve conflicts and get the work done.

Everything you do with your team—acquiring them, developing them, and managing them—depends on a good Staffing Management plan.

A common way of showing the timetable—or when people will work on what—is to use a resource histogram.

This is really important for telling the functional managers exactly who you'll need on your team, so they can provide the staff that you need to get the job done.

The resource histogram tells you the type and number of resources you need at any time. It's usually a <u>vertical bar chart</u>.

You need to make sure everyone on the team has the skills he or she needs to do the job.

A really important part of Human Resource Management is keeping your team motivated, and rewards tied to goals are a great way to do that.

You'll need to plan out exactly how your team members will roll off of your project so functional managers and other project managers will know if they're available for other projects.

Staffing Management Plan

Timetable – Resource Histogram

Training Needs

Each team member will be given one week as a ramp-up period, plus an additional three days of training for any new technology on the project. Mike will work with Amy and Brian to do a skills assessment for each team member to identify any other training requirements.

Recognition and Rewards

A reward system is in place to motivate the team. Each team member will be given a project-end bonus of $1,000 if the project ships on time. If all quality control standards are met, then each team member will be given an additional $500. There is a discretionary bonus line item in the budget of $2,500 to be used by Mike for additional motivation.

Release Criteria

Each team member will be released from the project according to the timetable, pending a review by Mike, Amy and Brian to verify that the work has been completed. No team member will be released until his or her deliverables are inspected and have passed all quality control procedures.

Answers on page 496.

Exercise

Read the **Staffing Management plan** on the facing page and answer these questions about the project.

1. How many designers, developers, and testers are needed in week #7 of the project?

.... designers   .... developers   .... testers

2. Who is responsible for verifying that each team member has the skills appropriate to the project?

....

3. Rewards should always be tied to performance goals in order to motivate the team. What performance goal has been set for the team, and what reward will each team member receive if it's achieved?

....
....

there are no Dumb Questions

Q: I still don't get the resource histogram. Am I supposed to make this myself, or does it come from somewhere?

A: You need to come up with the histogram yourself when you put together the Staffing Management plan. Since you're managing the project, you're the only one who knows when each person is needed on the project. Remember all of the activities that you came up with, back when you were building the schedule in the Chapter 6? Well, each of those activities had resource requirements, right? That means you know exactly what resources you'll need at any time in your project! That's why the activity resource requirements are an input to Plan Human Resource Management—you need the schedule and the activities in order to figure out the timetable. The histogram is the easiest way to show that information.

Q: Is that RACI chart really necessary?

A: Yes, definitely! Sometimes people split up responsibilities in ways that aren't immediately obvious just from people's titles or the names of their roles on the project—that's one of the big advantages of a matrix organization. RACI charts help everyone figure out their assignments. Mike might have Brian's senior developers sit in on Amy's design meetings, even though they don't usually do that. He'd put that in the RACI matrix to show everyone that's now part of their jobs for the project.

Q: Once I know what roles need to be filled on my project, how do I actually get the team on board?

A: That's what the next process is all about! It's called **Acquire Project Team**, and it's where you actually staff your project. Of course, you don't staff it during the planning phase. You have to wait until the project work begins, which is why it's in the *Executing* process group.

The hardest part about staffing your project is negotiating with the functional managers. The best resources are the ones that are in demand, which means your negotiating skills will be very important when it comes time to staff your project team.

Get the team together

Your Human Resource Management plan is in place, your project is ready to roll, and now it's time to begin the actual project work! You need your team, and the way you bring them on board is the **Acquire Project Team** process.

This is where you negotiate with functional managers for your project team members. You need the right people for the project, and you've done all the prep work to figure out who you need and when you need them. So now it's time to go get your team!

Executing process group

Watch it!

Beware of the halo effect!

That's when you put someone in a position they can't handle, just because they're good at another job.

Take a minute and think about why you need each of these inputs.

Inputs

Human Resource Management plan

Don't be thrown by the word "acquisition"—even though this is the Acquire Project Team process, this particular technique only refers to acquiring resources from outside.

Tools

There aren't any surprises here—if you've staffed a project team, you've done all of these things.

Negotiation is the most important tool in this process. There are resources that you need for your project, but they don't report to you. So you need to negotiate with the functional managers—and maybe even other project managers—for their time.

Virtual teams are when your team members don't all work in the same location. This is really useful when you're relying on consultants and contractors for outsourced work. Instead of meeting in person, they'll use phone, email, instant messaging, and online collaboration tools to work together.

Preassignment is when you can actually build the assignments into your Staffing Management plan. Sometimes you have resources who are guaranteed to you when you start the project, so you don't need to negotiate for them.

Acquisition means going outside of your company to contractors and consultants to staff your team.

Multicriteria decision analysis means looking at a bunch of factors when deciding on who should be part of your project team. Sometimes teams will use tools that help them weigh factors like cost, skills, knowledge, and availability when deciding on project team needs.

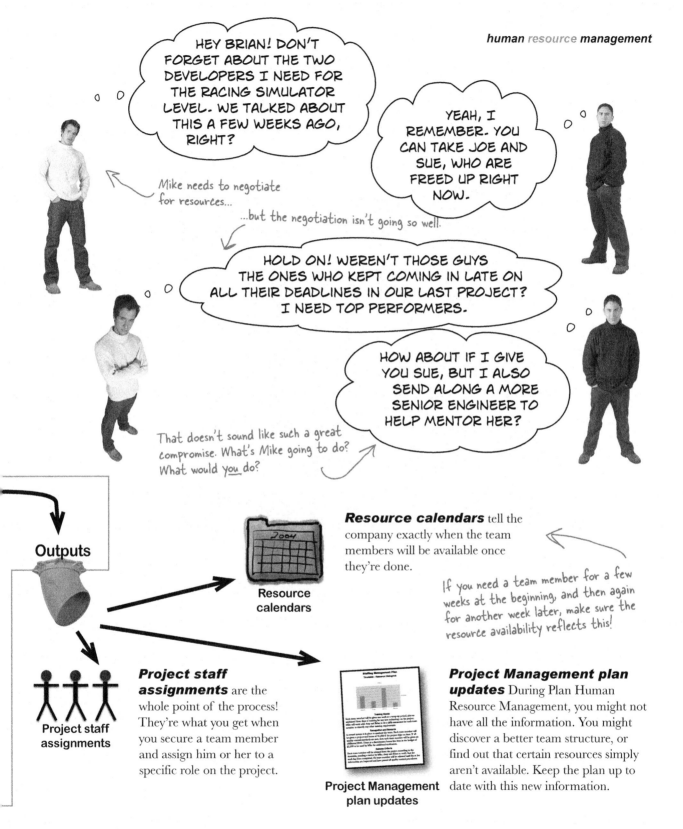

HEY BRIAN! DON'T FORGET ABOUT THE TWO DEVELOPERS I NEED FOR THE RACING SIMULATOR LEVEL. WE TALKED ABOUT THIS A FEW WEEKS AGO, RIGHT?

YEAH, I REMEMBER. YOU CAN TAKE JOE AND SUE, WHO ARE FREED UP RIGHT NOW.

Mike needs to negotiate for resources...

...but the negotiation isn't going so well.

HOLD ON! WEREN'T THOSE GUYS THE ONES WHO KEPT COMING IN LATE ON ALL THEIR DEADLINES IN OUR LAST PROJECT? I NEED TOP PERFORMERS.

HOW ABOUT IF I GIVE YOU SUE, BUT I ALSO SEND ALONG A MORE SENIOR ENGINEER TO HELP MENTOR HER?

That doesn't sound like such a great compromise. What's Mike going to do? What would you do?

Outputs

Resource calendars tell the company exactly when the team members will be available once they're done.

Resource calendars

If you need a team member for a few weeks at the beginning, and then again for another week later, make sure the resource availability reflects this!

Project staff assignments are the whole point of the process! They're what you get when you secure a team member and assign him or her to a specific role on the project.

Project staff assignments

Project Management plan updates During Plan Human Resource Management, you might not have all the information. You might discover a better team structure, or find out that certain resources simply aren't available. Keep the plan up to date with this new information.

Project Management plan updates

Human Resourcecross

Take some time to sit back and give your right brain something to do. It's your standard crossword; all of the solution words are from this chapter.

Answers on page 497.

Across

1. The resource _____ is a vertical bar chart in the Staffing Management plan that tells you the type and number of resources you need.

5. This is a great tool for communicating roles and responsibilities.

9. The main output of the Acquire Project Team process is project staff _____.

10. The _____ effect causes people with technical expertise to be put in positions for which they're unqualified.

11. The first process you perform in this knowledge area is Plan Human Resource _____.

12. This kind of chart tells you how team members relate to one another in your company.

13. The Manage Project Team process is part of the _____. process group.

Down

2. A great way to motivate your team.

3. What you provide for a human resource in order to ensure he or she has the skills necessary to do the project.

4. You use _____ teams when your team members don't all work in the same location.

6. _____ availability is an output that describes when each team member will be available to your project.

7. The most important tool in Acquire Project Team.

8. The _____ organization chart shows only the people assigned to your team, including consultants and subcontractors.

Cubicle conversation

> I'M STARTING TO GET A LITTLE CONCERNED WITH THE PACE OF OUR DESIGN WORK. IS EVERYTHING OK?

Looks like Mike's negotiation went well in the end! So now he's got his team...but can he get them motivated?

Amy: I haven't noticed anything. What's up?

Mike: Well, maybe it's nothing, but a couple of the design team members have been missing some deadlines. Nothing major, but it's starting to concern me.

Amy: Well, OK. I can keep an eye on them.

Mike: That's not all. One of them hasn't been replying to emails at all, and another scheduled a vacation right in the middle of a huge deadline week. I think we may have a real motivation problem.

Amy: You're right, that sounds pretty bad. What can I do about it?

Mike: Well, I built a discretionary bonus budget into the plan.

Amy: Right, that $2,500. But should we really be talking about giving bonuses? I thought these were underperformers. Shouldn't we reward only good behavior?

Mike: Well, right, but if we tie the bonus to meeting an aggressive deadline or high quality standards, it might help get them energized again.

Amy: We can give it a shot, but I'm skeptical.

Do you think Mike's idea will work? Why is it a good idea to make the bonus contingent on meeting specific goals? Can this plan backfire?

Develop your project team

The **Develop Project Team** process is the most important one in Human Resource Management. It's the one where you make sure your team is motivated and well managed—and those are some of the most important things that project managers do! You do it throughout the *entire* Executing phase of the project, because you need to keep your team moving toward the goal.

Executing
process
group

Get the team involved in planning—the more they feel like they're in control, the better they feel about the project!

Motivation

☐ One of your most important jobs as project manager is keeping the team motivated and constantly monitoring them to make sure they stay motivated.

☐ A really effective way to motivate your team is to set up a reward system. But make sure that they understand exactly what they're being rewarded for—and it *must* be fair, or it could backfire!

☐ Training is another great way to keep a team motivated. When people feel that they're growing professionally, they stay more involved and get more excited by their work.

This makes it more challenging to stay on top of the team and make sure the work is getting done.

Management

☐ When the project is being planned, you're directing everything—but by the time it's executing, the project manager is more of a coach and a facilitator.

☐ That's why it's really important for a project manager to have "soft skills"—you need to really understand what makes your team members tick, and help with their problems.

☐ A really good way to make sure that your project team sticks together is to establish **ground rules** for your project, which set a standard for how everyone works together.

This is one of the tools and techniques for Develop Project Team.

You develop your project team by keeping them motivated, and you do this all the way through your entire project.

Develop the team with your management skills

How do you keep your team motivated and up and running? With the tools and techniques for **Develop Project Team**, that's how. When you're working with your team, you need to be a leader. That means setting the rules for how people interact with one another, making sure they have the skills they need, setting up a good working environment, and keeping them motivated.

The inputs to Develop Project Team are the outputs you just created.

Project staff assignments

Resource calendars

Human Resource Management plan

Inputs

Tools

Discussing ground rules with the team can be really valuable, because it helps everyone see what's important to their teammates!

Recognition and rewards are the best way to keep your team motivated!

Interpersonal skills are all about using soft skills to help the people on your team solve problems.

Ground rules help you prevent problems between team members, and let you establish working conditions that everyone on the team can live with.

Team-building activities are important throughout your entire project. You're responsible for keeping the team together!

Training is a really important part of developing your team. If you've got a team member who doesn't have the skills to do the job, you need to get him trained…and it's up to you to plan enough of the project's time and budget to make sure it happens!

For example, you might have a rule where everyone always emails the team when they take a day off.

Colocation is the opposite of virtual teams. When you have all of your team located in the same room, you can increase communication and help them build a sense of community. Sometimes that room is called a **war room**.

Personnel assessment tools are used to figure out how your team approaches the work and how they like to work together. These tools include things like focus groups and surveys used to determine your team's style of working and interacting.

Your interpersonal skills can make a big difference for your team

Knowing all of the tools and techniques in the *PMBOK Guide* will help you learn a lot about your project, but the way you help your team to get the job done is just as important as the steps you take to get it done.

The team was happy to work on CGW III from the beginning. But when Mike told them that the company saw the game's success as responsible for 70% of their revenue, it really showed them how important the work was.

Leadership is all about giving the team a goal to shoot for and helping them to see the value in the work they are doing. It's not enough to have a team know the end product that they're building; they need to **understand the value that that product is going to bring to the company**. A project manager needs to constantly remind the team of the vision they're working toward, and make decisions to help keep the team on track toward it.

Team building involves helping your team learn to depend on and trust one another. As a project manager, you're responsible for helping the team come to an understanding about how they'll communicate and stay motivated when things go wrong. If you're open about your decision-making processes and communicate often about what you're doing, you can **help your team to bond**. Some people think of team building as going out for pizzas after work, but it's more about how you lead, and how you help to create an environment where your team members can trust one another.

Motivation demonstrates to your team the value that the project has for them. It includes making sure that people are compensated and rewarded financially for their work. But that's not the only facet of team motivation that you need to be concerned with. Your team also needs to know how the tasks they're doing contribute to project success, and what's in it for them. Motivating your team is about **helping them to be satisfied with the job they're doing**, recognizing them when they do a good job, and keeping them challenged with new and different problems.

Communication is a constant concern when you're leading a team. It's not enough to make the best decisions to get your project done; you've got to make sure that everybody in the team knows why you're making them, and feels like you're being **open and honest** about what's motivating every decision you make. If the people on your team feel like they're always getting the information they need from you and that they're never in the dark, they'll be able to trust you and one another more.

When Mike agreed to consolidate the code reviews for two features into one review, he forgot to tell the team about it. When the first code review was cancelled, the team was confused. They thought Mike didn't care about the quality of the product they were making, and the misunderstanding was really hard on them.

Influencing is all about using your relationships with the people on your team to get them to cooperate in making good decisions for the project. When you lead by example, you show your team how you want them to behave by doing it. It may seem subtle, but the way you work as a project manager can **set the standard for your teammates**. Collaborating with your team on the best way of working through your project is a really effective way of making sure that the team members gel, and know that they can rely on one another.

Mike made sure he came to work early as often as he could. After a while he noticed that everybody on the team was doing the same thing.

Political and cultural awareness means knowing the people on your team and understanding their backgrounds. Since projects sometimes span more than one culture, it's important to take the time to **understand the similarities and differences in the working environments** across the project team. It's equally important to communicate with your team members and understand what motivates them.

Decision making is how you handle the issues that come up when you're working through your project. There are a few basic techniques for decision making that you'll use:

Command: Sometimes you'll just make a decision and inform your team about it. You'll decide, and then team will do what you say.

Consultation: Sometimes you'll talk your decision over with your team and ask for their opinions before you decide.

Consensus: Another way of making decisions is to talk about a few options with your team and get everybody to agree on one of them before you decide.

Coin flip: Another way to make a decision is to just randomly choose one of the options.

Negotiation helps the people on your team come to an agreement about how to work together. It's important when you're negotiating to **listen to both parties** and to make sure that you **make it clear when concessions are made**. That should get everyone to see both sides of the issue and know that you're negotiating a fair resolution to it.

Trust building, coaching, and conflict management are also important in managing the interpersonal dynamics on your team. You'll remember trust building and coaching from Chapter 1. We'll talk about conflict management in just a minute.

Lead the team with your management skills

You've seen tools and techniques that help you set up a great environment for your team to succeed, but you need more than that to get them through a tough project. You need **leadership skills**, those "soft skills" you use to influence your team and keep them directed toward the project's goals.

> WAIT A SECOND! HOW CAN A MULTIPLE-CHOICE TEST QUIZ ME ON MY LEADERSHIP SKILLS?

You use leadership skills throughout the entire project! But they're most important in Develop Project Team because that's where you lead your team through their work.

You're right, it can't. What the PMP exam *can* quiz you on is your **knowledge** of leadership skills. There's been a lot of research on how people wield power in companies. The PMP exam concentrates on research done by two social psychology researchers named French and Raven who came up with five different kinds of power that people use to influence others.

The five kinds of power

There are five kinds of power that a project manager typically uses on a project. The first is called **legitimate power**, which is what you use when you assign work to someone who reports to you.

When you're someone's boss, you have legitimate power to tell them what to do. But when you work in a matrix organization, you don't have direct reports! So you'll need to use the other kinds of power to influence your team.

Reward power is what you have when you can award a bonus or another kind of reward in order to motivate team members. Always make sure that rewards are **fair**—you don't want to single out one person who is eligible for a reward without giving others a chance at it! And rewards work best when they're tied to specific goals or project priorities.

Making everyone compete for one single reward isn't fair—it's actually demotivating to force people to compete for an arbitrary prize.

Expert power means that the team respects you for your expertise in a specific area, and gives you credibility because of that. Think about it: a team of programmers is more likely to respect you and do what you ask them to do if they know that you're an accomplished software engineer!

> EVERYONE ON THE TEAM WILL GET AN EXTRA $500 BONUS IF WE MEET ALL OF THE QUALITY STANDARDS IN THE SPEC!

Reward and expert power are the most effective kinds of power that a project manager can use.

> I SPENT A FEW YEARS LEADING A GAME DESIGN TEAM, AND I GOT THE BEST RESULTS WHEN WE HELD JOINT DESIGN MEETINGS WITH THE PROGRAMMERS.

Referent power means that people admire you, are loyal to you, and want to do what you do because you're part of the cool crowd. Often, a project manager might wield referent power because he's trusted by people in authority, and others perceive him as associated with success.

Punishment power is exactly what it sounds like—you correct a team member for poor behavior. **Always remember to do this one-on-one and in private!** Punishing someone in front of peers or superiors is extremely embarrassing, and will be really counterproductive.

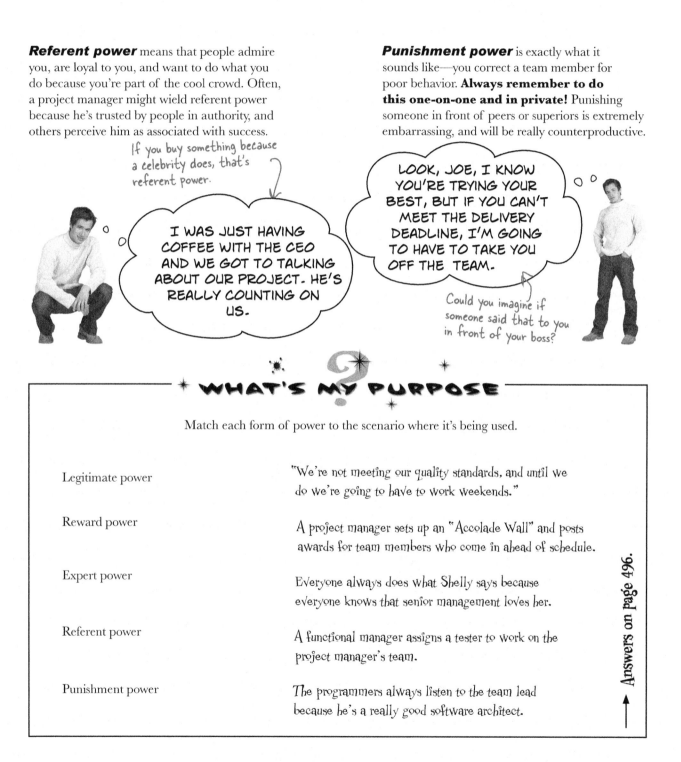

If you buy something because a celebrity does, that's referent power.

I WAS JUST HAVING COFFEE WITH THE CEO AND WE GOT TO TALKING ABOUT OUR PROJECT. HE'S REALLY COUNTING ON US.

LOOK, JOE, I KNOW YOU'RE TRYING YOUR BEST, BUT IF YOU CAN'T MEET THE DELIVERY DEADLINE, I'M GOING TO HAVE TO TAKE YOU OFF THE TEAM.

Could you imagine if someone said that to you in front of your boss?

WHAT'S MY PURPOSE

Match each form of power to the scenario where it's being used.

Legitimate power

Reward power

Expert power

Referent power

Punishment power

"We're not meeting our quality standards, and until we do we're going to have to work weekends."

A project manager sets up an "Accolade Wall" and posts awards for team members who come in ahead of schedule.

Everyone always does what Shelly says because everyone knows that senior management loves her.

A functional manager assigns a tester to work on the project manager's team.

The programmers always listen to the team lead because he's a really good software architect.

Answers on page 496.

Motivate your team

This stuff is all part of recognition and rewards—one of the tools and techniques for Develop Project Team.

No matter how good your soft skills are, if your team has a lousy work environment, they're going to have a hard time getting the project done. Luckily, there's been research done over the years to figure out exactly what makes a good working environment. For the PMP exam, you'll be expected to be familiar with the most popular theories of motivation and organization.

A "hygiene factor" is something like a paycheck or status—stuff that people need in order to do the job. If people don't have this stuff, it's really hard to motivate them!

You might see this in a question about "Maslow's theory," or it might show up on the exam as "Hierarchy of Needs" or "Maslow's Hierarchy."

Maslow's Hierarchy of Needs says that
people have needs, and until the lower ones are satisfied they won't even begin to think about the higher ones.

Herzberg's Motivation-Hygiene Theory

Sure, you love being a project manager. But would you do the job if you weren't getting a paycheck? Of course not!

Maslow says you can't achieve the higher needs until you're comfortable with the lower ones...

...you need to feel safe and accepted before you can make a higher contribution.

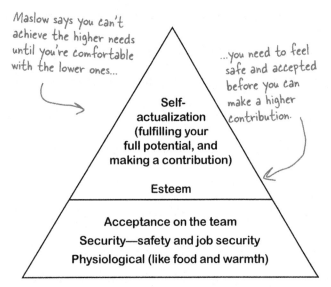

What Herzberg figured out was that you need things like good working conditions, a satisfying personal life, and good relations with your boss and coworkers—stuff he called "hygiene factors." They don't motivate you, but you need them before you can be motivated. Until you have them, you don't really care about "motivation factors" like achievement, recognition, personal growth, or career advancement.

Herzberg says that people need the stuff they normally expect out of a job—like hot coffee—before you can get them motivated about achievement and personal growth.

McGregor's Theory X and Theory Y

McGregor tells us that there are two kinds of managers: ones who assume that everyone on the team is selfish and unmotivated, and ones who trust their team to do a good job. He calls the kind of manager who distrusts the team a "Theory X" manager, and the kind who trusts them a "Theory Y" manager. **You could get exam questions where the answer could be "Theory X" or "Theory Y"—or both!**

A Theory X manager will micromanage the team, looking over everyone's shoulder all the time and making them feel like they aren't trusted.

It's much better—and easier—to be a Theory Y manager. If you trust the team to do their jobs, they won't let you down!

Two more theories that might appear on the PMP exam—although they're not nearly as common as the others.

Expectancy Theory says that you need to give people an expectation of a reward in order to motivate them—but this works only if that award is achievable. If everyone knows the award is either worthless or impossible to achieve, it will actually demotivate them!

McClelland's Achievement Theory says that people need to be motivated. Achievement is when someone performs well and is recognized for it. Power means he or she has a lot of control or influence in the company. And someone feels a strong sense of affiliation from being a part of a working team and having good relationships with coworkers.

Exercise Solution

Each of the following scenarios demonstrates one of the motivational theories at work. Write down which theory each scenario describes.

1. Bob is a programmer on the team, but he doesn't really feel like he's "one of the guys." He doesn't really have a lot of control over the work he's assigned. Recently, Bob put in a long weekend to get his work done, but nobody really seemed to take notice.

....

2. There was a break-in at the office, and now people are really jittery. Plus, the heating system has been broken for weeks, and it's freezing! No wonder nobody's getting any work done.

....

3. Eric's a functional manager, but his team seems to move really slowly. It turns out that everyone who reports to him has to hand him their work first, before they can give it to anyone else. He goes through it line by line, which sometimes takes hours! He doesn't trust his team to release anything he hasn't seen.

....

4. Joe's a functional manager, and his team is very efficient. He spot-checks their work, but for the most part he sets realistic performance goals and trusts them to meet it—he only pulls people aside if he finds that there's a specific problem that has to be corrected.

....

5. A project manager is having a lot of trouble motivating the team. He tries setting up rewards and a good working environment. But the team remains difficult to motivate—mostly because their paychecks all bounced last week, and everyone is angry at the CEO because they didn't get bonuses.

....

Answers on page 498.

Stages of team development

There's a process for a team to get from a group of strangers to a group that creates something good together, and that's what the stages of team development are all about.

Every team goes through these stages during a project.

Forming: People are still trying to figure out their roles in the group; they tend to work independently, but are trying to get along.

Storming: As the team learns more about the project, members form opinions about how the work should be done. This can lead to temper flare-ups in the beginning, when people disagree about how to approach the project.

Norming: As the team learns more about the other members, they begin to adjust their own work habits to help out one another and the team as a whole. Here's where the individuals on the team start learning to trust one another.

Performing: Once everyone understands the problem and what the others are capable of doing, they start acting as a cohesive unit and being efficient. Now the team is working like a well-oiled machine.

Adjourning: When the work is close to completion, the team starts dealing with the fact that the project is going to be closing soon.

Researcher Bruce Tuckman came up with these five stages as a model for team decision making.

Although this is the normal progression, it's possible that the team can get stuck in any one of the stages. One big contribution you can make, as the project manager, is to help the team get through the initial Storming phase, and into Norming and Performing. It's important to keep in mind that people have a tough time creating team bonds initially, and to try to use your soft skills to help the team to progress through the stages quickly.

BRAIN POWER

How does knowing the five stages of team development change the decisions that you'll make in handling conflicts on your team?

Exercise

Each of the following scenarios demonstrates one of the stages of team development. Write down which stage each scenario describes.

1. Joe and Tom are both programmers on the Global Contracting project. They disagree on the overall architecture for the software they're building, and frequently get into shouting matches over it. Joe thinks Tom's design is too short-sighted and can't be reused. Tom thinks Joe's design is too complicated and probably won't work. They're at a point right now where they're barely talking to each other.

2. Joan and Bob are great at handling the constant scope changes on the Business Intelligence project. Whenever the stakeholders request changes, they shepherd them through the change control process and make sure the team doesn't get bothered with them unless it's absolutely necessary. That leaves Darrel and Roger to focus on building the main product. Everybody is focusing on their area and doing a great job. It seems like it's all just clicking for the group.

3. Derek just got to the team, and he's really reserved. Folks on the team aren't quite sure what to make of him. Eveybody's polite, but it seems like some people are a little threatened by him.

4. Now that the product has shipped, the team is meeting to document all of their lessons learned and write up project evaluations.

5. Danny just realized that Janet is really good at developing web services. He's starting to think of ways to make sure that she gets all of the web service development work and Doug gets all of the client software work. Doug seems really happy about this too—he seems to really enjoy building Windows applications.

⟶ Answers on page 500.

How's the team doing?

There are two outputs of Develop Project Team. One is the **team performance assessment**. Developing the project team means working with them to keep everyone motivated, and training them to improve their skills. The other is **updates to your company's enterprise environmental factors**, to update your company's personnel records.

Outputs

> You'll need to keep track of how well the team is performing, so when the team has problems you'll have a good baseline to compare against.

> The project manager should look at how the team's skill set has improved, and make sure it's documented here.

> Has the team performance improved? Are the motivational techniques working? If so, that goes here!

> You can measure how motivated and happy the team is by keeping an eye on the turnover rate.

Cows Gone Wild IV
Team Performance Assessment

Competencies / Skills improvements
Developers: attended three-day training course on new vector graphics coding techniques. Designers: brought in industrial design professor from Ivy College to hold seminar on design techniques.

Team Performance
There's been a marked improvement in team cohesion, and it's resulted in a lower defect rate. We've awarded 50% of our $2,500 bonus budget.

Turnover Rate
Two designers and one developer have left the team, which is an improvement from CGW III.

BULLET POINTS: AIMING FOR THE EXAM

- Project managers use their **general management skills** ("soft skills") to motivate and lead the team.

- In a matrix organization, the project manager doesn't have **legitimate** power, because the team doesn't directly report to the project manager.

- The most effective forms of power are **reward power**, where the project manager sets up rewards and recognition for the team, and **expert power**, which means the team respects the project manager's technical expertise.

- **Referent power** is power that's based on identifying with or admiring the power holder.

- **Punishment power** is the least effective form of power. The project manager should never punish a team member in front of peers or managers!

- Project managers should be familiar with modern **theories of motivation and management**.

- **McGregor's Theories X and Y** state that there are poor Theory X managers who don't trust their teams, and good Theory Y managers who do.

- **Maslow's Hierarchy of Needs** is the theory that says that people can't achieve "self-actualization" (full potential) or esteem (feeling good and important) until lower needs like safety and security are met.

- **Herzberg's Theory** says that it's difficult to motivate people unless hygiene factors like a paycheck and job security are already in place.

- **Expectancy Theory** holds that people only respond to rewards that are tied to goals they feel they have a realistic chance of achieving.

- Bruce Tuckman's five stages of team development are **forming** (the team still finding their roles), **storming** (the team forming opinions), **norming** (adjusting work habits to help the team), **performing** (working like a well-oiled machine), and **adjourning** (closing down the project).

Cubicle conversation

> WE'VE GOT A PROBLEM, MIKE. MY TEAM NEEDS THE 3-D MODEL DESIGNS AND TEXTURE MAPS FROM THE DESIGN TEAM, BUT WE'RE NOT GETTING THEM.

Amy: Look, we've been over this, Brian. My team is working on level design, and that's the priority right now.

Brian: Mike, is that true?

Mike: I checked the schedule, and all three of those things are part of the current activity. It's not really clear which one of them is the priority.

Brian: Look, my team will be halted if we don't get those models and textures.

Amy: Come on, Brian. You guys have a whole bunch of unit tests that you can write, and I know you're a week behind on code reviews. Can't you just work on those in the meantime?

Brian: My team's been reviewing code for two weeks now. They need a break!

Amy: Aha! So it's not *really* that you're going to fall behind if you don't get the textures immediately.

Brian: Well, no, but I'll be dealing with a team that has motivation problems. And I'm the one who has to clean up that mess!

Mike: OK, hold on, guys. Let's see if we can work this out.

Amy: I don't see what there is to work out. He's being unreasonable.

⚛ BRAIN POWER

It looks like Brian and Amy are having a serious conflict, and it could have a big impact on the project if Mike doesn't get it under control! What usually causes conflicts in projects, and what can the project manager do about it when those conflicts happen?

Managing your team means solving problems

Wouldn't it be great if your team members never had any conflicts? Well, we all know that conflicts are a fact of life in any project. A good project manager knows how to handle conflicts so they don't delay or damage the project. And that's what the **Manage Project Team** process is about.

Executing process group

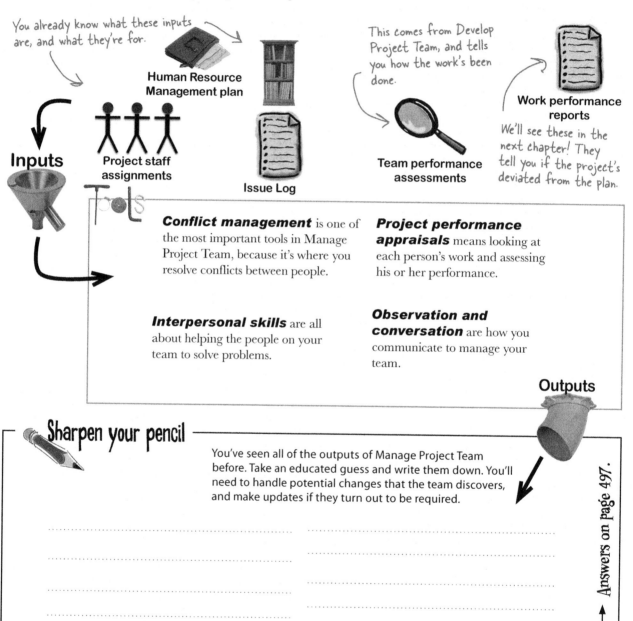

You already know what these inputs are, and what they're for.

Human Resource Management plan

This comes from Develop Project Team, and tells you how the work's been done.

Work performance reports

We'll see these in the next chapter! They tell you if the project's deviated from the plan.

Inputs

Project staff assignments

Issue Log

Team performance assessments

Tools

Conflict management is one of the most important tools in Manage Project Team, because it's where you resolve conflicts between people.

Project performance appraisals means looking at each person's work and assessing his or her performance.

Interpersonal skills are all about helping the people on your team to solve problems.

Observation and conversation are how you communicate to manage your team.

Outputs

Sharpen your pencil

You've seen all of the outputs of Manage Project Team before. Take an educated guess and write them down. You'll need to handle potential changes that the team discovers, and make updates if they turn out to be required.

Answers on page 497.

....

....

....

....

Conflict management up close

It's probably no surprise that over half of conflicts come from priorities, schedules, and people. That's why so many of the processes you're learning about are focused on preventing conflicts. Ground rules, good planning practices, and pretty much anything that has to do with communication are all there to prevent the most common reasons that conflicts happen.

Some of the common reasons that conflicts happen

Resources are scarce—that's why you have to negotiate for them. Have you ever been in a situation where there's a "good" conference room, or top-performing team member, or even that photocopy machine that always seems to be in use? Well, that's a scarce resource. No wonder resources cause so many conflicts.

These three things are the source of over 50% of all conflicts!

Priorities mean one project or person is more important than another, and gets more budget, resources, time, prestige, or other perks. If the company's priorities aren't crystal clear, then conflicts are definitely going to happen.

Over <u>half</u> of all conflicts are caused by resources, priorities, and schedules.

Schedules decide who gets what, when. Have you ever had a client, sponsor, or stakeholder get upset because your project won't come in as early as he or she wanted it to? Then you've had a conflict over schedules.

Some more sources of conflict

Personalities are always clashing. Sometimes two people just don't get along, and you're going to have to find a way to make them work together in order to get your project done.

Cost disagreements seem to come up a lot, especially where contracts are involved. Even when the price is agreed upon up front, buyer's remorse will set in, and it will lead to issues.

Technical opinions are definitely a reason that conflicts happen, because it's really hard to get an expert to change his mind…so when two of them disagree, watch out!

BRAIN POWER

What's the best way to deal with a conflict between two people on your project team?

How to resolve a conflict

When you're managing a project, you depend on people to get the work done. But when they have any sort of conflict, your project can grind to a halt…and you're the one who has to face the music when it causes delays and costs money! Since you're on the hook when a conflict threatens your project, **you're the one who has to resolve it**. Luckily, there are some techniques for getting your conflicts resolved.

The best way to resolve a conflict is to <u>confront the problem</u>: do your research, figure out what's behind it, and <u>fix the root cause</u>.

Confronting—or problem solving—is the most effective way to resolve a conflict.
When you confront the source of the conflict head-on and work with everyone to find a solution that actually fixes the reason that conflicts happen, then the problem is most likely to go away and never come back!

> OK, LET'S SEE IF WE CAN GET TO THE BOTTOM OF THIS.

The first thing you do whenever you face a problem is do your research and gather all the information so you can make an informed decision.

Compromise sounds good, doesn't it? But hold on a second—when two people compromise, it means that each person gives up something. That's why a lot of people call a compromise a "lose-lose" solution.

> LOOK, EACH OF YOU IS GOING TO HAVE TO GIVE UP SOMETHING IF WE WANT TO GET BACK ON TRACK.

You should always try to confront the problem first—you should forge a compromise only after you've tried every possible way to solve the real problem.

Collaborating means working with other people to make sure that their viewpoints and perspectives are taken into account. It's a great way to get a real commitment from everyone.

> AMY CAME UP WITH A GOOD IDEA, AND BRIAN EXPANDED ON IT. IT SOUNDS LIKE WE'VE GOT A REAL PLAN HERE!

Smoothing is what you're doing when you try to play down the problem and make it seem like it's not so bad. It's a temporary solution, but sometimes you need to do it to keep tempers from flaring and give people some space to step back and really figure out what's going on.

Forcing means putting your foot down and making a decision. One person wins, one person loses, and that's the end of that.

You should really try to avoid forcing and withdrawal if you can.

Withdrawal doesn't do much good for anyone. It's when people get so frustrated, angry, or disgusted that they just walk away from the argument. It's almost always counterproductive. If someone withdraws from a problem before it's resolved, it won't go away—and your project will suffer.

Who's really being unreasonable here?

BUT CONFRONTING SOUNDS LIKE A BAD THING! SHOULDN'T I AVOID CONFRONTATION?

"Confronting" is another way of saying "problem solving." Any time two people have a conflict, you need to step back and figure out what's actually causing the problem. That's how you "confront" it—by finding and fixing the underlying issue!

No! Confrontation is just another name for problem solving, because you solve a problem by confronting it head-on, doing your research, and fixing whatever is causing it. If you always remember to:

Confront the problem

...it will really help you through a bunch of questions on the exam!

Fix the root cause

Confront the Problem

wash the dog

Make it Stick

Exercise

Take a look at each of these attempts to resolve a conflict and figure out which conflict resolution technique is being used.

1. "I don't really have time for this—let's just do it your way and forget I ever brought up the problem."

   ....

2. "Look Sue, Joe's already filled me in on your issue. I've considered his position, and I've decided that he's right, so I don't need to hear anymore about it."

   ....

3. "Hold on a second, let's all sit down and figure out what the real problem is."

   ....

4. "Joe, you've got a solid case, but Sue really brings up some good points. If you just make two little concessions, and Sue gives up one of her points, we'll all be good."

   ....

5. "You guys are almost entirely in agreement—you just differ on one little point! I'll bet we'll be laughing about this next week."

   ....

6. "I don't really have time to deal with this right now. Just figure it out and get back to me."

   ....

7. "I know this problem seems really big, but I'll bet if we take a long, hard look at it, we can figure out how to fix it permanently."

   ....

⟶ Answers on page 499.

there are no
Dumb Questions

Q: How do I know what form of power to use?

A: You should always try to use expert power or reward power if you can. Expert power is effective because people naturally follow leadership from someone they respect. And reward power is also good because rewards help people motivate themselves.

When you use referent power, you're appealing to a really important psychological tool: the fact that when you like someone or she likes you, you're much more likely to influence her. And when you use punishment, you have to be very careful because it can be highly demotivating to the team. When you use it, always be careful not to punish someone in front of the team or other managers in your company. That can be embarrassing for the person, and just makes you look vindictive. Remember, your goal is to get your project back on track, not to put someone in his place!

Q: It sounds like compromise is a bad thing. But I've been told that when people are fighting, I should always look for a middle ground!

A: Yes, as little kids a lot of us were told that we should always look for a compromise. And that probably is the right thing to do on the playground. But when you're managing a project, you're judged by the success of your final product, not by how happy your team is. When you forge a compromise instead of really figuring out what's causing the problem, you're usually taking the easy way out.

Q: I'm still not quite clear about all of that storming and norming stuff. Do I need to know that to run a project?

A: Yes, you do! When Bruce Tuckman published his pioneering research about group development in 1965, he was looking for a model to describe how teams face their challenges, tackle their problems, find solutions to those problems, and deliver results. Since then, it's become the foundation for a lot of modern thinking about how teams form and work. More importantly, if you learn to recognize how teams evolve over the course of a project, it will actually help you in real life when you run your projects. If you understand how group dynamics work, you'll have a much better idea of what's causing conflicts and problems on your team, and you can help everyone work through those problems. Sometimes knowing that groups go through these patterns helps you keep perspective...and realize that it's normal—even healthy!—to have conflicts every now and then.

> **Try to avoid using punishment. When you do have to punish someone, make sure to do it in private, and not in front of peers or other managers.**

BULLET POINTS: AIMING FOR THE EXAM

- **Resources, schedules, and priorities** cause 50% of project problems and conflicts. Personality conflicts are actually the least likely cause.

- The best way to solve a problem is to **confront** it, which means doing your research, figuring out what's causing the problem, and fixing it.

- **Withdrawal** happens when someone gives up and walks away from the problem, usually because he's frustrated or disgusted. If you see a team member doing this, it's a warning sign that something's wrong.

- Don't be fooled by questions that make it sound like "confronting" is a bad thing. **Confronting is just another word for problem solving**.

- **Smoothing** is minimizing the problem, and it can help cool people off while you figure out how to solve it.

- You should **compromise only if you can't confront** the problem.

- **Forcing** means making a decision by simply picking one side. It's a really ineffective way to solve problems.

The Cows Gone Wild IV team ROCKS!

The odds were against Mike—he had to fight for a whole new team, keep them motivated, and solve some pretty serious problems. But he followed his plan, got a great team together, kept them on track, and got the product out the door!

Mike made sure everyone had a great working environment....

...he set up rewards that kept the team motivated...

...there were conflicts and arguments, but he confronted each of them and got the project back on track...

...and the team built the best Cows Gone Wild game ever!

COWS GONE
WILD IV
The Milk Man
Cometh

PHEW! THAT WAS A WHOLE LOT OF WORK, BUT WE GOT A GREAT PRODUCT OUT THE DOOR!

Question Clinic: The "have-a-meeting" question

THERE ARE A WHOLE LOT OF QUESTIONS ON THE EXAM THAT GIVE YOU A SITUATION WHERE THERE'S A CONFLICT, AN ISSUE, OR EVEN A CRISIS, AND ASK YOU WHAT TO DO FIRST. THE TRICK IS THAT *IN ALL OF THESE CASES, ONE OF THE OPTIONS IS TO HAVE A MEETING.* SOUNDS ODD, RIGHT? BUT THIS IS ACTUALLY REALLY IMPORTANT FOR A PROJECT MANAGER TO KNOW! THAT'S BECAUSE YOU NEED TO GATHER INFORMATION FROM OTHER PEOPLE BEFORE YOU MAKE A DECISION.

Don't be fooled—even though this asks about conflict, that doesn't mean it's asking you for a conflict resolution technique.

It's not always team members who have conflicts. You could have an unhappy client who has a complaint about you or your team members... and that client could be right.

Sounds like these guys are right, and the other person is wrong... right? Well, maybe not.

198. Three people on your project team are having conflicts about priorities. A junior team member wants to do the activities out of order, while two senior members want to follow the schedule that you had originally put together. What's the first step in resolving this conflict?

A. Tell everyone to work out the problem among themselves.

B. Tell the junior member that you should always follow the schedule.

C. Tell them to keep to the original schedule.

D. Meet with all three people and get all the information.

Never push off your management responsibilities on the team.

That's not true! What if the schedule has a problem and needs change control? The junior team member could be right.

You shouldn't make a unilateral decision without understanding the conflict.

This is the right answer. Get all the facts <u>before</u> you make any move.

REMEMBER HOW YOU ALWAYS LOOK AT THE IMPACT OF A CHANGE BEFORE YOU DECIDE WHETHER OR NOT TO MAKE IT? WELL, THIS IS THE SAME IDEA! YOU ALWAYS WANT TO LOOK AT ALL THE FACTS BEFORE YOU MAKE A MOVE.

HEAD LIBS

Fill in the blanks to come up with your own "have-a-meeting" question!

You're managing _____ when _____
(description of a project)   (two people with a conflict)
come to you with a disagreement about _____ . One team member
(source of disagreement)
says _____ , while the other says _____ .
(one idea about how to resolve it)   (a different idea about how to solve it)
What's the first thing that you do?

A. _____
(make a unilateral decision)

B. _____
(side with one person)

C. _____
(side with the other person)

D. _____
(have a meeting)

Here's an additional "have-a-meeting" exercise to help get you used to this kind of question.

How many different ways can you say "Have a meeting"?

Fill in a few more.

Gather information from everyone involved.   _____

Talk to the people involved directly.   _____

Make sure you know everything you need
about the situation.   _____

Don't make a move until you've got all the
information.   _____

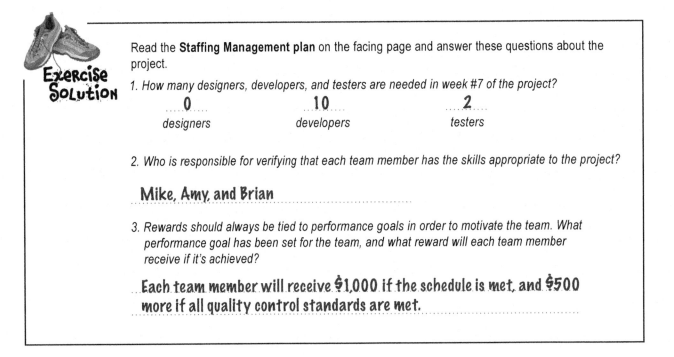

Exercise Solution

Read the **Staffing Management plan** on the facing page and answer these questions about the project.

1. How many designers, developers, and testers are needed in week #7 of the project?

     0   **10**   **2**

   designers   developers   testers

2. Who is responsible for verifying that each team member has the skills appropriate to the project?

   Mike, Amy, and Brian

3. Rewards should always be tied to performance goals in order to motivate the team. What performance goal has been set for the team, and what reward will each team member receive if it's achieved?

   Each team member will receive $1,000 if the schedule is met, and $500 more if all quality control standards are met.

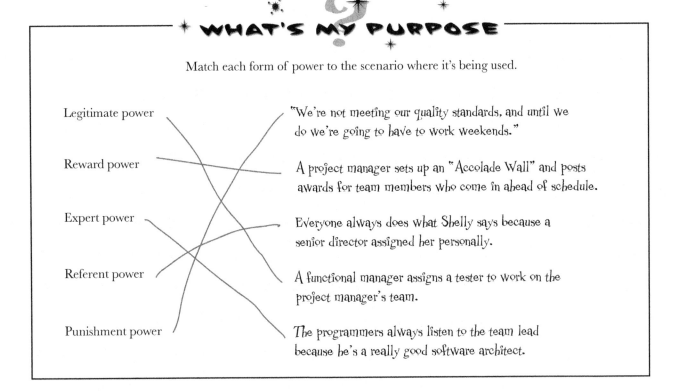

WHAT'S MY PURPOSE

Match each form of power to the scenario where it's being used.

Legitimate power   "We're not meeting our quality standards, and until we do we're going to have to work weekends."

Reward power   A project manager sets up an "Accolade Wall" and posts awards for team members who come in ahead of schedule.

Expert power   Everyone always does what Shelly says because a senior director assigned her personally.

Referent power   A functional manager assigns a tester to work on the project manager's team.

Punishment power   The programmers always listen to the team lead because he's a really good software architect.

Human Resourcecross Solution

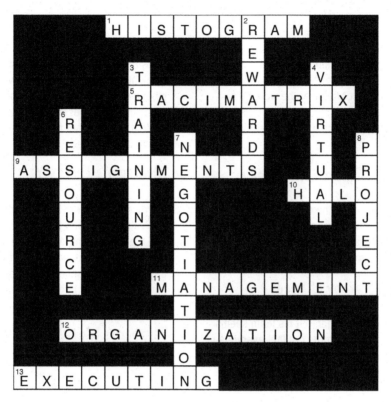

Crossword solution grid:

Across:
1. HISTOGRAM
5. RACIMATRIX
9. ASSIGNMENTS
10. HALO
11. MANAGEMENT
12. ORGANIZATION
13. EXECUTING

Down:
2. REWARD
3. TRAINING
4. VIRTURU
6. RESOURCE
7. NGOTIO
8. PROJECE

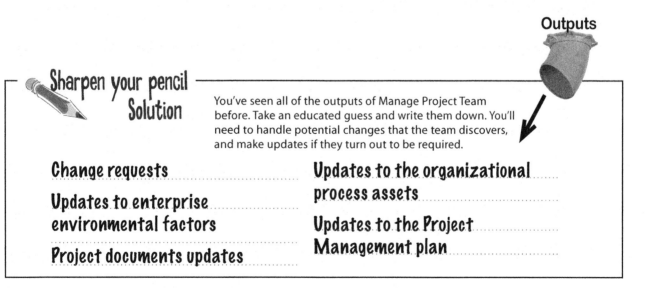

Sharpen your pencil
Solution

You've seen all of the outputs of Manage Project Team before. Take an educated guess and write them down. You'll need to handle potential changes that the team discovers, and make updates if they turn out to be required.

Outputs

Change requests

Updates to enterprise environmental factors

Project documents updates

Updates to the organizational process assets

Updates to the Project Management plan

Exercise Solution

Each of the following scenarios demonstrates one of the motivational theories at work. Write down which theory each scenario describes.

1. Bob is a programmer on the team, but he doesn't really feel like he's "one of the guys." He doesn't really have a lot of control over the work he's assigned. Recently, Bob put in a long weekend to get his work done, but nobody really seemed to take notice.

McClelland's Achievement Theory

2. There was a break-in at the office, and now people are really jittery. Plus, the heating system has been broken for weeks, and it's freezing! No wonder nobody's getting any work done.

Maslow's Hierarchy of Needs

3. Eric's a functional manager, but his team seems to move really slowly. It turns out that everyone who reports to him has to hand him their work first, before they can give it to anyone else. He goes through it line by line, which sometimes takes hours! He doesn't trust his team to release anything he hasn't seen.

McGregor's Theory X

4. Joe's a functional manager, and his team is very efficient. He spot-checks their work, but for the most part he sets realistic performance goals and trusts them to meet them—he only pulls people aside if he finds that there's a specific problem that has to be corrected.

McGregor's Theory Y

5. A project manager is having a lot of trouble motivating the team. He tries setting up rewards and a good working environment. But the team remains difficult to motivate—mostly because their paychecks all bounced last week, and everyone is angry at the CEO because they didn't get bonuses.

Herzberg's Motivation-Hygiene Theory

Exercise Solution

Take a look at each of these attempts to resolve a conflict and figure out which conflict resolution technique is being used.

1. "I don't really have time for this—let's just do it your way and forget I ever brought up the problem."

Withdrawal

2. "Look Sue, Joe's already filled me in on your issue. I've considered his position, and I've decided that he's right, so I don't need to hear any more about it."

Forcing

3. "Hold on a second, let's all sit down and figure out what the real problem is."

Confronting (or problem solving)

4. "Joe, you've got a solid case, but Sue really brings up some good points. If you just make two little concessions, and Sue gives up one of her points, we'll all be good."

Compromise

5. "You guys are almost entirely in agreement—you just differ on one little point! I'll bet we'll be laughing about this next week."

Smoothing

6. "I don't really have time to deal with this right now. Just figure it out and get back to me."

Withdrawal

7. "I know this problem seems really big, but I'll bet if we take a long, hard look at it, we can figure out how to fix it permanently."

Confronting (or problem solving)

ExeRCiSe SoLutiON

Each of the following scenarios demonstrates one of the stages of team development. Write down which stage each scenario describes.

1. Joe and Tom are both programmers on the Global Contracting project. They disagree on the overall architecture for the software they're building, and frequently get into shouting matches over it. Joe thinks Tom's design is too short-sighted and can't be reused. Tom thinks Joe's design is too complicated and probably won't work. They're at a point right now where they're barely talking to each other.

Storming

2. Joan and Bob are great at handling the constant scope changes on the Business Intelligence project. Whenever the stakeholders request changes, they shepherd them through the change control process and make sure the team doesn't get bothered with them unless it's absolutely necessary. That leaves Darrel and Roger to focus on building the main product. Everybody is focusing on their area and doing a great job. It seems like it's all just clicking for the group.

Performing

3. Derek just got to the team, and he's really reserved. Folks on the team aren't quite sure what to make of him. Eveybody's polite, but it seems like some people are a little threatened by him.

Forming

4. Now that the product has shipped, the team is meeting to document all of their lessons learned and write up project evaluations.

Adjourning

5. Danny just realized that Janet is really good at developing web services. He's starting to think of ways to make sure that she gets all of the web service development work and Doug gets all of the client software work. Doug seems really happy about this too—he seems to really enjoy building Windows applications.

Norming

Exam Questions

1. A RACI matrix is one way to show roles and responsibilities on your project. What does RACI stand for?

    A.   Responsible, Approve, Consult, Identify

    B.   Responsible, Accountable, Consulted, Informed

    C.   Retain, Approve, Confirm, Inform

    D.   Responsible, Accountable, Confirm, Inform

2. Everybody does what Tom says because he and the president of the company are golfing buddies. What kind of power does he hold over the team?

    A.   Legitimate

    B.   Reward

    C.   Punishment

    D.   Referent

3. What's the most effective approach to conflict resolution?

    A.   Smoothing

    B.   Confronting

    C.   Compromise

    D.   Withdrawal

4. Two of your team members are having a disagreement over which technical solution to use. What's the first thing that you should do in this situation?

    A.   Consult the technical documents.

    B.   Tell the team members to work out the problem themselves.

    C.   Ask the team members to write up a change request.

    D.   Meet with the team members and figure out what's causing the disagreement.

5. Joe is a project manager on a large software project. Very late in his project, the customer asked for a huge change and wouldn't give him any more time to complete the project. At a weekly status meeting, the client demanded that the project be finished on time. Joe told the client that he wasn't going to do any more status meetings until the client was ready to be reasonable about the situation. Which conflict resolution technique was he using?

    A.   Forcing

    B.   Compromise

    C.   Withdrawal

    D.   Confronting

Exam Questions

6. You've just completed your resource histogram. What process are you in?

    A.   Acquire Project Team

    B.   Develop Project Team

    C.   Plan Human Resource Management

    D.   Manage Project Team

7. Which of the following describes Maslow's Hierarchy of Needs?

    A.   You can't be good at your job if you don't have a nice office.

    B.   You need to feel safe and accepted to want to be good at your job.

    C.   Your boss's needs are more important than yours.

    D.   The company's needs are most important, then the boss's, then the employee's.

8. Jim and Sue are arguing about which approach to take with the project. Sue makes some good points, but Jim gets frustrated and storms out of the room. What conflict resolution technique did Jim demonstrate?

    A.   Withdrawal

    B.   Confronting

    C.   Forcing

    D.   Smoothing

9. Tina is a project manager who micromanages her team. She reviews every document they produce and watches when they come and go from the office. Which kind of manager is she?

    A.   Theory X

    B.   Theory Y

    C.   Theory Z

    D.   McGregor manager

10. Which of the following is NOT one of the top sources of conflict on projects?

    A.   Resources

    B.   Technical opinions

    C.   Salaries

    D.   Priorities

11. Which of the following is an example of the "halo effect"?

    A.   When a project manager is good, the team is good, too

    B.   The tendency to promote people who are good at technical jobs into managerial positions

    C.   When a project manager picks a star on the team and always rewards that person

    D.   When a technical person does such a good job that no one can find fault with her

Exam Questions

12. You are working on a construction project that is running slightly behind schedule. You ask the team to put in a few extra hours on their shifts over the next few weeks to make up the time. To make sure everyone feels motivated to do the extra work, you set up a $1,500 bonus for everyone on the team who works the extra hours if the deadline is met. What kind of power are you using?

  A. Legitimate

  B. Reward

  C. Expert

  D. Referent

13. Two team members are having an argument over priorities in your project. One thinks that you should write everything down before you start doing any work, while the other thinks you can do the work while you finish the documentation. You sit both of them down and listen to their argument. Then you decide that you will write most of it down first but will start doing the work when you are 80% done with the documentation. What conflict resolution technique are you using?

  A. Forcing

  B. Confronting

  C. Smoothing

  D. Compromise

14. What is a war room?

  A. A place where managers make decisions

  B. A room set aside for conflict management

  C. A room where a team can sit together and get closer communication

  D. A conflict resolution technique

15. You are writing a performance assessment for your team. Which process are you in?

  A. Develop Project Team

  B. Acquire Project Team

  C. Manage Project Team

  D. Plan Human Resource Management

16. You are working in a matrix organization. You don't have legitimate power over your team. Why?

  A. They don't report to you.

  B. They don't trust you.

  C. They don't know whether or not they will succeed.

  D. You haven't set up a good bonus system.

Exam Questions

17. Tom is using an organization chart to figure out how he'll staff his project. What process is he performing?

    A.   Plan Human Resource Management

    B.   Acquire Project Team

    C.   Develop Project Team

    D.   Manage Project Team

18. You're a project manager on an industrial design project. You've set up a reward system, but you're surprised to find out that the team is actually less motivated than before. You realize that it's because your rewards are impossible to achieve, so the team doesn't expect to ever get them. What motivational theory does this demonstrate?

    A.   Herzberg's Hygiene Theory

    B.   Maslow's Hierarchy of Needs

    C.   MacGregor's Theory of X and Y

    D.   Expectancy Theory

19. You're managing a software project when two of your programmers come to you with a disagreement over which feature to work on next. You listen to the first programmer, but rather than thinking through the situation and gathering all the information, you decide to go with his idea. Which conflict resolution technique did you use?

    A.   Compromise

    B.   Forcing

    C.   Confronting

    D.   Smoothing

20. Your client comes to you with a serious problem in one of the deliverables that will cause the final product to be unacceptable. Your team members look at his complaint and feel that it's not justifiable, and that the product really does meet its requirements. What's the first thing that you do?

    A.   Confront the situation by making the change that needs to be made in order to satisfy the client.

    B.   Explain to the client that the solution really is acceptable.

    C.   Work with the client and team members to fully understand the problem before making a decision.

    D.   Write up a change request and send it to the change control board.

Exam Questions ~~Answers~~

1. Answer: B

When you think about how you organize the work on your project, the RACI chart makes sense. Being **responsible** for a specific task or area of work means you're the one who's on the hook if it doesn't get done. Being **accountable** means you might not be doing it directly, but you have influence over it. Some people need to be **consulted** but don't get involved in the work, while others should just be kept **informed** of status.

2. Answer: D

Did you choose punishment? People might be afraid of punishment from the president of the company if they don't agree with Tom. But since Tom isn't the one who would punish them, it's referent power.

The power is here is referent. People are reacting to Tom's relationship to the president of the company, not his own authority.

3. Answer: B

Confronting does sound like it would be negative, but it just means solving the problem. If you actually solve the problem, there's no more reason for people to fight at all. That's always the best way to deal with a conflict. Any of the other options could lead to more problems later.

4. Answer: D

This is a classic "have-a-meeting" question! You should always gather the information you need before you make any kind of decision.

5. Answer: C

Joe decided that the best tactic was to refuse to talk to the client anymore—that's withdrawing. It's also probably not going to solve the problem.

6. Answer: C

You create the histogram as part of the Staffing Management plan. It's the main output of the Plan Human Resource Management process.

7. Answer: B

Maslow's Hierarchy of Needs says that your safety and acceptance are a prerequisite for your being able to do your best.

Answers

Exam ~~Questions~~

8. Answer: A

Jim took his ball and went home.
That's withdrawal.

> IT SEEMS LIKE JIM AND SUE HAD A CONFRONTATION, RIGHT? BUT THAT'S NOT WHAT "CONFRONTING" MEANS HERE! IT REALLY MEANS PROBLEM SOLVING.

9. Answer: A

A micromanager is a Theory X manager.
Tina believes that all employees need to be
watched very closely, or they will make mistakes.

10. Answer: C

You definitely need to know what causes conflicts on projects. Resources, technical opinions, priorities, and personalities all cause people to have conflicts, and there's a good chance you'll get a question on that!

11. Answer: B

Just because someone is good at a technical job, it doesn't mean he will be good at management. The jobs require very different skills.

12. Answer: B

You are motivating the work by offering a reward for it. People might be motivated by the bonus to put in the extra time even if they would not have been motivated by the deadline alone.

13. Answer: D

Both of them had to give something up, so that's a compromise.

14. Answer: C

War rooms are part of colocation. It's a way to keep your entire team in one room so they don't have any communication gaps.

Answers

Exam ~~Questions~~

15. Answer: A

Developing the team is where you evaluate performance and set up motivational factors. Manage Project Team is where you solve conflicts.

16. Answer: A

In matrix organizations, team members usually report to their functional managers. A project manager never has legitimate power over the team in those situations.

Don't forget that there are two org charts—one for the company, and one for the project.

17. Answer: A

Tom's project is at the very beginning—he's using the organization chart as a tool to figure out who's going to be assigned to his team.

18. Answer: D

Expectancy Theory says that people get motivated only by rewards that they can achieve, and that are fair. If you set up a reward system that selects people who don't deserve rewards, or that has rewards that are unattainable, then it will backfire and cause people to resent their jobs.

19. Answer: B

Whenever you choose one side over another without thinking or actually finding the root cause of the problem, you're forcing a solution on it. This is *not* a good way to solve problems!

20. Answer: C

Any time there's any sort of conflict, the first thing you need to do is gather all the information. And that's especially true when there's a disagreement between the client and the team! You'd better have your facts straight in such a charged situation.

> HE SHOULD HAVE SAT DOWN WITH BOTH PROGRAMMERS AND FIGURED OUT WHAT THE REAL PROBLEM WAS. EVEN IF THE SOLUTION ISN'T PERFECT, AT LEAST IT'S MORE FAIR.

10 Communications management

Getting the word out

GOOD NEWS, GERTRUDE! GET THE POT ROAST ON THE STOVE...I JUST CHECKED THE PERFORMANCE REPORT, AND I'LL MAKE IT HOME IN TIME FOR DINNER.

Communications management is about keeping everybody in the loop.
Have you ever tried talking to someone in a really loud, crowded room? That's what running a project is like if you don't get a handle on communications. Luckily, there's **Communications Management**, which is the knowledge area that gets everyone talking about the work that's being done, so that they all **stay on the same page**. That way, everyone has the information they need to **resolve any issues** and keep the project **moving forward**.

Party at the Head First Lounge!

Jeff and Charles want to launch their new Head First Lounge, so they're going to have a party for the grand opening. They're thinking of all of the things they need to arrange: the DJ, the hors d'oeuvres, the drinks, hula dancing. They need to start contacting caterers, DJs, and suppliers to make sure it all goes off without a hitch.

But something's not right

When Jeff called the caterer and the DJ to request everything he wanted for the party, his old staticky phone made it hard for everybody to understand what he was asking. Sometimes their taste for retro furniture can make things a little difficult.

1 MAKE SURE YOU BRING THE GOOD WINE!

This phone is old and staticky!!

2 HI JEFF. THIS IS THE CATERER. WE GOT YOUR MESSAGE. WE'LL BE SURE TO BRING PORK RINDS.

Oops! The caterer couldn't make out what Jeff was saying.

3 MAN, I'VE GOTTA CALL THEM BACK AND MAKE SURE THEY BRING WINE, NOT RINDS!

4 I WONDER IF I NEED TO CONTACT ANYONE OTHER THAN THE CATERER ABOUT THIS? IS THERE SOMETHING WRONG WITH THE PHONE? HOW CAN WE BE SURE THIS WON'T HAPPEN AGAIN?

⚛ BRAIN POWER

What can Jeff and Charles do to get a handle on their communication problems?

Anatomy of communication

When you communicate with your team, you need to **encode** your message into a phone call, a document, an IM chat, or sometimes even a different language for them to understand. Your team then **decodes** that message so they can get its content. If something happens to your message along the way (static on the phone line, your printer inserts garbage characters, your Internet connection is spotty, or your translation isn't very good), then your team might not get the intended message. The kind of interference that can alter your message is called **noise**.

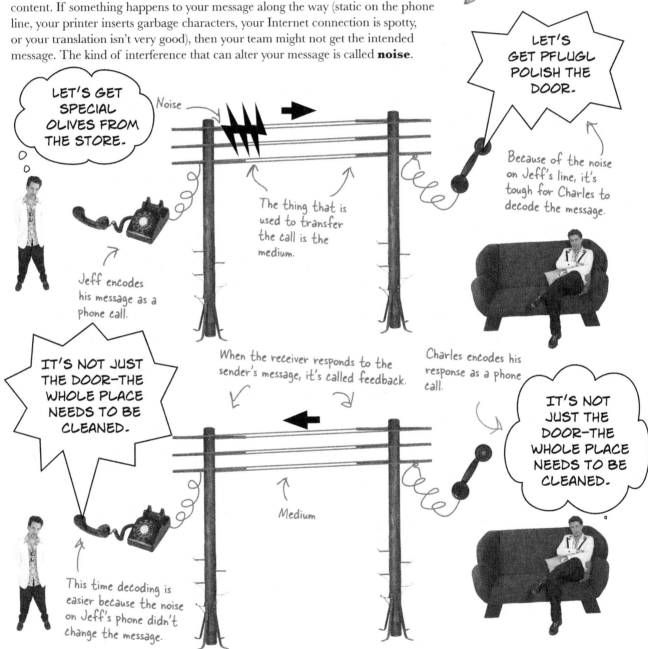

When you're talking about messages, encoding, decoding, and noise, you're talking about a <u>communications</u> model.

LET'S GET PFLUGL POLISH THE DOOR.

Because of the noise on Jeff's line, it's tough for Charles to decode the message.

LET'S GET SPECIAL OLIVES FROM THE STORE.

Noise

The thing that is used to transfer the call is the medium.

Jeff encodes his message as a phone call.

IT'S NOT JUST THE DOOR—THE WHOLE PLACE NEEDS TO BE CLEANED.

When the receiver responds to the sender's message, it's called feedback.

Charles encodes his response as a phone call.

IT'S NOT JUST THE DOOR—THE WHOLE PLACE NEEDS TO BE CLEANED.

Medium

This time decoding is easier because the noise on Jeff's phone didn't change the message.

Match each communication element to what it does.

Acknowledge

Getting the information from one
person to the other

Transmit message

Letting the sender know that the
message was received

Feedback/response

Modifying a message that has been
sent so that it can be understood

Encoding

An answer to a message

Decoding

Modifying a message so that it can be
sent

$\longrightarrow$ Answers on page 540.

Get a handle on communication

Any kind of communication can have interference. The wrong person can get the message; noise can garble the transmission; you can make mistakes decoding or encoding the message. It turns out that 90% of a project manager's job is communication, which is why there's a whole knowledge area devoted to it. The **Communications Management** processes are here to help you avoid these common kinds of errors, through planning and careful tracking of stakeholder communications on your project. Just like every other knowledge group we've covered so far, it all starts with a plan.

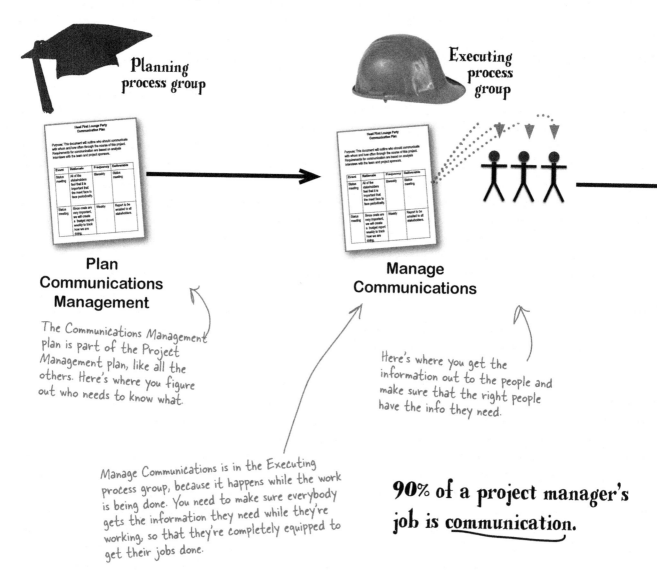

Planning process group

Plan Communications Management

Executing process group

Manage Communications

The Communications Management plan is part of the Project Management plan, like all the others. Here's where you figure out who needs to know what.

Here's where you get the information out to the people and make sure that the right people have the info they need.

Manage Communications is in the Executing process group, because it happens while the work is being done. You need to make sure everybody gets the information they need while they're working, so that they're completely equipped to get their jobs done.

90% of a project manager's job is <u>communication</u>.

It's not enough to plan and manage the communications on your project. You need to make sure that everybody who has a stake in your project is getting accurate reports of how it's going so they can make good decisions—that's what the **Control Communications** process is all about. You use it to monitor the data your project is producing, and control how it is presented to your stakeholders.

Monitoring
& Controlling
process group

Performance report

The team is 5% over budget
and 1 day behind schedule.

Status ... ing fine
and ...

Control
Communications

This is where you turn all of that **work performance data** (like how long it actually took the team to complete tasks, and actual costs of doing the work so far) into **work performance information** (like forecasted completion dates, and budget forecasts) that your stakeholders will use to stay informed.

Control Communications is in the Monitoring and Controlling process group. You need to constantly monitor and always stay in control of all of the communication that goes on throughout the project, whether it's to communicate your team's performance, or to keep stakeholders up to date on the project.

Communications Management makes sure <u>everybody</u> gets the <u>right message</u> at the <u>right time</u>.

Exercise

This is the Plan Communications Management process. You've seen a lot of planning processes now. Can you fill in the inputs and outputs for this one?

Tools

Communication requirements analysis means figuring out what kind of communication your stakeholders need from the project so that they can make good decisions. Your project will produce a lot of information; you don't want to overwhelm every member of your project team with all of it. Your job here is to figure out what all of them feel they need to stay informed and to be able to do their jobs properly.

Here's an example: Jeff and Charles will definitely care about the cost of the overall catering contract, but they don't need to talk to the caterer's butcher, liquor supplier, grocer, or other companies they work with.

Communication models demonstrate how the various people associated with your project send and receive their information. You've already learned about this—it's the **messages** you send, how you **encode** and **decode** the messages, the **medium** you use to transmit the messages, the **noise** that blocks the messages, and the **feedback** you get.

Inputs

....

....

This one is your company's culture and policies toward project communication.

Stakeholder register

Here's where your company keeps all of its templates and lessons learned.

....

You need to know who you're going to communicate with. You'll learn more about this in Chapter 13!

Here's where you plug in all of the planning you've done on your project so far.

Planning process group

Communication technology

has a major impact on how you can keep people in the loop. It's a lot easier for people to get information on their projects if it's all accessible through a website than it is if all of your information is passed around by paper memos. The technologies available to you will definitely figure into your plan of how you will keep everyone notified of project status and issues.

Communication methods are how

you actually share the information with your stakeholders. Communications can be **interactive**, where everyone exchanges information with one another. You can **push** information out to your stakeholders by sending out emails, memos, faxes, or other one-way communications. Or, if you need to get a lot of information out to people, they can **pull** it down themselves from intranet websites, e-learning courses, or libraries.

Meetings are always great

for helping your team to think about communication.

Outputs

Are you surprised at how much of this process you can fill in? Looks like you're getting the hang of this stuff!

There are several project documents that get updated when you're planning communications. Can you think of one of them?

Before you turn the page, take a minute and think of three examples of how you used each of these methods on your last project. That'll help you remember them for the exam!

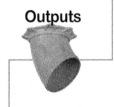

There are only two outputs. Can you guess what this one is?

Exercise
Solution

This is the Plan Communications Management process. You've seen a lot of planning processes now. Can yo fill in the inputs and outputs for this one?

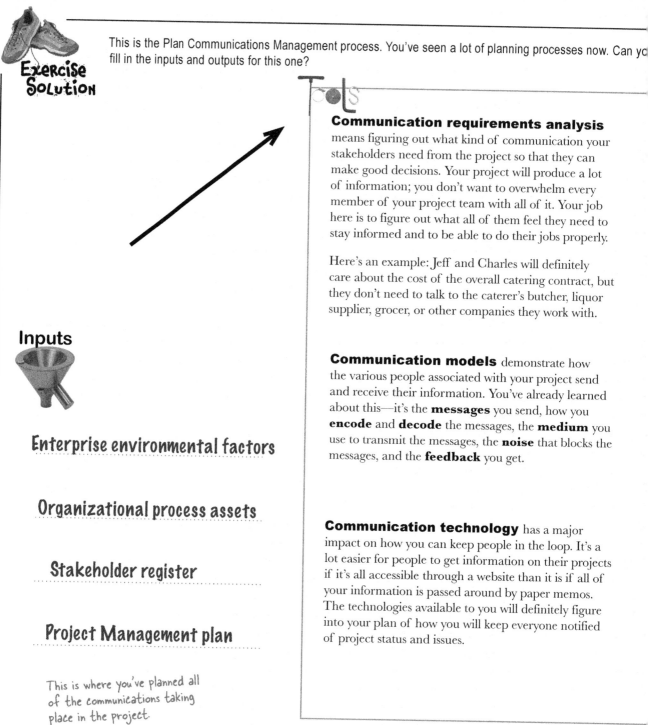

Tools

Communication requirements analysis
means figuring out what kind of communication your stakeholders need from the project so that they can make good decisions. Your project will produce a lot of information; you don't want to overwhelm every member of your project team with all of it. Your job here is to figure out what all of them feel they need to stay informed and to be able to do their jobs properly.

Here's an example: Jeff and Charles will definitely care about the cost of the overall catering contract, but they don't need to talk to the caterer's butcher, liquor supplier, grocer, or other companies they work with.

Communication models demonstrate how the various people associated with your project send and receive their information. You've already learned about this—it's the **messages** you send, how you **encode** and **decode** the messages, the **medium** you use to transmit the messages, the **noise** that blocks the messages, and the **feedback** you get.

Communication technology has a major impact on how you can keep people in the loop. It's a lot easier for people to get information on their projects if it's all accessible through a website than it is if all of your information is passed around by paper memos. The technologies available to you will definitely figure into your plan of how you will keep everyone notified of project status and issues.

Inputs

Enterprise environmental factors

Organizational process assets

Stakeholder register

Project Management plan

This is where you've planned all of the communications taking place in the project.

You'll usually need to update the project schedule, the stakeholder register, or the Stakeholder Management strategy when you plan communications for your project.

Communication methods are how you actually share the information with your stakeholders. Communications can be **interactive**, where everyone exchanges information with one another. You can **push** information out to your stakeholders by sending out emails, memos, faxes, or other one-way communications. Or, if you need to get a lot of information out to people, they can **pull** it down themselves from intranet websites, e-learning courses, or libraries.

Document updates

Planning process group

Meetings are always great for helping your team to think about communication.

Outputs

It's important that everyone involved understands why you are doing the meetings and reports you create.

The plan tells how you will distribute the information, to whom, and how often.

The PM's not always responsible for every communication. The plan makes it clear who communicates what on the project.

Head First Lounge Party
Communications Management Plan

Purpose: This document will outline who should communicate with whom and how often through the course of this project. Requirements for communication are based on analysis of interviews with the team and project sponsors.

Event	Rationale	Frequency	Deliverable
Status meeting	All of the stakeholders feel that it is important that they meet face to face periodically.	Biweekly	Meeting minutes to be emailed to all stakeholders. Archived in the document repository.
Budget report	Since costs are very important, we will create a budget report weekly to track how we are doing.	Weekly	Report to be emailed to all stakeholders.

Executing process group

Tell everyone what's going on

Once you have the Communications Management plan completed, it's time to make sure that everybody is getting the information that they need to help your project succeed. The **Manage Communications** process is all about making sure that the right information makes it to the right people.

Inputs

Communications Management plan
The Project Management plan includes the Communications Management plan you just created.

Work performance reports
This is your status information. This is the information you need to get out to your stakeholders.

Tools

Communication methods There are a lot of different ways to get a message across. For the test you will need to know four different kinds of communication, and when to use them.

❶ Formal written

Any time you're signing a legal document or preparing formal documentation for your project, that's formal written communication.

Any time you see anything that has to do with a contract, you should always use formal written communication.

Blueprints, specifications, and all other project documents are examples of formal written communication.

❷ Informal written

If you drop someone a quick email or leave her a memo or a sticky note, that's informal written communication.

❸ Formal verbal

If you ever have to give a presentation to update people on your project, that's formal verbal communication.

Speeches and prepared talks are formal. Meetings, hallway chats, and planning sessions are informal.

❹ Informal verbal

COME BY AT 7:00!

Just calling somebody up to chat about your project is informal verbal communication.

Exercise

Choose which kind of communication is being used in each situation.

1. You and your business analysts write a requirements specification for your project.

☐ Formal verbal   ☐ Informal verbal

☐ Formal written   ☐ Informal written

2. You call up a supplier for materials for your project to let him know that you are a week late, so he's got a little flexibility in his delivery schedule.

☐ Formal verbal   ☐ Informal verbal

☐ Formal written   ☐ Informal written

3. You present your project's status to your company's executive committee.

☐ Formal verbal   ☐ Informal verbal

☐ Formal written   ☐ Informal written

4. You send an email to some of your team members to get more information about an issue that has been identified on your project.

☐ Formal verbal   ☐ Informal verbal

☐ Formal written   ☐ Informal written

5. You leave a voicemail message for your test team lead following up on an issue she found.

☐ Formal verbal   ☐ Informal verbal

☐ Formal written   ☐ Informal written

6. You IM with your team members.

☐ Formal verbal   ☐ Informal verbal

☐ Formal written   ☐ Informal written

7. You prepare an RFP (request for proposals) for vendors to determine which of them will get a chance to contract a new project with your company.

☐ Formal verbal   ☐ Informal verbal

☐ Formal written   ☐ Informal written

Hint: We haven't talked about RFPs yet, but you don't need to know what they are to answer this question.

→ Answers on page 541.

Watch it!

Be careful about when you use different kinds of communication.

*Any time you need to get a message to a client or sponsor, you use **formal** communication. Meetings are always **informal verbal**, even if the meeting is to say something really important. And any project document—like a Project Management plan, a requirements specification, or especially a contract—is always **formal written**.*

Get the message?

Communication is about more than just what you write and say. Your facial expressions, gestures, tone of voice, and the context you are in have a lot to do with whether or not people will understand you. **Effective communication** takes the way you act and sound into account. Most of the communication on your project takes place during the Manage Communication process, so you need to know how to communicate effectively. Here are the important aspects to effective communication:

Nonverbal communication means your gestures, facial expressions, and physical appearance while you are communicating your message. Imagine what Jeff and Charles would think of the caterer if he negotiated the contract for their party while wearing a chicken suit. They probably wouldn't take him very seriously. You don't always think about it, but the way you behave can say more than your words when you are trying to get your message across.

When you're communicating with other people, you actually do <u>more nonverbal</u> *communication than verbal!*

Paralingual communication is the tone and pitch of your voice when you're talking to people. If you sound anxious or upset, that will have an impact on the way people take the news you are giving. You use paralingual communication all the time—it's a really important part of how you communicate. When your tone of voice makes it clear you're really excited about something, or if you're speaking sarcastically, that's paralingual communication in action.

If someone has dread in his voice when he tells you about a promotion, you get a much different impression than if he'd emailed you about it.

Feedback is when you respond to communication. The best way to be sure people know you are listening to them is to give lots of feedback. Some ways of giving feedback are summarizing their main points back to them, letting them know that you agree with them, or asking questions for clarification. When you give a lot of feedback to someone who is speaking, that's called **active listening**.

Like effective communication, effective listening is about taking everything the speaker says and does into consideration and asking questions when you don't understand.

> SO IT'S NOT ENOUGH TO SAY THE RIGHT THING. YOU NEED TO SAY IT THE RIGHT WAY, TOO.

That's why <u>active</u> *listening is an important part of communication.*

You do most of the project communication when you're performing the Manage Communications process.

Exercise

Jeff and Charles are interviewing new bartenders to help with the expanded space. Choose which kind of communication is being used in each situation.

1. One applicant came in 30 minutes late and was dressed unprofessionally. The guys knew that he would not be a good fit for the position.

   ☐ Paralingual   ☐ Nonverbal

   ☐ Feedback

2. Charles asked an applicant about her background. Her tone of voice was really sarcastic, and he got the impression she didn't take the job seriously. Charles and Jeff decided to pass on her, too.

   ☐ Paralingual   ☐ Nonverbal

   ☐ Feedback

3. Charles asked the next applicant if he knew how to make a sidecar. He said "A sidecar? Sure. It's one part brandy or cognac, one part Cointreau, and one part lemon juice."

   ☐ Paralingual   ☐ Nonverbal

   ☐ Feedback

4. Then the applicant told them about his background as a bartender for other retro clubs. As he spoke, he made eye contact with them and made sure to confirm agreement with them.

   ☐ Paralingual   ☐ Nonverbal

   ☐ Feedback

⟶ Answers on page 540.

WE FOUND A BARTENDER FOR THE NEW SPACE!

THIS NEW GUY WILL BE GREAT. HE HAS AWESOME COMMUNICATION SKILLS.

More Manage Communications tools

The tools in this process area are all about getting information from your team and making sure that the information makes it to the people who need it. You'll start your project with a kickoff meeting to get everyone on the same page, and follow your Communications Management plan as your project progresses. As you learn more about your project, you write down decisions you make and everything you learn on the project as lessons learned.

Communication methods are the specific methods you use to distribute information to your team...and you've already learned about them!

Communication models are important in the Manage Communications process, just like they are in planning. You already know about those too!

Information management systems are how you get the information your team needs to do the job. You might have an inbox where everyone puts their status information. If it's printed out on paper, you're doing **hard copy document distribution**. You could also use **electronic communication**. For example, you might use email, or you could have a software application that gathers information about your project and saves it to a database so that you can make your reports. Or your company might have **electronic tools for project management**, like a timesheet system for tracking hours spent on a project or a budgeting system for tracking expenditures. All of those are information gathering and retrieval systems, because the data they produce will be used to make decisions about your project.

Performance reporting is all about gathering information on how your team is progressing through the project. You might create **status reports** that show how close you are to your baseline schedule and highlight issues that your team has run into along the way. You'll always want to keep everybody informed on how your project is tracking risks, any changes that might come up that weren't planned for, and forecasts of what's coming up next for the team.

Communication technology is a tool that you use to get the message out. If you need to get a message to someone urgently, it might be hard to wait for a face-to face meeting. You might choose to use email, phone, or a ticketing system to communicate. There are a lot of factors that influence your decision to use a particular technology when communicating other than urgency, including availability, how easy the technology is to use, whether or not the team can meet to face-to-face because of where they work, and how confidential the information you're communicating is.

Outputs

Project communications

Throughout your project, you're creating status reports, presentations, and many other communications to keep your project stakeholders informed. It makes sense that all of these would be outputs of the Manage Communications process.

Project Management plan updates

As your project progresses, you'll make changes to the Project Management plan as new information is available. All of those project plan updates help to communicate what's going on in your project.

Project document updates

We've seen in other processes that keeping the project documentation updated is a big part of keeping everyone on the same page. Those project document updates are likewise a big part of how your project is communicated to all of the project stakeholders.

Organizational process asset updates

You've used lessons learned from all of the other projects your company has done as you've planned out your work. Here's where you get a chance to give your project's experience back to the company and to help future project managers learn from what's happened on your project.

Organizational
process assets

One of your most important outputs

Lessons learned are all of the corrective and preventive actions that you have had to take on your project, and anything you have learned along the way. And one of the most valuable things you'll do for future project managers is **write them down and add them to your company's organizational process asset library**. That way, other people can learn from your experience.

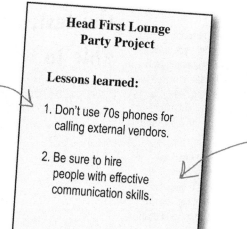

Next time they plan a party, Jeff and Charles won't run into the misunderstandings that they had on this one.

**Head First Lounge
Party Project**

Lessons learned:

1. Don't use 70s phones for calling external vendors.

2. Be sure to hire people with effective communication skills.

It's important to write down the good things you learned on the project, too. That way, you can be sure to repeat your successes next time.

there are no
Dumb Questions

Q: What do I do with lessons learned after I write them?

A: The great thing about lessons learned is that you get to help other project managers with them. You add them to your company's organizational process asset library, and other project managers then use them for planning their projects.

Since Jeff and Charles learned that they shouldn't use their retro phones for planning parties, no one should ever have to deal with that problem when planning a party for Jeff and Charles again. They wrote down the lesson they learned and filed it away for future planning efforts.

Q: I still don't get the different types of communication.

A: When you think about it, they are pretty easy to remember. You have formal and informal communication, and verbal and written communication types. The four different ways you can mix those up are all of the communication types. Think of informal verbal as phone calls between different team members. Formal verbal is giving a presentation. Informal written is sending out notes, emails, or memos. Formal written is when you have to write specifications or other formal project documentation.

For the test, you need to be able to identify which is which. If you just think of these examples, it should be a snap for you.

Q: Now, who's decoding, who's encoding, and where does feedback come from?

A: Think of encoding as making your message ready for other people to hear or read. If you write a book, you are encoding your message into words on pages. The person who buys the book needs to read it to decode it. The same is true for a presentation. When you present, you encode your thoughts into presentation images and text. The people who are listening to your presentation need to read the text, hear your voice, and see the visuals to decode it.

Feedback is all about the person who decodes the message letting the person who encoded it know that she received it. In the case of a book, this could be a reader sending a question or a note to the author or writing a review of it on a website. In a presentation, it could be as simple as nodding your head that you understand what's being said.

Q: Do I have to know everything that will be communicated to build a plan?

A: No. As you learn more about the project, you can always update the plan to include new information as you learn it. Pretty much all of the planning processes allow for progressive elaboration. You plan as much as you can up front, and then put all changes through change control from then on. So, if you find something new, put in a change request and update the plan when it's approved.

> **There are only four communication types; formal written, informal written, formal verbal, and informal verbal. For the test, you need to be able to tell which is which.**

Let everyone know how the project's going

You spend a lot of time collecting valuable information about how your projects are doing. So what do you do with it? You *communicate* it. And that's what the **Control Communications** process is for: taking the information you gathered about how work is being done and distributing it to the stakeholders who need to make decisions about the project.

Remember, the team members are all stakeholders, too—and this information is especially important to them!

You created this when you were executing the project—it was where you reported how the project work was going. Now you're using it to report the performance of the team to the stakeholders and the rest of the company.

Work performance data

It all starts with work performance data

You create one of the most important outputs of your entire project when the team is doing the project work in Direct and Manage Execution. **Work performance data** tells you the status of each deliverable in the project, what the team's accomplished, and all of the information you need to know in order to figure out how your project's going. But you're not the only one who needs this—your team members and stakeholders need to know what's going on, so they can adjust their work and correct problems early on.

Whenever you hear back from a team member about how the job is going, that's work performance data.

HEY, WE'RE STILL WAITING ON THE BAND TO CONFIRM. AND DIDN'T THE CATERER HAVE SOME QUESTIONS?

I HAVE NO IDEA! I HAVEN'T HEARD BACK FROM ANYONE!

Take a close look at the work being done

Work performance information isn't the only information you need to figure out how the project is going. There are a whole lot of outputs from the Executing processes that you need to look at if you really want to get a clear picture of your project.

Control Communications takes the outputs from the Executing process in Manage Communications and turns them into work performance information.

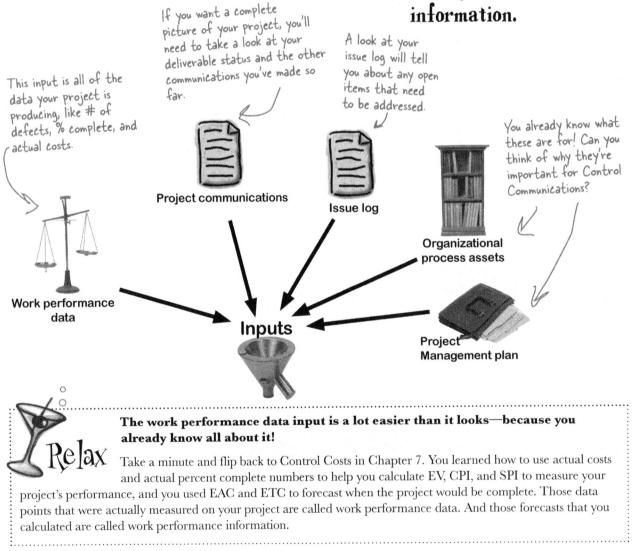

If you want a complete picture of your project, you'll need to take a look at your deliverable status and the other communications you've made so far.

A look at your issue log will tell you about any open items that need to be addressed.

This input is all of the data your project is producing, like # of defects, % complete, and actual costs.

You already know what these are for! Can you think of why they're important for Control Communications?

Project communications

Issue log

Organizational process assets

Work performance data

Inputs

Project Management plan

The work performance data input is a lot easier than it looks—because you already know all about it!

Relax

Take a minute and flip back to Control Costs in Chapter 7. You learned how to use actual costs and actual percent complete numbers to help you calculate EV, CPI, and SPI to measure your project's performance, and you used EAC and ETC to forecast when the project would be complete. Those data points that were actually measured on your project are called work performance data. And those forecasts that you calculated are called work performance information.

Sharpen your pencil

Tools

Control Communications is one of those *PMBOK Guide* processes that's really familiar to a lot of project managers. Can you figure out what each of its **tools and techniques** is for just from the name?

Information management systems

Expert judgment

Meetings

Sharpen your pencil
Solution

Control Communications is one of those *PMBOK Guide* processes that's really familiar to a lot of project managers. Can you figure out what each of its **tools and techniques** is for just from the name?

Information management systems

Here's where all of the project communication can be found. You'll find all of the current progress reports, risk and issue logs, and other project documents here.

Expert judgment

You might want to rely on expertise from your stakeholders, consultants, your PMO, or others to determine the right information to communicate about your project.

Meetings

You can hold meetings with your team members to get everyone on the same page about the progress reports on your project.

Now you can get the word out

Now that you've gathered up all the information about how the project's being done, it's time to get it out to the people who need it. The **outputs from Control Communications** shouldn't be particularly surprising… you're just packaging up the information you collected and turning it all into stuff that's easy to distribute to all the stakeholders. You've got three outputs from the process:

Work performance information is the most important output of the process—which shouldn't be a surprise, since the process is called Control Communications. Your performance reports tell everyone exactly how the project is doing, and how far off it is from its time, cost, and scope baselines. These include **forecasts**, which are what you turn your EAC and ETC numbers into. That way, everyone has a good idea of when the project is going to finish.

Organizational process asset updates need to be added—especially your **lessons learned**. There are always a lot of lessons to be learned when you're gathering this kind of project information.

Change requests happen when you do Control Communications. What do you do if you find out that your forecasts have your project coming in too late or over budget? You put the **change request** in as soon as possible. And if you need the project to change course, you'll need to **recommend corrective actions** to the team.

Project Management plan updates need to be done to make sure your plan reflects your project's current status.

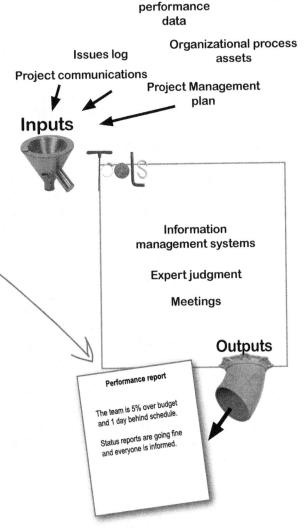

Work performance data

Issues log

Organizational process assets

Project communications

Project Management plan

Inputs

Tools

Information management systems

Expert judgment

Meetings

Outputs

Performance report

The team is 5% over budget and 1 day behind schedule.

Status reports are going fine and everyone is informed.

Project documents updates could mean updates to performance reports, issue logs, or forecasts.

BRAIN POWER

Control Communications is about more than just telling people how the project is doing. It's also about finding problems. What kind of problems are you likely to uncover when you sit down with stakeholders and put together your work performance information and forecasts?

Communicationcross

Take some time to sit back and give your right brain
something to do. It's your standard crossword; all of
the solution words are from this chapter.

Across

2. When you move a message from one person to another,
you _____ the message.

3. One thing you need to consider when choosing a
communication technology is the_____ of the
communication.

8. _____ listening is when a listener uses both verbal
and nonverbal clues like nodding or repeating the listener's
words to communicate that the message has been received.

9. The Control Communications process turns work
performance data into work performance_____

10. When your stakeholders get information from an intranet
website, you're using this communication model.

12. You can use the ETC and EAC calculations from Cost
Management to create _____.

Down

1. This kind of communication includes vocal but nonverbal
signals, such as changing the pitch and tone of voice.

4. According to the *PMBOK Guide*, _____percent of
project management is communication.

5. A conversation in a hallway is an example of_____
verbal communication.

6. A contract is always _____ communication.

7. When you receive a message and tell the sender that you
got it, you _____ the communicaiton.

11. The communication model you use when you send email
announcements.

→ Answers on page 542.

People aren't talking!

There's so much information floating around on any project, and if you're not careful it won't get to the people who need it. That's why so much of your job is communication—if you don't stay on top of all of it, your project can run into some serious trouble!

THE FORECAST LOOKED GOOD, AND EVERYONE WAS ON TOP OF THEIR JOBS. WE THOUGHT WE WERE ON TRACK FOR FRIDAY NIGHT. THEN ALL THESE PROBLEMS CAME UP...

Problems

※ The caterer's serving food that doesn't go with the drinks or theme.

※ The DJ and the band want to set up in the same place.

※ All the guests are telling us they like different food.

※ Has anyone even talked to the neighbors about the noise?

※ Three people are bringing friends, but nobody told the caterer.

HOW ARE WE GONNA GET A HANDLE ON THIS?

⚛ BRAIN POWER

What's causing all of these problems? Will better communication help?

Count the channels of communication

How many people need to talk to one another? Well, Jeff and Charles need to talk. But what about the DJ and the band? They wanted to set up their equipment in the same place—it looks like they need to talk, too. And the bartender needs to coordinate with the caterer. Wow, this is starting to get complicated. A good project manager needs to get a handle on all this communication, because it's really easy to lose track of it. That's why you need to know how to **count the channels of communication** on any project.

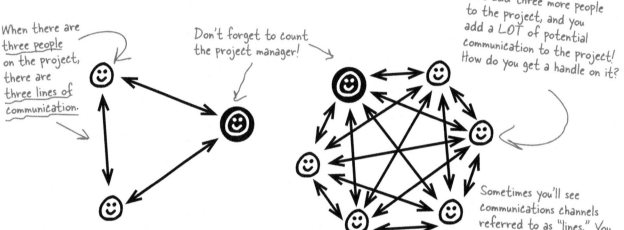

When there are three people on the project, there are three lines of communication.

Don't forget to count the project manager!

But add three more people to the project, and you add a LOT of potential communication to the project! How do you get a handle on it?

Sometimes you'll see communications channels referred to as "lines." You might see it either way on the exam, so we'll use both terms here to get you used to them.

Counting communication lines the easy way

It would be really easy to get overwhelmed if you tried to count all the lines of communication by hand. Luckily, there's a really easy way to do it by using a simple formula. Take the total number of people on the project—including the project manager—and call that number **n**. Then all you need to do is plug that number into this simple formula:

You'll need to know this formula on the PMP exam. Just keep using it, though, and you'll get it down in no time.

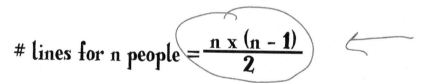

$$\text{\# lines for n people} = \frac{n \times (n-1)}{2}$$

So, how many more lines of communication were added when three more people joined the three-person project above? You know there were **three lines** to start with. So now just figure out *how many lines* there are *for 6 people*:

$$\text{\# lines for 6 people} = \frac{6 \times (6-1)}{2} = (6 \times 5) \div 2 = 15$$

When you added three more people to the three-person project—that had three lines of communication—the new team has 15 lines. So you **added 12 channels of communication**.

Sharpen your pencil

You'll need to know how to calculate the number of lines of communication for the exam…but don't worry, it's really easy once you get a little practice.

1. You're managing a project with five people on the team, plus one additional stakeholder—the sponsor. Draw in all the channels of communication on this picture.

Don't forget the project manager. There are six people on the team, but the total number of people who need to communicate is seven people, because the PM needs to communicate with the team members and sponsor.

2. Wow, that was a lot of work. Luckily, you won't need to do that again. Now do it the easy way: use the formula to figure out how many lines of communication there are for seven people.

$$\text{\# lines for } \underline{\hspace{1cm}} \text{ people} = \frac{\underline{\hspace{1cm}} \times (\underline{\hspace{1cm}} - 1)}{2} = (\underline{\hspace{1cm}} \times \underline{\hspace{1cm}}) \div 2 = \underline{\hspace{1cm}}$$

3. OK, now let's say that you've added two team members and two more stakeholders, so there are now 11 people on the project who need to communicate with one another. How many lines did you add?

First figure out how many lines there are for 11 people:

$$\text{\# lines for } \underline{\hspace{1cm}} \text{ people} = \frac{\underline{\hspace{1cm}} \times (\underline{\hspace{1cm}} - 1)}{2} = (\underline{\hspace{1cm}} \times \underline{\hspace{1cm}}) \div 2 = \underline{\hspace{1cm}}$$

So how many lines were added when four people joined the seven-person project?

lines added = # lines for 11 people − # lines for 7 people

$$= \underline{\hspace{1cm}} - \underline{\hspace{1cm}} = \underline{\hspace{1cm}}$$

⟶ Answers on page 543.

Q: Some of those communication skills seem like the same thing. What's the difference between active and effective listening?

A: Some of the communications ideas do have names that are a little confusing. But don't worry, they're really easy concepts for you to understand.

Active listening just means when you're listening to something, you keep alert and take specific actions that help make sure you understand. It includes both effective listening and feedback. Effective listening is a way that you do active listening—it means paying attention to both verbal and nonverbal communication. Feedback means doing things like repeating back the words that you were told in order to make sure you understood them, and giving your own nonverbal cues to show the speaker that you got the message.

Q: OK, so what about nonverbal and paralingual communication? Aren't those the same thing?

A: They *are* very similar, but they're not exactly the same. Nonverbal communication is any kind of communication that doesn't use words. That includes things like changing your body language, making eye contact, and using gestures. Paralingual communication is a kind of nonverbal communication—it's changing your tone of voice or intonation, finding ways to communicate things above and beyond just the words that you're saying. For example, the same words mean very different things if you say them sarcastically than if you say them in a normal tone of voice.

Q: Why is all that stuff about different kinds of communication important?

A: It's important because 90% of project management is communication, so if you want to be the best project manager that you can be, you need to constantly work to improve your communication skills!

Q: Should I always have a kickoff meeting?

A: Yes, absolutely! You should always have a kickoff meeting for every project. Not only that, but if you're running the kind of project with several phases, and you go through all of the process groups for each phase, then you should have a separate kickoff meeting for each new phase. Kickoff meetings also help you define who's responsible for various communications. Kickoff meetings are really important, because they give the team a chance to meet face-to-face, and give you the opportunity to make sure that everyone really understands all of the ways they can communicate with one another. That's a great way to head off a lot of potential project problems!

Q: Why do I need to be able to calculate the number of lines of communication?

A: It may seem like the lines of communication formula is something arbitrary that you just need to memorize for the exam, but it's actually pretty useful.

Let's say that you have a project with a whole lot of people on it. You set up a good communication system in your Communication Management plan, but you want to make sure that you really included every line in it, because if you missed one then you could run into communications problems down the line. So what do you do? Well, one thing you can do to check your work is to calculate the total number

of lines of communication in your project, and then make sure that every one of those lines is represented somewhere in your communications plan. It's a little more work up front, but it could really save you a lot of effort down the line!

Q: I spent all that time working on performance reports. What do I do with them once I'm done with them?

A: The same thing you do with any information that you generate on your project. You add them to your organizational process assets!

Think back to how you came up with your estimates in Time Management and Cost Management. You spent a lot of time doing analogous estimation, right? That's where you use performance from past projects to come up with a rough, top-down estimate for your new project. Well, where do you think the performance information from those past projects came from? You got them from your organizational process assets. And how did they end up there? Project managers from those past projects took their performance reports and added them. So you should add your performance reports, too. That way, project managers on future projects can use your project when they need to look up historical data.

You should add all of your performance reports to the organizational process assets so that project managers on future projects can use them as historical information.

It's party time!

The Head First Lounge party is a big hit! Everything came together beautifully, and Jeff and Charles are the new downtown sensation!

Question Clinic: The calculation question

YOU'LL RUN ACROSS A BUNCH OF QUESTIONS ON THE EXAM ASKING YOU TO USE SOME OF THE FORMULAS THAT YOU LEARNED. LUCKILY, THESE ARE SOME OF THE EASIEST QUESTIONS THAT YOU CAN ANSWER.

This is the wrong answer you'd get if you calculate the number of lines of communication if you include the team and two sponsors, but forget to include the project manager.

This wrong answer is the number of lines of communication BEFORE the team size was increased. You have 13 people (10 team members, 2 client sponsors, and you), so the number of lines is $13 \times 12 \div 2 = 78$.

12. You're managing a project with 2 client sponsors, and you have a 10-person team reporting to you. You've been given a budget increase, which allowed you to increase your team size by 30%. How many lines of communication were added?

A. 66
B. 78
C. 42
D. 120

This wrong answer is the number of lines of communication AFTER the team size was increased by 30%. You have 16 people (13 team members, 2 client sponsors, and you), so the number of lines is $16 \times 15 \div 2 = 120$.

Aha! Here's the right answer. Take the number of lines for 16 people and subtract the number of lines for 13 people: $120 - 78 = 42$.

WHEN YOU SIT DOWN TO TAKE AN EXAM AT A COMPUTER TESTING CENTER, YOU'LL BE GIVEN SCRATCH PAPER. YOU'LL ALSO HAVE 15 MINUTES TO GO THROUGH A TUTORIAL THAT SHOWS YOU HOW TO USE THE EXAM SYSTEM. BEFORE YOU FINISH THE TUTORIAL, TAKE A MINUTE AND WRITE DOWN ALL OF THE FORMULAS. WRITE DOWN THE EARNED VALUE FORMULAS AND THE FORMULA TO CALCULATE THE LINES OF COMMUNICATION ON THE SCRATCH PAPER. THAT WILL MAKE ANY CALCULATION QUESTION EASY.

HEAD LIBS

Try coming up with your own calculation question! But this time, try using one of the earned value formulas from Chapter 7.

You are managing a _____ project.
(kind of project)

You have _____ , _____ ,
(a value needed for the calculation)   (another value needed for the calculation)

and_____ .
(an irrelevant value that is NOT needed for the calculation)

Calculate _____ for your project.
(name of a formula)

A. _____
(the answer you'd get if you plug the wrong value into the formula)

B. _____
(the answer you'd get if you used the wrong formula)

C. _____
(the correct answer)

D. _____
(a totally bizarre answer that comes out of nowhere)

WHAT'S MY PURPOSE

Match each communication element to what it does

Acknowledge

Transmit message

Feedback/ Response

Encoding

Decoding

Getting the information from one person to the other

Letting the sender know that the message was received

Modifying a message that has been sent so that it can be understood

An answer to a message

Modifying a message so that it can be sent

Exercise Solution

Jeff and Charles are interviewing new bartenders to help with the expanded space. Choose which kind of communication is being used in each situation.

1. One applicant came in 30 minutes late, and was dressed unprofessionally. The guys knew that he would not be a good fit for the position.

☐ Paralingual   ☒ Nonverbal

☐ Feedback

The candidate repeated the question. That's a great example of feedback.

3. Charles asked the next applicant if he knew how to make a sidecar. He said "A sidecar? Sure. It's one part brandy or cognac, one part Cointreau, and one part lemon juice."

☐ Paralingual   ☐ Nonverbal

☒ Feedback

2. Charles asked an applicant about her background. Her tone of voice was really sarcastic, and he got the impression she didn't take the job seriously. Charles and Jeff decided to pass on her too.

☒ Paralingual   ☐ Nonverbal

☐ Feedback

4. Then the applicant told them about his background as a bartender for other retro clubs. As he spoke, he made eye contact with them and made sure to confirm agreement with them.

☐ Paralingual   ☒ Nonverbal

☐ Feedback

Exercise Solution

Choose which kind of communication is being used in each situation.

1. You and your business analysts write a requirements specification for your project.

☐ Formal verbal   ☐ Informal verbal

☒ Formal written   ☐ Informal written

2. You call up a supplier for materials for your project to let him know that you are a week late, so he's got a little flexibility in his delivery schedule.

☐ Formal verbal   ☒ Informal verbal

☐ Formal written   ☐ Informal written

3. You present your project's status to your company's executive committee.

☒ Formal verbal   ☐ Informal verbal

☐ Formal written   ☐ Informal written

4. You send an email to some of your team members to get more information about an issue that has been identified on your project.

☐ Formal verbal   ☐ Informal verbal

☐ Formal written   ☒ Informal written

5. You leave a voicemail message for your test team lead following up on an issue he found.

☐ Formal verbal   ☒ Informal verbal

☐ Formal written   ☐ Informal written

6. You IM with your team members.

☐ Formal verbal   ☐ Informal verbal

☐ Formal written   ☒ Informal written

7. You prepare an RFP (request for proposals) for vendors to determine which of them will get a chance to contract a new project with your company.

☐ Formal verbal   ☐ Informal verbal

☒ Formal written   ☐ Informal written

Anything that has to do with a contract is always formal written.

Communicationcross

Take some time to sit back and give your right brain something to do. It's your standard crossword; all of the solution words are from this chapter.

Sharpen your pencil
Solution

You'll need to know how to calculate the number of lines of communication for the exam…but don't worry, it's really easy once you get a little practice.

1. You're managing a project with five people on the team, plus one additional stakeholder—the sponsor. Draw in all the channels of communication on this picture.

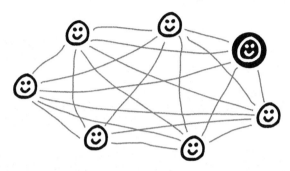

2. Wow, that was a lot of work. Luckily, you won't need to do that again. Now do it the easy way: use the formula to figure out how many lines of communication there are for seven people.

$$\# \text{ lines for } 7 \text{ people} = \frac{7 \times (7 - 1)}{2} = (7 \times 6) \div 2 = 21$$

3. OK, now let's say that you've added two team members and two more stakeholders, so there are now 11 people on the project who need to communicate with one another. How many lines did you add?

First figure out how many lines there are for 11 people:

$$\# \text{ lines for } 11 \text{ people} = \frac{11 \times (11 - 1)}{2} = (11 \times 10) \div 2 = 55$$

So how many lines were added when four people joined the seven-person project?

$$\# \text{ lines added} = \# \text{ lines for } 11 \text{ people} - \# \text{ lines for } 7 \text{ people}$$
$$= 55 - 21 = 34$$

Exam Questions

1. Keith, the project manager of a large publishing project, sends an invoice to his client. Which communication type is he using?

   A. Formal verbal

   B. Formal written

   C. Informal written

   D. Informal verbal

2. Which of the following is NOT an input to the Plan Communications Management process?

   A. Enterprise environmental factors

   B. Organizational process assets

   C. Information gathering techniques

   D. Project management plan

3. You take over for a project manager who has left the company and realize that the team is talking directly to the customer and having status meetings only when there are problems. The programming team has one idea about the goals of the project, and the testing team has another. Which document is the FIRST one that you should create to solve this problem?

   A. Communications Management plan

   B. Status report

   C. Meeting agenda

   D. Performance report

4. You ask one of your stakeholders how things are going on her part of the project and she says, "things are *fine*" in a sarcastic tone. Which is the BEST way to describe the kind of communication that she used?

   A. Feedback

   B. Active listening

   C. Nonverbal

   D. Paralingual

5. You're managing an industrial design project. You created a Communications Management plan, and now the team is working on the project. You've been communicating with your team, and now you're looking at the work performance data to evaluate the performance of the project. Which of the following BEST describes the next thing you should do?

   A. Use formal written communication to inform the client of the project status.

   B. Compare the work performance data against the time, cost, and scope baselines and look for deviations.

   C. Update the organizational process assets with your lessons learned.

   D. Hold a status meeting.

Exam Questions

6. You have five people working on your team, a sponsor within your company, and a client, all of whom need to be kept informed of your project's progress. How many lines of communication are there?

    A.   28

    B.   21

    C.   19

    D.   31

7. Which of the following is NOT an example of active listening?

    A.   Nodding your head in agreement while someone is talking

    B.   Restating what has been said to be sure you understand it

    C.   Asking questions for clarification

    D.   Multitasking by checking your email during a conversation

8. Sue sent a message to Jim using the company's voicemail system. When he received it, Jim called her back. Which of the following is true?

    A.   Sue encoded the voicemail; Jim decoded it, and then encoded his feedback message.

    B.   Sue decoded her voicemail message; Jim encoded his phone call and decoded the feedback.

    C.   Jim sent feedback to Sue, who encoded it.

    D.   Sue decoded her voicemail message and Jim encoded his feedback.

9. You're managing a construction project. Suddenly the customer asks for some major changes to the blueprints. You need to talk to him about this. What's the BEST form of communication to use?

    A.   Informal written

    B.   Informal verbal

    C.   Formal written

    D.   Formal verbal

10. Kyle is the project manager of a project that has teams distributed in many different places. In order to make sure that they all get the right message, he needs to make sure that his project plan is translated into Spanish, Hindi, French, and German. What is Kyle doing when he has his communications translated?

    A.   Encoding

    B.   Decoding

    C.   Active listening

    D.   Effective listening

Exam Questions

11. There are 15 people on a project (including the project manager). How many lines of communication are there?

    A. 105

    B. 112

    C. 113

    D. 52

12. Which communication process is in the Monitoring and Controlling process group?

    A. Manage Communications

    B. None of the communications processes

    C. Plan Communications Management

    D. Control Communications

13. You're working at a major conglomerate. You have a 24-person team working for you on a project with 5 major sponsors. The company announces layoffs, and your team is reduced to half its size. How many lines of communication are on your new, smaller team?

    A. 66

    B. 153

    C. 276

    D. 406

14. You've consulted your earned value calculations to find out the EAC and ETC of your project. Which of the following is the BEST place to put that information?

    A. Work performance information

    B. Forecasts

    C. Quality control measurements

    D. Lessons learned

15. Which of the following is an example of noise?

    A. An email that's sent to the wrong person

    B. A project manager who doesn't notice an important clause in a contract

    C. Garbled text and smudges that make a fax of a photocopy hard to read

    D. When the team is not paying attention during a status meeting

Exam ~~Questions~~ Answers

1. Answer: B

Any communication that can be used for legal purposes is considered formal written communication. An invoice is a formal document.

See the word "technique"? That's a good indication that it's a tool and not an input.

2. Answer: C

Information gathering techniques are not part of Plan Communications Management.

3. Answer: A

The Communications Management plan is the first thing you need to create in this situation. It will help you organize the meetings that are taking place and get everyone on the same page. The Communications Management plan will help you to streamline communications so that the customer can use you as a single point of contact, too.

> I GET IT! YOU CAN'T DO ANY COMMUNICATIONS UNLESS YOU'VE GOT A GOOD COMMUNICATIONS MANAGEMENT PLAN.

4. Answer: D

Paralingual communication happens when additional information is conveyed by the tone or pitch of your voice. It's when you use more than just words to communicate.

5. Answer: B

When you look at work performance data, you're in the Control Communications process. And what do you do with the work performance data? You compare it against the baselines to see if your project is on track! If it isn't, that's when you want to get the word out as quickly as possible.

A lot of people choose B here. Don't forget to include yourself! Look out for questions like this on the exam too.

6. Answer: A

The formula for lines of communication is n x (n − 1) ÷ 2. In this problem there were seven people named, plus you. (8 x 7) ÷ 2 = 28.

Answers

Exam ~~Questions~~

7. Answer: D

All of the other options show the speaker that you understand what is being said. That's <u>active listening.</u>

Active listening sometimes means saying things like "I agree," or "can you explain that a little further?" ↗

8. Answer: A

This question is just asking if you know the definitions of encode, decode, and feedback. Encoding is making a message ready for other people to understand, while decoding it involves receiving the message and understanding it. Feedback means letting the sender know that you got the message.

↙ *Any time you see anything about a formal document in communication with a client, it's formal written.*

9. Answer: C

Any time you are communicating with the customer about the scope of your project, it's a good idea to use formal written communication.

10. Answer: A

He has to encode his message so that others will understand it.

11. Answer: A

$(15 \times 14) \div 2 = 105$. This one is just asking if you know the formula $n \times (n-1) \div 2$.

12. Answer: D

Control Communications is the only Monitoring and Controlling process in Communications Management.

Answers

Exam Questions

13. Answer: B

There are now 12 team members, 5 sponsors, and a project manager. That gives you 18 people. Use the formula: n x (n – 1) ÷ 2 to calculate this: 18 x 17 ÷ 2 = 153.

Did you get one of the other answers? Make sure you included the five sponsors and the project manager!

14. Answer: B

The idea behind forecasts is that you are using the earned value calculations that forecast the completion of the project to set everyone's expectations. That's why you use EAC (which helps you estimate your project's total cost) and ETC (which gives you a good idea of how much more money you think you'll spend between now and when it ends).

15. Answer: C

There are plenty of ways that communication can go wrong. When you send email to the wrong person, your communication had trouble—but that's **not** noise. Noise is the specific thing that interferes with the communication. In this case, the garbled text is a great example of noise.

OH, I GET IT. I ALREADY CAME UP WITH GOOD COST AND TIME FORECASTS USING EAC AND ETC. NOW I CAN PACKAGE THEM UP AS FORECASTS AND SHARE THEM WITH THE TEAM.

11 Project risk management

Planning for the unknown

Even the most carefully planned project can run into trouble. No matter how well you plan, your project can always run into **unexpected problems**. Team members get sick or quit, resources that you were depending on turn out to be unavailable—even the weather can throw you for a loop. So does that mean that you're helpless against unknown problems? No! You can use *risk planning* to identify potential problems that could cause trouble for your project, **analyze** how likely they'll be to occur, take action to **prevent** the risks you can avoid, and **minimize** the ones that you can't.

What's a risk?

There are no guarantees on any project! Even the simplest activity can run into unexpected problems. Any time there's anything that ***might*** occur on your project and change the outcome of a project activity, we call that a **risk**. A risk can be an event (like a fire), or it can be a condition (like an important part being unavailable). Either way, it's something that may or may not happen…but if it does, you will be forced to change the way you and your team work on the project.

If your project requires that you stand on the edge of a cliff, then there's a risk that you could fall.

If it's very windy out or the ground is slippery and uneven, then falling is more likely.

**A risk is any uncertain event or condition that might affect your project.
Not all risks are negative.**

Not all risks are negative

Some events (like finding an easier way to do an activity) or conditions (like lower prices for certain materials) can help your project! When this happens, we call it an **opportunity**…but it's still handled just like a risk.

How you deal with risk

When you're planning your project, risks are still uncertain: they haven't happened yet. But eventually, some of the risks that you plan for *do* happen. And that's when you have to deal with them. There are four basic ways to handle a risk:

 Avoid

The best thing that you can do with a risk is avoid it—if you can prevent it from happening, it definitely won't hurt your project.

The easiest way to avoid this risk is to walk away from the cliff...but that may not be an option on this project.

2 Mitigate

If you can't avoid the risk, you can mitigate it. This means taking some sort of action that will cause it to do as little damage to your project as possible.

3 Transfer

One effective way to deal with a risk is to pay someone else to accept it for you. The most common way to do this is to buy insurance.

4 Accept

When you can't avoid, mitigate, or transfer a risk, then you have to accept it. But even when you accept a risk, at least you've looked at the alternatives and you know what will happen if it occurs.

If you can't avoid the risk, and there's nothing you can do to reduce its impact, then accepting it is your only choice.

LOOKS LIKE FALLING IS THE BEST OPTION.

Plan Risk Management

By now, you should have a pretty good feel for how each of the planning processes work. The past few knowledge areas started out with their own planning process, and Risk Management is no different. You start with the **Plan Risk Management** process, which should look very familiar to you.

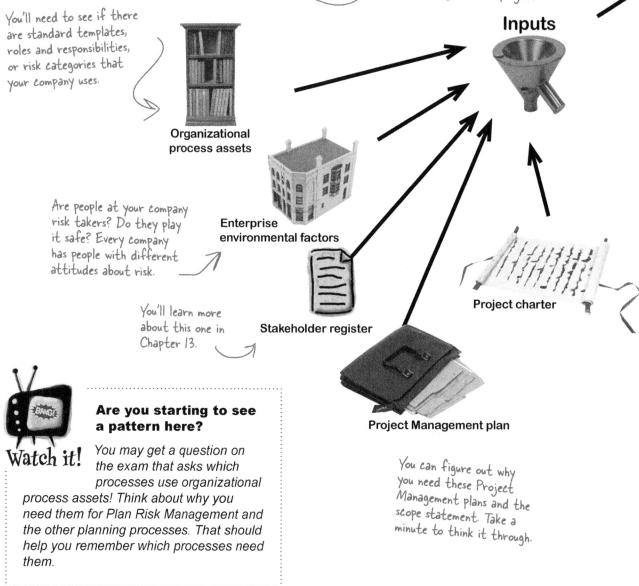

Planning process group

By the time a risk actually occurs on your project, it's too late to do anything about it. That's why you need to plan for risks from the beginning and keep coming back to do more planning throughout the project.

Inputs

You'll need to see if there are standard templates, roles and responsibilities, or risk categories that your company uses.

Organizational process assets

Are people at your company risk takers? Do they play it safe? Every company has people with different attitudes about risk.

Enterprise environmental factors

You'll learn more about this one in Chapter 13.

Stakeholder register

Project charter

Project Management plan

Are you starting to see a pattern here?

Watch it! *You may get a question on the exam that asks which processes use organizational process assets! Think about why you need them for Plan Risk Management and the other planning processes. That should help you remember which processes need them.*

You can figure out why you need these Project Management plans and the scope statement. Take a minute to think it through.

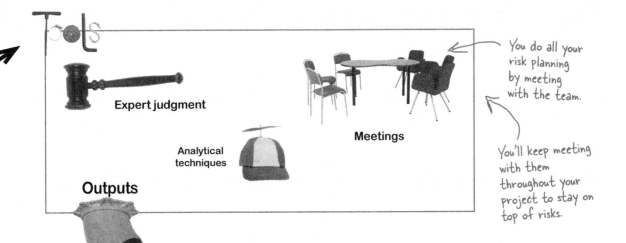

Expert judgment

Analytical techniques

Meetings

Outputs

You do all your risk planning by meeting with the team.

You'll keep meeting with them throughout your project to stay on top of risks.

The Risk Management plan is the only output

It tells you how you're going to handle risk on your project—which you probably guessed, since that's what management plans do. It says how you'll assess risk on the project, who's responsible for doing it, and how often you'll do risk planning (since you'll have to meet about risk planning with your team throughout the project).

The plan has parts that are really useful for managing risk:

The Risk Management plan is your guide to identifying and analyzing risks on your project.

Risk Management plan

It tells you who identifies and analyzes the risks, how they do it, and how often it happens.

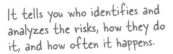

❏ It has a bunch of **risk categories** that you'll use to classify your risks. Some risks are technical, like a component that might turn out to be difficult to use. Others are external, like changes in the market or even problems with the weather. Risk categories help you to build a **risk breakdown structure (RBS)**.

❏ You'll need to describe the methods and approach you'll use for identifying and classifying risks on your project. This section of the document is called the **methodology**.

❏ It's important to come up with a plan to help you figure out how big a risk's impact is and how likely a risk is to happen. The impact tells you how much damage the risk will cause to your project. A lot of projects classify impact on a scale from minimal to severe, or from very low to very high. This section of the document is called the **definitions of probability and impact**.

Use a risk breakdown structure to categorize risks

You should build guidelines for risk categories into your Risk Management plan, and the easiest way to do that is to use a **risk breakdown structure** (RBS). Notice how it looks a lot like a WBS? It's a similar idea—you come up with major risk categories, and then decompose them into more detailed ones.

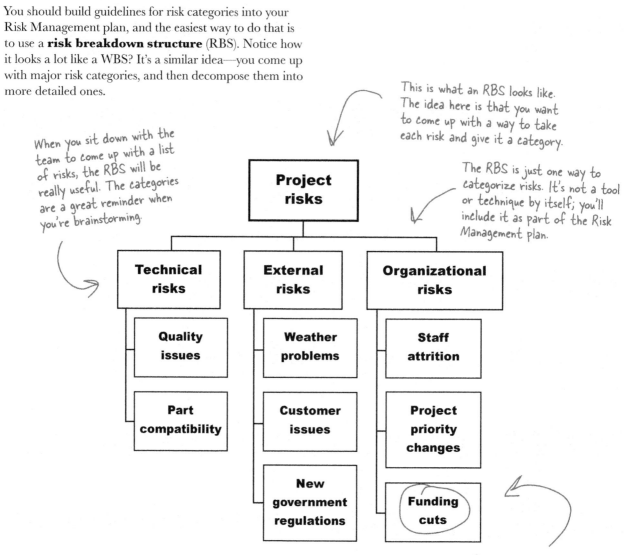

This is what an RBS looks like. The idea here is that you want to come up with a way to take each risk and give it a category.

When you sit down with the team to come up with a list of risks, the RBS will be really useful. The categories are a great reminder when you're brainstorming.

The RBS is just one way to categorize risks. It's not a tool or technique by itself; you'll include it as part of the Risk Management plan.

Once you come up with a list of risks, you'll label each risk with one of these categories. That will make it easier to figure out how to deal with the risks later.

Project risks

Technical risks
- **Quality issues**
- **Part compatibility**

External risks
- **Weather problems**
- **Customer issues**
- **New government regulations**

Organizational risks
- **Staff attrition**
- **Project priority changes**
- **Funding cuts**

Sharpen your pencil

Take a look at how each of these project risks is handled and figure out if the risk is being avoided, mitigated, transferred, or accepted.

1. Stormy weather and high winds could cause very slippery conditions, so you put up a tent and wear slip-resistant footwear to keep from losing your footing.

   ☐ Avoided   ☐ Mitigated

   ☐ Transferred   ☐ Accepted

2. You buy a surge protector to make sure a lightning strike won't blow out all of your equipment.

   ☐ Avoided   ☐ Mitigated

   ☐ Transferred   ☐ Accepted

3. Flooding could cause serious damage to your equipment, so you buy an insurance policy that covers flood damage.

   ☐ Avoided   ☐ Mitigated

   ☐ Transferred   ☐ Accepted

4. The manufacturer issues a warning that the safety equipment you are using has a small but nonzero probability of failure under the conditions that you'll be facing. You replace it with more appropriate equipment.

   ☐ Avoided   ☐ Mitigated

   ☐ Transferred   ☐ Accepted

5. A mud slide would be very damaging to your project, but there's nothing you can do about it.

   ☐ Avoided   ☐ Mitigated

   ☐ Transferred   ☐ Accepted

6. A team member discovers that the location you planned on using is in a county that is considering regulations that could be expensive to comply with. You work with a surveying team to find a new location.

   ☐ Avoided   ☐ Mitigated

   ☐ Transferred   ☐ Accepted

7. Surrounding geological features could interfere with your communications equipment, so you bring a flare gun and rescue beacon in case it fails.

   ☐ Avoided   ☐ Mitigated

   ☐ Transferred   ☐ Accepted

Answers: 1—Mitigated 2—Mitigated 3—Transferred 4—Avoided 5—Accepted 6—Avoided 7—Mitigated

Anatomy of a risk

Once you're done with Plan Risk Management, there are four more Risk Management processes that will help you and your team come up with the list of risks for your project, analyze how they could affect it, and plan how you and your team will respond if any of the risks materialize when you're executing it.

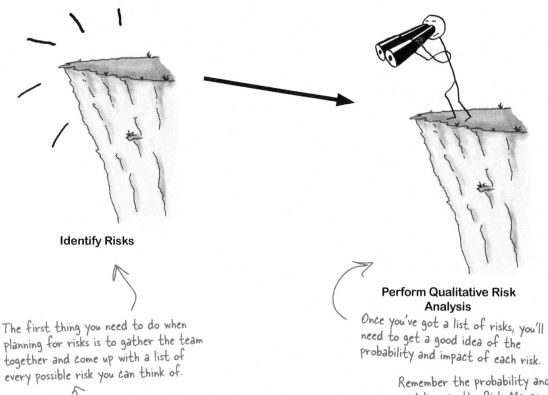

Identify Risks

The first thing you need to do when planning for risks is to gather the team together and come up with a list of every possible risk you can think of.

The RBS you created during Plan Risk Management will make it a lot easier to do this.

Perform Qualitative Risk Analysis

Once you've got a list of risks, you'll need to get a good idea of the probability and impact of each risk.

Remember the probability and impact guidelines in the Risk Management plan? This is where you use them to assign a probability and impact to each risk!

There are two more Risk Management processes. You already saw Plan Risk Management. There's also a Monitoring and Controlling process called Control Risks that you use when a risk actually materializes.

All <u>four</u> of these Risk Management processes are in the Planning process group—you need to plan for your project's risks before you start executing the project.

By the time you get here, you've got a list of risks, with a probability and impact assigned to each. That's a great starting point, but sometimes you need more information if you want to make good decisions...

All that's left now is to plan responses to each risk! This is where you decide whether to avoid, mitigate, transfer, or accept...and how you'll do it!

Perform Quantitative Risk Analysis
You can make better decisions with more precise information. That's what this process is about—assigning numerical values for the probability and impact of each risk.

Plan Risk Responses

BRAIN POWER

Some teams do Perform Qualitative Risk Analysis first, while others start with Perform Quantitative Risk Analysis. Some only do one or the other. Can you think of reasons they might do this?

What could happen to your project?

You can't plan for risks until you've figured out which ones you're likely to run into. That's why the next Risk Management process is **Identify Risks**. The idea is that you want to figure out every possible risk that might affect your project. Don't worry about how unlikely the risk is, or how bad the impact would be—you'll figure that stuff out later.

Planning process group

This will include all of your **activity cost estimates** and your **activity duration estimates**, which are also inputs.

You'll also need your **Stakeholder Register**, as well as **Procurement Documents** and other **Project Documents**.

Identify Risks

You should look at lessons learned from past projects to see what went wrong.

Cost and Schedule Management plans

Risk Management plan

Quality and Human Resource Management plans

Organizational process assets

Inputs

The Scope Baseline

Scope baseline

Enterprise environmental factors

T∘∘ls

The tools and techniques are all about gathering information from people and making sure it's right.

The goal of all of the risk planning processes is to produce the risk register. That's your main weapon against risk.

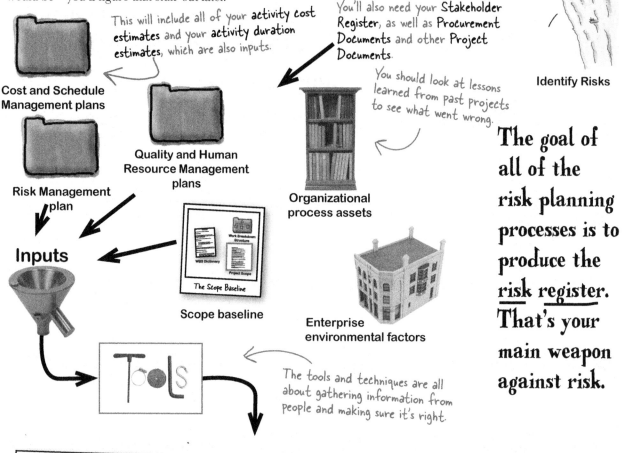

Identified Risks	Potential Responses	Root Causes
Landslide caused by loose gravel and dirt on the nearby mountain	Put up barrier or dig trench	Geological data review found loose topsoil nearby
High winds can lead to cliff disaster	Reinforce tent stakes; obtain weatherproof equipment	National weather service predicts 35% chance of high winds
Truck rental is unavailable	Pay to reserve equipment at a second company	Higher than expected demand for equipment in the area this season
Equipment failure during project	No responses were found by the team	Recent industry report cites higher-than-expected failure rates for critical equipment

Risk register

The risk register is the only output—and it's the most important part of Risk Management. It's a list of all of the risks and some initial ideas about how you'd respond to them.

Information-gathering techniques for Identify Risks

You probably already guessed that the goal of Identify Risks is to identify risks—seems pretty obvious, right? And the most important way to identify those risks is to gather information from the team. That's why the first—and most important—technique in Identify Risks is called **information gathering techniques**. These are time-tested and effective ways to get information from your team, stakeholders, and anyone else who might have information on risks.

Four useful information gathering techniques

There are a lot of different ways that you can find risks on your project. But there are only a few that you're most likely to use—and those are the ones that you will run across on the exam.

Brainstorming is the first thing you should do with your team. Get them all together in a room, and start pumping out ideas. Brainstorming sessions always have a **facilitator** to lead the team and help turn their ideas into a list of risks.

The team usually comes up with risks that have to do with building the product, while the sponsor or someone who would use the product will think about how it could end up being difficult to use.

The facilitator is really important—without her, it's just a disorderly meeting with no clear goal.

Interviews are a really important part of identifying risk. Try to find everyone who might have an opinion and ask them about what could cause trouble on the project. The sponsor or client will think about the project in a very different way than the project team.

The Delphi technique is a way to get opinions and ideas from experts. This is another technique that uses a facilitator, but instead of gathering team members in a room, the facilitator sends questionnaires to experts asking about important project risks. The facilitator will then take those answers and circulate them all to the experts—but each expert is kept **anonymous** so that they can give honest feedback.

The Delphi technique is always anonymous. People will give more honest opinions if they know their names won't be attached to them.

Root cause identification is analyzing each risk and figuring out what's actually behind it. Even though falling off of the cliff and having your tent blow away are two separate risks, when you take a closer look you might find that they're both caused by the **same thing**: high winds, which is the root cause for both of them. So you know that if you get high winds, you need to be on the lookout for *both* risks!

BRAIN POWER

What's the big difference between brainstorming and the Delphi technique? Can you think of a situation where one would be more useful than the other?

Tools

More Identify Risks techniques

Even though gathering information is the biggest part of Identify Risks, it's not the only part of it. There are other tools and techniques that you'll use to make sure that the risk register you put together lists as many risks as possible. The more you know about risk going into the project, the better you'll handle surprises when they happen. And that's what these tools and techniques are for—looking far and wide to get every risk possible.

Documentation reviews are when you look at plans, requirements, documents from your organizational process assets, and any other relevant documents that you can find to squeeze every possible risk out of them.

The RBS you created in Plan Risk Management is a good place to start for this. You can use all the risks you categorized in it as a jumping-off point.

Assumptions analysis is what you're doing when you look over your project's assumptions. Remember how important assumptions were when you were estimating the project? Well, now it's time to look back at the assumptions you made and make sure that they really are things you can assume about the project. Wrong assumptions are definitely a risk.

The team made assumptions during planning to deal with incomplete information...and there's a risk that each assumption could turn out to be wrong.

Checklist analysis means using checklists that you developed specifically to help you find risks. Your checklist might remind you to check certain assumptions, talk to certain people, or review documents you might have overlooked.

SWOT analysis lets you analyze strengths, weaknesses, opportunities, and threats. You'll start by brainstorming strengths and weaknesses, and then examine the strengths to find opportunities, and you'll look at the weaknesses to come up with threats to the project.

Expert judgment lets you rely on past experience to identify risks.

Diagramming techniques should be pretty familiar to you already. You can use the Ishikawa or fishbone diagrams from Quality Management to help you find the root cause of a risk, just like you did for a defect. You can also use flowcharts to see how parts of your system interact—any place where they get complex or uncertain is a good source of risks.

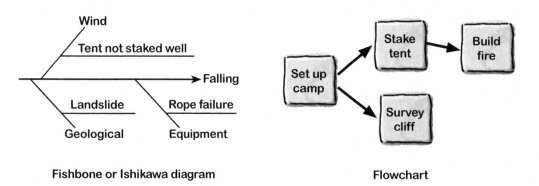

Fishbone or Ishikawa diagram

Flowchart

Exercise

Read each of these scenarios and identify which tool or technique is being used. If a scenario uses an information-gathering technique, specify which one.

1. Your project requires that you set up a campsite on the edge of a cliff. You gather your team members—including a geologist, a meteorologist, a tracker, and three campsite workers—and lead them in a directed discussion where they identify as many risks as possible.

2. You look through your company's asset library and discover that two previous projects involved setting up camp in this area. You look through the lessons learned to figure out what went wrong, and what could have been avoided through better planning.

3. You've sent a questionnaire to a park ranger and engineers at tent and hiking equipment companies to gather their opinions on the risk of falling off of a cliff. You remove their names from their responses, copy them, and send them back to everyone to get their feedback.

4. You've identified a risk that is very complex, so you identify the root cause. You use an Ishikawa diagram to gain insight into it.

5. You've reviewed your estimates and find that you had assumed that seasonal weather patterns would hold. If they change, then it could cause serious problems with the project.

6. You meet individually with many different people: the sponsor, stakeholders, team members, and experts. You ask each of them detailed questions about what they think could go wrong on the project.

1. Information-gathering techniques—Brainstorming   2. Documentation reviews
3. Information-gathering techniques—Delphi technique   4. Diagramming techniques
5. Assumptions analysis   6. Information-gathering techniques—Interviews

Where to look for risks

A good way to understand risks for the exam is to know where they come from. If you start thinking about how you find risks on your project, it will help you figure out how to handle them.

Here are a few things to keep in mind when you're looking for risks:

 1. RESOURCES ARE A GOOD PLACE TO START.

Have you ever been promised a person, equipment, conference room, or some other resource, only to be told at the last minute that the resource you were depending on wasn't available? What about having a critical team member get sick or leave the company at the worst possible time? Check your list of resources. If a resource might not be available to you when you need it, then that's a risk.

2. THE CRITICAL PATH IS FULL OF RISKS.

Remember the critical path method from Chapter 6? Well, an activity on the critical path is a lot riskier than an activity with plenty of float, because any delay in that activity will delay the project.

If an activity that's not on the critical path has a really small float, that means a small problem could easily cause it to become critical—which could lead to big delays in your project.

 3. "WHEN YOU ASSUME..."

Have you ever heard that old saying about what happens when you assume? At the beginning of the project, your team had to make a bunch of assumptions in order to do your estimates. But some of those assumptions may not actually be true, even though you needed to make them for the sake of the estimate. It's a good thing you wrote them down—now it's time to go back and look at that list. If you find some of them that are likely to be false, then you've found a risk.

4. LOOK OUTSIDE YOUR PROJECT.

Is there a new rule, regulation, or law being passed that might affect your project? A new union contract being negotiated? Could the price of a critical component suddenly jump? There are plenty of things outside of your project that are risks—and if you identify them now, you can plan for them and not be caught off guard.

Finding risks means talking to your team and being creative. Risks can be anywhere.

These areas are a good start, but there are plenty of other places on your project where you can find risks. Can you think of some of them?

Outputs

Now put it in the risk register

The point of the Identify Risks process is to…well, identify risks. But what does that really give you? You need to know enough about each risk to analyze it and make good decisions about how to handle it. So when you're doing interviews, leading brainstorming sessions, analyzing assumptions, gathering expert opinions with the Delphi technique, and using the other Identify Risks tools and techniques, you're gathering exactly the things you need to add to the risk register.

Each risk that you and the team come up with should go here.

It's a good idea for your Identify Risks meetings to include a discussion of how to respond to the risks, but you'll really dive into this later in the Plan Risk Responses process.

This is where the results of your root cause analysis go.

Identified Risks	Potential Responses	Root Causes
Landslide caused by loose gravel and dirt on the nearby mountain	Put up barrier or dig trench	Geological data review found loose topsoil nearby
High winds can lead to cliff disaster	Reinforce tent stakes; obtain weatherproof equipment	National weather service predicts 35% chance of high winds
Truck rental is unavailable	Pay to reserve equipment at a second company	Higher than expected demand for equipment in the area this season
Equipment failure during project	No responses were found by the team	Recent industry report cites higher-than-expected failure rates for critical equipment

Risk register

You might discover new risk categories, like Equipment. If you do, you'll go back to the RBS and add them.

You'll get a chance to come up with more complete responses later.

Some risks do not have an obvious response.

You already created the Risk Management plan in the last process. Now you're going back and updating it by adding the risk register.

The risk register is built into the Risk Management plan. Updates to the risk register are the only output of the Identify Risks process.

Rank your risks

It's not enough to know that risks are out there. You can identify risks all day long,
and there's really no limit to the number of risks you can think of. But some of them
are likely to occur, while others are very improbable. It's the ones that have much
better odds of happening that you really want to plan for.

Besides, some risks will cause a whole lot of damage to your project if they happen,
while others will barely make a scratch…and you care much more about the risks
that will have a big impact. That's why you need the next Risk Management process,
Perform Qualitative Risk Analysis—so you can look at each risk and figure out
how likely it is and how big its impact will be.

**Perform
Qualitative Risk
Analysis**

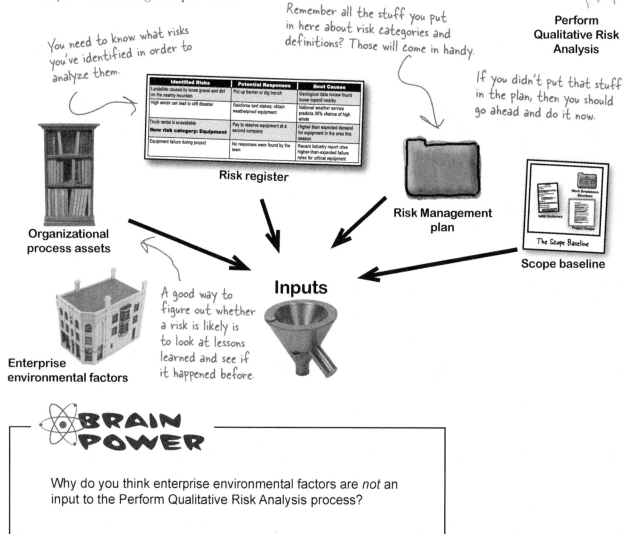

*You need to know what risks
you've identified in order to
analyze them.*

*Remember all the stuff you put
in here about risk categories and
definitions? Those will come in handy.*

*If you didn't put that stuff
in the plan, then you should
go ahead and do it now.*

Identified Risks	Potential Responses	Root Causes
Landslide caused by loose gravel and dirt on the nearby mountain	Put up barrier or dig trench	Geological data review found loose topsoil nearby
High winds can lead to cliff disaster	Reinforce tent stakes; obtain weatherproof equipment	National weather service predicts 35% chance of high winds
Truck rental is unavailable **New risk category: Equipment**	Pay to reserve equipment at a second company	Higher than expected demand for equipment in the area this season
Equipment failure during project	No responses were found by the team	Recent industry report cites higher-than-expected failure rates for critical equipment

Risk register

**Risk Management
plan**

The Scope Baseline

Scope baseline

**Organizational
process assets**

Inputs

*A good way to
figure out whether
a risk is likely is
to look at lessons
learned and see if
it happened before.*

**Enterprise
environmental factors**

BRAIN POWER

Why do you think enterprise environmental factors are *not* an
input to the Perform Qualitative Risk Analysis process?

Examine each risk in the register

Not all risks are created equal. Some of them are really likely to happen, while others are almost impossible. One risk will cause a catastrophe on your project if it happens; another will just waste a few minutes of someone's time.

Risk data quality assessment means making sure that the information you're using in your risk assessment is accurate. Sometimes it makes sense to bring in outside experts to check out the validity of your risk assessment data. Sometimes you can even confirm the quality of the data on your own, by checking some sample of it against other data sources.

Risk urgency assessment is checking out how soon you're going to need to take care of a particular risk. If a risk is going to happen soon, you'd better have a plan for how to deal with it soon, too.

Expert judgment definitely comes in handy when you're assessing risks. Who better to help you come up with things that might go wrong than experts who have been through similar projects before?

Risk probability and impact assessment is one of the best ways to be sure that you're handling your risks properly by examining how likely they are to happen, and how bad (or good) it will be if they do. This process helps you assign a probability to the likelihood of a risk occurring, and then figure out the actual cost (or impact) if it does happen. You can use these values to figure out which of your risks need a pretty solid mitigation plan, and which can be monitored as the project goes on.

Probability and impact matrix is a table where all of your risks are plotted out according to the values you assign. It's a good way of looking at the data so you can more easily make judgments about which risks require response. The ones with the higher numbers are more likely to happen and will have a bigger impact on your project if they do. So you'd better figure out how to handle those.

Risk categorization is all about grouping your risks so that you can come up with a better strategy for dealing with them. You might group them by the phase of the project where you'll see them, or by the source of the risk. Or you could come up with a bunch of additional categories that would help you to organize your responses better and be ready for the risk if it should happen.

Creating risk categories can help you deal with whole groups of risks in one response plan.

Perform Qualitative Risk Analysis helps you prioritize each risk and figure out its <u>probability</u> and <u>impact</u>.

Sometimes you'll find that some risks have obviously low probability and impact, so you won't put them in the main section of your register. Instead, you can add them to a separate section called the <u>watchlist</u>, which is just a list of risks. It'll include risks you don't want to forget about, but which you don't need to track as closely. You'll check your watchlist from time to time to keep an eye on things.

Probability	P&I				
.9	.09	.27	.45	.63	.81
.7	.07	.21	.35	.49	.63
.5	.05	.15	.25	.35	.45
.3	.03	.09	.15	.21	.27
.1	.01	.03	.05	.07	.09
Impact	.1	.3	.5	.7	.9

Sharpen your pencil

Here are some facts about the cliff project that were uncovered during qualitative analysis. Update the risk register on the facing page with the appropriate information.

Risk	Probability	Impact
1. Landslide	.1	.9
2. Winds	.7	.9
3. No truck	.3	.7
4. Storms	.5	.3
5. Supplies	.1	.5
6. Illness	.1	.7

During the Perform Qualitative Risk Analysis sessions, the team assigned a probability and impact number to each of the risks on the facing page.

Prob. & impact matrix

Probability					
.9	.09	.27	.45	.63	.89
.7	.07	21	.35	49	63
.5	.05	.15	25	35	45
.3	.03	.09	.15	.21	27
.1	.01	.03	.05	.07	.09
	.1	.3	.5	.7	.9

Impact

This gives you a good picture of the threshold the company has set for evaluating risks.

You can figure out the priority of each risk based on its probability and impact. Low-priority risks have no shading, medium ones are light gray, and high ones are dark gray.

1. The organizational process assets at your company set a high-priority risk as any risk with a probability and impact score higher than 0.20. Medium-priority risks are those between 0.10 and 0.19, and low-priority are those between 0 and 0.09. Low-priority risks can be monitored on a watchlist, but high and medium ones must have a response strategy.

   Fill in the missing values in the Priority and Probability columns in the risk register on the right, using the Probability and Impact matrix to figure out which ones are low, medium, or high. For example, we filled in "High" under Priority for row #3 by looking up risk ("No truck") in the first table, finding the probability and impact values, and then using the Probability and Impact matrix. The probability is .3 and the impact is .7, so you can find the corresponding box in the matrix. Since it's dark gray, its priority is high.

2. After analyzing your data, you came up with three risk categories for the project: natural, equipment, and human.
   Fill in the missing values in the Category column of the risk register with either "Natural," "Equipment," or "Human." We started you out by filling in a few of them.

3. For this particular project, you'll need the equipment at the start of the project, so any equipment risks are considered high urgency. Natural and human risks are all medium urgency, except for ones that have to do with storms, which you consider low urgency for this project because of limited mitigation potential.
   Figure out the whether the urgency for each risk is low, medium, or high and fill in the Urgency column in the risk register.

It's OK for some responses to be blank—you'll fill them in later during the Plan Risk Responses process.

	Identified risks	Potential response	Root cause	Category	Priority	Urgency
1.	Landslide caused by loose gravel and dirt on the nearby mountain	Put up barrier or dig trench	Geological data review found loose topsoil nearby	Natural	low	~~low~~ medium
2.	High winds can lead to cliff disaster	Reinforce tent stakes; obtain weatherproof equipment	National Weather Service predicts 35% chance of high winds	Natural	high	Medium
3.	Truck rental is unavailable	Secure other transport	Higher-than-expected demand for equipment this season	Equipment	High	high
4.	Storms predicted through the first two weeks of project schedule time	Create reserves to account for time lost due to storms	El Niño weather pattern	Natural	Medium	Low
5.	Supply shortage if we don't accurately predict food needs		Nearest store is 30 miles away	Equipment	low	~~low~~ ~~medium~~ Hi
6.	If someone gets sick, it could be a problem getting medical care	Bring a doctor with us on the project	Nearest hospital is 50 miles away	Human	low	~~low~~ medium

Outputs

Qualitative analysis helps you figure out which risks are most important to your project's success. When you've finished your analysis, you should have a risk register that tells you a lot more about what could go wrong.

The only output of Perform Qualitative Risk Analysis is project documents updates—including updates to the risk register.

	Identified risks	Potential response	Root cause	Category	Priority	Urgency
1.	Landslide caused by loose gravel and dirt on the nearby mountain	Put up barrier or dig trench	Geological data review found loose topsoil nearby	Natural	Low	Medium
2.	High winds can lead to cliff disaster	Reinforce tent stakes; obtain weatherproof equipment	National Weather Service predicts 35% chance of high winds	Natural	High	Medium
3.	Truck rental is unavailable		Higher-than-expected demand for equipment this season	Equipment	High	High
4.	Storms predicted through the first two weeks of project schedule time	Create reserves to account for time lost due to storms	El Niño weather pattern	Natural	Medium	Low
5.	Supply shortage if we don't accurately predict food needs		Nearest store is 30 miles away	Equipment	Low	High
6.	If someone gets sick, it could be a problem getting medical care	Bring a doctor with us on the project	Nearest hospital is 50 miles away	Human	Low	Medium

there are no
Dumb Questions

Q: Who does Perform Qualitative Risk Analysis?

A: The whole team needs to work on it together. The more of your team members who are helping to think of possible risks, the better off your plan will be. Everybody can work together to think of different risks to their particular part of the work, and that should give an accurate picture of what could happen on the project.

Q: What if people disagree on how to rank risks?

A: There are a lot of ways to think about risks. If a risk has a large impact on your part of the project or your goals, you can bet that it will seem more important to you than the stuff that affects other people in the group. The best way to keep the right perspective is to keep everybody on the team evaluating risks based on how they affect the overall project goals. If everyone focuses on the effect each risk will have on your project's constraints, risks will get ranked in the order that is best for everybody.

Q: Where do the categories come from?

A: You can create categories however you want. Usually, people categorize risks in ways that help them come up with response strategies. Some people use project phase. That way, they can come up with a risk mitigation plan for each phase of a project, and they can cut down on the information they need to manage throughout. Some people like to use the source of the risk as a category. If you do that, you can find mitigation plans that can help you deal with each source separately. That might come in handy if you are dealing with a bunch of different contractors or suppliers and you want to manage the risks associated with each separately.

Q: How do I know if I've got all the risks?

A: Unfortunately, you never know the answer to that one. That's why it's important to keep monitoring your risk register throughout the project. It's important that you are constantly updating it and that you never let it sit and collect dust. You should be looking for risks throughout all phases of your project, not just when you're starting out.

Q: I'm still not clear on the difference between the Delphi technique and brainstorming.

A: It's easy to get those two confused because both are about people sitting and thinking of risks. Delphi is a technique where you ask experts (who may or may not be team members) to give their opinion anonymously, and then you evaluate those opinions. Brainstorming is just you and your team sitting in a room thinking of risks.

Q: What's the point in even tracking low-priority risks? Why have a watchlist at all?

A: Actually, watchlists are just a list of all of the risks that you want to monitor as the project goes on. You might be watching them to see if conditions change and make them more likely to happen. By keeping a watchlist, you make sure that all of the risks that seem low priority when you are doing your analysis get caught before they cause serious damage if they become more likely later in the project.

The conditions that cause a risk are called **triggers**. So, say you have a plan set up to deal with storms, and you know that you might track a trigger for lightning damage, such as a thunderstorm. If there's no thunderstorm, it's really unlikely that you will see lightning damage, but once the storm has started, the chance for the risk to occur skyrockets.

Q: I still don't get the difference between priority and urgency.

A: Priority tells you how important a risk is, while urgency tells you when you need to deal with it. Some risks could be high priority but low urgency, which means that they're really important, but not time-critical. For example, you might know that a certain supplier that provides critical equipment will go out of business in six months, and you absolutely need to find a new supplier. But you have six months to do it. Finding a new supplier is a high priority, because your project will fail if it's not taken care of. But it's not urgent—even if it takes you four months to find a new supplier, nothing bad will happen.

Qualitative vs. quantitative analysis

Let's say you're a fitness trainer, and your specialty is helping millionaires get ready for major endurance trials. You get paid the same for each job, but the catch is that you get paid only if they succeed. Which of these clients would you take on?

Running a marathon   vs.   Climbing Mount Everest

One client wants you to help him train so that he can finish a marathon. He doesn't have to win, just get to the finish line.

Another client wants you to help him get to the top of Mount Everest. He won't be satisfied unless he gets to the summit.

It's much more likely that you can get even an out-of-shape millionaire to finish a marathon than it is that you can get him to climb Mount Everest successfully.

In fact, since the 1950s, 10,000 people have attempted to climb Mount Everest, and only 1,200 have succeeded. 200 have died. Your qualitative analysis probably told you that the climbing project would be the *riskier* of the two. But having the numbers to back up that judgment is what quantitative analysis is all about.

Perform Quantitative Risk Analysis

Once you've identified risks and ranked them according to the team's assessment, you need to take your analysis a little further and make sure that the numbers back you up. Sometimes you'll find that your initial assessment needs to be updated when you look into it further.

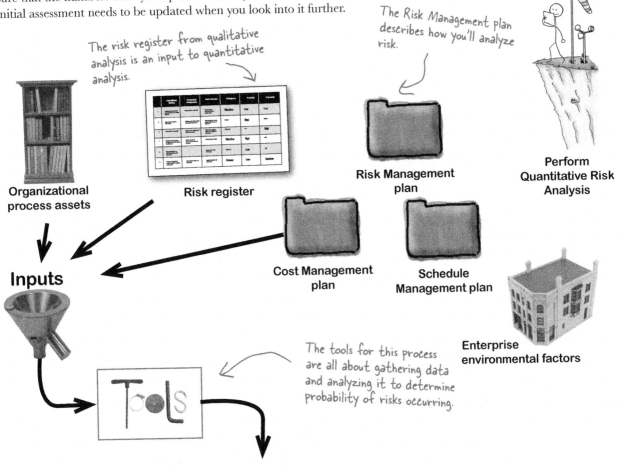

Planning process group

The risk register from qualitative analysis is an input to quantitative analysis.

The Risk Management plan describes how you'll analyze risk.

The tools for this process are all about gathering data and analyzing it to determine probability of risks occurring.

Once you're done analyzing, you update the risk register with the data you've gathered.

Organizational process assets

Risk register

Risk Management plan

Perform Quantitative Risk Analysis

Inputs

Cost Management plan

Schedule Management plan

Enterprise environmental factors

Tools

Risk register updates

First gather the data...

Quantitative tools are broken down into three categories: the ones that help you get more information about risks, the ones that help you to analyze the information you have, and expert judgment to help you put it all together. The tools for gathering data focus on gathering numbers about the risks you have already identified and ranked. These tools are called **data gathering and representation techniques**.

Interviewing
Sometimes the best way to get hard data about your risks is to interview people who understand them. In a risk interview, you might focus on getting three-point cost estimates so that you can come up with a budget range that will help you mitigate risks later. Another good reason to interview is to establish ranges of probability and impact, and document the reasons for the estimates on both sides of the range.

Probability distribution
Sometimes taking a look at your time and cost estimate ranges in terms of their distribution will help you generate more data about them. You probably remember these distribution curves from your probability and statistics classes in school. Don't worry, you won't be asked to remember the formal definition of probability distributions or even to be able to create them. You just need to know that they are another way of gathering data for quantitative analysis.

Beta Distribution   **Triangular Distribution**

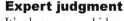

Expert judgment
It's always a good idea to contact the experts if you have access to them. People who have a good handle on statistics or risk analysis in general can be helpful when you are doing quantitative analysis. Also, it's great to hear from anybody who has a lot of experience with the kind of project you are creating, too.

...then analyze it

Now that you have all the data you can get about your risk register, it's time to analyze that information. Most of the tools for analyzing risk data are about figuring out how much the risk will end up costing you. These tools are called **quantitative risk analysis and modeling techniques**.

Sensitivity analysis is all about looking at the effect one variable might have if you could completely isolate it. You might look at the cost of a windstorm on human safety, equipment loss, and tent stability without taking into account other issues that might accompany the windstorm (like rain damage or possible debris from nearby campsites). People generally use tornado diagrams to look at a project's sensitivity to just one risk factor.

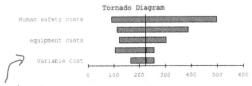

The tornado diagram lets you look at just one uncertain factor while assuming that all other data will stay where you expect it to.

Expected monetary value analysis lets you examine costs of all of the paths you might take through the project (depending on which risks occur) and assign a monetary value to each decision. So, if it costs $100 to survey the cliff and $20 to stake your tent, choosing to stake your tent *after* you've looked at the cliff has an expected monetary value of $120.

The main method of expected monetary value analysis you need to know for the test is **decision tree analysis.** For decision tree analysis, you just diagram out all of the decisions you think you will need to make to deal with risks. Then you add up all that you would need to spend to make each decision.

We'll talk about this in a couple of pages...

THE WIND'S BLOWING AT 31 MPH SE. I WEIGH 153 LBS. ACCORDING TO MY SIMULATIONS, I HAVE A 28.3% CHANCE OF FALLING OFF THE CLIFF IN THESE CONDITIONS.

Quantitative analysis means taking measurements and coming up with exact numbers to describe your risks.

Modeling and simulation. It's also a good idea to run your project risks through modeling programs if you can. Monte Carlo analysis is one tool that can randomize the outcomes of your risks and the probabilities of them occurring to help you get a better sense of how to handle the risks you have identified.

This is the same technique you learned about back in Chapter 6.

Monte Carlo analysis lets you run a lot of simulations to come up with data about what could happen on your project.

Calculate the expected monetary value of your risks

OK, so you know the probability and impact of each risk. How does that really help you plan? Well, it turns out that if you have good numbers for those things, you can actually figure out how much those risks are going to cost your project. You can do that by calculating the **expected monetary value** (or EMV) of each risk:

You can find these in your risk register.

❶ Start with the probability and impact of each risk.

Risk	Probability	Impact
High winds	35%	cost $48 to replace equipment
Mudslide	5%	lose $750 in damage costs
Wind generator is usable	15%	save $800 in battery costs
Truck rental unavailable	10%	cost $350 for last-minute rental

❷ Take the first risk and multiply the probability by the impact. For opportunities, use a positive cost. For threats, use a negative one. Then do the same for the rest of the risks.

The wind generator risk is an opportunity because you'll save money if it happens. So when you do the EMV calculation, you use a positive number for the impact.

High winds: 35% x –$48 = –$16.80

Even though the impact of a mudslide is big, the probability is low so the EMV is small.

Mudslide: 5% x –$750 = –$37.50

Wind generator: 15% x $800 = $120.00

Truck rental: 10% x –$350 = –$35.00

❸ Now that you've calculated the EMV for each of the risks, you can add them up to find the total EMV for all of them.

EMV = –$16.80 + –$37.50 + $120.00 + –$35.00 = $30.70

If you add $30.70 to the budget, then it should be enough to account for these risks.

Sharpen your pencil

You'll need to know how to do EMV calculations for the test. Give them a shot now—they're pretty easy once you get the hang of them.

Take a look at this table of risks.

Risk	Probability	Impact
Navigation equipment failure	15%	costs $300 due to getting lost
Unseasonably warm weather	8%	saves $500 in excavation costs
Wild animals eat rations	10%	costs $100 for replacements

1. Calculate the EMV for each of these three risks.

2. If these are the only risks on the project, calculate the total EMV.

3. The latest weather report came out, and there is now a 20% chance of unseasonably warm weather. What's the new EMV for the project?

4. Now the cost of replacement rations goes up to $150. What's the new EMV for the project?

→ Answers on page 598.

Decision tree analysis uses EMV to help you make choices   T☉☉Ls

There's another way to do EMV—you can do it visually using something called a **decision tree**. This decision tree shows the hidden costs of whether or not you buy a heavier tent. The tent is more expensive—it costs $350, while the lighter tent costs $130. But the heavier tent has better protection against the wind, so if there are high winds, your equipment isn't damaged.

If you buy a heavy tent, then it protects your equipment better, but it'll cost more. You figure that if there are high winds, you'll lose $953 worth of equipment with a light tent, but only $48 worth if you have a heavy one. If there are low winds, then you'll only lose $15 worth with a light tent and $10 worth with a heavy tent.

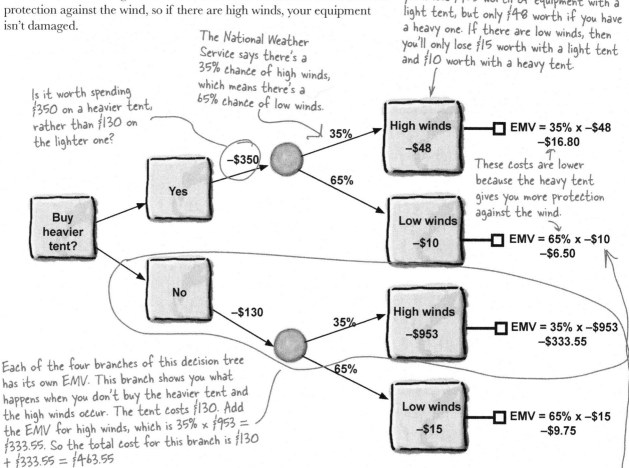

The National Weather Service says there's a 35% chance of high winds, which means there's a 65% chance of low winds.

Is it worth spending $350 on a heavier tent, rather than $130 on the lighter one?

These costs are lower because the heavy tent gives you more protection against the wind.

Each of the four branches of this decision tree has its own EMV. This branch shows you what happens when you don't buy the heavier tent and the high winds occur. The tent costs $130. Add the EMV for high winds, which is 35% × $953 = $333.55. So the total cost for this branch is $130 + $333.55 = $463.55

What's the EMV—or how much it's likely to cost you—of choosing the heavier tent?

If we add the EMV for high winds plus the EMV for low winds to the cost of the tent, we'll figure out the "real" cost of choosing the heavier tent. So that's –$16.80 + –$6.50 + –$350 = –$373.30.

This is just the EMV of the low winds if you buy the heavier tent. The probability of low winds is 65%, and the cost is $10. So it's just like the other EMV calculations: 65% × –$10 = –$6.50.

Compare that with the EMV of choosing the lighter tent. Which decision makes sense?

We can do the same thing for the bottom two branches of the tree. The "cheaper" tent costs –$130 + –$333.55 + –$9.75 = $473.30. So it's actually more expensive!

Exercise

Looking at the decision tree on the facing page, see if you can figure out the expected monetary value depending on the decisions the team makes.

⤷ Hint: Figure out the new EMV for each branch— that will tell you if the decision makes sense.

1. You hear a weather report that says there's now a 45% chance of high winds. Does it still make sense to buy the heavier tent?

2. If you don't buy the heavier tent, then you have room to take along a wind generator that can power your equipment, and that will save you $1,100 in portable batteries if there's a heavy wind. If there's still a 45% chance of high winds, does it still make sense to buy the heavier tent?

⟶ Answers on page 599.

This is an opportunity. So it should have a POSITIVE value when you do the EMV calculation.

there are no Dumb Questions

Q: I still don't get this Monte Carlo stuff. What's the deal?

A: All you really need to know about Monte Carlo analysis for the test is that it's a way that you can model out random data using software. In real life, though, it's a really cool way of trying to see what could happen on your project if risks do occur. Sometimes modeling out the data you already have about your project helps you to better see the real impact of a risk if it did happen.

Q: I can figure out how much the risk costs using EMV, or I can do it with decision tree analysis. Why do I need two ways to do this?

A: That's a good question. If you take a really careful look at how you do decision tree analysis, you might notice something... it's actually doing exactly the same thing as EMV. It turns out that those two techniques are really similar, except that EMV does it using numbers and decision tree analysis spells out the same calculation using a picture.

Q: I understand that EMV and decision trees are related, but I still don't exactly see how.

A: It turns out that there are a lot of EMV techniques, and decision tree analysis is just one of them. But it's the one you need to know for the test, because it's the one that

helps you make decisions by figuring out the EMV for each option. You can bet that you'll see a question or two that asks you to calculate the EMV for a project based on decision tree like the one on the facing page. As long as you remember that risks are negative numbers and that opportunities are positive ones, you should do fine.

Q: So are both quantitative analysis and qualitative analysis really just concerned with figuring out the impact of risks?

A: That's right. Qualitative analysis focuses on the impact as the team judges it in planning. Quantitative analysis focuses on getting the hard numbers to back up those judgments.

Update the risk register based on your quantitative analysis results

When you've finished gathering data about the risks, you change your priorities, urgency ratings and categories (if necessary), and you update your risk register. Sometimes modeling out your potential responses to risk helps you to find a more effective way to deal with them. That's why the only output of the **Perform Quantitative Risk Analysis** is **Project Documents Updates**.

Outputs

Analysis showed us that this would be the most expensive risk if it were to occur. So it got upgraded to a high priority.

	Identified risks	Potential response	Root cause	Category	Priority	Urgency
1.	Landslide caused by loose gravel and dirt on the nearby mountain	Put up barrier or dig trench	Geological data review found loose topsoil nearby	Natural	High	Medium
2.	High winds can lead to cliff disaster	Reinforce tent stakes; obtain weatherproof equipment	National Weather Service predicts 35% chance of high winds	Natural	High	Medium
3.	Truck rental is unavailable	Pay to reserve equipment at a second company	Higher-than-expected demand for equipment this season	Equipment	High	High
4.	Storms predicted through the first two weeks of project schedule time	Create reserves to account for time lost due to storms	El Niño weather pattern	Natural	Medium	Low
5.	Supply shortage if we don't accurately predict food needs		Nearest store is 30 miles away	Equipment	Low	High
6.	If someone gets sick, it could be a problem getting medical care	Bring a doctor with us on the project	Nearest hospital is 50 miles away	Human	Low	Low

This one got downgraded when quantitative analysis showed that it was not very likely to happen on such a short-term project.

BULLET POINTS: AIMING FOR THE EXAM

- The main output of all of the Risk Management planning processes is **updated project documents**, and the main document that gets updated is the **risk register**.

- The first step in Risk Management is **Identify Risks**, where you work with the whole team to figure out what risks could affect your project.

- Qualitative and quantitative analysis are all about **ranking risks** based on their probability and impact.

- Qualitative analysis is where you take the **categories** in your risk plan and **assign** them to each of the risks that you've identified.

- Quantitative analysis focuses on **gathering numbers** to help evaluate risks and **making the best decisions** about how to handle them.

- **Decision tree analysis** is one kind of **expected monetary value** analysis. It focuses on adding up all of the costs of decisions being made on a project so that you can see the overall value of risk responses.

- To calculate EMV, be sure to **treat all negative risks as negative numbers** and all **opportunities as positive ones**. Then add up all of the numbers on your decision tree.

- Don't forget **watchlists**. They let you monitor lower-priority risks so that you can see if triggers for those risks occur and you need to treat them as higher priorities.

- All of the processes in Risk Management are **Planning or Monitoring and Controlling processes**. There are **no Executing** processes here. Since the goal is to plan for risks, there is no need to focus on actually doing the work. By then, it's too late to plan for risks.

Your risk register should include both <u>threats</u> and <u>opportunities</u>. Opportunities have <u>positive</u> impact values, while threats have <u>negative</u> ones. Don't forget the plus or minus sign when you're calculating EMV.

 BRAIN POWER

How would you handle the risks listed in the risk register so far?

How do you respond to a risk?

After all that analysis, it's time to figure out what you're going to do if a risk occurs. Maybe you'll be able to keep a reserve of money to handle the cost of the most likely risks. Maybe there's some planning you can do from the beginning to be sure that you avoid it. You might even find a way to transfer some of the risk with an insurance policy.

However you decide to deal with each individual risk, you'll update your risk responses in the risk register to show your decisions when you're done. When you're done with **Plan Risk Responses**, you should be able to tell your change control board what your response plans are and who will be in charge of them so they can use them to evaluate changes.

Planning
process group

**Plan Risk
Responses**

Plan Risk Responses is figuring out what you'll do if risks happen.

You've updated your risk register as part of all of your analysis so far. It should contain everything you know about the risks facing your project, and even some preliminary responses you might have thought of along the way.

You might consult the Risk Management plan to figure out who is responsible for various activities in the Plan Risk Responses process, and for guidelines to help you prioritize your risks.

Risk register

Risk Management plan

Notice that organizational process assets aren't here. You can't use a template for this one. It's all about figuring out the responses that make sense for your project's SPECIFIC risks.

Inputs

It isn't always so bad

Remember the strategies for handling negative risks—avoid, mitigate, transfer, and accept—from earlier? Well, there are strategies for handling positive risks, too. The difference is that **strategies for positive risks** are all about how you can try to get the most out of them. The strategies for handling negative and positive risks are the tools and techniques for the Plan Risk Responses process.

The strategies for negative risks are also tools and techniques for this process. They're the ones you already learned: avoid, mitigate, transfer, and accept. Acceptance is a technique for both negative and positive risks.

❶ Exploit

This is when you do everything you can to make sure that you take advantage of an opportunity. You could assign your best resources to it. Or you could allocate more than enough funds to be sure that you get the most out of it.

❷ Share

Sometimes it's harder to take advantage of an opportunity on your own. Then you might call in another company to share in it with you.

❸ Enhance

This is when you try to make the opportunity more probable by influencing its triggers. If getting a picture of a rare bird is important, then you might bring more food that it's attracted to.

❹ Accept

Just like accepting a negative risk, sometimes an opportunity just falls in your lap. The best thing to do in that case is to just accept it!

Response planning can even find more risks

Secondary risks come from a response you have to another risk. If you dig a trench to stop landslides from taking out your camp, it's possible for someone to fall into the trench and get hurt.

Residual risks remain after your risk responses have been implemented. So even though you reinforce your tent stakes and get weatherproof gear, there's still a chance that winds could destroy your camp if they are strong enough.

I GET IT. SO, I HAVE TO GO BACK AND ANALYZE SECONDARY RISKS. BUT RESIDUAL RISKS JUST SIT THERE, SO I CAN DEAL WITH THEM LATER.

Which risk response technique is being used in these situations? Match each technique to its scenario.

Mitigate

If the weather's good, then there's a chance you could see a meteor shower. If the team gets a photo that wins the meteor photo contest, you can get extra funding. You have your team stay up all night with their telescopes and cameras ready.

Avoid

You hear that it's going to rain for the first three days of your trip, so you bring waterproof tents and indoor projects for the team to work on in the meantime.

Accept

You read that there's a major bear problem in the spring on the cliff where you are planning to work. You change your project start date to happen in the fall.

Transfer

On your way up the cliff, you meet another team that is looking to survey the area. You offer to do half of the surveying work while they do the other half and then trade your findings with each other.

Exploit

There's a high probability of water damage to some of your equipment, so you buy insurance to avoid losses.

Share

There's always the chance that someone could make a mistake and fall off the cliff. No matter how much you plan for the unexpected, sometimes mistakes happen.

Enhance

About 10 years ago a really rare bird, the black-throated blue warbler, was seen on this cliff. If you could get a picture of it, it would be worth a lot of money. So, you bring special seeds that you have read are really attractive to this bird, and you set up lookout points around the cliff with cameras ready to get the shot.

⟶ Answers on page 600.

Add risk responses to the register

It's time to add—you guessed it—more updates to project documents, including the risk register. All of your risk responses will be tracked through change control. Changes that you need to make to the plan will get evaluated based on your risk responses, too. It's even possible that some of your risk responses will need to be added into your contract.

Every risk needs to have one person who owns the response plan.

	Identified risks	Response strategy	Root cause	Risk owner	Cat	Priority	Urgency
1.	Landslide caused by loose gravel and dirt on the nearby mountain	Put up barrier or dig trench	Geological data review found loose topsoil nearby	**Joe S.**	Natural	High	Medium
2.	High winds can lead to cliff disaster	Reinforce tent stakes; obtain weatherproof equipment	National Weather Service predicts 35% chance of high winds	**Tanya T.**	Natural	High	Medium
3.	Truck rental is unavailable	Pay to reserve equipment at a second company	Higher-than-expected demand for equipment this season	**Joe S.**	Equipment	High	High
4.	Storms predicted through the first two weeks of project schedule time	**Buy storm insurance in case the equipment is damaged**	El Niño weather pattern	**Michael R.**	Natural	Medium	Low
5.	Supply shortage if we don't accurately predict food needs		Nearest store is 30 miles away	**James S.**	Equipment	Low	High
6.	If someone gets sick, it could be a problem getting medical care	Bring a doctor with us on the project	Nearest hospital is 50 miles away	**Tanya T.**	Human	Low	Low
7.	Someone could fall in the landslide trench	Set up a trench patrol to make sure no one gets hurt	Dig trench for landslides	**Joe S.**	Human	Low	Low

During Plan Risk Responses, the team agreed to buy insurance for this one.

The PM plan needs to be updated so that integrated change control can include the risk responses.

Project Management plan updates

Project documents updates

Risk Management Exposed

This week's interview:
Stick figure who hangs out on cliffs

Head First: We've seen you hanging out on cliffs for a while now. Apparently, you've also been paying people to stand on the cliff for you, or getting a friend to hold a trampoline at the foot of the cliff; we've even seen you jump off of it. So now that I've finally got a chance to interview you, I want to ask the question on everyone's mind: "Are you insane? Why do you spend so much time up there?"

Stick Figure: First off, let me dispel a few myths that are flying around out there about me. I'm not crazy, and I'm not trying to get myself killed! Before Risk Management entered my life I, like you, would never have dreamed of doing this kind of thing.

Head First: OK, but I'm a little skeptical about your so-called "Risk Management." Are you trying to say that because of Risk Management you don't have to worry about the obvious dangers of being up there?

Stick Figure: No. Of course not! That's not the point at all. Risk Management means you sit down and make a list of all of the things that could go wrong. (And even all the things that could go right.) Then you really try to think of the best way to deal with anything unexpected.

Head First: So you're doing this Risk Management stuff to make it less dangerous for you?

Stick Figure: Yes, exactly! By the time I'm standing up there on that cliff, I've really thought my way through pretty much everything that might happen up there. I've thought through it both qualitatively and quantitatively.

Head First: Quantitatively?

Stick Figure: Yes. You don't think I'd go up there without knowing the wind speed, do you? Chance of landslides? Storms? The weight of everything I'm carrying? How likely I am to fall in weather conditions? I think about all of that and I measure it. Then I sit down and come up with risk response strategies.

Head First: OK, so you have strategies. Then what?

Stick Figure: Then I constantly monitor my risks while I'm on the cliff. If anything changes, I check to see if it might trigger any of the risks I've come up with. Sometimes I even discover new risks while I'm up there. When I do, I just add them to the list and work on coming up with responses for them.

Head First: I see. So you're constantly updating your list of risks.

Stick Figure: Yes! We call it a **risk register**. Whenever I have new information, I put it there. It means that I can actually hang out on these cliffs with a lot of confidence. Because, while you can't guarantee that nothing will go wrong, you can be prepared for whatever comes your way.

Head First: That's a lot of work. Does it really make a difference?

Stick Figure: Absolutely! I'd never be able to sleep at night knowing that I could fall off the cliff at any time. But I've planned for the risks, and I've taken steps to stay safe…and I sleep like a baby.

You can't plan for every risk at the start of the project

Monitoring
& Controlling
process group

Even the best planning can't predict everything—there's always a chance that a new risk could crop up that you hadn't thought about. That's why you need to constantly monitor how your project is doing compared to your risk register. If a new risk happens, you have a good chance of catching it before it causes serious trouble. When it comes to risk, the earlier you can react, the better for everybody. And that's what the **Control Risks** process is all about.

THERE HAVE BEEN REPORTS OF BEARS CAUSING PROBLEMS FOR PEOPLE AROUND HERE LATELY. BE CAREFUL OUT HERE.

The park ranger's come by to let you know about some recent bear sightings on this cliff.

The risk register doesn't say anything about handling bears. Looks like this is a new risk altogether...

Control Risks is another change control process

Risk responses are treated just like changes. You monitor the project in every status meeting to see how the risks in the risk register are affecting it. If you need to implement a risk response, you take it to your change control board, because it amounts to a change that will affect your project constraints.

You compare all of your actual data to your plans using the risk register and the PM plan.

Project Management plan

Risk register

You should review status reports, metrics, and other work outputs to see if risks are happening.

Work performance data

Work performance reports

Inputs

You should keep monitoring your risks at every meeting until the project is closed.

That's why the tools and techniques include status meetings.

Weekly Meeting Agenda

1. Review work performance information and earned value numbers

2. Examine performance reports.

   • What have people gotten done?

3. Review risk register.

   • Have any risks occurred that require implementation of a risk response strategy?

   • Have we identified any new risks that need to be analyzed?

Risk monitoring should be done at _every_ status meeting.

How to control your risks

Controlling risks means keeping your finger on the pulse of the project. If you are constantly reviewing all of the data your project is producing, you will be able to react quickly if a new risk is uncovered, or if it looks like one of your response strategies needs to spring into action. Without careful monitoring, even your best plans won't get implemented in time to save your project if a risk happens.

Risk reassessment

You should have some regularly scheduled reassessment meetings to go over all of the information you have to date and see if your risk register still holds true. In a reassessment, your main goal is to find any new risks that have come up. That's why it's important to reassess your risk register every so often, and be sure that all of the risks in it are still the right ones.

Variance and trend analysis

Comparing the actual project performance to the plan is a great way to tell if a risk might be happening. If you find that you're significantly over budget or behind schedule, a risk could have cropped up that you didn't take into account. Looking for trends in your defects or schedule variance, for example, might show patterns that indicate that risks have occurred before you would have found that out on your own.

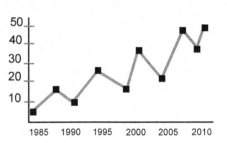

Reserve analysis

Just like you keep running tabs on your budget, you should always know how much money you have set aside for risk response. As you spend it, be sure to subtract it so you know if you have enough to cover all of your remaining risks. If you start to see that your reserves are running low and there are still a lot of risks being identified, you might be in trouble. Keeping tabs on your reserves means that you will always know if you need to set aside more funds or make different choices about how to handle risks as they come up.

Sometimes this kind of reserve is called a "contingency"—because its use is contingent on a certain risk happening.

More control risk tools and techniques

There are just a few more tools in the **Control Risks** process. They're all focused on finding new risks if they crop up, dealing with changes to the risks you've already planned for, and responding quickly to risks you know how to handle.

Tools

Never stop looking for new risks and adapting your strategies for dealing with them.

Risk audits are when you have an outside party come in and take a look at your risk response strategies to judge how effective they are. Sometimes risk audits will point out better ways of handling a specific risk so that you can change your response strategy going forward.

Auditors will also look at how effective your overall processes for risk planning are.

Technical performance measurement means comparing the performance of your project with its planned performance. So if you expected to hit a specific milestone, you could check performance information on your product at that time to see if it measured up to the plan. If not, that might indicate that there are risks you didn't plan for.

Meetings are the most important way to keep the team up to date on risk planning—so important that **they should happen throughout the entire project**. The more you talk about risks with the team, the better. Every single status meeting should have risk review on the agenda. Status meetings are a really important way of noticing when things might go wrong, and of making sure that you implement your response strategy in time. It's also possible that you could come across a new opportunity by talking to the team.

Here are some risk monitoring and control activities. Can you determine which of the tools is being used in each one?

Exercise

1. At every milestone, you do a new round of Identify Risks and make sure that the risks in your risk register still apply to the project.

☐ Reassessment   ☐ Audit

☐ Technical performance measurement

☐ Trend analysis   ☐ Reserve analysis

2. You check to make sure that you have all of the features developed in your project that you had planned when you reach the "feature complete" milestone. When you find that you are missing one of the planned features, you realize that a new risk has shown up—you missed one of the required features in your functional specification.

☐ Reassessment   ☐ Audit

☐ Technical performance measurement

☐ Trend analysis   ☐ Reserve analysis

3. You take a look at the number of defects you have found in your project per phase and find that it is higher in your project than it has been in most other projects that the company is doing. You dig a little deeper and find some previously unplanned risks that have been causing trouble on your project.

☐ Reassessment   ☐ Audit

☐ Technical performance measurement

☐ Trend analysis   ☐ Reserve analysis

4. Your company sends a risk expert in to take a look at your risk response strategies. She finds that you are missing a few secondary risks that might be caused by the responses you have planned. So you update your risk register to include the secondary risks.

☐ Reassessment   ☐ Audit

☐ Technical performance measurement

☐ Trend analysis   ☐ Reserve analysis

5 You decide to implement a risk response that costs $4,000. You check to make sure that you have enough money to cover the rest of the risks that might happen from here on out in the project.

☐ Reassessment   ☐ Audit

☐ Technical performance measurement

☐ Trend analysis   ☐ Reserve analysis

Answers:
1—Reassessment
2—Technical performance measurement
3—Trend analysis
4—Audit
5—Reserve analysis

there are no Dumb Questions

Q: Why do I need to ask about risks at every status meeting?

A: Because a risk could crop up at any time, and you need to be prepared. The better you prepare for risks, the more secure your project is against the unknown. That's also why the triggers and watchlists are really important. When you meet with your team, you should figure out if a trigger for a risk response has happened. And you should check your watchlist to make sure none of your low-priority risks have materialized.

For the test, you need to know that status meetings aren't just a place for you to sit and ask each member of your team to tell you his or her status. Instead, you use them to figure out decisions that need to be made to keep the project on track or to head off any problems that might be coming up. In your status meetings, you need to discuss all of the issues that involve the whole team and come up with solutions to any new problems you encounter. So, it makes sense that you would use your status meetings to talk about your risk register and make sure that it is always up to date with the latest information.

Q: I still don't get trend analysis. How does it help me find risks?

A: It's easy to miss risks in your project—sometimes all the meetings in the world won't help your team see some of them. That's why a tool like trend analysis can be really useful. Remember the control chart from Chapter 8? This is really similar, and it's just as valuable. It's just a way to see if things are happening that you did not plan for.

Q: Hey, didn't you talk about risks back in the Time Management chapter too?

A: Wow—it's great that you remembered that! The main thing to remember about risks from Chapter 6 is that having a very long critical path or, even worse, multiple critical paths means you have a riskier project. The riskiest is when all of the activities are on the critical path. That means that a delay to even one activity can derail your whole project.

Q: Shouldn't I ask the sponsor about risks to the project?

A: Actually the best people to ask about risks are the project team itself. The sponsor knows why the project is needed and how much money is available for it, but from there, it's really up to the team to manage risks. Since you are the ones doing the work, it makes sense that you would have a better idea of what has gone wrong on similar projects and what might go wrong on this one. Identify Risks, Perform Qualitative and Quantitative Risk Analysis, and Plan Risk Responses are some of the most valuable contributions the team makes to the project. They can be the difference between making the sponsor happy and having to do a lot of apologizing.

Q: Why do we do risk audits?

A: Risk audits are when you have someone from outside your project come in and review your risk register—your risks and your risk responses—to make sure you got it right. The reason we do it is because risks are so important that getting a new set of eyes on them is worth the time.

Q: Hold on, didn't we already talk about reserves way back in the Cost Control chapter? Why is it coming up here?

A: That's right, back in Chapter 7 we talked about a **management reserve**, which is money set aside to handle any unknown costs that come up on the project. That's a different kind of reserve than the one for controlling risks. The kind of reserve used for risks is called a **contingency reserve**, because its use is *contingent* on a risk actually materializing.

Project managers sometimes talk about both kinds of reserves together, because they both have to show up on the same budget. When they do, you'll sometimes hear talk of "known unknowns" and "unknown unknowns." The management reserve is for unknown unknowns—things that you haven't planned for but could impact your project. The contingency reserve is for known unknowns, or risks that you know about and explicitly planned for and put in your risk register.

The better you prepare for risks, the more secure your project is against the unknown.

Sharpen your pencil

By now, you know what comes out of a typical Monitoring and Controlling process. Draw in the missing outputs for Control Risks.

Outputs

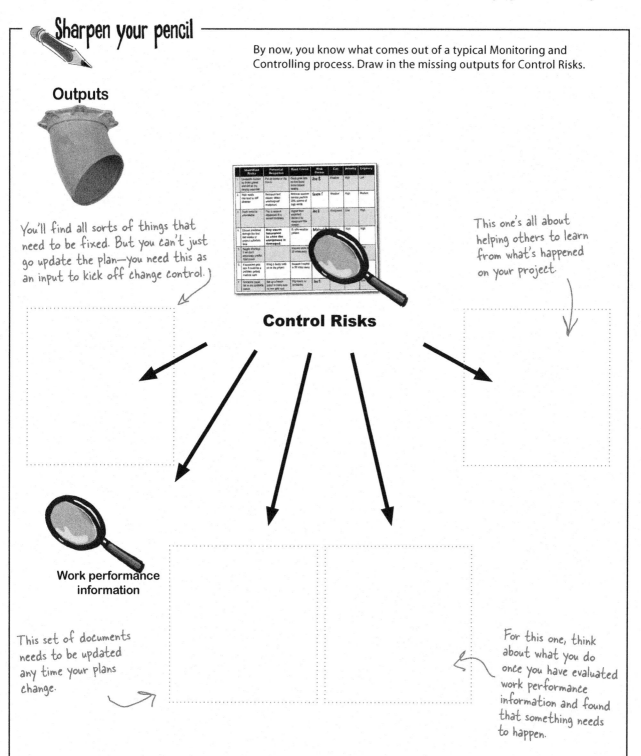

Control Risks

You'll find all sorts of things that need to be fixed. But you can't just go update the plan—you need this as an input to kick off change control.

This one's all about helping others to learn from what's happened on your project.

Work performance information

This set of documents needs to be updated any time your plans change.

For this one, think about what you do once you have evaluated work performance information and found that something needs to happen.

Sharpen your pencil
Solution

By now, you know what comes out of a typical Monitoring and Controlling process. Draw in the missing outputs for Control Risks.

Outputs

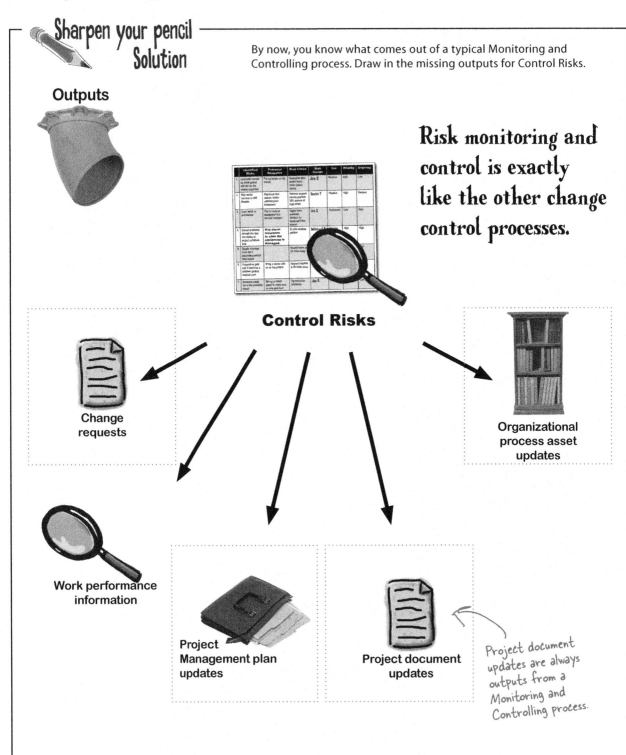

Risk monitoring and control is exactly like the other change control processes.

Control Risks

Change requests

Organizational process asset updates

Work performance information

Project Management plan updates

Project document updates

Project document updates are always outputs from a Monitoring and Controlling process.

* Note from the authors: We're not exactly sure why he feels his mission was accomplished after spraying a bear in the face and then jumping off of a cliff. But it seems to work!

Question Clinic: The "which-is-NOT" question

YOU'LL SEE SOME QUESTIONS ON THE EXAM THAT LIST INPUTS, OUTPUTS, TOOLS, OR CONCEPTS AND ASK YOU TO DETERMINE WHICH ONE OF THEM IS NOT PART OF THE GROUP. USUALLY, YOU CAN FIGURE THEM OUT BY GOING THROUGH THE ANSWER CHOICES ONE BY ONE AND ELIMINATING THE ONE THAT DOESN'T BELONG.

This is the one with tornado diagrams. It's got to be part of the Group.

This one is definitely a quantitative analysis technique. Multiplying probability with the value of positive and negative outcomes of the project is all about putting numbers to risk.

117. Which of the following is not a quantitative analysis technique?

A. Sensitivity analysis

B. Expected monetary value

C. Monte Carlo analysis

D. Reserve analysis

Remember reading something about Monte Carlo back in Chapter 6? It might be right...right? But wait, it's also a tool for using random numbers to model out possible risks on the project. It's definitely part of quantitative analysis.

D's definitely the right answer. It's about numbers, but it isn't concerned with assigning numbers to the risk. It's about keeping tabs on the contingency reserve when risks materialize, so it's a Monitoring and Controlling process. This has to be it!

TAKE YOUR TIME AND THINK YOUR WAY THROUGH IT. ALL OF THEM WILL HAVE SOMETHING IN COMMON BUT ONE. AS LONG AS YOU REMEMBER THE GROUP YOU'RE FITTING THEM INTO, YOU WON'T HAVE ANY TROUBLE.

Take your time answering which-is-NOT questions.

HEAD LIBS

Fill in the blanks to come up with your own "which-is-NOT" question!

Which of the following is NOT a _____ ?
(input, output, tool, process, or concept)

A. _____
(input, output, tool, or process that is in the group)

B. _____
(input, output, tool, or process that is in the group)

C. _____
(input, output, tool, or process that is in the group)

D. _____
(the right answer)

Sharpen your pencil
Solution

You'll need to know how to do EMV calculations for the test. Give them a shot now—they're pretty easy once you get the hang of them.

Take a look at this table of risks.

Risk	Probability	Impact
Navigation equipment failure	15%	costs $300 due to getting lost
Unseasonably warm weather	8%	save $500 in excavation costs
Wild animals eat rations	10%	costs $100 for replacements

1. Calculate the EMV for each of these three risks.

Navigation equipment failure: 15% x –$300 = –$45.00

Unseasonably warm weather: 8% x $500 = $40.00

Don't forget to use a positive value here because it's an opportunity, not a threat.

Wild animals eat rations: 10% x –$100 = –$10.00

2. If these are the only risks on the project, calculate the total EMV.

Total EMV = –$45.00 + $40.00 + –$10.00 = –$15.00

You get the total EMV by adding up the EMV for each risk.

3. The latest weather report came out, and there is now a 20% chance of unseasonably warm weather. What's the new EMV for the project?

Unseasonably warm weather: 20% x $500 = $100.00

The new total EMV = –$45.00 + $100.00 + –$10.00 = $45.00

The EMV is now positive, which means the project should cost less than you originally budgeted.

4. Now the cost of replacement rations goes up to $150. What's the new EMV for the project?

Wild animals eat rations: 10% x –$150 = –$15.00

The new total EMV = –$45.00 + $100.00 + –$15.00 = $40.00

When the probability of high
winds changed to 45%, then the
probability of low winds also
changed: to 55%.

Exercise Solution

1. You hear a weather report that says there's now a 45% chance of high winds. Does it still make sense to buy the heavier tent?

EMV of choosing the heavier tent: –$350 plus (45% x –$48) plus (55% x –$10) = –$377.10

EMV of choosing the lighter tent: –$130 plus (45% x –$953) plus (55% x –$15) = –$567.10

It still makes sense to choose the heavier tent.

2. If you don't buy the heavier tent, then you have room to take along a wind generator that can power your equipment, and that will save you $1,100 in portable batteries if there's a heavy wind. If there's still a 45% chance of high winds, does it still make sense to buy the heavier tent?

EMV of choosing the heavier tent: –$350 plus (45% x –$48) plus (55% x –$10) = –$377.10

EMV of choosing the lighter tent: –$130 plus (45% x $147) plus (55% x –$15) = –$72.10

Now it makes sense to choose the lighter tent.

So where did this $147 come from? Well, if
there's a heavy wind, then the generator turns
this into an opportunity. You'll still see $953 in
equipment damage, but that's offset by the
$1,100 in savings for portable batteries. That
puts you ahead by $147—but only if there's a
heavy wind!

✦ WHAT'S MY ❓ PURPOSE

Which risk response technique is being used in these situations? Match each technique to its scenario.

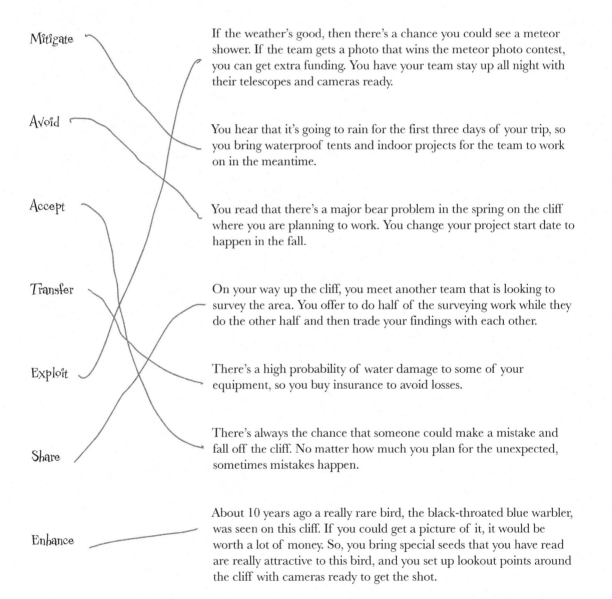

Mitigate

Avoid

Accept

Transfer

Exploit

Share

Enhance

If the weather's good, then there's a chance you could see a meteor shower. If the team gets a photo that wins the meteor photo contest, you can get extra funding. You have your team stay up all night with their telescopes and cameras ready.

You hear that it's going to rain for the first three days of your trip, so you bring waterproof tents and indoor projects for the team to work on in the meantime.

You read that there's a major bear problem in the spring on the cliff where you are planning to work. You change your project start date to happen in the fall.

On your way up the cliff, you meet another team that is looking to survey the area. You offer to do half of the surveying work while they do the other half and then trade your findings with each other.

There's a high probability of water damage to some of your equipment, so you buy insurance to avoid losses.

There's always the chance that someone could make a mistake and fall off the cliff. No matter how much you plan for the unexpected, sometimes mistakes happen.

About 10 years ago a really rare bird, the black-throated blue warbler, was seen on this cliff. If you could get a picture of it, it would be worth a lot of money. So, you bring special seeds that you have read are really attractive to this bird, and you set up lookout points around the cliff with cameras ready to get the shot.

Exam Questions

1. The project manager for a construction project discovers that the local city council may change the building code to allow adjoining properties to combine their sewage systems. She knows that a competitor is about to break ground in the adjacent lot and contacts him to discuss the possibility of having both projects save costs by building a sewage system for the two projects.

This is an example of which strategy?

- A. Mitigate
- B. Share
- C. Accept
- D. Exploit

2. Which of the following is NOT a risk response technique?

- A. Exploit
- B. Transfer
- C. Mitigate
- D. Confront

3. You are using an RBS to manage your risk categories. What process are you performing?

- A. Plan Risk Management
- B. Identify Risks
- C. Perform Qualitative Risk Analysis
- D. Perform Quantitative Risk Analysis

4. Which of the following is used to monitor low-priority risks?

- A. Triggers
- B. Watchlists
- C. Probability and Impact matrix
- D. Monte Carlo analysis

Exam Questions

5. You're managing a construction project. There's a 30% chance that weather will cause a three-day delay, costing $12,000. There's also a 20% chance that the price of your building materials will drop, which will save $5,000. What's the total EMV for both of these?

    A.   −$3,600

    B. $1,000

    C.   −$2,600

    D. $4,600

6. Joe is the project manager of a large software project. When it's time to identify risks on his project, he contacts a team of experts and has them all come up with a list and send it in anonymously. What technique is Joe using?

    A.   SWOT

    B.   Ishikawa diagramming

    C.   Delphi

    D.   Brainstorming

7. Susan is the project manager on a construction project. When she hears that her project has run into a snag due to weeks of bad weather on the job site, she says "No problem, we have insurance that covers cost overruns due to weather." What risk response strategy did she use?

    A.   Exploit

    B.   Transfer

    C.   Mitigate

    D.   Avoid

8. You're performing Identify Risks on a software project. Two of your team members have spent half of the meeting arguing about whether or not a particular risk is likely to happen on the project. You decide to table the discussion, but you're concerned that your team's motivation is at risk. The next item on the agenda is a discussion of a potential opportunity on the project in which you may be able to purchase a component for much less than it would cost to build.

Which of the following is NOT a valid way to respond to an opportunity?

    A.   Exploit

    B.   Transfer

    C.   Share

    D.   Enhance

Exam Questions

9. Risks that are caused by the response to another risk are called:

   A. Residual risks

   B. Secondary risks

   C. Cumulative risks

   D. Mitigated risks

10. What's the main output of the Risk Management processes?

    A. The Risk Management plan

    B. The risk breakdown structure

    C. Work performance information

    D. The risk register and project documents updates

11. Tom is a project manager for an accounting project. His company wants to streamline its payroll system. The project is intended to reduce errors in the accounts payable system and has a 70% chance of saving the company $200,000 over the next year. It has a 30% chance of costing the company $100,000.

    What's the project's EMV?

    A. $170,000

    B. $110,000

    C. $200,000

    D. $100,000

12. What's the difference between management reserves and contingency reserves?

    A. Management reserves are used to handle known unknowns, while contingency reserves are used to handle unknown unknowns.

    B. Management reserves are used to handle unknown unknowns, while contingency reserves are used to handle known unknowns.

    C. Management reserves are used to handle high-priority risks, while contingency reserves are used to handle low-priority risks.

    D. Management reserves are used to handle low-priority risks, while contingency reserves are used to handle high-priority risks.

Exam Questions

13. How often should a project manager discuss risks with the team?

    A.   At every milestone

    B.   Every day

    C.   Twice

    D.   At every status meeting

14. Which of the following should NOT be in the risk register?

    A.   Watchlists of low-priority risks

    B.   Relative ranking of project risks

    C.   Root causes of each risk

    D.   Probability and impact matrix

15. Which of the following is NOT true about Risk Management?

    A.   The project manager is the only person responsible for identifying risks

    B.   All known risks should be added to the risk register

    C.   Risks should be discussed at every team meeting

    D.   Risks should be analyzed for impact and priority

16. You're managing a project to remodel a kitchen. You find out from your supplier that there's a 50% chance that the model of oven that you planned to use may be discontinued, and you'll have to go with one that costs $650 more. What's the EMV of that risk?

    A. $650

    B.   –$650

    C. $325

    D.   –$325

17. Which risk analysis tool is used to model your risks by running simulations that calculate random outcomes and probabilities?

    A.   Monte Carlo analysis

    B.   Sensitivity analysis

    C.   EMV analysis

    D.   Delphi technique

Exam Questions

18. A construction project manager has a meeting with the team foreman, who tells him that there's a good chance that a general strike will delay the project. They brainstorm to try to find a way to handle it, but in the end decide that if there's a strike, there is no useful way to minimize the impact to the project. This is an example of which risk response strategy?

    A.   Mitigate

    B.   Avoid

    C.   Transfer

    D.   Accept

19. You're managing a project to fulfill a military contract. Your project team is assembled, and work has begun. Your government project officer informs you that a supplier that you depend on has lost the contract to supply a critical part. You consult your risk register and discover that you did not plan for this. What's the BEST way to handle this situation?

    A.   Consult the Probability and Impact matrix

    B.   Perform Quantitative and Perform Qualitative Risk Analysis

    C.   Recommend preventive actions

    D.   Look for a new supplier for the part

20. Which of the following BEST describes risk audits?

    A.   The project manager reviews each risk on the risk register with the team

    B.   A senior manager audits your work and decides whether you're doing a good job

    C.   An external auditor reviews the risk response strategies for each risk

    D.   An external auditor reviews the project work to make sure the team isn't introducing a new risk

Answers

Exam ~~Questions~~

1. Answer: B

Sharing is when a project manager figures out a way to use an opportunity to help not just her project but another project or person as well.

← It's OK to share an opportunity with a competitor—that's a win—win situation.

2. Answer: D

Confronting is a conflict resolution technique.

3. Answer: A

You use an RBS to figure out and organize your risk categories even before you start to identify them. Then you decompose the categories into individual risks as part of Identify Risks.

4. Answer: B

Your risk register should include watchlists of low-priority risks, and you should review those risks at every status meeting to make sure that none of them have occurred.

When you're calculating EMV, negative risks give you negative numbers.

5. Answer: C

The expected monetary value (or EMV) of the weather risk is the probability (30%) times the cost ($12,000), but don't forget that since it's a risk, that number should be negative. So its EMV is 30% x –$12,000 = –$3,600. The building materials opportunity has an EMV of 20% x $5,000 = $1,000. Add them up and you get –$3,600 + $1,000 = –$2,600.

Common sense would tell you that the answer is D, but brainstorming doesn't have to be anonymous. So, it's got to be Delphi.

6. Answer: C

Using the Delphi technique, experts supply their opinions of risks for your project anonymously so that they each get a chance to think about the project without influencing each other.

Make it Stick

EMV
Expected
Monetary
value

Negative
risks =
Negative
numbers

wash
the
dog

Answers

Exam ~~Questions~~

7. Answer: B

Wow, did you see that huge red herring?

Susan bought an insurance policy to cover cost overruns due to weather. She transferred the risk from her company to the insurance company.

8. Answer: B

You wouldn't want to transfer an opportunity to someone else! You always want to find a way to use that opportunity for the good of the project. That's why the response strategies for opportunities are all about figuring out ways to use the opportunity to improve your project (or another, in the case of sharing).

9. Answer: B

A secondary risk is a risk that could happen because of your response to another risk.

10. Answer: D

The processes of Risk Management are organized around creating the risk register, and updating it as part of project documents updates.

11. Answer: B

The key to this one is to remember that the money the project makes is positive, and the money it will cost is negative.

$200,000 x 0.70 = $140,000 savings, and $100,000 x 0.30 = -$30,000 expenses. Add them together and you get $110,000.

That's why it's useful to figure out the EMV for a risk—so you know how big your contingency reserve should be.

12. Answer: B

Contingency reserves are calculated during Perform Quantitative Risk Analysis based on the risks you've identified. You can think of a risk as a "known unknown"—an uncertain event that you know about, but which may not happen—and you can add contingency reserves to your budget in order to handle them. Management reserves are part of Cost Management—you use them to build a reserve into your budget for any unknown events that happen.

Answers

Exam ~~Questions~~

Answers

13. Answer: D

Risk monitoring and response is so important that you should go through your risk register at every status meeting!

14. Answer: D

The Probability and Impact matrix is a tool that you use to analyze risks. You might find it in your Project Management plan, but it's not included in the risk register.

15. Answer: A

It's really important that you get the entire team involved in the Identify Risks process. The more people who look for risks, the more likely it is that you'll find the ones that will actually occur on your project.

16. Answer: D

Even though this looks a little wordy, it's just another EMV question. The probability of the risk is 50%, and the cost is –$650, so multiply the two and you get –$325.

17. Answer: A

This is just the definition of Monte Carlo analysis. That's where you use a computer simulation to see what different random probability and impact values do to your project.

18. Answer: D

There are some risks that you just can't do anything about. When that happens, you have to accept them. But at least you can warn your stakeholders about the risk, so nobody is caught off guard.

Exam ~~Questions~~ *Answers*

19. Answer: D

You've got an unplanned event that's happened on your project. Is that a risk? No. It's a project problem, and you need to solve that problem. Your Probability and Impact matrix won't help, because the probability of this happening is 100%—it's already happened. No amount of risk planning will prevent or mitigate the risk. And there's no sense in trying to take preventive actions, because there's no way you can prevent it. So the best you can do is start looking for a new part supplier.

20. Answer: C

It's a good idea to bring in someone from outside of your project to review your risks. The auditor can make sure that each risk response is appropriate and really addresses the root causes of each risk.

> I SEE—THIS WASN'T A RISK AT ALL, IT WAS JUST A PROBLEM THAT CAME UP DURING THE PROJECT. I BET BETTER RISK PLANNING MIGHT HAVE HELPED THE TEAM PREPARE FOR THIS!

12 Procurement management

Getting some help

Some jobs are just too big for your company to do on its

own. Even when the job isn't too big, it may just be that you don't have the expertise or equipment to do it. When that happens, you need to use **Procurement Management** to find another company to **do the work for you**. If you find the **right seller**, choose the **right kind of relationship**, and make sure that the **goals of the contract are met**, you'll get the job done, and your project will be a success.

Victim of her own success

Kate's last project went really well. In fact, maybe a little too well. The company's customer base grew so much that now the IT department's technical support staff is overwhelmed. Customers who call up looking for technical support have to spend a long time on hold, and that's not good for the company.

Calling in the cavalry

WE'VE GOT A NEW PROJECT FOR YOU, KATE. WE FIGURE IT'S GOING TO TAKE ABOUT 18 MONTHS TO RAMP UP THE NEW TECH SUPPORT CALL CENTER. CAN YOU HANDLE IT?

Kate: No problem. The hard part will be figuring out how to manage the transition. Are we going to try to expand the team immediately, or call in a supplier to help us out?

Ben: Whoa, hold on there! Is going outside the company even an option?

Kate: Look, our tech support team is already at full capacity, and it'll take months to upgrade the facilities to handle more people...not to mention to hire and train the staff. We may be able to handle it ourselves, but there's a good chance that the easiest way to get the job done is to go outside our company to find a vendor to do the work.

Ben: But isn't it kind of risky thinking about working with another company? I mean, what if it goes out of business during our project? Or what if it costs too much?

Kate: Well, we'll need to make sure that we answer those questions. But this isn't the first time our company's brought on a contractor like this. The legal department has done this kind of thing before. I'll set up a meeting with somebody over there and see if they can help us out.

Ben: OK, you can follow up on that. But I'm still not sure about this.

Sometimes you need to hire an outside company to do some of your project work. That's called <u>procurement</u>, and the outside company is called the <u>seller</u>.

> WAIT A MINUTE. I'M NOT A LAWYER. WHY SHOULD I HAVE TO KNOW A BUNCH OF DETAILS ABOUT CONTRACTING?

You need to be involved because it's your project, and you're responsible for it.

One of the most common mistakes people make on the exam (and in real life) is to assume that if another company is selling products or services for your project and they don't deliver, it's not your problem. After all, you've got a contract with the company, right? So if they don't deliver, they won't get paid.

Well, it's not that simple. Yes, there are plenty of sellers who fail to deliver on their agreements. But for each seller that doesn't deliver, there's a frustrated project manager whose project ran into trouble because of it. That's why a lot of the Procurement Management tools and techniques are focused on selecting the *right* seller and communicating exactly what you'll need to the people doing the work.

Watch it!

The PMP exam is based on contracting laws and customs in the United States.

Are you used to working in a country that ISN'T the U.S.? Then you should be especially careful about these processes. You may be used to working with agreements in a way that isn't exactly the same as how they'll work on the exam questions. Luckily, the U.S. government publishes a lot of information on contracting at http://www.acquisition.gov/. Take a look at the site if you want a little more background.

Agreement Process Magnets

There are four Procurement Management processes. They're pretty easy to understand—you can probably guess which ones are which from their descriptions. Connect the description of each process with its name, and then try to guess which process group it's in.

Descriptions of each process	**Process names**	**Process groups**
Plan out what you'll purchase, and how and when you will need the contracts to be negotiated for your project.	....	....
Decide on the seller (or sellers) you are going to work with, and finalize and sign the contract.	....	....
Keep tabs on the contract. Make sure your company is getting what you paid for.	....	....
Confirm that the work was done right and that all obligations are fulfilled on both sides.	....	....

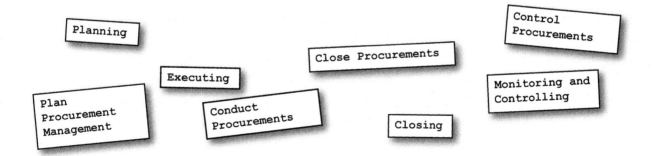

Planning

Control Procurements

Executing

Close Procurements

Plan Procurement Management

Conduct Procurements

Monitoring and Controlling

Closing

Agreement Process Magnets Solutions

There are four Procurement Management processes. They're pretty easy to understand—you can probably guess which ones are which from their descriptions. Connect the description of each process with its name, and then try to guess which process group it's in.

Descriptions of each process	Process names	Process groups

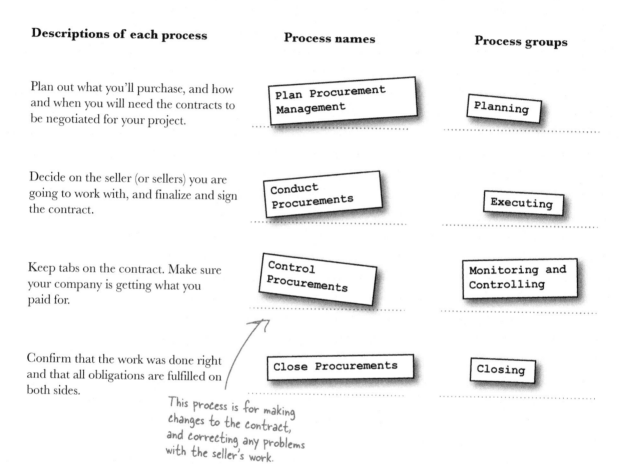

Plan out what you'll purchase, and how and when you will need the contracts to be negotiated for your project.

Plan Procurement Management

Planning

Decide on the seller (or sellers) you are going to work with, and finalize and sign the contract.

Conduct Procurements

Executing

Keep tabs on the contract. Make sure your company is getting what you paid for.

Control Procurements

Monitoring and Controlling

Confirm that the work was done right and that all obligations are fulfilled on both sides.

Close Procurements

Closing

This process is for making changes to the contract, and correcting any problems with the seller's work.

Ask the legal expert

> HI KATE. I'M STEVE FROM LEGAL. BEN SAID YOU NEEDED TO TALK TO ME—DO YOU HAVE A MINUTE?

Thanks for coming by, Steve. We're looking for a contractor to handle tech support while we bring on more people in our call center. How do we normally handle this stuff?

Here's how it usually works. I'll actually write the contract and do the negotiation. But before I do that, I'll need to sit down with you to understand what the contract has to accomplish.

Kate: So I'm not involved at all?

Steve: Oh, you're definitely involved. You need to help with the negotiations, because you're the only person who really understands what we're trying to accomplish with the contract.

Kate: OK, that makes sense. So when do we get started?

Steve: Well, not so fast. We need to be really sure that the way we pick our vendors is absolutely fair. We've got some company guidelines that you'll need to follow. And once we've got the contract signed and the work is under way, we'll need to meet to make sure the contract is really being followed. And if there's a problem and we need to negotiate a change to the contract, you'll need me to do it.

Kate: OK, I can handle that. So should I start working on something to send out to sellers?

Steve: Not quite. Before we even get started with all of that, are you sure we really need to contract this work?

BRAIN POWER

What should Kate do to figure out if it's really a good idea to contract the work?

Anatomy of an agreement

Procurement is pretty intuitive, and the **four Procurement Management processes** follow a really sensible order. First you plan what you need to contract; then you plan how you'll do it. Next, you send out your contract requirements to sellers. They bid for the chance to work with you. You pick the best one, and then you sign the contract with them. Once the work begins, you monitor it to make sure the contract is being followed. When the work is done, you close out the contract and fill out all the paperwork.

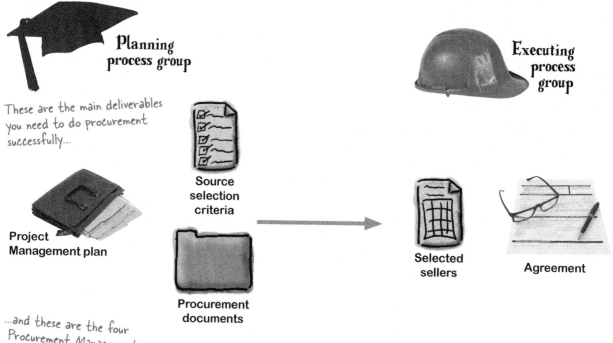

Planning process group

These are the main deliverables you need to do procurement successfully...

Source selection criteria

Project Management plan

Procurement documents

Executing process group

Selected sellers

Agreement

...and these are the four Procurement Management processes that you use to create those deliverables.

Plan Procurement Management Here's where you take a close look at your needs, to be sure that you really need to create a contract. You figure out what kinds of contracts make sense for your project, and you try to define all of the parts of your project that will be contracted out.

You'll need to plan out each individual contract for the project work and work out how you'll manage it. That means figuring out what metrics it will need to meet to be considered successful, how you'll pick a seller, and how you'll administer the contract once the work is happening.

Conduct Procurements

This process is all about getting the word out to potential agreement or contract partners about the project and how they can help you. You hold **bidder conferences** and find qualified sellers that can do the work.

Next, you evaluate all of the responses to your procurement documents and find the seller that suits your needs the best. When you find them, you sign the contract, and then the work can begin.

You can have several contracts for a single project

The first Procurement Management process is **Plan Procurement Management**. It's a familiar planning process, and you use it to plan out all of your procurement activities for the project. The other three processes are done for every contract. Here's an example. Say you're managing a construction project, and you've got one contract with an electrician and another one with a plumber. That means you'll go through those three processes two separate times, once for each contractor.

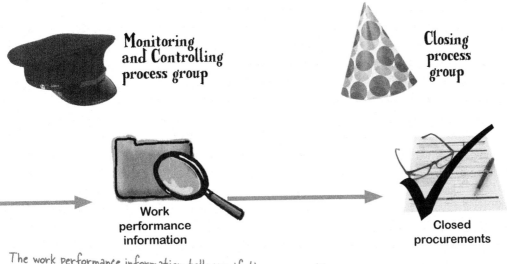

Monitoring
and Controlling
process group

Closing
process
group

Work
performance
information

Closed
procurements

The work performance information tells you if there are problems with your agreement—so Control Procurements is just like any other Monitoring and Controlling process.

Control Procurements

When the contract is under way, you stay on top of the work and make sure the contract is adhered to. You monitor what the contractor is producing and make sure everything is running smoothly. Occasionally, you'll need to make changes to the contract. Here's where you'll find and request those changes.

Close Procurements

When the work is done, you'll close your contract out. You'll make sure that the product that is produced meets the criteria for the contract, and that the contractor gets paid.

Start with a plan for the whole project

Planning process group

You need to think about all of the work that you will contract out for your project before you do anything else. The **Plan Procurement Management process** is all about figuring that out, and writing up a plan for how you'll do it.

Your cost estimates and your schedule play a big part in how you plan out your procurement.

Inputs

Past contracts can be a useful resource in setting up a new procurement.

Project Management plan

Organizational process assets

Activity cost estimates

Schedule

Tools

Your resource requirements and requirements document tell you a lot about the work you need to do and how this contract will fit into it.

Activity resource requirements

Requirements documentation

Make or buy analysis means figuring out whether or not you should be contracting the work or doing it yourself. It could also mean deciding whether to build your own solution to your problem or buy one that is already available. Most of the same factors that help you make every other major project decision will help you with this one. How much does it cost to build it versus buy it? How will this decision affect the scope of your project? How about your project schedule? Do you have time to do the work and still meet your commitments? As you plan out what you will and won't contract, you need to have thought through your reasoning pretty carefully.

There are some resources (like heavy equipment) that your company can buy, rent, or lease depending on the situation. You'll need to examine leasing versus buying costs and determine the best way to go forward.

You need to consider the level of expertise your company has with the product or work you're thinking of contracting, as well as the job and contracting markets you operate in.

Enterprise environmental factors

Contracting adds an extra dimension of risk to your project because your seller will have different management and policies. So managing risks is especially important!

Risk register

Stakeholder register

This plan will have:

- The planned delivery dates for the work or products you are contracting
- The company's standard documents you will use
- The contract types you plan to use, and any metrics that will be used to measure the contractor's performance
- Any constraints or assumptions you need to know about all of the contracts you plan to create for your project

Procurement Management plan

Expert judgment means asking someone who's made the same kind of decision before to help you look at all the information you have for your project and make the right decision. Experts can be really helpful in evaluating technology, or providing insight into how your work might be done in different sourcing scenarios.

Market research You will want to check out reviews of possible vendors to work with. Sometimes procurement teams will go to conferences or read published reports that evaluate vendors doing similar contracts to help make decisions.

Meetings help your team put their heads together and make sure they're covering all of the project's needs when setting up a procurement.

This is just a list of the work that will be contracted. This statement of work will be given to potential contracting partners later.

Procurement statement of work

After doing your make or buy analysis, you write down what you learned so that other people understand your rationale.

Make or buy decisions

You'll use this output to help you find the sellers that will do the work.

Procurement documents

And you'll use this one to help you figure out which seller you want to hire.

Outputs

Change requests

Project document updates

Source selection criteria

> I STILL DON'T BUY IT. WHY SHOULD I GO OUTSIDE OF MY COMPANY? WHY CAN'T I JUST HAVE MY TEAM DO ALL THE WORK?

Because sometimes it's not worth having your team do part of the job.

If your company needed to renovate your office, would you hire the carpenter, electrician, and builders? Would you buy the power tools, cement mixer, trucks, and ladders? Of course not. You'd hire a contractor to do the work, because it would cost too much to buy all that stuff for one job, and you wouldn't want to hire people just for the job and then fire them when it was done. Well, the same goes for a lot of jobs on your projects. You don't always want to have your company build everything. There are a lot of jobs where you want to hire a **seller**.

There are a lot of words for the company you're hiring: contractor, consultant, external company...but for the PMP exam, you'll typically see the term "seller."

It's natural to feel a little nervous about this contracting stuff.

Relax

A lot of project managers have only ever worked with teams inside their own companies. All this talk of contracts, lawyers, proposals, bids, and conferences can be intimidating if you've never seen it before. But don't worry. Managing a project with a contractor is really similar to managing one that uses your company's employees. There are just a few new tools and techniques that you need to learn...but they're not hard, and you'll definitely get the hang of them really quickly.

Make or Buy Magnets

Figure out whether or not Kate and Ben should contract out the tech support work by organizing these facts about the project into make or buy columns. The first few have been done for you.

This <u>really</u> is how a lot of people handle make or buy decisions—looking at all of the information you have for the project and using it to determine whether the facts line up under "Make" or "Buy."

Make

Buy

> The next big product release is six months away.

> We think the procurement process will take around three months, and ramping up staff in the call center will take eight months.

> Training the contractor's employees will be less valuable because we won't be able to use their knowledge when the contract is up.

> It might be hard to control the quality of the contractor's work.

> The contract team can have a staff trained and ready within a month from the signed contract.

> Our estimate is that it will cost around $30,000 per month to hire an additional 10 people and reduce wait time to 10 minutes per call. The cheapest contract for this is around $40,000 per month.

> The cost for equipment and training for a 10-person team is around $50,000. Contracts could be drawn up to cut that cost down a lot.

> Contracting companies who specialize in tech support have access to a lot of information and best practices that could make the project go more smoothly.

Make or Buy Magnets Solutions

Figure out whether or not Kate and Ben should contract out the tech support work by organizing these facts about the project into make or buy columns. The first few have been done for you.

Make

Buy

There's no way they are going to be able to support even more customers with a new product in six months if they don't have the staff then.

> It might be hard to control the quality of the contractor's work.

> Our estimate is that it will cost around $30,000 per month to hire an additional 10 people and reduce wait time to 10 minutes per call. The cheapest contract for this is around $40,000 per month.

> Training the contractor's employees will be less valuable because we won't be able to use their knowledge when the contract is up.

> The next big product release is six months away.

> We think the procurement process will take around three months and ramping up staff in the call center will take eight months.

> The cost for equipment and training for a 10-person team is around $50,000. Contracts could be drawn up to cut that cost down a lot.

> The contract team can have a staff trained and ready within a month from the signed contract.

> Contracting companies who specialize in tech support have access to a lot of information and best practices that could make the project go more smoothly.

Even though the staff costs will be higher with the contractor, not having to pay for equipment and training could offset the higher labor cost.

Sometimes contractors can bring their expertise from running lots of similar projects and make everything run more smoothly than it would if you do it yourself.

The decision is made

Doing make or buy analysis just means understanding the reasons for the contract and deciding whether or not to contract out the work. Once you've done that, if you still think contracting is an option, then you should have a good idea of what you need to get out of the contracting process.

STEVE, I'VE TAKEN A REALLY CLOSE LOOK AT THIS, AND I THINK IT MAKES SENSE TO FIND A SELLER TO HANDLE SOME OF OUR TECHNICAL SUPPORT FOR US. IF THEY TAKE OVER THE SECOND SHIFT FOR A FEW MONTHS, THAT'LL GIVE US TIME TO RAMP UP OUR OWN TEAM AND ADD PHONE CAPACITY.

SOUNDS GOOD. WHAT KIND OF CONTRACT ARE WE LOOKING AT?

Types of contractual agreements

If you want to see a really thorough overview of the different types of contracts, check out the U.S. Federal Acquisition Regulation website:

http://www.acquisition.gov/far/

It's a good idea to know a little bit about the most commonly used contract types. They can help you come up with a contract that will give both you and the seller the best chance of success.

Fixed price contracts

Some PMP exam questions might just refer to a contract type by its acronym (FP, CPFF, etc.).

Fixed price (FP) means that you are going to pay one amount regardless of how much it costs the contractor to do the work. A fixed-price contract only makes sense in cases where the scope is very well known. If there are any changes to the amount of work to be done, the seller doesn't get paid any more to do it.

Fixed price plus incentive fee (FPIF) means that you are going to pay a fixed price for the contract and give a bonus based on some performance goal. You might set up a contract where the team gets a $50,000 bonus if they manage to deliver an acceptable product before the contracted date. If the fixed-price contract does not include a fee, it's often referred to as a **firm fixed-price (FFP)** contract.

Cost-reimbursable contracts

Don't worry about trying to cram these into your head right now— you'll get a lot of practice with them throughout the chapter.

Costs plus fixed fee (CPFF) means what it says. You pay the seller back for the costs involved in doing the work, plus you agree to an amount that you will pay on top of that.

Costs plus award fee (CPAF) is similar to the CPFF contract, except that instead of paying a fee on top of the costs, you agree to pay a fee based on the buyer's evaluation of the seller's performance.

Costs plus incentive fee (CPIF) means you'll reimburse costs on the project and pay a fee if some performance goals are met. Kate could set up her project using this contract type by suggesting that the team will get a $50,000 bonus if they keep the average wait time for the calls down to seven minutes per customer for over a month. If she were on a CPIF contract, she would pay the team their costs for doing the work, and also the $50,000 bonus when they met that goal.

Time and materials

A lot of people say that the T&M contract is a lot like a combination of a cost-plus and fixed-price contract, because you pay a fixed price per hour for labor, but on top of that you pay for costs like in a cost-plus contract.

Time and materials (T&M) is used in labor contracts. It means that you will pay a rate for each of the people working on your project plus their materials costs. The "time" part means that the buyer pays a fixed rate for labor—usually a certain number of dollars per hour. And the "materials" part means that the buyer also pays for materials, equipment, office space, administrative overhead costs, and anything else that has to be paid for. The seller typically purchases those things and bills the buyer for them. This is a really good contract to use if you don't know exactly how long your contract will last, because it protects both the buyer and seller.

Even if your project has several contracts, they don't all have to be the same type. That's why you need to administer each one separately.

Sharpen your pencil

This is a tough one—take your time and <u>think</u> about each kind of contract.

There are advantages and disadvantages to every kind of contract. Different kinds of contracts carry different risks to both the buyer and seller. Can you think of some of them?

Here's a hint: FP contracts don't have much risk for the buyer.

Firm fixed price (FFP)

Risks to the buyer ....

Risks to the seller ....

Fixed price plus incentive fee (FPIF)

Risks to the buyer ....

Risks to the seller ....

Cost plus fixed fee (CPFF)

Risks to the buyer ....

Risks to the seller ....

CPAF contracts are really risky for the seller, not the buyer. Can you figure out why?

Cost plus award fee (CPAF)

Risks to the buyer ....

Risks to the seller ....

Cost plus incentive fee (CPIF)

Risks to the buyer ....

Risks to the seller ....

Time and materials (T&M)

Risks to the buyer ....

Risks to the seller ....

Sharpen your pencil
Solution

There are advantages and disadvantages to every kind of contract. Different kinds of contracts carry different risks to both the buyer and seller. Can you think of some of them?

There are a lot of right answers—even if yours aren't here, it doesn't mean that they're wrong.

A fixed-price contract has a lot more risk for the seller than the buyer.

Firm fixed price (FFP)

Risks to the buyer — **The only risk is if the seller doesn't deliver because of costs.**

Risks to the seller — **Unexpected costs could be bigger than the contract itself.**

Fixed price plus incentive fee (FPIF)

Risks to the buyer — **There's still not much risk to the buyer in fixed-price contracts.**

Risks to the seller — **The seller still has the same risks as FP, but may make more.**

CPFF contracts have risks for both the buyer and the seller.

Cost plus fixed fee (CPFF)

Risks to the buyer — **If the costs are too high, the buyer will have to pay a lot more.**

Risks to the seller — **A fixed fee on top of costs might not be worth it for the seller.**

CPAF contracts are risky for the sellers, because the buyer subjectively determines their performance, and can decide they didn't perform well enough.

Cost plus award fee (CPAF)

Risks to the buyer — **There aren't many risks to the buyer.**

Risks to the seller — **If the buyer determines the seller underperformed, he can withhold the fee.**

Incentive fees are a really good way to reduce the risk to the buyer on a cost-plus contract.

Cost plus incentive fee (CPIF)

Risks to the buyer — **There's still a risk of cost overruns, but it's not as bad.**

Risks to the seller — **The incentive fee isn't guaranteed, so it might not be paid.**

A lot of T&M contracts include a "cost not to exceed" clause to make sure this doesn't happen. If the contract doesn't have this, it can get really risky for the buyer!

Time and materials (T&M)

Risks to the buyer — **If costs are too high, the contract could get expensive.**

Risks to the seller — **The contract might not cover high overhead costs.**

Take a minute and try to figure out why the T&M contract is a really good choice if you don't know how long the job will last.

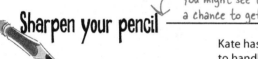

More about contracts

There are just a few more things you need to know about any contract to do procurement work.

Every contract needs to outline the work to be done and the payment for that work.

- You might see an exam question that mentions "consideration"—that's just another word for the payment.

- Remember in Risk Management how you used insurance to transfer risk to another company? You did that using a special kind of contract called an insurance policy.

- You might get a question that asks about ***force majeure***. This is a kind of clause that you'll see in a contract. It says that if something like a war, riot, or natural disaster happens, you're excused from the terms of the contract.

Always pay attention to the point of total assumption.

- The **point of total assumption** is the point at which the seller assumes the costs. In a fixed-price contract, this is the point where the costs have gotten so large that the seller basically runs out of money from the contract and has to start paying the costs.

You should always make sure both the buyer and seller are satisfied.

- When you negotiate a contract, you should make sure that the buyer and the seller **both** feel comfortable with the terms of the contract. You don't want the people at the seller's company to feel like they got a raw deal—after all, you're depending on them to do good work for your project.

Sharpen your pencil

You might see this kind of question about whether to make or buy. Here's a chance to get a little more practice with making contract decisions.

Kate has **18 months** to build up the capacity her company needs to handle all the technical support calls. See if you can figure out whether it's a better deal for Kate to make or buy.

1. If they handle the extra work within the company instead of finding a seller, it will cost an extra $35,000 in overtime and $11,000 in training costs in total, on top of the $4,400 per person per month for the five-person team needed to do the extra support work. What's the total cost of keeping the work within the company?

2. Kate and Ben talked to a few companies and estimate that it will cost $20,000 per month to hire another company to do the work, but they'll also need to spend $44,000 in setup costs. What will contracting the work cost?

3. So does it make more sense to make or buy? Why?

Answers on page 650.

Figure out how you'll sort out potential sellers

The two big outputs of Plan Procurement are the **procurement documents** and **source selection criteria.** The procurement documents are what you'll use to find potential sellers who want your business. The **source selection criteria** are what you'll use to figure out which sellers you want to use.

**Procurement
documents**

A big part of Procurement Management is making sure that both the buyers and sellers are treated fairly. Writing out source selection criteria beforehand is a good way to make sure each seller gets a fair shake.

**Source
selection
criteria**

You'd be amazed at how many sellers respond to bids that they have no business responding to. You definitely need to make sure the seller has the skill and capacity to do the work you need.

There are a bunch of different documents you might want to send to sellers who want to bid on your work.

You'll usually include the **procurement statement of work (SOW)** so that sellers know exactly what work is involved.

An **invitation for bid (IFB)** is a document that tells sellers that you want them to submit proposals. You'll also hear of people using **requests for information (RFI)** and **requests for proposals (RFP)**.

There's another kind of invitation—an **invitation for quote (IFQ)**. This is a way to tell sellers that you want them to give you a quote on a fixed-price contract to do the work.

A **purchase order** is something you'll send out to a seller who you know that you want to work with. It's an agreement to pay for certain goods or services.

In some cases you'll want to allow for more flexibility in your contract. If you're hiring a seller to build something for you that you've never built before, you'll often encourage them to help you set the scope instead of locking it down.

Decide in advance on how you want to select the sellers.

There are a lot of ways you can select a potential seller. Figuring out if a seller is appropriate for your work is something that takes a lot of talking and thinking—and there's no single, one-size-fits-all way of selecting sellers. But there are some things that you should definitely look for in any seller:

- Can the seller actually do the work you need done?

- How much will the seller charge?

- Can the seller cover any costs and expenses necessary to do the job?

- Are there subcontractors involved that you need to know about?

- Does the seller really understand everything in the SOW and contract?

- Is the seller's project management capability up to the task?

You always put together procurement documents and source selection criteria before you start talking to individual sellers who want your business.

Contract Magnets

Which of the magnets are part of the procurement documents, and which of them are part of the source selection criteria?

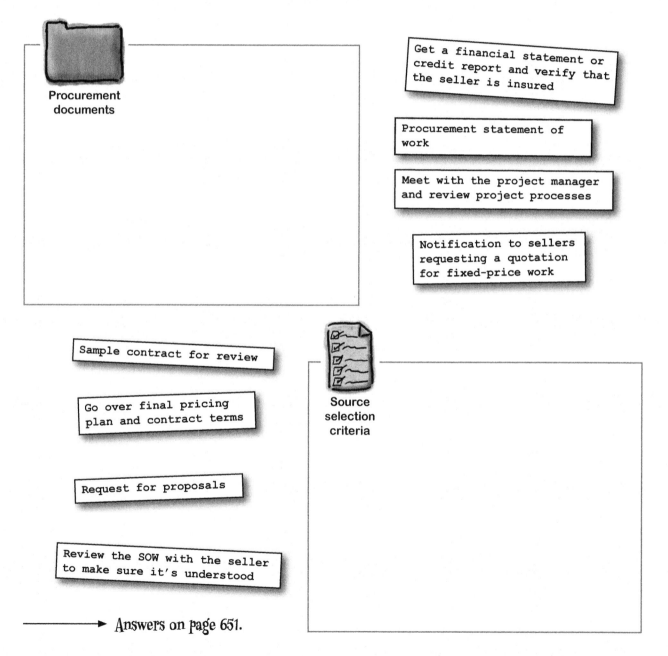

Procurement documents

> Get a financial statement or credit report and verify that the seller is insured

> Procurement statement of work

> Meet with the project manager and review project processes

> Notification to sellers requesting a quotation for fixed-price work

> Sample contract for review

> Go over final pricing plan and contract terms

> Request for proposals

> Review the SOW with the seller to make sure it's understood

Source selection criteria

⟶ Answers on page 651.

WHAT'S MY PURPOSE

Here are some of the proposals that Kate is evaluating. Match each one up to the contract type that's most important for the proposal.

The buyer will pay for the cost of phone service, rent on the facilities, and employees, plus an additional $2,500 per month.

Fixed price

The buyer will pay the seller a total of $285,000 for 18 months of technical support services.

Fixed-price incentive fee

The buyer will pay for the cost of phone service and rent on the facilities, plus $4,500 per month for employees' time. Costs will not exceed $14,500 per month.

Cost plus fixed fee

The buyer will pay for the cost of phone service, rent on the facilities, and employees. An additional $2,750 will be awarded each month that the seller provides an average of 10 issues resolved per person per day and an average wait time of under 3 minutes.

Cost plus incentive fee

The buyer will pay for the cost of phone service, rent on the facilities, and employees. An additional $5,000 will be paid for exceptional performance.

Cost plus award fee

The buyer will pay the seller a total of $285,000 for 18 months of technical support services. An additional $2,750 will be awarded each month that the seller provides an average of 10 issues resolved per person per day and an average wait time of under 3 minutes.

Time and materials

⟶ **Answers on page 638.**

Get in touch with potential sellers

The next step in procurement is pretty straightforward. You use the **Conduct Procurements process** to, well, get the word out to sellers and see what kind of responses you get. Once you narrow down your list of sellers to a few who look like they might be good candidates, you evaluate all of their responses against your source selection criteria and choose the vendor you're going to work with. All that's left to do after that is to get it all on paper...and then you award the contract!

Use outputs from the Plan Procurement Management process to find the right seller

When you perform the Conduct Procurements process, you'll start with some of the outputs you created in Plan Procurement Management. Here's how you'll use them:

Make or buy decisions

The make or buy decisions you made will come in handy because they'll tell you what you need to find a contractor to help out with and what you'll do yourself.

Source selection criteria

Use the source selection criteria to evaluate the sellers that respond to you. By evaluating all of your sellers using the same criteria, you'll be sure that you evaluate everyone fairly and find the right seller for your company.

Procurement documents

Procurement documents will have all of the information that you'll actually give to potential sellers to help them bid on your contract. Two of the most commonly used procurement documents are the RFI and the RFP.

Procurement statement of work

The procurement statement of work is where you write out all of the work that needs to be done by a contractor. It tells you the scope of the work that you're going to contract to another company.

RFI: Request for information documents are sent to potential sellers to ask for information about their capability to do the work.

RFP: Request for proposal is when you give a seller the opportunity to examine your procurement documents and write up a proposal of how they'd do the work.

Pick a partner

You've figured out what services you want to procure, and you've gone out and found a list of potential sellers. Now it's time to choose one of them to do the project work—and that's exactly what you do in **Conduct Procurements**.

Inputs

Procurement documents

Project documents

Source selection criteria

Seller proposals

Procurement statement of work

Make or buy decisions

Organizational process assets

Procurement Management plan

Tools

Analytical techniques

You'll need to determine whether or not the vendor is capable of completing the project in the required time frame and understand how well it can deliver to your budget. If you check out its track record on past projects, you'll have a better sense of how it'll perform.

Advertising

Sometimes the best way to get in touch with sellers is to place an ad. Also, sometimes you are **required** to take out an ad (like for some government-funded projects), in order to give all sellers full notice.

Independent estimates

A lot of the time, you don't have the expertise in your company to figure out whether or not a seller is quoting you a fair price. That's why you'll often turn to a third party to come up with an estimate of what the work should cost.

Bidder conferences

It's really important that you make sure all of the bidders can compete in a fair, unbiased way. And the best way to do that is to get them all in a room together, so that they can ask questions about your contract. That way, you don't give any one seller an advantage by providing inside information that the other sellers don't have access to.

Expert judgment

Here's another case where it's really good to bring in someone from outside your project to help evaluate each proposal. You should bring in someone with a lot of specific expertise in the work being done to make sure the seller is up to the job.

The whole point of the Conduct Procurements process is, well, to conduct procurements...and here they are. Along with the contract, this is the most important output of the process.

Selected sellers

Finally! Everyone's signed on the dotted line, and you've got your contract.

Agreements

Procurement negotiation

When you send out a package of procurement documents to a potential seller, it usually has some information about the contract you want to sign: the type of contract, some of the terms, maybe some rough calculations and estimates of the total costs, and other numbers. But not all sellers will want to sign that particular contract, even ones who you'll eventually want to work with. That's why you need to negotiate the terms of the contract.

This is one of those times where your company's lawyers will probably do most of the talking—but that doesn't mean you're not a critical part of the process. Your job is to provide the expertise and in-depth understanding. After all, you're the one who understands your project and your project's needs better than anyone else.

Proposal evaluation techniques

You're going to have to work closely with the seller to figure out if his proposal really is appropriate for the work. You need to be very careful before you choose someone to do the work. That's what this tool is for—it's a kind of "catch-all" that's there to remind you that there's no single way to evaluate a proposal. You need to look at the whole picture—the seller, your needs, and the job.

In the course of selecting a source, you might find changes that need to be made to your requirements or other documents.

Outputs

Project documents updates

If your contract makes some of the seller's resources available to you, you'll need this in order to update your Staffing Management plan.

Resource calendars

Any time you do negotiations, you usually end up making some adjustments to your plans—so you'll need to use these for change control.

Change requests

Project Management plan updates

BRAIN POWER

When Kate selects a seller, she'll need to help her company's legal team negotiate the terms of the contract. Which type of contract do you think is right for Kate's project?

Sharpen your pencil

Kate is putting out an RFP to find a seller to provide technical support for her company. Can you figure out which Conduct Procurements tool she's using?

1. Kate works with her company's seller evaluation committee, which follows a documented, formal evaluation review process to determine which seller should be selected for the contract.

....

2. Kate contacts an IT trade journal and places a classified ad to try to find sellers.

....

3. The CEO's brother-in-law runs a company that's bidding on the contract. Kate needs to make sure he gets fair—not preferential—treatment. She doesn't want to give him an unfair advantage, but she also doesn't want to exclude him from the bidding process. So she gathers representatives from all sellers into a room where they can ask questions about the contract out in the open and hear the responses to each question.

....

4. Kate's company takes part in an equal-opportunity program in which seller companies owned by minorities must be given notice of any RFPs. She uses a website approved by the program to find seller companies with performance data from similar projects.

....

→ Answers on page 650.

there are no Dumb Questions

Q: Do I always need to hold a bidder conference whenever I do procurement?

A: No, you don't always need a bidder conference. Sometimes your company has a preferred supplier who you always deal with, so you don't have to advertise for sellers. And sometimes there's a **sole source** for a particular service or part—there may only be one company that provides it. In that case, advertising and bidder conferences would be pointless.

The bidder conference has two goals. The first is to make sure that you answer all of the questions from potential sellers. But the other is to make sure that all potential sellers are treated equally and have access to the same information.

Q: I'm still not clear on why I'd want to use a cost-plus contract.

A: One of the best reasons to use a cost-plus contract is to make sure that the seller you're working with doesn't end up getting a raw deal. A fixed-price contract can be pretty risky for a seller. When the seller uses a cost-plus contract—like a cost plus incentive fee, or cost plus fixed fee—it means that there's a built-in guarantee that the seller won't have to swallow cost overruns. If you're reasonably certain that the costs can be contained, or if you set up a good incentive system, then a cost plus contract can be a really good one for making sure that both the buyer and seller are treated fairly.

Q: Why all this talk about treating the seller fairly? I'm trying to get the best deal I can. Doesn't that mean I should try to get as many concessions from sellers as possible?

A: One of the most important parts of procurement is that both the buyer and seller should feel like they're getting a good deal. You should never expect a seller to have to take on a bad contract. After all, you're depending on the seller to deliver a necessary piece of your project. That's why the goal in any procurement should always be for the buyer and seller to both feel like they were treated fairly.

Q: How do I use organizational process assets to find a seller?

A: When you're conducting procurements, you need to actually find sellers to do the work. Many companies have a list of sellers who they consider qualified to work for them because of past project performance. You'll usually find a list like this on file, and that's what the **qualified seller list** is.

Sharpen your pencil

Kate needs to use most of the tools and techniques in Conduct Procurements. Which technique is Kate using in each of these scenarios?

1. This is the first time that Kate's company has contracted out technical support services, so she hires a consultant to help her and the legal team estimate a fair price for the contract.

....

2. Kate sets up criteria for each seller. Before sellers can submit a bid, they must show that they have handled technical support contracts before and have facilities that can handle over 150 simultaneous calls.

....

3. The CIO and the director of the IT department at Kate's company spent a lot of time setting up the company's existing technical support department, so Kate meets with them to get their technical opinions.

....

4. Kate sent a notification out to all potential sellers who responded to her RFP informing them that she was calling a meeting with all of them. She made sure that all prospective sellers had a clear understanding of the work that needed to be done, and she answered all questions from each seller out in the open.

....

5. Kate reviews all of the sellers' past performance on similar projects to determine whether or not there might be risks if she were to award the contract to any of them.

....

6. Kate and her company's legal team sit down with the sellers and work out the terms of the contract. There's a lot of back and forth, but they settle on an agreement that everyone is comfortable with.

....

WHAT'S MY PURPOSE

Here are some of the proposals that Kate is evaluating. Match each one up to the contract type that's most important for the proposal.

The buyer will pay for the cost of phone service, rent on the facilities, and employees, plus an additional $2,500 per month.

The contract lays out the costs, and then adds a dollar amount fee on top of that. That's a fixed fee, so it's a CPFF contract.

Fixed price

The buyer will pay the seller a total of $285,000 for 18 months of technical support services.

Since a preset price will be paid, this is a fixed price (or lump sum) contract.

Fixed-price incentive fee

The buyer will pay for the cost of phone service and rent on the facilities, plus $4,500 per month for employees' time. Costs will not exceed $14,500 per month.

A lot of T&M contracts will have a "not to exceed" clause to limit risk for the buyer.

Cost plus fixed fee

The buyer will pay for the cost of phone service, rent on the facilities, and employees. An additional $2,750 will be awarded each month that the seller provides an average of 10 issues resolved per person per day and an average wait time of under 3 minutes.

Notice how the incentive fee was tied to specific quality measurements. That's a great way to motivate the seller to do a good job.

Cost plus incentive fee

The buyer will pay for the cost of phone service, rent on the facilities, and employees. An additional $5,000 will be paid for exceptional performance.

This is the same agreement from the fixed-price contract, but it's got the incentive fee from the CPIF contract. So it's fixed-price incentive fee.

Cost plus award fee

The buyer will pay the seller a total of $285,000 for 18 months of technical support services. An additional $2,750 will be awarded each month that the seller provides an average of 10 issues resolved per person per day and an average wait time of under 3 minutes.

Time and materials

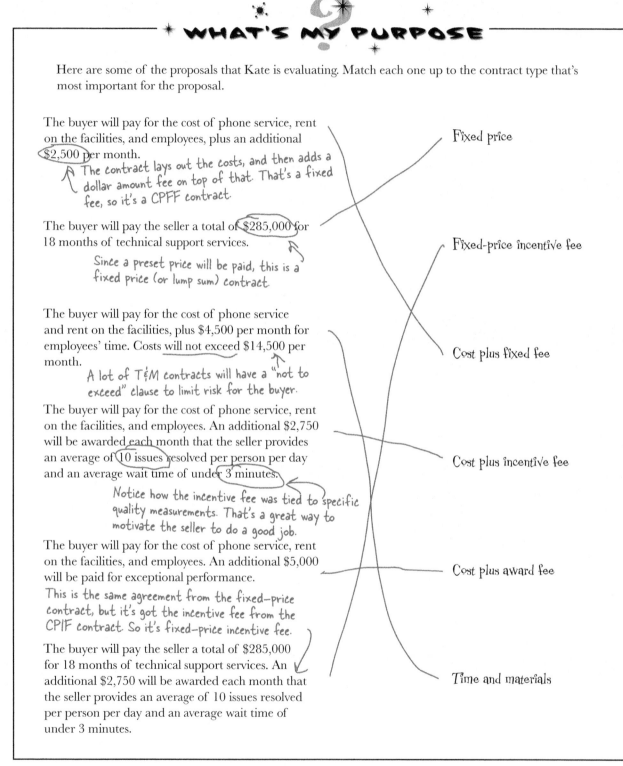

Sharpen your pencil
Solution

Kate needs to use most of the tools and techniques in Conduct Procurements. Which technique is Kate using in each of these scenarios?

1. This is the first time that Kate's company has contracted out technical support services, so she hires a consultant to help her and the legal team estimate a fair price for the contract.

Independent estimates

It's often hard to come up with a fair price yourself, because the skills you need to do that are usually the same skills that you need to do the job. Sometimes you don't have those skills in your company, which could be why you looked for a seller in the first place.

2. Kate sets up criteria for each seller. Before sellers can submit a bid, they must show that they have handled technical support contracts before and have facilities that can handle over 150 simultaneous calls.

Proposal evaluation techniques

When you screen out potential sellers, it makes the job of selecting a seller a lot easier.

3. The CIO and the director of the IT department at Kate's company spent a lot of time setting up the company's existing technical support department, so Kate meets with them to get their technical opinions.

Expert judgment

You've seen a whole lot of other processes that have this same technique. Expert judgment always means getting an opinion from someone outside your project.

4. Kate sent a notification out to all potential sellers who responded to her RFP informing them that she was calling a meeting with all of them. She made sure that all prospective sellers had a clear understanding of the work that needed to be done, and she answered all questions from each seller out in the open.

Bidder conferences

5. Kate reviews all of the sellers' past performance on similar projects to determine whether or not there might be risks if she were to award the contract to each of them.

Analytical techniques

6. Kate and her company's legal team sit down with the sellers and work out the terms of the contract. There's a lot of back and forth, but they settle on an agreement that everyone is comfortable with.

Procurement negotiations

Project managers don't usually do the negotiation themselves. They'll get involved and provide expertise and knowledge, but usually rely on a lawyer or legal department to work out the actual terms of the contract.

Two months later...

Kate's procurement project had been going really well…or so she thought. But it turns out there's a problem.

> KATE, THE CEO CALLED ME AT 3 A.M. THERE'S A JANITOR'S STRIKE AT THE SELLER'S TECHNICAL SUPPORT OFFICE, AND THAT'S CAUSING ALL SORTS OF HAVOC. NOW OUR WAIT TIMES ARE EVEN LONGER THAN THEY WERE THREE MONTHS AGO. WHAT ARE YOU GOING TO DO ABOUT THIS?

Kate never even thought to ask about the janitors union when the legal team was negotiating the contract.

Watch it!

Keep an eye out for questions that ask about unions, even when they don't have to do with contracts or Procurement Management.

When you work with a union, even if it's through a seller, then the union contract (also called a **collective bargaining agreement**) *can have an impact on your project. That means you need to consider the union itself a stakeholder, and when you do your planning you need to make sure any union rules and agreements are considered as constraints.*

BRAIN POWER

What could Kate have done to prevent this problem? Could she have detected it sooner? What should she do now?

Monitoring
& Controlling
process group

Keep an eye on the contract

You wouldn't just start off a project and then assume everything would go perfectly, would you? Well, you can't do that with a contract either. That's why you use the **Control Procurements process**.

The idea behind the Control Procurements process is that staying on top of the work that the seller is doing is more difficult than working with your own project. That's because when you hire a seller to take over part of your project, the team who's doing the work doesn't report to you. That's why **the first three inputs are especially important**. The approved change requests are the way that you change the terms of the contract if something goes wrong, and the work performance data and work performance reports are how the seller tells you how the project is going.

The tools and techniques for the Control Procurements process are on the next page.

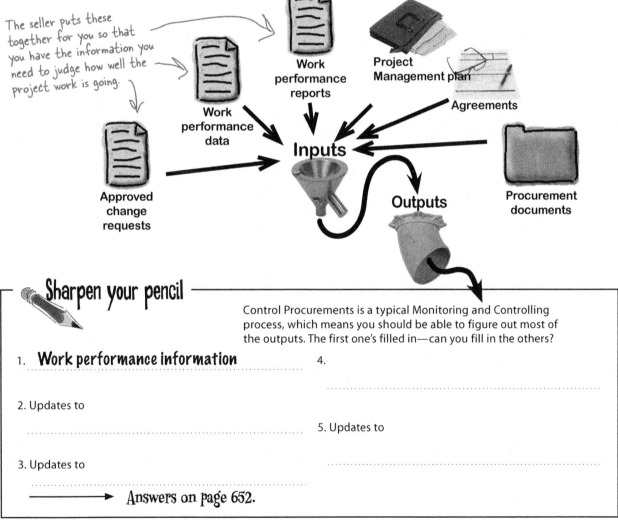

The seller puts these together for you so that you have the information you need to judge how well the project work is going.

Work performance reports

Project Management plan

Work performance data

Agreements

Inputs

Approved change requests

Outputs

Procurement documents

Sharpen your pencil

Control Procurements is a typical Monitoring and Controlling process, which means you should be able to figure out most of the outputs. The first one's filled in—can you fill in the others?

1. **Work performance information** ....

4. ....

2. Updates to ....

5. Updates to ....

3. Updates to ....

→ Answers on page 652.

Stay on top of the seller

Tools

The **tools and techniques for Control Procurements** are all there to help you work with the seller. Some of them help you look for any potential problems with the seller and make changes to correct them. Others help you with the day-to-day administration work that you need to do in order to keep your project running.

Tools and techniques to keep your project running

There are seven tools and techniques for the Control Procurements process.

Payment systems

Your partner won't be very happy if you don't pay. The payment system is how your company pays its sellers. It's usually established by an Accounting or Accounts Receivable department.

Records management system

There are a lot of records produced by a typical contract: invoices, receipts, communications, memos, emails, instructions, clarifications, etc. You'll need to put a system in place to manage them.

Tools and techniques to find and fix problems

You can conduct a performance review to get the data you need for your performance report.

Claims administration

When there's a dispute between a buyer and a seller, that's called a **claim**. Most contracts have some language that explains exactly how claims should be resolved—and since it's in the contract, it's legally binding, and both the buyer and seller need to follow it.

Performance reporting

The easiest way for you to keep track of the contract work being done is to write up performance reports. These are exactly like the performance reports that you saw earlier in the book—you'll use them to monitor the project work and report on the progress to your company's management.

Procurement performance reviews

Most contracts lay out certain standards for how well the seller should do the job. Is the seller doing all the work that was agreed to? Is the work being done on time? The buyer has the right to make sure this is happening, and the way to do this is to go over the performance of the seller's team.

Inspections and audits

This tool is how the buyer makes sure that the product that the seller produces is up to snuff. This is where you'll check up on the actual product or service that the project is producing to make sure that it meets your needs and the terms of the contract.

Contract change control system

This is just like all of the other change control systems that you've seen already. It's a set of procedures that are set up to handle changes in the contract. You might have a different one for every contract in your project.

Buyer-conducted performance reviews let buyers check all of the work that the sellers are doing.

Exercise

Which of the tools and techniques from Control Procurements should Kate use for each situation?

1. An important client calls technical support, but ends up spending two hours waiting on hold. Kate doesn't find out until the seller calls her directly. She needs a better way to manage information about how the seller is performing.

....

2. The CEO's mother calls technical support, but spends two hours waiting for him to answer. Kate needs to make sure the seller is delivering the quality it promised.

....

3. Kate gets a call from the Accounting department about a duplicate invoice that was accidentally paid twice.

....

4. According to the statement of work, the seller is supposed to have weekly training sessions with technical support staff, but Kate isn't sure they're being conducted as often as they should be.

....

5. A manager at the seller says that it's not responsible for training sessions, but Kate thinks it is.

....

Exercise Solution

Which of the tools and techniques from Control Procurements should Kate use for each situation?

1. An important client calls technical support, but ends up spending two hours waiting on hold. Kate doesn't find out until the seller calls her directly. She needs a better way to manage information about how the seller is performing.

   Records management system

   A records management system can help Kate by giving her a place to store all the reports from the seller.

2. The CEO's mother calls technical support, but spends two hours waiting for them to answer. Kate needs to make sure the seller is delivering the quality it promised.

   Inspections and audits

   You use inspections and audits when you want to review the quality of the product or service being produced.

3. Kate gets a call from the Accounting department about a duplicate invoice that was accidentally paid twice.

   Payment system

4. According to the statement of work, the seller is supposed to have weekly training sessions with technical support staff, but Kate isn't sure they're being conducted as often as they should be.

   Procurement performance review

   If you need to check whether work is being done well, you can use a procurement performance review.

5. A manager at the seller says that it's not responsible for training sessions, but Kate thinks it is.

   Claims administration

there are no Dumb Questions

Q: Should I only care about unions when I'm working with contracts?

A: Unions come up in procurement and contracts whenever a seller has an existing contract with a union. That contract is called a **collective bargaining agreement**, and if that agreement impacts the work that the seller is going to do for you, then you need to make sure that your legal department considers it when they work out the terms of the contract.

But unions are also important when you're doing Human Resource Management. If your company has a collective bargaining agreement with a union, then you need to consider the terms of that contract as **external constraints** to your project plan. Here's an example: let's say you're managing a construction project, and your workers are all union members. Then you need to make sure that you consider any overtime rules and other restrictions on resource availability when you put together your team, your budget, and your plan.

Whenever you see "inspection" or "audit," it means that you're looking at the products that the seller delivered to see if they meet your standards.

Q: Once a contract is signed, does that mean it's never allowed to change?

A: No. This confuses some people, because when you sign a contract, it's legally binding—which means you must abide by the terms of the contract. But that doesn't mean those terms can't change. If both the buyer and the seller agree to make a change to the contract, then they have every right to do so. That's why you have a contract change control system—so you can make sure these changes are made properly.

But you can't always assume that you have the ability to change a contract that you're not happy with. Once your company has agreed to a contract, then you're absolutely required to meet its terms and complete your side of it. If you want to make a change to it, you need to negotiate that change, and it's possible that the seller won't agree to it—just like you have every right to refuse an unreasonable change that the seller requests.

Q: Does the type of contract make a difference in how changes are handled?

A: No, it doesn't. While the type of contract definitely affects a lot of things, changes are always handled the same way. You always use the contract change control system to handle the changes.

That's why the contract change control system is so important. It tells you the exact rules that you need to go through in order to make a change to a contract. No contract is perfect, and most of the time there are little tweaks that both the buyer and seller want to make. This gives them the tools they need to make only the changes that they need, without either team agreeing to a change that they don't want included in the contract.

Q: I still don't get the difference between a performance review and an audit.

A: The difference is that performance reviews are about the **work**, while inspections and audits are about the **deliverables and products**.

You'll use a performance review when you want to make sure that the team at the seller is doing every activity that they should. For example, if you have a contract that requires the seller to perform certain quality control or project management tasks, you might conduct a performance review where you observe the team and verify that they do those tasks. On the other hand, if you want to make sure that the products that the team is producing meet your requirements and standards, you'll send out an auditor to inspect the products that the seller is making to verify that they meet the requirements.

Q: So do project managers usually get involved in contract negotiations?

A: Project managers don't usually do the negotiating themselves, but they do often get involved in contract negotiations. Remember, nobody knows more about the project than the project manager—you know what work needs to be done, what requirements the product must meet, and what kind of budget you need to stay within. So even though a lawyer or legal department will do the actual negotiation, they won't know if the seller is capable of doing the job without the project manager's help.

Close the contract when the work is done

When the seller's work is done, it's time to close the contract, and that's when you use the **Close Procurements** process. Even if your contract ends disastrously (or in court), you still need to close out the contract so that you can make sure all of your company's responsibilities are taken care of—and that you learn from the experience.

There's only one other process in the Closing group. Take a minute and flip back to Chapter 4 to refresh your memory.

Closing process group

Inputs

Project Management plan

Procurement documents

The Close Procurements process is one of the two processes in the Closing process group, and it's the very last Procurement Management process that you perform.

Tools

Procurement audits

Once you've closed out the contract, you go over everything that happened on the project to figure out the lessons learned and look for anything that went right or wrong.

Records management system

When you were working with the legal team to put the contract together, you looked at information about past contracts. So where did it come from? Past project managers stored their contracts and other documents in the records management system.

Procurement negotiations

You need to make sure that all of the terms of the contract have been met and there are no outstanding claims on it. If the buyer or the seller have outstanding claims from the relationship, they need to get resolved, sometimes through legal arbitration or, in the worst-case scenario, in court.

Outputs

The way you close out a contract is by giving notice—a formal written one—that the contract is complete... and instructions for doing that should be part of the contract terms.

Closed procurements

Organizational process asset updates

Any lessons you learned from the procurement audits should be added here, along with any documentation and a copy of the formal acceptance that you gave to the seller.

Contractcross

Give your right brain something to do. It's your standard crossword; all of the solution words are from this chapter.

Across

1. The kind of contract where the buyer pays a lump sum.

6. In a T&M contract, the buyer pays for _____, which includes equipment, office space, administrative overhead costs, and anything else (other than labor) that has to be paid for.

9. The procurement _____ of work defines the portion of the project's scope that the seller will work on for the contract.

12. The potential sellers submit seller _____ to a buyer to explain how the contract will be fulfilled.

14. An invitation for _____ is a document that asks sellers for the price of the work.

16. The ___ contract has high risk to the seller because the seller must cover costs that go beyond the price.

17. This kind of analysis is determine whether to procure a service or stay within the company.

19. The buyer selects the _____ selection criteria before contacting potential sellers.

20. This kind of fee is used in cost-plus contracts to encourage the seller to increase performance.

21. The _____ management system stores all of the documents, communications, and information relating to the contract.

Down

2. Even if the buyer fails to deliver on a contract and it has to be terminated early, you still need to perform the _____ Procurements process.

3. A force _____ clause protects both the buyer and seller from things like war and natural disasters preventing the completion of the contract.

4. An invitation for _____ is a document that tells sellers you want them to submit proposals.

5. Some organizations maintain a _____ seller list of sellers pre-screened based on past experience with them.

7. _____ documents contain all of the information the seller wants to communicate to potential buyers.

8. A buyer is sometimes required to use _____ in order to announce a project to all potential sellers.

10. A disagreement between the buyer and seller is called a _____.

11. The contract-related process in the Monitoring and Controlling group is called _____ Procurements.

13. The company or organization that's performing services for the contract.

15. In this kind of contract, the seller is paid the costs as well as a fee that's determined based on the buyer's evaluation of performance.

18. The company or organization that's procuring services.

→ Answers on page 652.

Kate closes the contract

The 18-month contract's ready to close! The seller did a great job handling technical support, and that gave Kate and Ben the time they needed to ramp up their own company's team and facilities.

WELL, THE CONTRACT'S DONE, AND WE'VE MOVED TECH SUPPORT BACK INTO THE COMPANY.

AND IT GAVE US ENOUGH TIME TO TRAIN OUR OWN TEAM TO TAKE OVER. GREAT WORK!

HELLO, THIS IS TECHNICAL SUPPORT. HOW CAN I HELP YOU?

Question Clinic: BYO questions

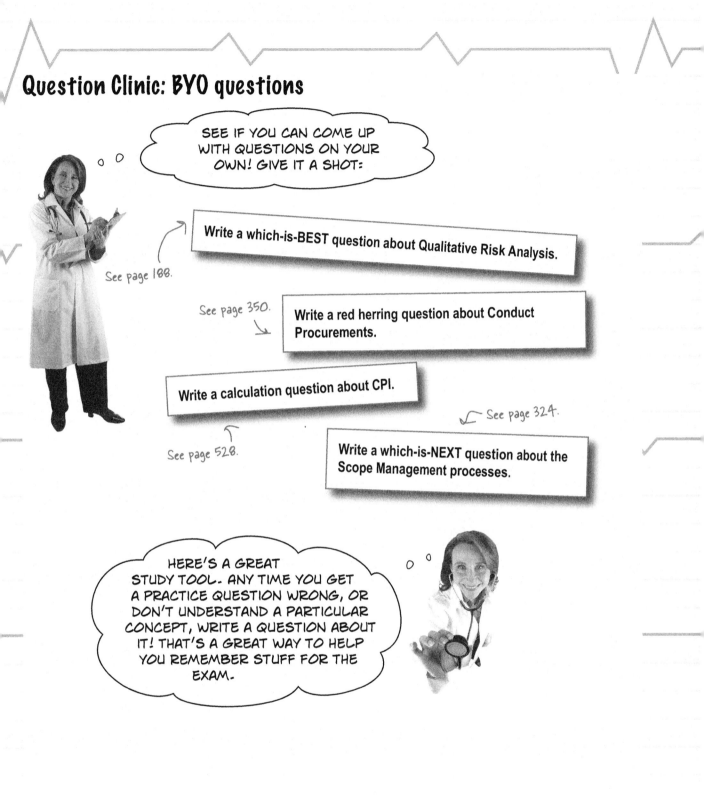

SEE IF YOU CAN COME UP
WITH QUESTIONS ON YOUR
OWN! GIVE IT A SHOT:

Write a which-is-BEST question about Qualitative Risk Analysis.

See page 188.

See page 350.

Write a red herring question about Conduct Procurements.

Write a calculation question about CPI.

See page 324.

See page 528.

Write a which-is-NEXT question about the Scope Management processes.

HERE'S A GREAT
STUDY TOOL. ANY TIME YOU GET
A PRACTICE QUESTION WRONG, OR
DON'T UNDERSTAND A PARTICULAR
CONCEPT, WRITE A QUESTION ABOUT
IT! THAT'S A GREAT WAY TO HELP
YOU REMEMBER STUFF FOR THE
EXAM.

exercise *solutions*

Sharpen your pencil
Solution

Kate has **18 months** to build up the capacity her company needs to handle all the technical support calls. See if you can figure out whether it's a better deal for Kate to make or buy.

1. If they handle the extra work within the company instead of finding a seller, it will cost an extra $35,000 in overtime and $11,000 in training costs in total, on top of the $4,400 per person per month for the five-person team needed to do the extra support work. What's the total cost of keeping the work within the company?

The total cost for keeping the work is the monthly cost ($4,400 per person × 5 people × 18 months = $396,000) plus the extra costs ($35,000 overtime and $11,000 training costs). $396,000 + $35,000 + $11,000 = $442,000 total costs for keeping the work inside the company ("making").

2. Kate and Ben talked to a few companies and estimate that it will cost $20,000 per month to hire another company to do the work, but they'll also need to spend $44,000 in setup costs. What will contracting the work cost?

The cost for hiring another company to do the work is $20,000 per month × 18 months plus the $44,000 in setup costs. ($20,000 × 18) + $44,000 = $404,000 total costs for contracting out the work ("buying").

3. So does it make more sense to make or buy? Why?

In this case, it makes more sense to buy because the costs of making ($442,000) are greater than the costs of buying ($404,000).

Sharpen your pencil
Solution

Kate is putting out an RFP to find a seller to provide technical support for her company. Can you figure out which Conduct Procurements tool she's using?

1. Kate checks works with her company's seller evaluation committee, which follows a documented, formal evaluation review process to determine which seller should be selected for the contract.

Proposal evaluation techniques

2. Kate contacts an IT trade journal and places a classified ad to try to find sellers.

Advertising

3. The CEO's brother-in-law runs a company that's bidding on the contract. Kate needs to make sure he gets fair—not preferential—treatment. She doesn't want to give him an unfair advantage, but she also doesn't want to exclude him from the bidding process. So she gathers representatives from all sellers into a room where they can ask questions about the contract out in the open and hear the responses to each question.

Bidder conference

4. Kate's company takes part in an equal-opportunity program in which seller companies owned by minorities must be given notice of any RFPs. She uses a website approved by the program to find seller companies with performance data from similar projects.

Analytical techniques

Contract Magnets Solution

Which of the magnets are part of the procurement documents, and which of them are part of the source selection criteria?

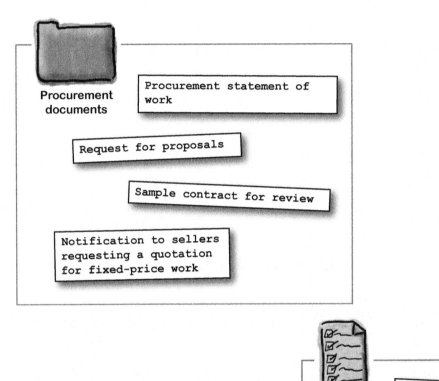

Procurement documents

> Procurement statement of work

> Request for proposals

> Sample contract for review

> Notification to sellers requesting a quotation for fixed-price work

Source selection criteria

> Review the SOW with the seller to make sure it's understood

> Meet with the project manager and review project processes

> Go over final pricing plan and contract terms

> Get a financial statement or credit report and verify that the seller is insured

Sharpen your pencil
Solution

Control Procurements is a typical Monitoring and Controlling process, which means you should be able to figure out most of the outputs. The first one's filled in—can you fill in the others?

1. **Work performance information**

4. **Change requests**

2. Updates to
   Organizational process assets

5. Updates to
   Project Management Plan

3. Updates to
   Project documents

Contractcross Solutions

Give your right brain something to do. It's your standard crossword; all of the solution words are from this chapter.

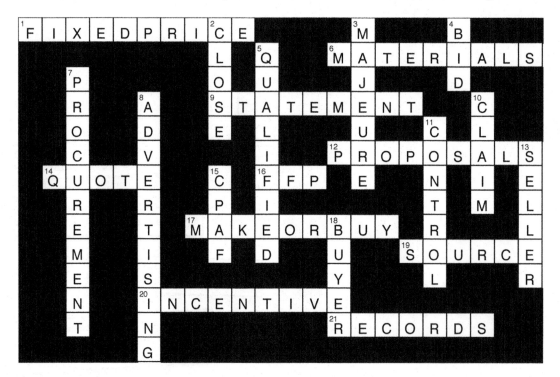

Exam Questions

1. Tom is a project manager for a software company. He is contracting a long-term software project with an external company. That company charges him $20/hour per employee and $300 overhead per month. What kind of contract is he using?

- A. FP
- B. CPAF
- C. CR
- D. T&M

2. Which of the following is NOT true about bidder conferences?

- A. All potential sellers should meet separately with the buyer.
- B. Potential sellers should ask questions in an open forum so other sellers can hear the answers.
- C. Bidder conferences are a good way to make sure sellers are treated fairly.
- D. All sellers are given the same procurement documents.

3. You work for a seller that is bidding on a contract. Which type of contract has the MOST risk for your company?

- A. CPIF
- B. T&M
- C. FP
- D. CPAF

4. Which of the following BEST describes the "point of total assumption" for a contract?

- A. The point in a cost-plus contract where the buyer assumes that the seller will need to be paid
- B. The total cost of a T&M contract
- C. The point in a fixed-price contract where the seller has to assume all costs going forward
- D. The total number of resources required for a contract

5. You're trying to decide whether or not to contract out a construction job. To do it within your company, you will have to hire an engineer for $35,000 and pay a construction team $15,000 per week. A contractor quotes you a price of $19,000 per week, and your expert agrees that you won't find a lower price than that. The job will take 16 weeks. What's the BEST way to proceed?

- A. Pay the contractor to do the job.
- B. Select a T&M contract.
- C. Don't contract out the work; hire the engineer and pay the construction team to do the work.
- D. Make sure the contract has a force majeure clause.

Exam Questions

6. You're managing a project that might have to contract out work, and you're comparing the relative advantages and disadvantages of finding a seller versus having your company do the work itself. Which process are you in?

    A.   Plan Procurement Management

    B.   Plan Contracting

    C.   Conduct Procurements

    D.   Request Seller Responses

7. You're using a qualified seller list. Which process are you in?

    A.   Plan Procurement Management

    B.   Plan Contracting

    C.   Conduct Procurements

    D.   Request Seller Responses

8. You've been contracted by an industrial design firm to manage its contracting. Your client asks you to take over the negotiations for an important contract to design a new remote-control lighting system. You've narrowed it down to one seller, and now you're working with the legal department at the buyer to negotiate the terms of the contract. Which of the following BEST describes your goal?

    A.   You want to get the best deal for your client by making sure the seller's price is as low as possible, no matter what it costs them.

    B.   You want to get a fair deal for both the buyer and the seller.

    C.   You want to make sure that the seller gets as much money as possible.

    D.   You want to prolong the negotiation so that you earn a higher fee.

9. You've been contracted by a construction company to manage its contracting. It has a choice of either buying an excavator or renting it. To buy it, the company would have to pay $105,000, but owning it will require approximately $10,000 in maintenance costs per year. The price to rent the excavator is $5,000 per month, with a one-time service charge of $2,000. What's the minimum number of months the company needs to use the excavator in order for it to make sense to buy it rather than rent?

    A.   8 months

    B.   16 months

    C.   21 months

    D.   25 months

10. Which of the following contracts has the MOST risk for the buyer?

    A.   FP

    B.   CPAF

    C.   CPIF

    D.   T&M

Exam Questions

11. You're managing a project that is difficult to estimate, so you don't have a good idea of when the project will end. Which of the following contracts is BEST?

    A. FP

    B. CPAF

    C. CPIF

    D. T&M

12. You're looking for a seller to do work for your project. When do you send out an RFP?

    A. After you create the procurement documents, but before you select the seller

    B. Before you plan contracting, but after you Plan Procurement Management

    C. After the bidder conference, but before you select the seller

    D. During Control Procurements

13. You're creating source selection criteria for your contract. What process are you in?

    A. Conduct Procurements

    B. Control Procurements

    C. Close Procurements

    D. Plan Procurement Management

14. You're managing a project when you and the seller both agree that you need to have the seller add more resources to the project in order to finish on time. The number of resources is written into the contract. What's the BEST way to proceed?

    A. Your project will be late because you can't change the contract once it's signed.

    B. You need to convince the buyer to sign a new contract.

    C. You need to use the contract change control system to make the change to the contract.

    D. You need to use claims administration to resolve the issue.

15. Which of the following BEST explains the difference between a seller audit during Control Procurements and a procurement audit during Close Procurements?

    A. The seller audit reviews the products being created, while the procurement audit reviews how well the seller is doing the job.

    B. The procurement audit reviews the products being created, while the seller audit reviews how well the seller is doing the job.

    C. The seller audit reviews the products being created, while the procurement audit is used to examine successes and failures and gather lessons learned.

    D. The procurement audit reviews the products being created, while the seller audit is used to examine successes and failures and gather lessons learned.

Exam ~~Questions~~ Answers

1. Answer: D

This contract is a time and materials contract. It's charging a rate for labor and overhead for materials.

Eliminating the wrong answers works really well with questions like this.

2. Answer: A

One of the most important things about a bidder conference is that no one seller is given better access to the buyer. They should all have the same opportunity to gather information, so that no single seller is given preferential treatment.

Sellers should meet in the same room, and any time one of them asks the question, everyone else should hear the answer.

3. Answer: C

A fixed-price contract is the riskiest sort of contract for the seller. That's because there's one price for the whole contract, no matter what happens. So if it turns out that there's a lot more work than expected, or the price of parts or materials goes up, then the seller has to eat the costs.

4. Answer: C

This is just the definition of the point of total assumption.

5. Answer: C

This is a simple make-or-buy decision, so you can work out the math. The contractor's quote of $19,000 per week for a 16-week job means that buying will cost you $19,000 x 16 = $304,000. On the other hand, if you decide to keep the work in-house, then it will cost you $35,000 for the engineer, plus $15,000 per week for 16 weeks: $35,000 + (16 x $15,000) = $275,000. It will be cheaper to make it rather than buy it!

OH, I GET IT. I JUST HAVE TO FIGURE OUT HOW MUCH IT COSTS TO MAKE IT OR BUY IT, AND CHOOSE THE LOWER NUMBER!

Exam ~~Questions~~

Answers

---

6. Answer: A

This question describes make-or-buy analysis, which is part of the Plan Procurement Management process.

That makes sense. You can't start contracting until you figure out whether or not you should.

7. Answer: C

One of the most important things that you do when you're finding sellers during the Conduct Procurements process is to select the sellers that will do the work. And the qualified seller list is an input that you use for that.

Your company should already have a qualified seller list on file.

8. Answer: B

One of the most important parts of Procurement Management is that both the buyer and the seller want to feel like they're getting a good deal. Every procurement should be a win-win situation for both parties!

9. Answer: D

This may look like a tough problem, but it's actually pretty easy. Just figure out how much the rental would cost you for each of the answers:

A.	8 months	8 months x $5,000 per month + $2,000 service charge = $42,000
B.	16 months	16 months x $5,000 per month + $2,000 service charge = $82,000
C.	21 months	21 months x $5,000 per month + $2,000 service charge = $107,000
D.	25 months	25 months x $5,000 per month + $2,000 service charge = $127,000

Now look at what the excavator would cost for 25 months. It would cost $105,000 plus $20,000 for the maintenance costs, for a total of $125,000. So at 25 months, the excavator is worth buying—but before that, it makes more sense to rent.

If this seems a little out of place, remember that renting equipment is a kind of contract, and the same kind of make-or-buy decision is necessary.

10. Answer: D

The time and materials (T&M) contract is the riskiest one for the buyer, because if the project costs are much higher than the original estimates, the buyer has to swallow them, while the seller keeps getting paid for the time worked.

~~Exam Questions~~ Answers

11. Answer: D

Both cost-plus and fixed-price contracts are based on the idea that you know how long the contract is going to last. A seller would only agree to a fixed-price contract if there's a good idea of how much it's going to cost. And a cost-plus contract will hurt the buyer if it goes over. Only the time and materials contract will give both the buyer and seller a fair deal if neither has a good idea of how long the work will take.

That's the only time you really should use a T&M contract.

12. Answer: A

Contracting is a pretty linear process—first you plan the contract, then you put together a package of procurement documents to send to potential sellers, and then you select a seller and start the work. So you send out a request for proposals after you've put together the procurement document package so that you can select a seller for the job.

13. Answer: D

You put together the source selection criteria as part of the Plan Procurement Management process. That way, you can use the criteria when you're looking at the responses you get from sellers.

14. Answer: C

You can always change a contract, as long as both the buyer and the seller agree to it. When you do that, you need to use the contract change control system—just like with any other change.

This is not what claims administration is for. Since the buyer and seller agree, there is no claim.

15. Answer: C

It's easy to get mixed up with all of these audits, but if you think about how they're used, it gets less confusing. When you're performing Control Procurements, the most important part of your job is to figure out if the products that the seller is producing meet your requirements. By the time closure happens, the products have all been completed—if there was a problem, you should have caught it during Control Procurements. All you can do now is come up with any lessons learned so that you can avoid mistakes in the future.

13 Stakeholder management

Keeping everyone engaged

BUT IT WASN'T UNTIL I MAPPED OUT ALL OF THE PARTY GUESTS IN A POWER AND INTEREST GRID THAT I REALLY KNEW HOW TO KEEP THEM SATISFIED.

Project management is about knowing your audience.

If you don't get a handle on the people who are affected by your project, you might discover that they have needs you aren't meeting. If your project is going to be successful, you've got to satisfy your stakeholders. Luckily, there's the **Stakeholder Management** knowledge area, which you can use to understand your stakeholders and figure out what they need. Once you really understand how important those needs are to your project, it's a lot easier to **keep everyone satisfied**.

Party at the Head First Lounge (again)!

Jeff and Charles had a great time upgrading the lounge in
Chapter 10. They've been doing so well since they upgraded
that they want to have another party to ring in the summer.

Not everybody is thrilled

There were a few people who weren't quite as enthused about the opening party for Head First Lounge as Jeff and Charles were. Can you use the four **Stakeholder Management** processes to help them get their party on track?

> THAT WAS THE BEST PARTY IN TOWN, AND WE DIDNT KNOW ABOUT IT UNTIL THE NEXT DAY.

> WHAT MAKES HF LOUNGE BETTER THAN EVERY OTHER DOWNTOWN PARTY SPOT?

Mike writes a blog on downtown nightlife.

Mark and Laura are really into the downtown nighife scene.

> THE MUSIC LAST TIME WAS TOO COMMERCIAL. I'VE GOT TO PLAY SOMETHING MORE INTERESTING IF PEOPLE ARE GOING TO REMEMBER ME.

> THE LAST PARTY GOT A LITTLE LOUD. I'M JUST HOPING THEY'LL KEEP IT DOWN IN THE FUTURE.

Adam lives next door to HF Lounge.

Tom's been DJing for a couple of years at HF Lounge and he's really starting to take off.

⚛ BRAIN POWER

What can Jeff and Charles do to get a handle on their stakeholder problems?

Understanding your stakeholders

When you think about it, there are a lot of people who have an interest in your project. That obviously includes the sponsor who's paying for it, the team who's making it, and the people who will support it. But there are people who aren't so obvious who have a stake in your project as well. If you don't pay attention to all of your stakeholders, you could find that you don't meet their needs, and that can cause your project to run off the rails. The **Stakeholder Management** processes are here to help you figure out who your stakeholders are, plan how you'll keep them engaged, and manage your project to keep them satisfied.

Initiating process group

Planning process group

Stakeholder Management Plan

Stakeholder	Unaware	Resistant	*****	Supportive	Leading
Jeff -HFL employee				Current	Desired
Charlie-HFL employee				Current	Desired
DJ Tom-HFL employee			Current	Desired	
Adam Neighbor	Current		Desired		
Mike the Blogger		Current	Desired		
Mark and Laura party guests	Current		Desired		

Stakeholder Communication Requirements

Stakeholder	Type of information required	Frequency
Jeff -HFL employee	Status meetings Issue Reviews	Daily
Charlie-HFL employee	Status meetings Issue Reviews	Daily
DJ Tom-HFL employee	Status meetings Issue Reviews	Weekly
Adam Neighbor	Party announcement	1 month prior to party
Mike the Blogger	Party announcement Special invitation	1 Month prior to party Follow up one week prior to party
Mark and Laura party guests	Party announcement Special invitation	1 Month prior to party

Identify Stakeholders

You need to spend a little time figuring out who your stakeholders are before you can do anything else. That's why this one is an Initiating process.

Plan Stakeholder Management

Here's where you figure out what your stakeholders' current level of engagement is, and plan how you'll get them to support your project.

Stakeholder Management makes sure you know who you need to engage to keep your project on track.

Stakeholder <u>requirements</u> and <u>expectations</u> sometimes change over the course of the project. Control Stakeholder Engagement makes sure you stay on top of those changes and adjust your plans accordingly.

Executing
process
group

Monitoring
and Controlling
process group

**Manage
Stakeholder
Engagement**

This process is all about working with your stakeholders to understand their needs and keep them involved in all of the decisions that affect them as the project progresses.

**Control
Stakeholder
Engagement**

This is where you keep track of all of the stakeholder relationships you've built and adjust your plans and actions to keep their needs in mind.

Find out who your stakeholders are

Initiating process group

One of the first things you need to do when you start a project is to figure out who your stakeholders are and what you need to do to keep them all in the loop. The **Identify Stakeholders** process is all about writing down your stakeholders' names along with their goals, expectations, and concerns in a document called the **stakeholder register.** Most projects succeed or fail based on how well the project manager knows and manages stakeholder expectations. Writing them down up front will help you to come up with a strategy to identify the people who could impact your project, but still need to be convinced of its value.

Knowing the way your company runs should help you to find the people who will be impacted by your project.

Enterprise environmental factors

The project charter will tell you who's funding and championing the project.

Project charter

Organizational process assets

Procurement documents

Any suppliers or vendors listed in contracts should be part of your stakeholder list.

Inputs

Tools

Stakeholder analysis is a critical tool in this process. You need to interview all of the stakeholders you can find for your project, and find out the value the project has for them. As you sit with stakeholders, you'll identify more people to interview.

During stakeholder analysis, you can divide your stakeholders into groups based on their level of involvement and need for communication. When you understand what motivates all of your stakeholders, you can come up with a strategy to make sure that they're told about the things that they find important, and that they're not bored with extraneous details.

Expert Judgment

Expert judgment in this process means talking to all of the experts on your project to identify more stakeholders, and learn more about the ones you've identified.

Meetings

Meetings are a great tool for getting everyone together to think through the stakeholders who might be impacted by your project.

BRAIN POWER

What effect would a resistant stakeholder have on your project? What about a neutral one?

Stake-hold-er, noun
A person who has an interest or concern about something.
Tom was a stakeholder in the Little League game since his son was playing in it.

The PMBOK Guide defines a stakeholder a little more specifically than its everyday term: "Individuals, groups, or organizations who may affect, be affected by, or perceive themselves to be affected by a decision, activity, or outcome of a project." [PMBOK Guide, 5th edition, page 394]

Outputs

The register should tell you what individual stakeholder, get out of the project so that you can help them to see the value in the project.

The classification tells you whether the stakeholder is internal or external, but also whether he's a supporter, resistor, or neutral participant in the project.

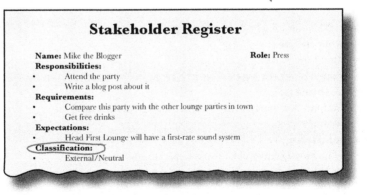

Stakeholder Register

Name: Mike the Blogger   **Role:** Press
Responsibilities:
- Attend the party
- Write a blog post about it

Requirements:
- Compare this party with the other lounge parties in town
- Get free drinks

Expectations:
- Head First Lounge will have a first-rate sound system

Classification:
- External/Neutral

**Flip the page to learn more about the stakeholder register.

Stakeholder analysis up close

When you get started on your project, the first thing you should do is examine the charter and any contract information you have to figure out who will be impacted by it. Once you have a preliminary list of stakeholders, you should sit down with each one of them and figure out their responsiblities, goals, expectations, and concerns. These interviews will be the basis for the stakeholder profiles in your stakeholder register. As you interview people, you'll likely find more stakeholders to include in the list.

It's useful to group stakeholders together, because stakeholders in a particular group tend to have similar needs and project interests.

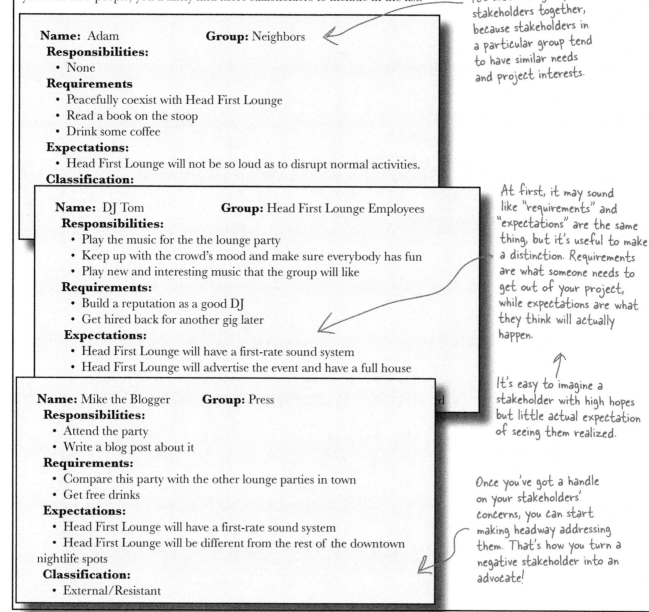

Name: Adam   **Group:** Neighbors
Responsibilities:
• None
Requirements
• Peacefully coexist with Head First Lounge
• Read a book on the stoop
• Drink some coffee
Expectations:
• Head First Lounge will not be so loud as to disrupt normal activities.
Classification:

Name: DJ Tom   **Group:** Head First Lounge Employees
Responsibilities:
• Play the music for the the lounge party
• Keep up with the crowd's mood and make sure everybody has fun
• Play new and interesting music that the group will like
Requirements:
• Build a reputation as a good DJ
• Get hired back for another gig later
Expectations:
• Head First Lounge will have a first-rate sound system
• Head First Lounge will advertise the event and have a full house

Name: Mike the Blogger   **Group:** Press
Responsibilities:
• Attend the party
• Write a blog post about it
Requirements:
• Compare this party with the other lounge parties in town
• Get free drinks
Expectations:
• Head First Lounge will have a first-rate sound system
• Head First Lounge will be different from the rest of the downtown nightlife spots
Classification:
• External/Resistant

At first, it may sound like "requirements" and "expectations" are the same thing, but it's useful to make a distinction. Requirements are what someone needs to get out of your project, while expectations are what they think will actually happen.

It's easy to imagine a stakeholder with high hopes but little actual expectation of seeing them realized.

Once you've got a handle on your stakeholders' concerns, you can start making headway addressing them. That's how you turn a negative stakeholder into an advocate!

One way to get a handle on how to communicate with your stakeholders is to create a **power/interest grid.** When you plot your stakeholders on a power/interest grid, you can determine who has high or low power to affect your project, and who has high or low interest. People with high power need to be kept satisfied, while people with high interest need to be kept informed. When a stakeholder has both, make sure you manage her expectations very closely!

People with **high power and low interest** need to be kept in the loop. You need these people to be **kept satisfied** with the project, even if they aren't interested in it.

The people who are **high power and high interest** are the decision makers who have the biggest impact on project success, so **closely manage** their expectations.

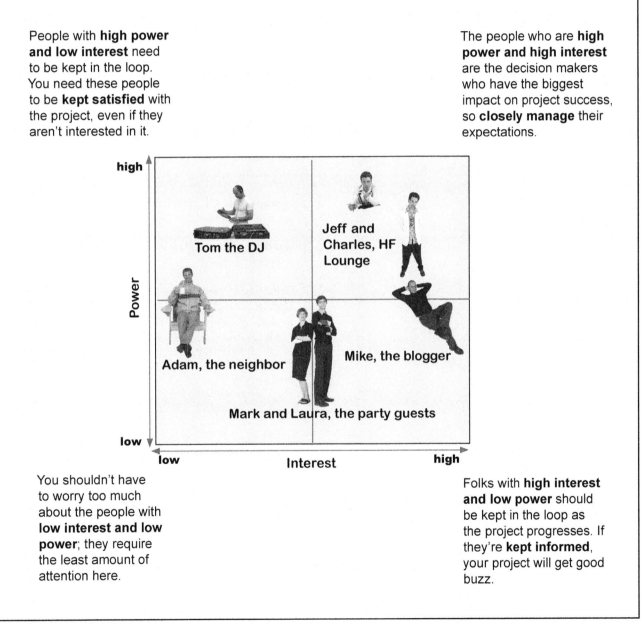

Tom the DJ

Jeff and Charles, HF Lounge

Adam, the neighbor

Mike, the blogger

Mark and Laura, the party guests

high / low — Power

low / high — Interest

You shouldn't have to worry too much about the people with **low interest and low power**; they require the least amount of attention here.

Folks with **high interest and low power** should be kept in the loop as the project progresses. If they're **kept informed**, your project will get good buzz.

Exercise

This is the Plan Stakeholder Management process. You've seen a lot of planning processes now. Can you fill in the inputs and outputs for this one?

Inputs

Tools

Analytical techniques means figuring out how engaged your stakeholders are today, and how engaged you want them to be as your project gets under way. The levels of engagement are:

Unaware: The stakeholder doesn't know that the project is happening.

Resistant: The stakeholder doesn't want the project or decision you're making to happen.

Neutral: The stakeholder is fine with the project or decision no matter how it turns out.

Supportive: The stakeholder wants your project or decision to succeed.

Leading: The stakeholder is actively helping the project to succeed.

When you're mapping out your Stakeholder Management plan, it helps to determine where all of your stakeholders are in relation to these classifications, and where you ultimately want them to be.

This one is your company's culture and policies toward project communication.

....

Here's where your company keeps all of its templates and lessons learned.

....

You need to know who you're going to communicate with.

....

Here's where you use all of the planning you've done for all of the knowledge areas in your project so far.

Planning
process group

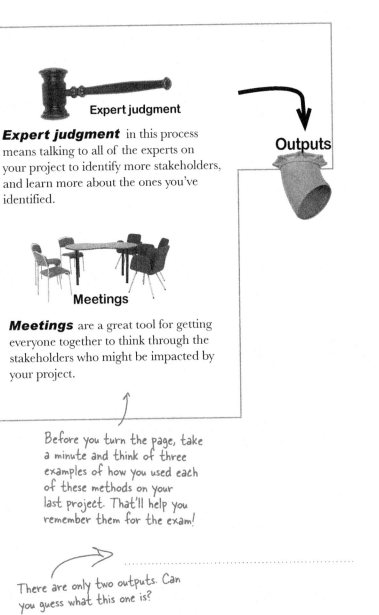

Expert judgment

Expert judgment in this process means talking to all of the experts on your project to identify more stakeholders, and learn more about the ones you've identified.

Outputs

Meetings

Meetings are a great tool for getting everyone together to think through the stakeholders who might be impacted by your project.

Are you surprised at how much of this process you can fill in? Looks like you're getting the hang of this stuff!

**Document
updates**

Before you turn the page, take a minute and think of three examples of how you used each of these methods on your last project. That'll help you remember them for the exam!

There are only two outputs. Can you guess what this one is?

There are several project documents that get updated when you're planning Stakeholder Management. Can you think of one of them?

Exercise Solution

This is the Plan Stakeholder Management process. You've seen a lot of planning processes now. Can you fill in the Inputs and Outputs for this one?

Tools

Analytical Techniques means figuring out how engaged your stakeholders are today, and how engaged you want them to be as your project gets underway. The levels of engagement are:

Unaware: The stakeholder doesn't know that the project is happening.

Resistant: The stakeholder doesn't want the project or decision you're making to happen.

Neutral: The stakeholder is fine with the project or decision no matter how it turns out.

Supportive: The stakeholder wants your project or decision to succeed.

Leading: The stakeholder is actively helping the project to succeed.

When you're mapping out your Stakeholder Management plan, it helps to determine where all of your stakeholders are in relation to these classifications, and where you ultimately want them to be.

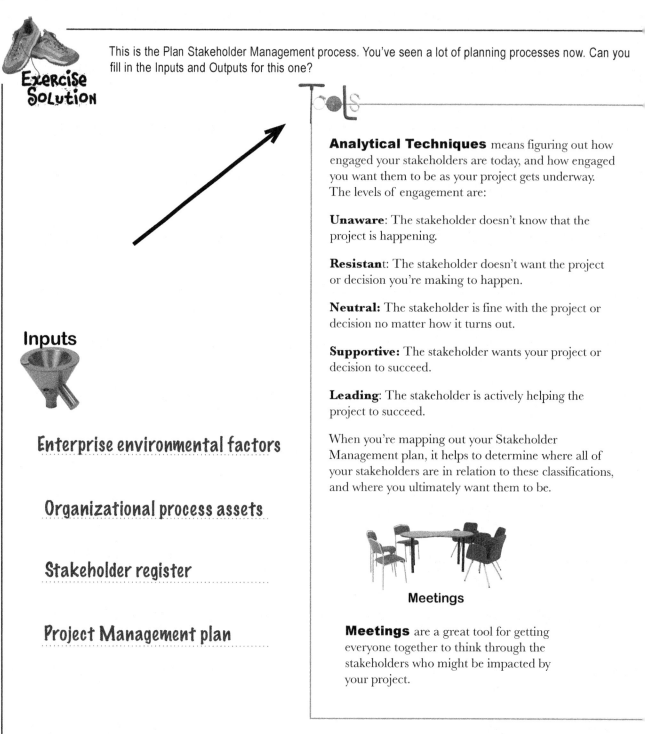

Meetings

Meetings are a great tool for getting everyone together to think through the stakeholders who might be impacted by your project.

Inputs

Enterprise environmental factors

Organizational process assets

Stakeholder register

Project Management plan

You'll usually need to update the project schedule and the stakeholder register when you plan Stakeholder Management for your project.

Expert judgment

Expert judgment in this process means talking to all of the experts on your project to identify more stakeholders, and learn more about the ones you've identified.

Document updates

Planning process group

Outputs

Think about everyone who's involved in the project, and try to come up with a plan for how to engage them.

The plan says how you will distribute the information, to whom, and how often.

Stakeholder Management Plan

Stakeholder	Unaware	Resistant	Neutral	Supportive	Leading
Jeff—HFL employee				Current	Desired
Charles—HFL employee				Current	Desired
DJ Tom—HFL employee			Current	Desired	
Adam—Neighbor	Current		Desired		
Mike—Blogger		Current		Desired	
Mark and Laura—party guests	Current			Desired	

Stakeholder Communication Requirements

Stakeholder	Type of information required	Frequency
Jeff—HFL employee	Status meetings Issue reviews	Daily
Charles—HFL employee	Status meetings Issue reviews	Daily
DJ Tom—HFL employee	Status meetings Issue reviews	Weekly
Adam—Neighbor	Party announcement	One month prior to party
Mike—Blogger	Party announcement Special invitation	One month prior to party Follow up one week prior to party
Mark and Laura—party guests	Party announcement Special invitation	One month prior to party

How engaged are your stakeholders?

It's not enough to know who your stakeholders are—you need to understand what motivates them, and what it will take to make the project a success for each of them. That's where the **Stakeholder Engagement Assessment matrix** comes in. Jeff and Charles sat down and worked to come up with one for their upcoming party. Here's what it looked like:

Stakeholder	Unaware	Resistant	Neutral	Supportive	Leading
Jeff—HFL employee				Current	Desired
Charles—HFL employee				Current	Desired
DJ Tom—HFL employee			Current	Desired	
Adam—Neighbor	Current		Desired		
Mike—Blogger		Current		Desired	
Mark and Laura—party guests	Current			Desired	

DO WE REALLY NEED ANOTHER DOWNTOWN PLACE TO HANG OUT?

HF LOUNGE: UNDERGROUND IS THE HOTTEST PARTY SPOT IN TOWN!

Turning around a negative stakeholder can take a lot of effort. You might need to pay special attention to the things that affect his opinion to do it. But usually your project will be better off for it, especially if that negative stakeholder has a lot of influence.

Choose which engagement level applies in each situation.

Exercise

1. Jeff and Charles haven't reached out to the caterer yet, but they hope to get his help at a discount.

☐ Unaware ☐ Resistant ☐ Neutral

☐ Supportive ☐ Leading

2. The sound engineers are already booked for that night. They're going to charge an extra fee if you want them to come help out.

☐ Unaware ☐ Resistant ☐ Neutral

☐ Supportive ☐ Leading

3. The liquor distributor usually has no trouble accommodating double and triple orders.

☐ Unaware ☐ Resistant ☐ Neutral

☐ Supportive ☐ Leading

4. The neighborhood business association is looking for a place to celebrate the past year's successful programs, and a Head First Lounge party sounds like just the thing for them.

☐ Unaware ☐ Resistant ☐ Neutral

☐ Supportive ☐ Leading

5. A local event magazine was so happy with the last Head First Lounge party that they've offered to partially sponsor this one.

☐ Unaware ☐ Resistant ☐ Neutral

☐ Supportive ☐ Leading

6. The neighbors haven't been told that HFL is planning another party. We probably ought to let them know way in advance.

☐ Unaware ☐ Resistant ☐ Neutral

☐ Supportive ☐ Leading

→ Answers on page 686.

⚛ BRAIN POWER

Think about a major project you've worked on where there were a large number of stakeholders. Where was their engagement level at the beginning of the project? Where did it end up?

there are no
Dumb Questions

Q: How do I figure out who all of my stakeholders are?

A: The short answer is: *look around*. You probably can name the majority of the stakeholders on your project right off the top of your head. You'll surely know who the sponsor is; you can point to her name right on the charter. Then there's the team that's doing the work; you'll know who they are because you work with them every day. From there, it gets a little harder. Any business partners (like trainers or support people for software packages) that your company has contracted to help out with the project are also stakeholders. Consultants or other vendors you might've contracted to help you deliver your product will also be stakeholders. Then you'll have to think about how the product of your project will affect the rest of your company. Will it change the way people work when it's complete? How will those people who need to change their work feel about the project if you asked them today? They're stakeholders, too. If you're thorough, the list might get pretty long, but it's much better to think about your stakeholders up front than it is to ignore them. A stakeholder you don't plan for today could cause a lot of turbulence in your project later.

Q: Explain the point of that power/interest grid again.

A: Different people have different perspectives on your project. Some of them will put a lot of time and effort into making your project succeed, while others will not have the spare cycles to give. And some people might even actively work against your project if they don't understand it. The power/interest grid is there to help you understand how you should approach the stakeholders for your project. If someone without a lot of power to influence your project doesn't have the time or the will to help you with it, that's less of a problem than when someone with a lot of power to affect your project is resistant to it. It's a tool to help you figure out the best approach to managing all of the stakeholders on your project. It will help you choose the right way to influence the people who can help you succeed.

Q: How do I turn around resistant stakeholders?

A: Many times they're resistant for a good reason. The best approach is really to try to understand why they're resistant and help them to see the benefits of your project. Many times, stakeholders who are resistant to change have good suggestions that can make the project better in the long run.

Q: What does it mean to have a stakeholder in a leading role?

A: When a stakeholder takes on a leading role, he is actively involved in making sure your project is a success. He may go to meetings and convince others to support the project, and help you to clear any obstacles that might jeopardize your project's goals. When a stakeholder has a leading role in the project, he has a stake in seeing it succeed. Leading stakeholders are willing to put time and energy into making sure that others support the project.

There are five engagement levels for stakeholders: unaware, resistant, neutral, supportive, and leading.

Managing stakeholder engagement means clearing up misunderstandings

As your project progresses, you'll need to check in with your stakeholders regularly so that misunderstandings don't develop. Your job is to help them to take part in the decisions the team is making, so that they can be supportive. When a stakeholder is resistant to change, you'll need to negotiate with her and understand her resistance so that you can take her perspective into account.

Sometimes a stakeholder you don't always talk to might have a good suggestion for improvement that can help the whole team. It's also possible that there are some facts about the project that the stakeholder hasn't considered, and you can help him to be better informed. The key to success in Stakeholder Management is being inclusive, and sharing information with everyone who's impacted by the project's outcome.

WHAT'S UP WITH THE SOUND ENGINEERS? ARE WE GOING TO BE ABLE TO GET THEM TO COME OUT ON THE NIGHT OF THE PARTY?

LAST I HEARD, THEY WERE CHARGING A BIG FEE. DID THEY EVER CONFIRM?

Jeff and Charles didn't realize that they scheduled their party on the same day as another music event that was happening in the city. The sound engineers needed to bring in extra help to handle both events, and that's why they were charging a bigger fee. Once you found out what the issue was and communicated it back to the Head First Lounge team, they understood the cost and approved it.

Exercise

The **Manage Stakeholder Engagement** process is a typical Executing process. You already know the inputs, tools and techniques, and outputs! See if you can figure out what they do just from their names. Write down a description for each of them, and then flip the page to see if you're right.

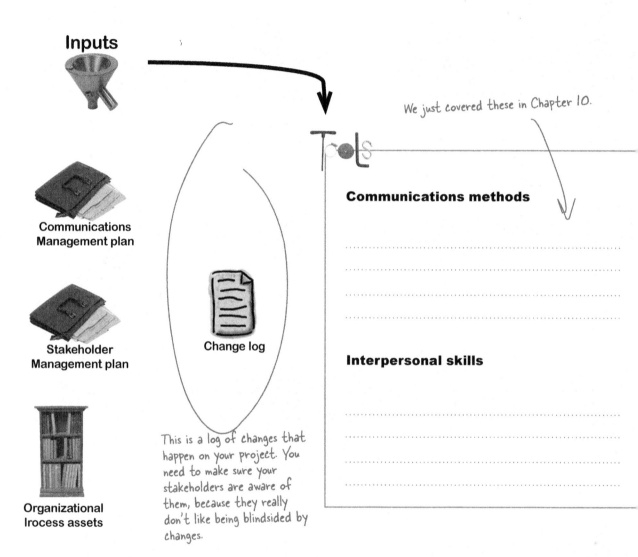

Inputs

Communications Management plan

Stakeholder Management plan

Change log

This is a log of changes that happen on your project. You need to make sure your stakeholders are aware of them, because they really don't like being blindsided by changes.

Organizational Irocess assets

We just covered these in Chapter 10.

T●●ls

Communications methods

....
....
....
....

Interpersonal skills

....
....
....
....

Executing
process
group

Issue log

....

....

....

....

Since the Manage Stakeholder
Engagement process is all about resolving
issues that the stakeholders experience,
the tools are focused on communicating
with the stakeholders about those issues.

Change requests

....

....

....

....

Management skills

....

....

....

....

Project document updates

....

....

....

....

Outputs

Organizational process asset updates

....

....

....

....

Project Management Plan updates

How will the outputs be used to
communicate with stakeholders?
Don't forget that every team
member is a stakeholder!

....

....

....

....

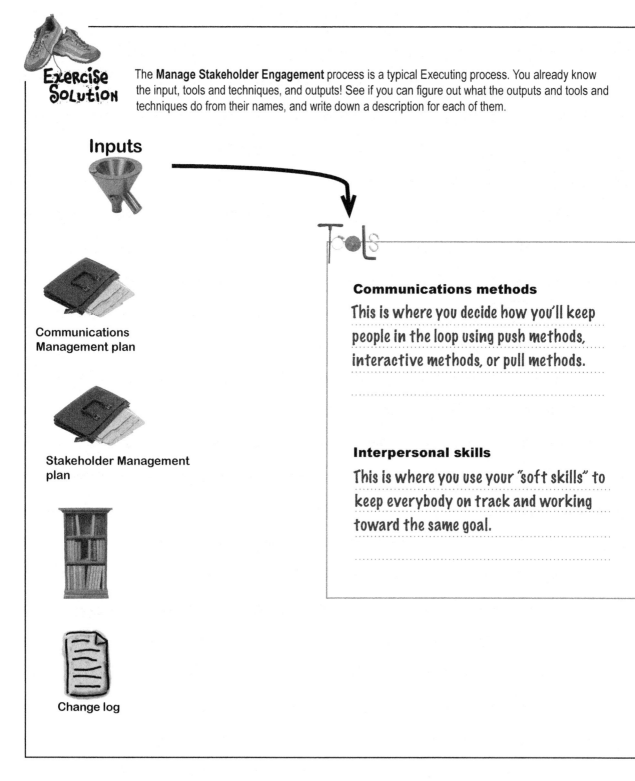

Exercise Solution

The **Manage Stakeholder Engagement** process is a typical Executing process. You already know the input, tools and techniques, and outputs! See if you can figure out what the outputs and tools and techniques do from their names, and write down a description for each of them.

Inputs

Communications Management plan

Stakeholder Management plan

Change log

Tools

Communications methods

This is where you decide how you'll keep people in the loop using push methods, interactive methods, or pull methods.

Interpersonal skills

This is where you use your "soft skills" to keep everybody on track and working toward the same goal.

Executing
process
group

Issue log

These are any issues that come up during the project that need to be shared with the stakeholders.

Change requests

These are any changes to the project plan or other documents that involve stakeholder communication.

Management skills

This is where you gather important information about your project and use it to make decisions about how to keep the team on track.

Project document updates

These are any updates to previously written project documents that come from the Manage Stakeholder Engagement process.

Outputs

Organizational process asset updates

Any lessons you learn from talking to stakeholders are added to the organizational process assets.

Project Management plan updates

Approved changes actually need to be made to the project plan.

Control your stakeholders' engagement

Now that you've got a great framework set up for managing the way your stakeholders interact with your project, you need to monitor those interactions to make sure that everybody stays in the loop. When you run into a problem or find a place where you might be able to bring the project closer to meeting a stakeholder's goal, you can make course corrections and changes to keep as many of your stakeholders satisfied as possible—and that's what the **Control Stakeholder Engagement** process is all about.

Monitoring and Controlling process group

Once you know what your stakeholders' requirements are, you can monitor how close or far away your project is from meeting them.

This is all of the data about how your project is progressing. You can use it to make forecasts and tell how close you are to meeting your stakeholders' goals.

You'll need to stay on top of all of the current issues for your project if you want keep everybody in the loop.

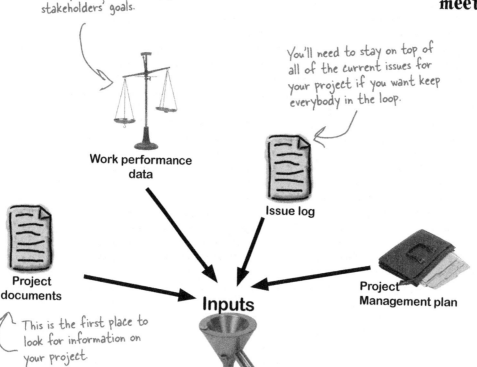

Work performance data

Issue log

Project documents

Inputs

Project Management plan

This is the first place to look for information on your project.

Sharpen your pencil

Can you figure out what each of the **Control Stakeholder Engagement** tools and techniques is for, just from the name?

Information management systems

Expert judgment

Meetings

Sharpen your pencil
Solution

Can you figure out what each of the **Control Stakeholder Engagement** tools and techniques is for, just from the name?

Tools

Information management systems

Stores reports on project performance, like cost, schedule, and scope reports.

Not everybody has a predefined information management system. Some teams provide the same information in spreadsheets or even emails.

Expert judgment

Getting all of your current stakeholders together to determine if there might be other stakeholders who haven't yet been identified.

Meetings

Status meetings keep all of the the stakeholders in the loop about project progress, and provide a place for everyone to share their opinions on how to keep the project on track.

Meetings are a great way to encourage an open exchange of ideas about how the project is going.

Now you can tell when you need to change the way you deal with stakeholders

Outputs

Now that you've taken a look at all of the data coming from your project, you're in a better position to tell if you need to make changes to the way you're managing stakeholder engagement. The **outputs from Control Stakeholder Engagement** are all about making changes to the documents you've been using to keep them engaged all along.

Work performance information is the analyzed work performance data, changed from raw data into reports that are understood in context and can be used to make project decisions. Where work performance data was the raw outputs of the various controlling processes, the work performance information is consolidated. Think of work performance data as your raw budget performance data (this month, we're over budget by $1,000) and work performance information as that data in context (the project budget forecast is currently $10,000 under budget, even though we're over by $1,000 this month).

Change requests happen as part of controlling stakeholder engagement. What do you do if you find out that one of your stakeholders' requirements has been missed? You put the **change request** in as soon as possible.

Project documents updates happen as you work with your stakeholders to monitor and control their engagement. You might find new stakeholders along the way who make you update your **stakeholder register**. You could also run into issues that need to be added to the **issue log**.

Project Management Plan updates might be needed as you work with your stakeholders to ensure successful delivery. Your stakeholders might request changes to your overall strategy for any of the knowledge areas you've planned as part of your Project Management plan. You'll need to go back and make updates to those plans if you want to keep everyone in the loop about the change in approach.

Organizational process asset updates need to be added—especially your **lessons learned**. There are always a lot of lessons to be learned when you're working with stakeholders to ensure project success.

It's party time!

The Head First Lounge party is a big hit! Thanks to your
Stakeholder Management skills, Jeff and Charles are the reigning
kings of downtown nightlife!

I'M SO GLAD I SPOKE UP, AND I'M PLAYING WHAT I WANT TONIGHT.

Tom the DJ told you that he wanted to play different music at the next party, and you took that suggestion to Jeff and Charles. Now he's playing what he wants!

THANKS TO THE EARLY INVITE, WE MADE IT TO THIS GREAT PARTY. THE MUSIC IS PERFECT.

By reaching out early to potential party-goers who'd missed the last event, you made sure the lounge was stocked with people who'd really appreciate the party.

HF LOUNGE: UNDERGROUND IS THE HOTTEST PARTY SPOT IN TOWN!

You paid special attention to Mike the Blogger, gave him a VIP invitation, and made sure he knew about the event way in advance. You knew he cared about sound systems, so you made it a point to show him the work the sound engineers had done for the event. Looks like it paid off!

Thanks to diligent planning, Adam the neighbor was given ample notice about the evening's events and took his family on a weekend vacation.

THANKS TO ALL OF OUR STAKEHOLDERS, WE DID IT AGAIN!

Jeff and Charles learned a lot about how to manage stakeholders from watching the way you managed this project.

Exercise

Choose which engagement level applies in each situation

1. Jeff and Charles haven't reached out to the caterer yet, but they hope to get his help at a discount.

☒ Unaware   ☐ Resistant   ☐ Neutral

☐ Supportive   ☐ Leading

2. The sound engineers are already booked for that night. They're going to charge an extra fee if you want them to come help out.

☐ Unaware   ☒ Resistant   ☐ Neutral

☐ Supportive   ☐ Leading

3. The liquor distributor usually has no trouble accommodating double and triple orders.

☐ Unaware   ☐ Resistant   ☒ Neutral

☐ Supportive   ☐ Leading

4. The neighborhood business association is looking for a place to celebrate the past year's successful programs, and a Head First Lounge party sounds like just the thing for them.

☐ Unaware   ☐ Resistant   ☐ Neutral

☒ Supportive   ☐ Leading

5. A local event magazine was so happy with the last Head First Lounge party that they've offered to partially sponsor this one.

☐ Unaware   ☐ Resistant   ☐ Neutral

☐ Supportive   ☒ Leading

6. The neighbors haven't been told that HFL is planning another party. We probably ought to let them know way in advance.

☒ Unaware   ☐ Resistant   ☐ Neutral

☐ Supportive   ☐ Leading

Exam Questions

1. Matt, the sponsor of a large publishing project, meets with all of the stakeholders on the project to ask for their support in an upcoming testing event. Which engagement level is he displaying?

    A.   Unaware

    B.   Resistant

    C.   Supportive

    D.   Leading

2. Which of the following is NOT an input to the Plan Stakeholder Management process?

    A.   Enterprise environmental factors

    B.   Organizational process assets

    C.   Work performance data

    D.   Stakeholder register

3. You take over for a project manager who has left the company, and realize that there are stakeholders in the project who haven't been included in any of the status meetings so far. Some upper managers think the project is not going to succeed, and others are actively thinking of cancelling it. Which document is the FIRST one that you should create to solve this problem?

    A.   Stakeholder register

    B.   Status report

    C.   Budget forecast

    D.   Performance report

4. In a monthly steering meeting, you ask one of the stakeholders on your project if she has reviewed the latest document updates you've made since the last meeting. She says, "I'm not working on that; I'm not even sure what it is." How would you BEST describe her level of engagement?

    A.   Unaware

    B.   Resistant

    C.   Supportive

    D.   Leading

5. You're managing a construction project. You created a stakeholder register and Stakeholder Management plan, and now the team is working on the project. You've been managing the work, and now you're looking at the work performance data to keep your stakeholders informed of the status of the project. You've discovered a change in the way your stakeholders look at the budget for your project. Which of the following BEST describes the next thing you should do?

    A.   Update the risk register to include any changes to risk strategy.

    B.   Compare the work performance information against the time, cost, and scope baselines and look for deviations.

    C.   Create a change request and update the issue log and Cost Management plan to reflect the approved change.

    D.   Hold a status meeting.

Exam Questions

6. Joe, a stakeholder on your project, has been plotted on the power/interest grid as high interest with low power. What's the BEST approach for managing his engagement?

  A. Make him responsible for a deliverable on the project.

  B. Keep him informed of all of the decisions that might affect the project's outcome.

  C. Closely manage his requirements and expectations.

  D. Ignore his requirements and expectations, since he doesn't have the power to affect the project's outcome.

7. Which Stakeholder Management process is in the Initating process group?

  A. Manage Stakeholder Engagement

  B. Identify Stakeholders

  C. Plan Stakeholder Management

  D. Register Stakeholders

8. Sue, the sponsor of the industrial design project you're managing, is plotted on your power/ interest grid as high power, high interest. Which is the BEST approach for managing her requirements and expectations?

  A. Keep her informed of all project decisions.

  B. Manage her requirements and expectations closely.

  C. Keep her satisfied by inviting her to all of the team meetings.

  D. Understand her goals and expectations, but don't do anything with them.

9. Which is NOT an input of the Identify Stakeholders process?

  A. Procurement documents

  B. Enterprise environmental factors

  C. Project charter

  D. Project Management plan

10. Kyle is the project manager of a project that has teams distributed in many different places. One of the stakeholders in his project has asked that all formal communications from the project be shared with all of the teams, regardless of their location. This is an example of:

  A. A stakeholder expectation

  B. A stakeholder goal

  C. A stakeholder requirement

  D. Decoding communication

Exam Questions

11. Which information is NOT included in the stakeholder register?

    A.   Stakeholder name and group

    B.   Stakeholder requirements

    C.   Stakeholder expectations

    D.   Stakeholder deliverables

12. Which Stakeholder Management process is in the Monitoring and Controlling process group?

    A.   Distribute Information

    B.   Manage Stakeholder Engagement

    C.   Plan Communications

    D.   Control Stakeholder Engagement

Answers

Exam ~~Questions~~

1. Answer: D

Since Matt is working to bring other stakeholders to support the project, he's in a leading engagement role.

2. Answer: C

Work performance data is an input of some Stakeholder Management processes, but not an input to Plan Stakeholder Management.

3. Answer: A

The stakeholder register is the first thing you need to create here. It looks like some of the upper managers who might be thinking of cancelling the project need to have their ideas taken into account. Once you've identified them as stakeholders, you can work to bring their perspective into account on your project and include them in project management decisions.

> RIGHT.
> YOU NEED TO KNOW WHO YOUR STAKEHOLDERS ARE BEFORE YOU CAN GET THEM INVOLVED.

4. Answer: A

It sounds like this stakeholder is completely unaware of her responsibilities on the project. The next step here is to spend some time bringing her up to speed on what's expected.

5. Answer: C

When you look at work performance data and discover a new stakeholder requirement, you're doing the Control Stakeholder Engagement process. Some of the outputs of that process are change requests, project document updates, and Project Management plan updates.

Once you recognize the process that's being described, try to think of the outputs of the process to figure out which answer is best.

6. Answer: B

Stakeholders in the low power/high interest quadrant of the power/interest grid need to be kept informed of all project decisions.

Exam ~~Questions~~ Answers

7. Answer: B

Identify Stakeholders is the only process in Stakeholder Management that is part of the Initiating process group.

Flip back to chapter 4 for a quick refresher on the Initiating process group, now that you've studied Stakeholder Management. It's a good way to review.

8. Answer: B

Stakeholders in the high power/high interest quadrant of the power/interest grid need to have their expectations and requirements closely managed by the project manager.

It can get confusing because Identify Stakeholders is discussed at the end of the PMBOK Guide, but it's part of the Initiating processes and done before you get to the Planning processes.

9. Answer: D

The Project Management plan is not an input to the Identify Stakeholders process. Since the Identify Stakeholders process is part of the Initiating process group, the stakeholders are identified as an input to the Planning processes that create the Project Management plan.

10. Answer: A

This is a stakeholder expectation. The stakeholder expects that the team will do as requested.

11. Answer: D

Not all stakeholders have deliverables on a project. Some are sponsors, vendors, or others who might be involved in the project but not actually producing deliverables. When a stakeholder does have deliverables that she's accountable for, she'll be documented in the Scope Management plan.

12. Answer: D

Control Stakeholder Engagement is the only Monitoring and Controlling process in Stakeholder Management.

Making good choices

It's not enough to just know your stuff. You need to make good choices to be good at your job. Everyone who has the PMP credential agrees to follow the **Project Management Institute Code of Ethics and Professional Conduct**, too. The Code helps you with **ethical decisions** that aren't really covered in the body of knowledge—and it's a big part of the PMP exam. Most of what you need to know is **really straightforward**, and with a little review, you'll do well.

Doing the right thing

You'll get some questions on the exam that give you situations that you might run into while managing your projects and then ask you what to do. Usually, there's a clear answer to these questions: *it's the one where you stick to your principles.* Questions will make the decisions tougher by offering rewards for doing the wrong thing (like money for taking a project shortcut), or they will make the infraction seem really small (like photocopying a copyrighted article out of a magazine). If you stick to the principles in the PMP Code of Professional Conduct regardless of the consequences, you'll **always** get the answers right.

> The PMP exam categorizes ethics and professional conduct as part of a project manager's cross-cutting knowledge and skills. This means that there will be questions about ethics and professional conduct scattered throughout the exam alongside questions about each of the process groups. You should expect to see questions about specific ethical situations around each of the process groups. When you come across a question about ethics and professional conduct, you're likely to see it in the context of initiating, planning, executing, or monitoring and controlling a project.

The main ideas

In general, there are a few kinds of problems that the code of ethics prepares you to deal with.

1. **Follow all laws and company policies.**

2. **Treat everybody fairly and respectfully.**

3. **Have respect for the environment and the community you're working in.**

4. **Give back to the project management community by writing, speaking, and sharing your experience with other project managers.**

5. **Keep learning and getting better and better at your job.**

6. **Respect other people's cultures.**

7. **Respect copyright laws.**

8. **Always be honest with everyone on the project.**

9. **If you find that another person has done something to damage the PMP credential in any way, you must report it to PMI.**

> So if you find out that someone has stolen questions from the PMP exam, cheated on the PMP exam, falsely claimed to have a PMP certification, or lied about anything related to the PMP certification process, then you MUST report that to PMI.

Ethics and professional responsibility questions make up **10%** of the exam. That's good news because these questions are really easy if you understand the ideas behind the PMP Code of Professional Conduct.

COME ON. IS THIS REALLY ON THE TEST? I KNOW HOW TO DO MY JOB. DO I REALLY NEED A MORALITY LESSON?

Being a PMP-certified project manager means that you know how to do your job and that you will do it <u>with</u> integrity.

It might seem like it doesn't really matter how you will handle these situations, but think about it from an employer's perspective for a minute. Because of the PMI Code of Ethics and Professional Conduct, employers know that when they hire a PMP-certified project manager, they are hiring someone who will follow company policies and do everything aboveboard and by the book. That means that you'll help to protect their company from litigation and deliver on what you promise, which is actually pretty important.

So you should definitely expect to see questions about ethics and professional responsibility on the exam. Not only that, but you won't necessarily see them as straightforward, black-and-white questions, either. Since the questions on this topic are combined into the other questions for the process groups, you're likely to get questions about situations that might occur on real projects. A question about ethics or professional responsibility might look at first like a question about, say, a particular tool or technique in planning. Keep your eye out for "red herring" questions that turn out to be about ethics and social responsibility. They'll lay out a situation that sounds like a normal project management problem, but requires you to use one of the principles in the PMI Code of Ethics and Professional Conduct.

Can you think of some situations where you might need to make decisions using these principles in your own projects?

Keep the cash?

A lot of ethics questions on the PMP exam concern bribery. It is never, under any circumstances, OK to accept a bribe—even if your company and customer might benefit from it somehow. And bribes aren't always cash. They can be anything from free trips to tickets to a ball game. Any time you're offered anything to change your opinion or the way you work, you must decline the offer and disclose it to your company.

In some countries, even though you may be "expected" to pay a bribe, it's not okay to do it—even if it's customary or culturally acceptable.

> KATE, YOU WERE SO GREAT TO WORK WITH. WE'D LIKE TO SEND YOU $1,000 AS A TOKEN OF OUR APPRECIATION.

> AWESOME. I'VE BEEN WANTING TO GO SHOPPING FOR A WHILE. AND WHAT ABOUT THAT VACATION? ACAPULCO, HERE WE COME!

The easy way

> I WOULD NEVER ACCEPT A GIFT LIKE THAT. DOING A GOOD JOB IS ITS OWN REWARD!

The right way

> I'M SORRY, I CAN'T ACCEPT THE GIFT. I REALLY APPRECIATE THE GESTURE, THOUGH.

Fly business class?

Any time there's a policy in your company, you need to follow it. Even if it seems like no harm will be done if you don't follow the policy, and even if you will be able to get away with it, you should not do it. And that goes double for laws—under no circumstances are you ever allowed to break a law, no matter how much good it "seems" to do you or your project.

And if you ever see someone in your company breaking the law, you need to report it to the authorities.

WE'VE GOT SOME EXTRA MONEY IN THE BUDGET AND YOU'RE REALLY DOING A GREAT JOB. I KNOW THE TRAVEL POLICY SAYS WE ALWAYS FLY COACH. BUT WE CAN AFFORD TO SPLURGE A BIT. WHY DON'T YOU BUY A BUSINESS TICKET THIS TIME?

DID YOU KNOW THAT THOSE CHAIRS GO INTO TOTALLY FLAT BEDS? THIS IS SO COOL. I'VE WORKED SO HARD, I'VE TOTALLY EARNED IT!

THERE'S NO EXCUSE FOR NOT FOLLOWING THE RULES. THE TRAVEL POLICY SAYS FLY COACH. NO EXCEPTIONS!

WOW, BEN. THAT'S REALLY NICE OF YOU. BUT THE ECONOMY FARE WILL BE FINE.

New software

When it comes to copyright, it's never OK to use anything without permission. Books, articles, music, software...you always need to ask before using it. For example, if you want to use some copyrighted music in a company presentation, you should write to the copyright owner and ask for permission.

HEY KATE, I JUST GOT A COPY OF THAT SCHEDULING SOFTWARE YOU WANTED. YOU CAN BORROW MY COPY AND INSTALL IT.

ABSOLUTELY! THIS WILL TOTALLY MAKE MY JOB ABOUT 100 TIMES FASTER, AND IT'S FREE? IT'S MY LUCKY DAY.

THAT SOFTWARE WAS CREATED BY A COMPANY THAT DESERVES TO BE PAID FOR THEIR WORK. IT'S JUST WRONG NOT TO BUY A LICENSED COPY.

THANKS FOR LETTING ME KNOW IT'S AVAILABLE. I'LL GO BUY A COPY.

Shortcuts

You might see a question or two that asks if you really need to follow all of the processes. Or you might be asked by your boss to keep certain facts about your project hidden from stakeholders or sponsors. You have a responsibility to make sure your projects are run properly, and to never withhold information from people who need it.

WE DON'T HAVE TIME FOR ALL OF THIS DOCUMENTATION. LET'S CUT OUT A COUPLE OF THESE PLANS TO KEEP OUR PROJECT ON SCHEDULE.

ALL RIGHT, LESS WORK FOR ME! LET'S FACE IT, IT'S NOT LIKE I HAVE ALL THE TIME IN THE WORLD FOR WRITING PLANS!

I WOULD NEVER DO A PROJECT WITHOUT FOLLOWING ALL OF THE 47 PROCESSES OUTLINED IN THE PMBOK GUIDE.

I KNOW WE DON'T HAVE MUCH TIME, BUT TAKING THIS SHORTCUT COULD ACTUALLY COST US MORE TIME THAN IT SAVES IN THE END.

A good price or a clean river?

Being responsible to the community is even more important than running a successful project. But it's more than being environmentally aware—you should also respect the cultures of everyone else in your community, and the community where your project work will be done.

That means even though languages, customs, holidays, and vacation policies might be different from country to country, you need to treat people the way they are accustomed to being treated.

WE JUST FOUND OUT THAT ONE OF OUR SUPPLIERS DUMPS HARMFUL CHEMICALS IN THE RIVER. THEY'VE ALWAYS GIVEN US GREAT RATES, AND OUR BUDGET WILL GO THROUGH THE ROOF IF WE SWITCH SUPPLIERS NOW. THE WHOLE THING GIVES ME A HEADACHE. WHAT SHOULD WE DO?

WE CAN'T LET THE PROJECT FAIL FOR A BUNCH OF STUPID FISH.

THE EARTH IS OUR HOME AND IS SO MUCH MORE IMPORTANT THAN THIS PROJECT. WE HAVE TO DO WHAT'S RIGHT...

BEN, I KNOW IT COULD CAUSE US PROBLEMS, BUT WE'RE GONNA HAVE TO FIND ANOTHER SUPPLIER.

We're not <u>all</u> angels

We know that the choices you make on your project are not always black and white. Remember that the questions on the exam are designed to test your knowledge of the PMP Code of Professional Conduct and how to apply it. A lot of situations you will run into in real life have a hundred circumstances around them that make these decisions a little tougher to make than the ones you see here. But if you know what the code would have you do, you're in a good position to evaluate those scenarios as well.

Seriously, it's a quick read—and it'll help you on the exam.

Now, go read the PMP Code of Professional Conduct before you take these exam questions. Go to this URL, and click on the "PMI Code of Ethics and Professional Conduct" link.

http://www.headfirstlabs.com/hfpmp

I MAY NOT BE THE LIFE OF THE PARTY, BUT THINK LIKE ME, AND YOU'LL NAIL THE ETHICS PART OF THE EXAM.

Exam Questions

1. You read a great article over the weekend, and you think your team could really benefit from it. What should you do?

   A.   Photocopy the article and give it to the team members.

   B.   Type up parts of the article and email it to the team.

   C.   Tell everyone that you thought of the ideas in the article yourself.

   D.   Buy a copy of the magazine for everyone.

2. You find out that a contractor that you're working with discriminates against women. The contractor is in another country, and it's normal in that country. What should you do?

   A.   Respect the contractor's culture and allow the discrimination to continue.

   B.   Refuse to work with the contractor, and find a new seller.

   C.   Submit a written request that the contractor no longer discriminate.

   D.   Meet with your boss and explain the situation.

3. You're a project manager at a construction company that's selling services to a client. You are working on a schedule and a budget when the CEO at the client demands that you do not produce those things. Instead, he wants you to begin work immediately. What the BEST thing that you can do?

   A.   Meet with the CEO to explain why the budget and schedule are necessary.

   B.   Stop work immediately and go into claims administration.

   C.   Don't produce the schedule and budget.

   D.   Ask the buyer to find another company to work with.

4. You're working on a project when the client demands that you take him out to lunch every week if you want to keep his business. What's the BEST thing to do?

   A.   Take the client out to lunch and charge it to your company.

   B.   Refuse to take the client out to lunch because it's a bribe.

   C.   Take the client out to lunch, but report him to his manager.

   D.   Report the incident to PMI.

5. You are working on one of the first financial projects your company has attempted, and you have learned a lot about how to manage the project along the way. Your company is targeting financial companies for new projects next year. What's the BEST thing for you to do?

   A.   Talk to your company about setting up some training sessions so that you can teach others what you have learned on your project.

   B.   Keep the information you've learned to yourself so that you'll be more valuable to the company in the next year.

   C.   Decide to specialize in financial contracts.

   D.   Focus on your work with the project and don't worry about helping other people to learn from the experience.

Exam Questions

6. You find out that you could save money by contracting with a seller in a country that has lax environmental protection rules. What should you do?

    A.   Continue to pay higher rates for a environmentally safe solution.

    B.   Take advantage of the cost savings.

    C.   Ask your boss to make the decision for you.

    D.   Demand that your current contractor match the price.

7. You overhear someone on your team using a racial slur. This person is a critical team member and you are worried that if he leaves your company it will cause project problems. What should you do?

    A.   Pretend you didn't hear it so that you don't cause problems.

    B.   Report the team member to his boss.

    C.   Bring it up at the next team meeting.

    D.   Meet in private with the team member and explain that racial slurs are unacceptable.

8. You've given a presentation for your local PMI chapter meeting. This is an example of what?

    A.   A PDU

    B.   Contributing to the Project Management Body of Knowledge

    C.   Donating to charity

    D.   Volunteering

9. You are about to hold a bidder conference, and a potential seller offers you great tickets to a baseball game for your favorite team. What should you do?

    A.   Go to the game with the seller but avoid talking about the contract.

    B.   Go to the game with seller and discuss the contract.

    C.   Go to the game, but make sure not to let him buy you anything because that would be a bribe.

    D.   Politely refuse the tickets.

10. Your company has sent out an RFP, and your brother wants to bid on it. What's the BEST thing for you to do?

    A.   Give your brother inside information to make sure that he has the best chance at getting the project.

    B.   Publicly disclose your relationship with him and excuse yourself from the selection process.

    C.   Recommend your brother but don't inform anyone of your relationship.

    D.   Don't tell anyone about your relationship, but be careful not to give your brother any advantage when evaluating all of the potential sellers.

Exam ~~Questions~~ Answers

1. Answer: D

You should never copy anything that's copyrighted. Make sure you always respect other people's intellectual property!

2. Answer: B

It's never OK to discriminate against women, minorities, or others. You should avoid doing business with anyone who does.

3. Answer: A

This is a difficult situation for any project manager. But you can't cut corners on the project management processes, and you certainly can't tell the client that you're refusing their business. The best thing you can do is meet with the CEO to explain why you need to follow the rules.

4. Answer: B

The client is demanding a bribe, and paying bribes is unethical. You should not do it. If your project requires you to bribe someone, then you shouldn't do business with that person.

5. Answer: A

This is called contributing to the Project Management Body of Knowledge.

You should always try to help other people learn about managing projects.

6. Answer: A

You should never contract work to a seller who pollutes the environment. Even though it costs more to use machinery that doesn't damage the environment, it's the right thing to do.

Answers

Exam Questions

7. Answer: D

You should make sure that your team always respects other people.

8. Answer: B

Any time you help share your knowledge with others, you are contributing to the Project Management Body of Knowledge, and that's something you should do as a certified project manager!

9. Answer: D

You have to refuse the tickets even if the game sounds like a lot of fun. The tickets amount to a bribe, and you shouldn't do anything that might influence your decision in awarding your contract.

10. Answer: B

You have to disclose the relationship. It's important to be up front and honest about any conflict of interest that could occur on your projects.

15 A little last-minute review

Check your knowledge

BOY, EXAM DAY IS MY FAVORITE DAY OF THE YEAR! I ONLY WISH WE COULD HAVE SCHOOL ALL SUMMER, TOO.

Wow, you sure covered a lot of ground in the last 14 chapters! Now it's time to take a look back and drill in some of the most important concepts that you learned. That'll keep it all fresh and give your brain a final workout for exam day!

Process Magnets

Can you put all of the processes in the right knowledge areas?
Give it a shot—and while you're at it, see if you can put them inside each
knowledge area in the order that they're typically performed on a project.

Integration	Scope	Time	Cost
1	1	1	1
2	2	2	2
3	3	3	3
4	4	4	4
5	5	5	**Quality**
6	6	6	1
		7	2
			3

Plan Scope
Management

Develop Project
Management Plan

Plan Schedule
Management

Estimate
Activity
Durations

Perform
Integrated
Change Control

Determine Budget

Estimate Costs

Develop
Project Team

Conduct
Procurements

Control Scope

Plan Risk
Management

Define Scope

Control Stakeholder
Engagement

Develop
Schedule

Acquire
Project Team

Plan Cost
Management

Perform
Quantitative
Risk Analysis

Plan
Communications
Management

Control Quality

Plan Procurement
Management

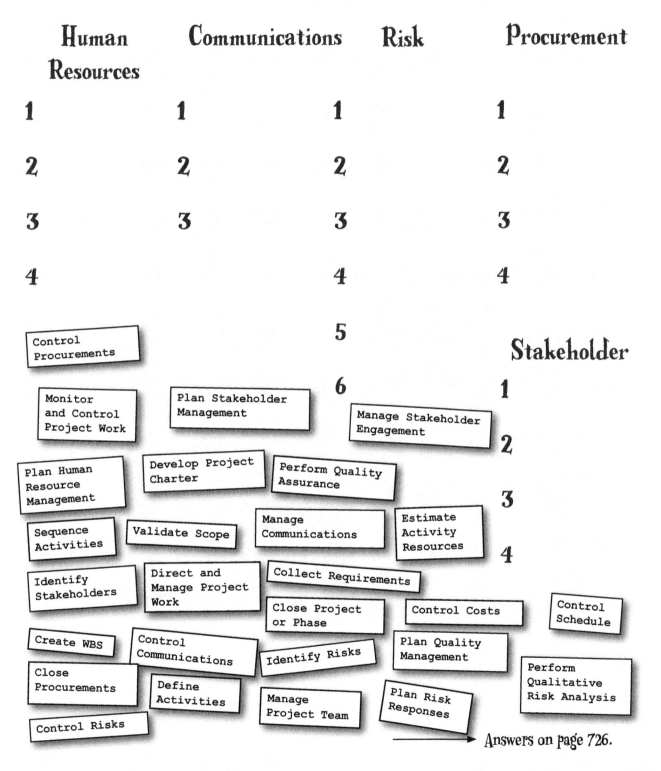

Human Resources	Communications	Risk	Procurement
1	1	1	1
2	2	2	2
3	3	3	3
4		4	4
		5	
		6	**Stakeholder**

Stakeholder

1
2
3
4

Control Procurements

Monitor and Control Project Work

Plan Stakeholder Management

Manage Stakeholder Engagement

Plan Human Resource Management

Develop Project Charter

Perform Quality Assurance

Sequence Activities

Validate Scope

Manage Communications

Estimate Activity Resources

Identify Stakeholders

Direct and Manage Project Work

Collect Requirements

Create WBS

Control Communications

Close Project or Phase

Control Costs

Control Schedule

Close Procurements

Identify Risks

Plan Quality Management

Perform Qualitative Risk Analysis

Define Activities

Manage Project Team

Plan Risk Responses

Control Risks

Answers on page 726.

Processcross

How well do you know the 47 processes in the PMBOK Guide? Let's find out!

Answers on page 728.

Across

2. The _____ Communications process gets the right information to the right people. It's where most of the communication on the project happens.

5. The Close Project or _____ process is where you make sure the project is finished, and all of the lessons learned are documented.

6. The document that authorizes you to do the work is built in the Develop Project _____ process.

Down

1. The _____ Stakeholder Management Process is where you plan all of the activities you'll do to keep stakeholders satisfied through the project.

2. You create a document that defines how you'll handle every aspect of the project in the Develop Project _____ Plan process.

3. The Perform _____ Risk Analysis is where you categorize each risk.

Across

7. You manage all communication for the people who are affected by your project in the Manage Stakeholder _____ process.

8. The _____ Risks process is where you create a risk register that contains a list of risks that might affect your project.

9. The _____ Procurements process is where you make sure all of the contracting activities are finished.

10. In the Determine _____ process, you add up all of your estimated costs and figure out how much money your project will spend in total.

13. The _____ Procurements process is the Monitoring and Controlling process for procurements, where you look for changes in your contracts.

14. The Collect _____ process is where you gather the needs of the stakeholders and document them.

16. The _____ Management knowledge area is all about figuring out your budget.

17. The _____ Project Team process is where you track your team members' performance, provide feedback, and resolve issues.

19. In the _____ Management knowledge area, you determine how long the work will take.

20. The _____ Management knowledge area helps you figure out the work that needs to be done for your project.

21. The _____ Procurements process is where you determine which sellers will do the work.

22. The _____ Scope process is where you make sure all of the work has been done, and you get formal approval from the stakeholders.

26. The Plan _____ _____ Management process is where you create a plan for how you will assign and manage your staff.

27. You track your work closely and manage your costs in the _____ Costs process.

28. The _____ process group is where you make sure the project starts out right.

29. In the _____ Activities process, you decompose each work package into a complete list of activities for the project.

32. The _____ process group is where you do the most work. It's where you build a document to guide you through each of the knowledge areas.

33. In the _____ Schedule process, you build a bar chart, milestone list, calendar, or other document out of all of your estimates.

38. The _____ and Controlling process group is concerned with finding and dealing with changes.

39. The Control Risks process is where you look for any new risks and changes to the risk _____.

40. The Control _____ process is where you look for changes to the planned dates for performing activities and meeting milestones.

46. The _____ Stakeholder Engagement process is the process of tracking stakeholder engagement through the project and making corrections and changes based on project information.

47. The _____ Scope process is where you write down exactly what the team will do to produce the product.

48. In the _____ and Control Project Work process, you constantly look for changes or problems that occur.

49. The Estimate Activity _____ process is where you generate an estimate of how long each activity will take.

50. You assign your team to the project in the _____ Project Team process.

52. The _____ process group is where you shut down the project.

53. The Plan _____ Management process is where you create a plan to ensure that your deliverables conform to requirements, and are fit for use.

Across

54. The _____ Communications process gets the right information to the right people. It's where most of the communication on the project happens.

55. The Estimate Activity _____ process is where you figure out what people, equipment, and other things that you'll need for the project, and when you'll need them.

Down

4. The _____ Activities process is where you put the list of activities in order and create network diagrams.

5. In the Plan _____ Management process, you decide what work you'll contract out to a seller.

6. In the _____ Quality process, your team looks for defects in deliverables.

11. In the Perform _____ Risk Analysis process, you assign numerical values to your risks in order to more accurately assess them.

12. You create a plan that tells you how you manage unexpected events in the Plan _____ Management process.

15. _____ Management is the knowledge area where you bring all of the work and project plans together.

18. Control _____ is where you figure out how your project is doing, and let everyone else know.

23. In the Perform Quality _____ process, you make sure the entire project and quality processes meet your company's quality standards.

24. The Create _____ process is where you create a graphical, hierarchical document that describes all work packages.

25. In the _____ Management knowledge area, you contract with sellers to do project work.

30. The _____ process group is where the team does the project work.

31. The Develop Project _____ process is where you keep your team motivated, and set goals and rewards for them.

34. _____ Management is the knowledge area where you figure out who's talking to whom, and how.

35. The _____ Management knowledge area is where you figure out who is impacted by your projects and work to keep them satisfied.

36. The Plan _____ Management process is where you'll figure out how you'll handle messages, channels of communication, meetings, and reporting.

37. The deliverables and work performance information are created in the Direct and Manage Project _____ process.

39. The Plan Risk _____ process is where you decide how your team will react to each risk, should it occur.

41. You work with sponsors and stakeholders to decide whether or not to make changes during the Perform _____ Change Control process.

42. The Human _____ Management knowledge area is where you put together and manage your team.

43. The _____ Management knowledge area is where you plan for the unknown.

44. The _____ Scope process is where you look for changes to scope, and make only those changes that are necessary.

45. You make sure your deliverables conform to their requirements using the _____ Management knowledge area processes.

51. The Estimate _____ process is where you figure out how much money you'll spend on each activity in the schedule.

Exercise

These questions are all about specific things that you're likely to see on the exam. They're drawn from many different knowledge areas. Take some time and try to answer all of them—remember, these are a little harder than questions you'll see on the exam, since they're not multiple choice!

1. What's it called when you bring your stakeholders together at the beginning of the project in order to figure out how everyone will communicate throughout the project?

....

2. What do you call the point in a fixed-price contract where the seller assumes the rest of the cost?

....

3. Which conflict resolution technique is most effective?

....

4. What's the range for a rough order of magnitude estimate?

....

5. Which contract type is best when you don't know the scope of the work?

....

6. What are you doing when you add resources to the critical path in order to shorten the schedule?

....

7. Which management theory states that employees can't be trusted and need to be constantly monitored?

....

8. What are the top three causes of conflict on projects?

....

9. Customer satisfaction is part of which knowledge area?

....

10. Which type of power is typically unavailable to project managers in a matrix organization?

....

11. Which form of communication is **always** necessary whenever you are performing procurement processes?

....

12. What's it called when you add up the costs of inspection, test planning, testing, rework (to repair defects discovered), and retesting?

....

13. What are the three characteristics of a project that differentiate it from a process?

....

14. Where would you find out details about a specific work package, such as an initial estimate or information about what account it should be billed against?

....

15. What do you do when you and your team can't identify a useful response to a risk that you've identified?

....

16. What do you decompose work packages into before you can build your schedule?

....

17. What's the float for any activity on the critical path?

....

18. What percentage of a project manager's time is spent communicating?

....

19. What should you do with the factors that cause change?

....

20. Which two types of estimate require historical information?

....

⟶ **Answers on page 729.**

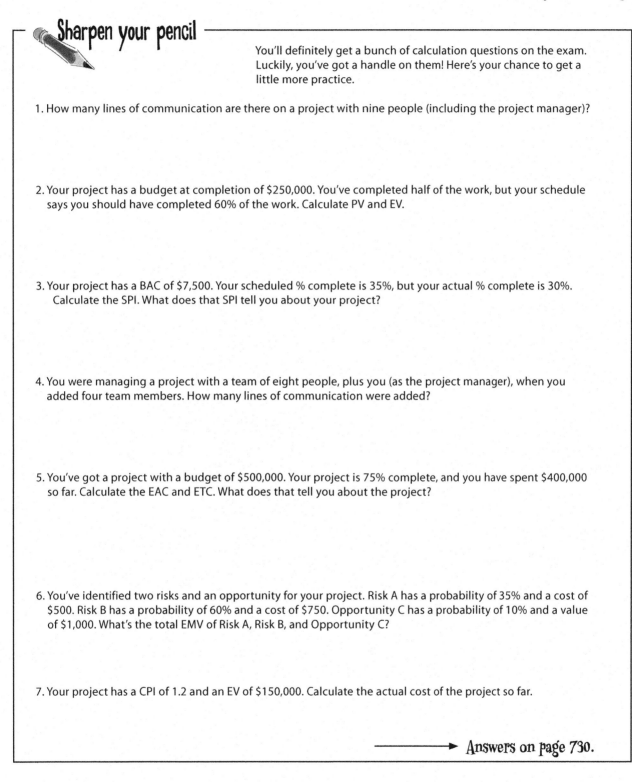

Sharpen your pencil

You'll definitely get a bunch of calculation questions on the exam. Luckily, you've got a handle on them! Here's your chance to get a little more practice.

1. How many lines of communication are there on a project with nine people (including the project manager)?

2. Your project has a budget at completion of $250,000. You've completed half of the work, but your schedule says you should have completed 60% of the work. Calculate PV and EV.

3. Your project has a BAC of $7,500. Your scheduled % complete is 35%, but your actual % complete is 30%. Calculate the SPI. What does that SPI tell you about your project?

4. You were managing a project with a team of eight people, plus you (as the project manager), when you added four team members. How many lines of communication were added?

5. You've got a project with a budget of $500,000. Your project is 75% complete, and you have spent $400,000 so far. Calculate the EAC and ETC. What does that tell you about the project?

6. You've identified two risks and an opportunity for your project. Risk A has a probability of 35% and a cost of $500. Risk B has a probability of 60% and a cost of $750. Opportunity C has a probability of 10% and a value of $1,000. What's the total EMV of Risk A, Risk B, and Opportunity C?

7. Your project has a CPI of 1.2 and an EV of $150,000. Calculate the actual cost of the project so far.

⟶ Answers on page 730.

A long-term relationship for your brain

Take a minute and think back over everything you've just learned. Does it seem a little…well, overwhelming? Don't worry, that's absolutely normal. You've got all of this information that's floating around in your brain, and your brain is still trying to organize it.

Your brain is an amazing machine, and it's really good at organizing information. Luckily, when you feed it so much new data, there are ways that you can help make it "stick." That's what you've been doing in this chapter. Your brain wants its new information to be categorized. That's why it helps to first review how the processes are organized, and then review what the processes do.

> I GET IT! THAT'S WHY I DID THE PROCESS MAGNETS FIRST, THEN THE PROCESS CROSSWORD, THEN THE CALCULATION QUESTIONS. WORKING THROUGH THESE THINGS SEPARATELY CAN HELP ME GET THEM INTO MY BRAIN.

Yes! Cognitive psychologists call it *chunking*, and it's a really effective way of getting information into your long-term memory. When you have a collection of things that are strongly associated with one another, it gives your brain a sort of "guideline" for storing it. And the weaker associations with the other "chunks" give it a bigger framework for managing this large amount of information, so that it's all mutually reinforcing.

Here's how to do this next section

This next section consists of a series of questions grouped together by knowledge area. To make this as effective as possible:

* **Make this the only PMP study activity you do today.**

* **Give yourself plenty of time to do it.**

* **Make sure you drink lots of water while you're answering the questions.**

* **As you're answering the questions, think about each answer and only mark down one response, even if you're not 100% sure.**

* **After you do each section, read through each question again.**

* **Don't look at the answers until you've gone through all of the knowledge areas.**

* **Make sure you get plenty of sleep the night after you do these questions.**

COGNITIVE PSYCHOLOGISTS RECOGNIZE THAT SLEEP PLAYS A REALLY IMPORTANT ROLE IN HELPING YOUR BRAIN ORGANIZE AND CONSOLIDATE INFORMATION INTO YOUR LONG-TERM MEMORY.

Integration Questions

1. You're managing a project for a military subcontractor to modify the software for a missile guidance system. You're planning the project, and need to take into account information about the company's operating environment. Which of the following is NOT an example of the factors you are looking at:

    A.   Marketplace conditions

    B.   The forecast for project completion, including ETC and TCPI

    C.   Government standards you need to comply with

    D.   The political climate that can affect your project

2. The CFO of a company tells you that you need to include a new feature in software that's being produced by a project you are managing. The deadline for the project is tight, and if you take on the extra work then the project will come in late. What BEST describes the first thing that you should do?

    A.   Update the Project Management plan to make sure the new feature is included in the project.

    B.   Tell the CFO that he needs to wait until you're working on the next version, and then submit a change request to the change control board.

    C.   Tell the CFO that the deadline is too tight, and the feature can't be included.

    D.   Evaluate the impact that the change will have on the project.

3. Which of the following BEST describes the role of the project sponsor?

    A.   Assigning work to the project team

    B.   Paying for the project

    C.   Politically supporting the project inside the organization

    D.   Defining the type of organization (matrix, functional, etc.)

4. Which of the following is NOT a part of the Project Management plan?

    A.   The lifecycle selected for the project

    B.   The level of implementation for each of the processes

    C.   The organization's staffing and retention guidelines

    D.   Techniques used to communicate with stakeholders

Scope Questions

5. You're managing an architecture project to design an extension on an existing building. One of the stakeholders has been adamant that the plans do not include an interior supporting wall, because she wants to be able to reconfigure the floorplan. Where do you document this information?

    A.   Project scope statement

    B.   Scope Management plan

    C.   Work breakdown structure

    D.   WBS dictionary

6. You're in the process of talking to the project stakeholders, figuring out what they need, and writing it down. Which of the following is a tool or technique that you would use?

    A.   Decomposition

    B.   Observations

    C.   Variance analysis

    D.   Inspection

7. A project manager is analyzing deliverables and subdividing them into smaller, more manageable components. This project manager is performing WHICH process?

    A.   Control Scope

    B.   Define Scope

    C.   Collect Requirements

    D.   Create WBS

8. A software project team lead is working with stakeholders to make sure that there's formal, documented acceptance of every one of the project deliverables. Which of the following BEST describes the work that she's doing?

    A.   Performing variance analysis on the cost baseline

    B.   Updating the traceability matrix

    C.   Structuring and organizing the WBS

    D.   Running the software and walking through it with the stakeholders

Time Questions

9. You're working on an IT project to set up a development environment, including designing and building a computer room, installing the operating systems and software, and performing a security evaluation. You need at least two weeks to order the hardware before you can configure it and install the operating systems. Which of the following BEST describes this relationship?

    A.   Lead

    B.   Lag

    C.   Finish-to-Start (FS)

    D.   Start-to-Start (SS)

10. You're planning an IT project to set up a development environment, including designing and building a computer room, installing the operating systems and software, and performing a security evaluation. Your project includes three different activities that involve three different network technicians splicing ends onto wires in order to build their own Ethernet cables, because that's less expensive than buying prepackaged ones. Every Ethernet cable must be tested with a qualification tester. That's an expensive piece of equipment, and there are only a few of them that must be shared among all of the technicians in the company. You need to plan your schedule based on the availability of the testing equipment. What's the BEST place to find that information?

    A.   Staffing requirements

    B.   Activity network diagram

    C.   Resource calendar

    D.   Activity resource requirements

11. You're working on a IT project to set up a development environment, including designing and building a computer room, installing the operating systems and software, and performing a security evaluation. Once the operating system on a machine is installed, it needs to be imaged and copied to three identical boxes. Which of the following BEST describes this relationship?

    A.   Lead

    B.   Lag

    C.   Finish-to-Start (FS)

    D.   Start-to-Start (SS)

12. You're working on a IT project to set up a development environment, including designing and building a computer room, installing the operating systems and software, and performing a security evaluation. Your team comes up with a best-case scenario for the activity that involves ordering and installing the equipment. If everything goes perfectly, they feel it will take five weeks. However, they think it's much more likely to take nine. A team member points out that on his last project, there was a major equipment delivery delay that cost the project an extra four weeks, and the rest of the team agrees that this is a possibility in a worst-case scenario. Use PERT analysis to calculate how long you should expect this activity to take.

    A.   5 weeks

    B.   9 weeks

    C.   12 weeks

    D.   13 weeks

Cost Questions

13. Which of the following BEST describes funding limit reconciliation?

    A.   Comparing your project's budget against the project's reserves

    B.   Comparing your project's planned expenditures against the funding constraints

    C.   Comparing your project's planned value against the actual costs

    D.   Comparing your project's net present value against the internal rate of return

14. Your project has a budget at completion (BAC) of $75,000, and you need to figure out if you're on track to meet it. You know that you've already spent $56,000, and you're 70% done with the project. If your project continues expenditures at the current rate, what's the lowest that you can allow your CPI to go before you've exceeded your project's budget?

    A.   .94

    B.   1.18

    C.   1.43

    D.   1.08

15. You're managing a construction project to install 7,500 light switches in a new high-rise building. You've installed 3,575 of them so far, and you've spent $153,500 of your total budget of $245,000. Which of the following is true?

    A.   The AC is $153,500, so I'm ahead of schedule.

    B.   The ETC is $168,995, so I'm within my budget.

    C.   The CV is $36,880, so I've exceeded my budget.

    D.   The CPI is .7597, so I've exceeded my budget.

16. You're managing a construction project, and you're estimating your project's activities using a spreadsheet that a consulting company created for you. You have to fill in the dry weight of the materials, the number of people required to do the work, the type of building you're working on, and other information about the project. Which BEST describes what you are doing?

    A.   Parametric estimation

    B.   Analogous estimation

    C.   Bottom-up estimation

    D.   Top-down estimation

Quality Questions

17. Which of the following is NOT considered when calculating cost of quality?

    A.   How much it costs to repair deliverables when they don't meet requirements

    B.   The cost of training the team to perform inspections

    C.   The cost of working with customers who find problems with the work delivered to them

    D.   The cost of gaining formal acceptance of project deliverables

18. Which of the following is NOT an example of quality assurance?

    A.   Examining the way deliverables are produced to see if processes are being followed

    B.   Examining deliverables to see if they meet requirements

    C.   Examining a group of deliverables to figure out why they all had the same defect

    D.   Examining the company's documentation on how processes are to be performed

19. Which of the following BEST describes a situation where statistical sampling is appropriate?

    A.   You just got a shipment of 50,000 parts, and you need to figure out if enough of them are within tolerances to be used on your project.

    B.   You need to figure out which defects are critical, and which can be delivered to your customer and repaired later.

    C.   You need to use the rule of seven on a control chart.

    D.   You need to determine if your project is ahead of schedule and within budget.

20. You're planning your project's quality activities. You know that as you go through your project, your team will find lots of ways to improve how your company does work in the future. You need a way to handle that information in a systematic manner. What document is BEST used to plan for this?

    A.   Quality checklist

    B.   Process Improvement plan

    C.   Quality Management plan

    D.   Quality metrics

Human Resource Questions

21. A project manager working in a projectized organization is putting a team together to perform a project. Which of the following is NOT a tool or technique that might be used to do this?

    A.   Colocation

    B.   Virtual teams

    C.   Preassignment

    D.   Negotiation

22. A project manager is building her team. She notices that team members are not collaborating efficiently, and she is concerned that this is leading to a destructive environment. Which of the following BEST describes the state that the team is in?

    A.   Norming

    B.   Forming

    C.   Storming

    D.   Performing

23. A project manager is running into trouble with his team, because they always come to meetings late. Which of the following is MOST LIKELY to be the cause of this problem?

    A.   The team resents the work that they're doing.

    B.   The project manager needs to schedule meetings later in the day.

    C.   The project manager was late to meetings himself, and this influenced the team.

    D.   The team is in the storming phase of team development.

24. A project manager for a software project is well regarded by people in her company as an expert programmer. She has an especially good reputation among the senior managers of the company. The CTO of the company tells her that any time she runs into trouble, he'll back up any decision that she makes—and everyone on the team admires the CTO, and they have a lot of loyalty to him. Which BEST describes the power the project manager is exerting?

    A.   Reward power

    B.   Expert power

    C.   Punishment power

    D.   Referent power

Communications Questions

25. Bob and Sue are stakeholders in your project. Sue is in the high-power, low-interest quadrant of the power/interest grid. Bob is in the low-power, high-interest quadrant. Which of the following BEST describes the approach you should take:

A. Sue needs to participate in every important meeting, while Bob needs to have his opinion heard in those meetings.

B. You need to make sure that Bob is on the change control board, but Sue requires only minimum effort.

C. You need to go out of your way to satisfy all of Sue's needs, while Bob has to be kept in the loop on important decisions.

D. Sue needs to be managed closely, and you must satisfy Bob by making sure he feels his needs are being met.

26. A project manager is analyzing the communications requirements for a project. There are six team members, four stakeholders, and two subcontractors. He needs to find the number of potential communication channels. How many channels are there?

A. 13

B. 55

C. 66

D. 78

27. You're about to close your project when you are surprised to get an email from someone you've never spoken to before. He is very angry, because he's directly impacted by your project, and there are specific things he needs from it and you aren't delivering those things. Which BEST describes the the communications process you are in?

A. You did not manage communications effectively.

B. You did not identify a project stakeholder.

C. You did not encode your communications.

D. You did not have a Stakeholder Management strategy.

28. You are in the process of making relevant information available to your project's stakeholders. Which of the following is NOT a tool or technique that you would use?

A. Hard copies of documents interoffice-mailed to stakeholders

B. Power/interest grid that includes all stakeholders

C. Conference calls with stakeholders

D. An online folder that contains project documents

Risk Questions

29. A project manager is analyzing project risks by using quantitative techniques to assign a numeric value to each of those risks. Which of the following tools and techniques is NOT used for this?

    A.   EMV analysis that uses a decision tree

    B.   Sensitivity analysis to determine which risks pose the biggest threat

    C.   A Probability and Impact matrix that assigns numeric values to each risk priority

    D.   A simulation that runs through many different project scenarios

30. One of your construction project team members warns you that your concrete supplier caused serious delays in his last project because he delivered the wrong kind of concrete. You discuss it with the team, and decide that you need to accept the possibility that this will happen. But you make back-up plans with an alternate provider by putting a down payment on an emergency shipment, just in case. This is an example of:

    A.   Avoidance

    B.   Transferrance

    C.   Mitigation

    D.   Acceptance

31. A project manager is forced to dip into his management reserve. Which of the following is the MOST LIKELY cause of this?

    A.   One of the risks on the risk register caused a budget overrun.

    B.   The project went over budget because the project manager miscalculated his forecast.

    C.   A risk that was never planned for occurred.

    D.   The team's estimates were incorrect.

32. Your software project team informs you that another team working at the company built a tool that will save three weeks on the project. You ask the other team's project manager to have his team members share the tool with your team members. Which BEST describes this strategy?

    A.   Exploiting

    B.   Sharing

    C.   Enhancing

    D.   Accepting

Procurement Questions

33. A project manager is in trouble because he took on a contract that overran its budget, and now he has to eat the cost out of his own budget. Which of the following BEST describes his contract?

    A.   Time and materials

    B.   Fixed price

    C.   Cost plus incentive fee

    D.   Cost plus percentage of cost

34. A project manager for a software project hires a subcontractor to build a module that will be used in the rest of the project. When it's time to integrate that module into the rest of the code, there are serious quality problems. The subcontractor claims that the module meets its requirements, but the project manager's lead developer says that the requirements have clearly been violated. The contract is unclear about how to handle the situation. It's clear that the contract needs to be amended to indicate a solution, but there is no agreement on the exact wording of the change. What BEST describes the next step for the project:

    A.   The buyer and seller must proceed to claims administration.

    B.   The buyer should file a lawsuit against the seller.

    C.   The buyer should conduct an audit of the seller.

    D.   The buyer and seller must adhere strictly to the wording of the contract.

35. A buyer and seller have a teaming agreement. Which of the following BEST describes their relationship?

    A.   The seller and buyer both have team members on every project team.

    B.   The buyer can dictate the structure of the seller's project teams.

    C.   The seller is free to dictate deliverables and contract terms.

    D.   The seller has input into project decisions, and representation in the buyer's management structure.

36. You're a project manager planning a project that requires that you hire a contractor. Before you can find sellers, you need to develop a document that defines the portion of the work that the contractor will do. Which of the following is NOT true about this document?

    A.   It's based on the project scope baseline.

    B.   It includes exact specifications for the deliverables the contract will produce.

    C.   Its terms are either fixed price or cost reimbursable.

    D.   It must completely define the work that the contractor must do.

Stakeholder Questions

37. Tom, a stakeholder in an IT project, has asked to be included in all of the status meetings and team communications. While he is not actually on the team or accountable for any deliverables, he cares about the outcome of the project and wants to be kept in the loop. Where would you plot him on a power/interest grid?

- A.   High power, high interest
- B.   High power, low interest
- C.   Low power, high interest
- D.   Low power, low interest

38. A project manager for an industrial design project is reviewing the work performance data that's being produced by his project team to determine whether or not changes need to be made in Stakeholder Management. When he finds a change, he writes up a request and incorporates all approved requests into the Project Management plan, project documents, and lessons learned. What process is he performing?

- A.   Plan Stakeholder Management
- B.   Identify Stakeholders
- C.   Manage Stakeholder Engagement
- D.   Control Stakeholder Engagement

39. Joanne, the sponsor for a software project, has been working with the senior management team in her company to cancel the project even though it's only just begun its planning processes. Which is the BEST classification for Joanne's engagement with the project?

- A.   Unaware
- B.   Resistant
- C.   Supportive
- D.   Leading

Great job! It looks like you're almost ready

If you've read all the chapters, done all the exercises, and taken all of the practice questions, then you have a solid grasp on the material for the PMP exam. You're *almost ready* to get certified! By the way, don't worry if you didn't get some of the questions on the past few pages. This was really hard stuff—some of it was even harder than the PMP exam. Remember, a great way to prepare is to write your own Question Clinic–style questions for anything that's giving you trouble.

So if you did get them, you should feel very proud of yourself!

Process Magnets Solutions

Can you put all of the processes in the right knowledge areas?
Give it a shot—and while you're at it, see if you can put them inside each
knowledge area in the order that they're typically performed on a project.

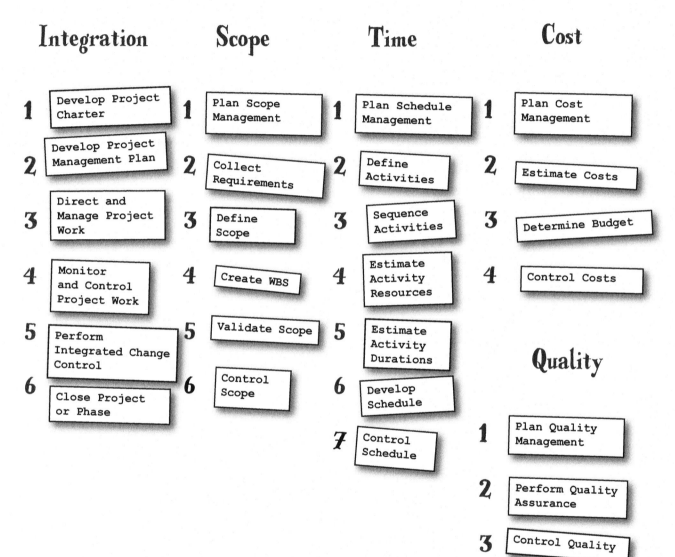

Integration

1. Develop Project Charter
2. Develop Project Management Plan
3. Direct and Manage Project Work
4. Monitor and Control Project Work
5. Perform Integrated Change Control
6. Close Project or Phase

Scope

1. Plan Scope Management
2. Collect Requirements
3. Define Scope
4. Create WBS
5. Validate Scope
6. Control Scope

Time

1. Plan Schedule Management
2. Define Activities
3. Sequence Activities
4. Estimate Activity Resources
5. Estimate Activity Durations
6. Develop Schedule
7. Control Schedule

Cost

1. Plan Cost Management
2. Estimate Costs
3. Determine Budget
4. Control Costs

Quality

1. Plan Quality Management
2. Perform Quality Assurance
3. Control Quality

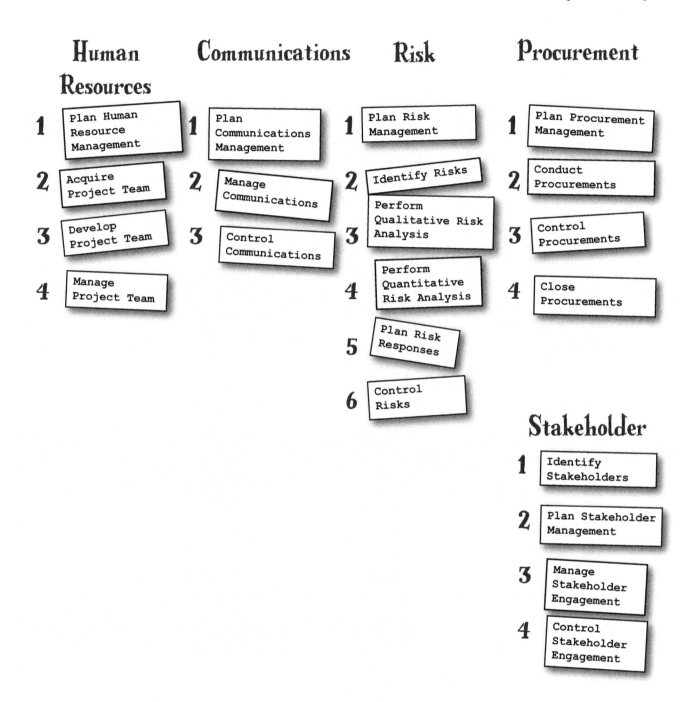

Human Resources

1 Plan Human Resource Management

2 Acquire Project Team

3 Develop Project Team

4 Manage Project Team

Communications

1 Plan Communications Management

2 Manage Communications

3 Control Communications

Risk

1 Plan Risk Management

2 Identify Risks

3 Perform Qualitative Risk Analysis

4 Perform Quantitative Risk Analysis

5 Plan Risk Responses

6 Control Risks

Procurement

1 Plan Procurement Management

2 Conduct Procurements

3 Control Procurements

4 Close Procurements

Stakeholder

1 Identify Stakeholders

2 Plan Stakeholder Management

3 Manage Stakeholder Engagement

4 Control Stakeholder Engagement

Processcross Solutions

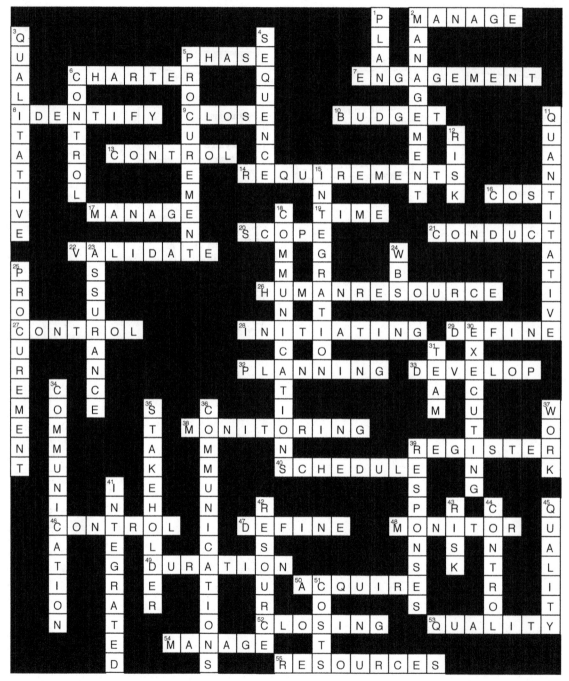

Exercise Solution

These questions are all about specific things that you're likely to see on the exam. They're drawn from many different knowledge areas. Take some time and try to answer all of them—remember, these are a little harder than questions you'll see on the exam, since they're not multiple choice!

1. What's it called when you bring your stakeholders together at the beginning of the project in order to figure out how everyone will communicate throughout the project?
   Kickoff meeting

2. What do you call the point in a fixed-price contract where the seller assumes the rest of the cost?
   Point of total assumption

3. Which conflict resolution technique is most effective?
   Problem solving or confronting

4. What's the range for a Rough Order of Magnitude estimate?
   –25% to +75%

5. Which contract type is best when you don't know the scope of the work?
   Time and materials

6. What are you doing when you add resources to the critical path in order to shorten the schedule?
   Crashing the schedule

7. Which management theory states that employees can't be trusted and need to be constantly monitored?
   McGregor's Theory X

8. What are the top three causes of conflict on projects?
   Resources, priorities, and schedules

9. Customer satisfaction is part of which knowledge area?
   Quality Management

10. Which type of power is typically unavailable to project managers in a matrixed organization?
    Lgitimate power

11. Which form of communication is *always* necessary whenever you are performing procurement processes?
    Formal written

12. What's it called when you add up the costs of inspection, test planning, testing, rework (to repair defects discovered), and retesting?
    Cost of quality

13. What are the three characteristics of a project that differentiate it from a process?
    Temporary, unique, and progressively elaborated

14. Where would you find out details about a specific work package, such as an initial estimate or information about what account it should be billed against?
    WBS dictionary

15. What do you do when you and your team can't identify a useful response to a risk that you've identified?
    Accept it

16. What do you decompose work packages into before you can build your schedule?
    Activities

17. What's the float for any activity on the critical path?
    Zero

18. What percentage of a project manager's time is spent communicating?
    90%

19. What should you do with the factors that cause change?
    Try to influence them

20. Which two types of estimate require historical information?
    Analogous and parametric

Sharpen your pencil
Solution

You'll definitely get a bunch of calculation questions on the exam. Luckily, you've got a handle on them! Here's your chance to get a little more practice.

1. How many lines of communication are there on a project with nine people (including the project manager)?

The formula for # lines of communication is: # lines = $n \times (n - 1) \div 2$

So # lines = $9 \times (9 - 1) \div 2 = 9 \times 8 \div 2 = 36$ lines of communication

2. Your project has a budget at completion of $250,000. You've completed half of the work, but your schedule says you should have completed 60% of the work. Calculate PV and EV.

The formulas are: PV = BAC × scheduled % complete   EV = BAC × actual % complete

So the answers are: PV = $250,000 × 60% = $150,000   EV = $250,000 × 50% = $125,000

3. Your project has a BAC of $7,500. Your scheduled % complete is 35%, but your actual % complete is 30%. Calculate the SPI. What does that SPI tell you about your project?

The formula is: SPI = EV ÷ PV.   So first calculate EV and PV: EV = $7,500 × 30% = $2,250
PV = $7,500 × 35% = $2,625

Now calculate SPI: SPI = $2,250 ÷ $2,625 = .86   Since SPI is lower than 1, your project is behind schedule.

4. You're managing a project with a team of eight people, plus you (as the project manager), when you added four team members. How many lines of communication were added?

We already figured out the team of 9 people in question 1—you have 36 lines of communication. The new team has four more team members, or a total of 13, so # lines = $13 \times (13 - 1) \div 2 = 13 \times 12 \div 2 = 78$ lines
So the number of lines added is 78 lines − 36 lines = 42 lines of communication

5. You've got a project with a budget of $500,000. Your project is 75% complete, and you have spent $400,000 so far. Calculate the EAC and ETC. What does that tell you about the project?

The formulas are: EV = BAC × % complete   CPI = EV ÷ AC   EAC = BAC ÷ CPI and ETC = EAC − AC
First calculate EV = $500,000 × 75% = $375,000, and then calculate CPI = $375,000 ÷ $400,000 = .94

Now you can calculate EAC = $500,000 ÷ .94 = $531,915 and ETC = $531,915 − $400,000 = $131,915
This means that you should expect to spend about $131,915 before the project ends.

6. You've identified two risks and an opportunity for your project. Risk A has a probability of 35% and a cost of $500. Risk B has a probability of 60% and a cost of $750. Opportunity C has a probability of 10% and a value of $1,000. What's the total EMV of Risk A, Risk B, and Opportunity C?

To calculate EMV, add up the individual probabilities multiplied by costs (negative) or values (positive).

EMV = (.35 × −$500) + (.60 × −$750) + (.10 × $1,000) = −$525

7. Your project has a CPI of 1.2 and an EV of $150,000. Calculate the actual cost of the project so far.

The formula is: CPI = EV ÷ AC.   First fill in what we know: 1.2 = $150,000 ÷ AC

Now flip it around: AC = $150,000 ÷ 1.2 = $125,000

Integration Questions ~~Answers~~

(Answers)

1. B.

Since this question is asking about the company's operating environment, it's really asking you to figure out which of the answers is not an enterprise environmental factor. Marketplace conditions, government standards, and political climate are enterprise environmental factors, but forecasts aren't—that's a Cost Management output.

2. D.

Whenever your project's scope changes, that means you need to put your project through change control. That doesn't necessarily mean that you need to have your team change the way they'll work, or reject the change outright. The first step in evaluating any change is understanding the impact that it will have on the project.

3. B.

The main role of the sponsor is to provide funding for the project. That's why the sponsor is an important stakeholder. However, the sponsor does not necessarily have a specific role on the project beyond paying for it.

That doesn't mean that the sponsor can't also be involved in those other things. A sponsor who's paying for a project also wants it to succeed, and will often fill other roles on the project as well.

4. C.

When you put together your Project Management plan, one of the first things that you do is select a project methodology (or lifecycle), and determine exactly how each of the processes will be implemented. Your Communications Management plan will definitely have specifics about how you communicate with your stakeholders, because that's a really important part of managing a project. But your Project Management plan doesn't typically include your company's policies. For example, guidelines about how your company hires and retains staff are usually set by an HR department, and they generally don't vary from project to project.

Staffing and retention guidelines are an example of one of your company's policies, and policies are enterprise environmental factors.

Scope ~~Questions~~ Answers

5. A.

This stakeholder wants to exclude something from the scope, and project exclusions—which identify what is out of scope for the project—should always be documented in the project scope statement. (You might also consider it a constraint—technically, it's both—but constraints are also documented in the project scope statement.)

This will also be a project requirement, so it will end up in a requirements document as well. But that wasn't one of the choices in the question!

6. B.

When you're talking to project stakeholders and documenting their needs, you're collecting requirements. "Observations" is a technique used in the Collect Requirements process. And that makes sense—you often need to observe people doing their jobs in order to figure out how they'll use your project's deliverables.

7. D.

When a project manager analyzes deliverables and subdivides them into smaller components, he is using decomposition, which is the only tool or technique in the Create WBS process.

8. D.

When a project team lead—who, in this case, is acting as the project manager—is gathering formal acceptance of deliverables, she is performing the Validate Scope process. That process has only one tool or technique—Inspection—and a walkthrough is a very common way to inspect a deliverable.

Time Questions Answers

9. A.

When one activity (such as ordering hardware) must take place a certain amount of time before another activity (installing the operating systems), that's called a lead.

10. C.

The resource calendar has information on which resources—which includes equipment and material, as well as people—are available at specific times. You'll typically use it to estimate resource utilization, which you'll need to do when you're building the schedule for this project.

Did you think that the resource calendar applied only to human resources? Any scarce piece of equipment is also a resource, and you can use a resource calendar to make sure it's available when your team needs it.

11. C.

A Finish-to-Start (FS) relationship is the most common sort of predecessor that you'll see in a project schedule. That's what you call it when one activity (imaging the machine) starts as soon as another one is finished (the operating system is installed).

12. B.

This question is asking you to apply the PERT analysis three-point estimate. The optimistic estimate is 5 weeks, the most likely is 9 weeks, and the pessimistic is 13, so the three-point estimate is (5 + 4 × 9 + 13) ÷ 6 = 9 weeks.

Answers

~~**Cost Questions**~~

13. B.

Funding limit reconciliation means checking the project's expenditures—how much you've already spent—against any limits that your company has on the commitments of funds. Companies typically don't allow project managers to throw unlimited amounts of money at the project, so this makes sure that the project can be done within the company's guidelines...so you can catch cost overruns before you've spent more than you're allowed to spend! This is how you know you haven't blown your budget—by comparing it against the hard limits set by your company.

14. B.

When you're asked to figure out the lower limit of your CPI to keep your project within its budget, you're being asked to calculate the to-complete performance index (TCPI). If you've got a BAC, an AC, and a % complete, then you know enough to compute the EAC, so you should use the formula TCPI = (EAC – EV) ÷ (EAC – AC):

EV = BAC × actual % complete = $75,000 x 70% = $52,500

AC = $56,000

So as long as your CPI doesn't go below 1.18 for the rest of the project, you'll still hit your budget.

CPI = EV ÷ AC = $52,500 ÷ $56,000 = 0.9375

EAC = BAC ÷ CPI = $80,000

TCPI = (EAC – EV) ÷ (EAC – AC) = ($80,000 - $52,500) ÷ ($80,000 – $56,000) = (1.18)

This looks like a lot of work, but once you break it down, it's all formulas that you know.

15. D.

The first step in figuring this out is to figure out the actual % complete, which you can do by figuring that you've installed 3,570 of the total 7,500 lightswitches, or 3,575 ÷ 7,500 = 47.6%. Then you can use the formulas, which show you that CPI is .7597, telling you that you've exceeded your budget:

Actual % complete = 47.6%

AC = $153,500 (but since this doesn't tell you anything about your schedule, answer A is incorrect)

EV = BAC × actual % complete = $245,000 x 47.6% = $116,620

CV = EV – AC = $116,620 - $153,500 = –$36,880 (since the CV is negative, you're below your budget, so answer C is incorrect; the CV in this case is negative)

CPI = EV ÷ AC = $116,620 ÷ $153,500 = .7597

EAC = BAC ÷ CPI = $245,000 ÷ .7597 = $322,495

ETC = EAC – AC = $322,495 – $153,500 = $168,995 (which doesn't actually tell you whether or not you're within your budget, so answer B is incorrect)

16. A.

A really common way of doing parametric estimation involves entering numbers into a spreadsheet that performs calculations based on historical data gathered from previous projects.

Quality ~~Questions~~ Answers

17. D.

Gaining formal acceptance of project deliverables is part of Scope Verification, which is NOT a quality activity, so it's not part of the cost of quality. All of the other answers involve costs incurred in either finding defects or dealing with them once they've been found, and that's all part of cost of quality.

18. B.

When you're performing quality assurance, that means that you're looking at the way that people are doing their jobs. Often, it means that you're taking a step back to look at the big picture—if many defects have the same root cause, if all of the processes are documented, and if they're being followed properly. However, if you're looking at individual deliverables, then you're doing inspection, which is part of quality control, not quality assurance.

A lot of the time, when you're performing quality assurance, you're looking at ongoing processes and not just projects.

19. A.

Statistical sampling helps you make decisions about a large number of items without having to look at every single one of them.

20. B.

The Process Improvement plan is a plan that you build as part of the Plan Quality Management process that helps you improve the way your company does its work. This is what you use to help your team think "outside" of your project, and look at the company's overall process or methodology for doing projects.

Answers

Human Resource Que~~stions~~

21. A.

Colocation means having most or all of your team members working in the same location. This is an important tool for running a team, but the question is asking about acquiring a team. Virtual teams, preassignment, and negotiation are all tools and techniques of Acquire Project Team.

22. C.

Many teams go through five stages of development: forming, storming, norming, performing, and adjourning. The storming stage occurs early on in team development, before the team members are really comfortable with one another or the work. During this stage, they often have trouble collaborating, and are not necessarily open to one another's ideas.

23. C.

One of the most important interpersonal skills that a project manager has is influencing, and leading by example is a very effective way to do this. But when a project manager sets a bad example, it is almost certain to be reflected in the behavior or attitude of the team.

Even if the team is in the storming phase of team development, they should still be expected to act in a professional manner—and this includes showing up to meetings on time.

24. D.

Referent power means that you have the power or ability to attract others and build loyalty. One effective way that people often wield referent power is to take advantage of the loyalty that the team already has to someone high up in the company—in this case, the CTO.

Communications Questions ~~Answers~~

Answers

25. C.

Since Sue is in the high-power, low-interest quadrant, she needs to be kept satisfied. This means that she has to feel that her needs are actually being met, but since she's not following the project on a day-to-day basis, the only way you can do that is by delivering a final product that meets those needs. Bob is low-power, high-interest, so he needs to be kept informed. This means that he needs to feel like he's constantly in the loop on important decisions, even if he won't actively be taking part in them.

26. D.

This is a basic "lines of communication" problem—you need to figure out how many people are communicating. In this case it's six team members, four stakeholders, two subcontractors...and don't forget the project manager! That's 6 + 4 + 2 + 1 = 13 people. The formula is $n(n - 1) \div 2 = 13(12) \div 2 = 78$.

27. B.

The definition of a stakeholder includes "anyone who is directly impacted by your project." That means this person is a stakeholder! Since you never spoke to this person, you failed to identify him as a stakeholder, and as a result you did not meet his needs.

Tools and techniques like a Stakeholder Management strategy or stakeholder register are great, but they don't work well if you haven't identified all of the project's stakeholders.

28. B.

When you're making relevant information available to your project's stakeholders, you're performing the Manage Communications process. The power/interest grid is a useful tool, but it's not part of Manage Communications.

Answers
Risk ~~Questions~~

29. C.

The question asked for "quantitative techniques," which means that it's asking about tools and techniques in the Perform Quantitative Risk Analysis process. Even though the Probability and Impact matrix assigns numbers to risks, it's not one of those tools and techniques—in fact, it's not a quantitative technique at all! It's a qualitative technique, because it's used for prioritization and categorization.

30. C.

This is an example of risk mitigation, because you are taking steps to deal with the problem just in case it happens.

Did the word "accept" throw you? Make sure that you always read the whole question, and don't just take a single word out of context!

31. C.

The management reserve is the part of the budget reserved for unplanned risks. If a risk was on the risk register, it was planned for, and the contingency reserve is used to pay for it. In this case, since the management reserve was used, that means the risk wasn't planned for at all.

32. A.

When the other team built that tool, that presented an opportunity. Asking the team to share the tool with you is your way of taking advantage of it. That's called exploiting the opportunity.

Even though the word "share" appeared in the question, since only one party—your team!—is benefiting from the action, you're exploiting the opportunity, not sharing it. Since your use of the tool doesn't benefit the other team at all, this isn't an opportunity for them.

Procurement Questions ~~Answers~~

Answers

33. B.

The fixed-price contract is the riskiest type of contract that a seller can take on. When the project's costs exceed the price of the contract, the seller has to pay for the overrun.

> This is usually bad for the buyer, too! Contracting works best when it's a win–win situation for both the buyer and the seller.

34 A.

Claims administration is what you do when there are contested changes to the contract, and the seller and buyer can't reach an agreement on the change. Any time you see a dispute or appeal between the buyer and seller and there's no clear resolution, that's where claims administration comes into play.

35. D.

When a buyer and seller have a teaming agreement, the seller acts as a partner with the buyer. It's not just a typical "here's the terms, you go off and do the work" relationship that a lot of sellers and buyers have. Instead, the buyer asks for the seller's input, and makes the seller an active part of the management of the project.

> When a seller works on a project with a teaming agreement, the team members from both the buyer's and seller's teams usually have a lot of mutual respect. That's why this is a really effective way to do procurement.

36. C.

A document that defines the portion of work the contractor will do is the procurement statement of work (SOW). It's a clear, concise, unambiguous, and complete document that describes the work that must be done, and can include specifications of deliverables. But it doesn't include contract terms—those terms are part of the contract itself.

37. C.

Tom is a low-power, high-interest stakeholder. The project manager on the project should work to keep him informed.

38. D.

The project manager is performing the Control Stakeholder Engagement process. The outputs of Control Stakeholder Engagement are work performance information, change requests, Project Management plan updates, project document updates, and organizational process asset updates. The question mentioned most of these outputs.

39. B.

Joanne is a resistant stakeholder. If she is attempting to cancel the project, she is not onboard with the benefits the project might bring.

16 Practice makes perfect

Practice PMP exam

I KNOW WE'RE SUPPOSED TO BE STUDYING, BUT I CAN'T STOP THINKING ABOUT FUDGE.

Bet you never thought you'd make it this far! It's been a long journey, but here you are, ready to review your knowledge and get ready for exam day. You've put a lot of new information about project management into your brain, and now it's time to see just how much of it stuck. That's why we put together this 200-question PMP practice exam for you, completely updated for the 2013 exam. It looks just like the one you're going to see when you take the real PMP exam. Now's your time to flex your mental muscle. So take a deep breath, get ready, and let's get started.

Exam Questions

1. A team member approaches you with a change that could cut your schedule down by a month. What is the first thing you should do?

    A.   Write up a change request and see if you can get it approved.

    B.   Make the change; it's going to save time and nobody will want the project to take longer than it should.

    C.   Figure out the impact on the scope of the work and the cost before you write up the change request.

    D.   Tell the team member that you've already communicated the deadline for the project, so you can't make any changes now.

2. You have been hired by a contractor who wants you to manage a construction project for one of his clients. The project team has been working for six weeks. You need to determine whether the team is ahead of or behind schedule. Which of the following tools and techniques is the BEST one for you to consult?

    A.   Work performance data

    B.   Project management software

    C.   Schedule change control system

    D.   Bottom-up estimating

3. Which of the following is not a tool or technique of the Control Quality process?

    A.   Inspection

    B.   Quality audits

    C.   Pareto charts

    D.   Statistical sampling

4. Brandi is a project manager on a software project. About halfway through development, her team found that they had not estimated enough time for some of the technical work they needed to do. She requested that the new work be added to the scope statement and that the time to do the work be added to the schedule. The change control board approved her change. What's her next step?

    A.   Update the scope and schedule baselines to reflect the approved change.

    B.   Start doing the work.

    C.   Gather performance metrics on the team's work so far.

    D.   Perform Quality Assurance.

5. You are initiating a project that has a virtual team. Half of your team members will be located in another country, where they are working for a subcontractor. The subcontractor's team members speak a different language than your team does. When you need to get information to them, it must first be translated. This is an example of:

    A.   Encoding

    B.   Transmission

    C.   Decoding

    D.   Acknowledgment

Exam Questions

6. Which of the following is NOT a source of information about specific project constraints and assumptions?

    A.   The Scope Management plan

    B.   Requirements documentation

    C.   The project scope statement

    D.   The scope baseline

7. When do you perform stakeholder analysis?

    A.   During the Initiating process group

    B.   When developing the project charter

    C.   When creating the Project Management plan

    D.   When putting changes through change control

8. Which of the following is NOT true of obtaining project plan approval?

    A.   Until you obtain plan approval, you don't need to put changes to it through change control.

    B.   Change control makes sure that only approved changes can make it into the approved plan.

    C.   Only one person needs to approve the Project Management plan and that's the project manager.

    D.   It's important for the entire team to buy into the Project Management plan for it to be successful.

9. When are the most expensive defects most likely to be introduced into a product?

    A.   When the product is being assembled

    B.   When the product is being designed

    C.   When the Quality Management plan is being written

    D.   When the product is being reviewed by the customers

Exam Questions

10. You are the project manager for a railroad construction project. Your sponsor has asked you for a forecast for the cost of project completion. The project has a total budget of $80,000 and CPI of .95. The project has spent $25,000 of its budget so far. How much more money do you plan to spend on the project?

  A. $59,210
  B. $80,000
  C. $84,210
  D. $109,210

11. Which of the following best describes decomposition?

  A. Waiting for a task to expire so that it can break down into smaller tasks
  B. Taking a deliverable and breaking it down into the smaller work packages so that it can be organized and planned
  C. Categorizing work packages
  D. Dividing work packages into deliverables that can be planned for

12. Which is the BEST definition of quality?

  A. A product made of very expensive materials
  B. A product made with a lot of care by the team who built it
  C. A product that satisfies the requirements of the people who pay for it
  D. A product that passes all of its tests

13. In which plan do you define the processes that will be used to keep people informed throughout the project?

  A. Staffing Management plan
  B. Project Management plan
  C. Schedule Management plan
  D. Communications Management plan

14. Which enterprise environmental factor defines how work is assigned to people?

  A. RACI matrix
  B. Project management information system (PMIS)
  C. Resource histogram
  D. Work authorization system

15. You are currently performing the Conduct Procurements process. You are considering two bids from companies on your qualified sellers list. Your project is on a tight budget, and you have been instructed by senior management to consider the cost over any other criteria. You used the company that submitted the lower

Exam Questions

bid on a previous project, and you were not happy with its work. The company that submitted the higher bid has a reputation for treating its clients well, flying project managers first-class, and giving them accommodations in five-star hotels. What is the BEST way to handle this situation?

- A. Select the company with the lowest bid.
- B. Give the manager at the company with the higher bid information that will allow him to tailor his bid so that it better meets your needs.
- C. Rewrite the RFP so that the company with the lowest bid is excluded.
- D. Select the company with the higher bid.

16. Which of the following shows roles and responsibilities on your project?

- A. Bar chart
- B. Resource histogram
- C. RACI matrix
- D. Human Resource Management plan

17. A project manager is working in a country where it is customary to pay the police for private protection services, and those costs have increased the AC. This has affected his earned value calculations. When he reviews the budget with his company's management, his supervisor tells him that in another country, those costs would be considered a bribe, and questions whether they should be added to the budget. What is the BEST way for the project manager to proceed?

- A. Do not pay the police for private protection services, because that would be a bribe.
- B. Pay the police for private protection services, because it is customary in the country they are operating in.
- C. Consult the Cost Management plan about payment.
- D. Initiate cost control to update the cost baseline.

18. Which conflict resolution technique is most effective?

- A. Withdrawal
- B. Compromise
- C. Smoothing
- D. Confronting

19. Which of the following is NOT an input to Control Quallity?

- A. Deliverables
- B. Work performance data
- C. Quality checklists
- D. Validated changes

Exam Questions

20. You have just delivered a product to your client for acceptance when you get a call that some features they were expecting are missing. What's the first thing you should do?

A. Get your team together and reprimand them for building a product that doesn't meet user expectations.

B. Tell the client that the product passed all of your internal quality inspections and scope verification processes, so it must be fine.

C. Tell the team to start building the missing features into the product right away.

D. Call a meeting with the client to understand exactly what is unacceptable in the product and try to figure out what went wrong along the way.

21. You are managing a software project. You are partway through the project, and your team has just delivered a preliminary version of part of the software. You are holding a weekly status meeting when one of the team members points out that an important stakeholder is running into a problem with one of the features of the current software. The team member feels that there is a risk that the stakeholder will ask for a change in that feature, even though that change would be out of scope of the current release—and if the stakeholder requests that change, there is a high probability that the change control board would approve the change. What is the BEST action to take next?

A. Mitigate the risk by asking a team member to get familiar with the feature of the software that might be changed.

B. Schedule a meeting with the stakeholder to discuss the risk.

C. Add the risk to the risk register and gather information about its probability and impact.

D. Add the risk to the issue log and revisit it when there is more information.

22. Tom is the project manager of an accounting project. He has just finished defining the scope for the project and is creating the WBS. He goes to his organizational process asset library and finds a WBS from a past project to use as a jumping-off point. Which of the following describes the asset that Tom is using?

A. Decomposition

B. Delphi technique

C. Brainstorming

D. Templates

23. Which of the following BEST describes the main purpose of the project charter?

A. It authorizes the project manager to work on the project.

B. It identifies the sponsor and describes his or her role on the project.

C. It contains a list of all activities to be performed.

D. It describes the initial scope of the work.

24. You are the project manager for a software development project. When you need to get staff from the manager of the QA department, he suggests a few test engineers with performance problems for your team. Which is the BEST response to this situation?

Exam Questions

    A.   Stop talking to the QA manager.

    B.   Call a meeting with the QA manager to try to figure out why he suggested those candidates and how the two of you can work together to find team members with suitable skills and interests for your team.

    C.   Tell the QA manager that the staffing problems are really no big deal, and you're sure that the two of you can eventually figure out the right answer together.

    D.   Tell the manager that you know which team members you want for your team and he needs to give them to you.

25. You are managing a construction project using a firm fixed price (FFP) contract. The contract is structured so that your company will be paid a fee of $85,000 to complete the work. There was a $15,000 overhead cost that your company had to cover. It's now three months into the project, and your costs have just exceeded $70,000. The project has now consumed the entire fee, and your company will now be forced to pay for all costs on the project from this point forward. What's the BEST way to describe this situation?

    A.   The project manager has overspent the budget.

    B.   The project is overdrawn.

    C.   The project has reached the point of total assumption.

    D.   The project has ceased to be a profit center for the company.

26. A project manager is reporting the final status of the closed contract to the stakeholders. Which form of communication is appropriate?

    A.   Informal written

    B.   Informal verbal

    C.   Formal written

    D.   Formal verbal

27. You are managing a software engineering project. While investigating the cause of a low SPI, you discover that your team is having trouble completing their object design tasks, which are on the critical path. One of your team members tells you that her friend at another company sent her a copy of a software package it owns that will help your team meet its deadline. Without that software package, your project will probably be late. But you don't have enough money in the budget to purchase it. What's the BEST way to handle this situation?

    A.   Tell the team member not to use the software, and accept that the project will be late.

    B.   Use the software so that your project comes in on time.

    C.   Purchase the software so that you have a licensed copy.

    D.   Find a way to add resources to the object design activity or move it off of the critical path.

Exam Questions

28. You are managing a project with a total budget of $450,000. According to the schedule, your team should have completed 45% of the work by now. But at the latest status meeting, the team reported that only 40% of the work has actually been completed. The team has spent $165,000 so far on the project. How would you BEST describe this project?

    A.   The project is ahead of schedule and within its budget.

    B.   The project is behind schedule and within its budget.

    C.   The project is ahead of schedule and over its budget.

    D.   The project is behind schedule and over its budget.

29. Which of the following is the correct order of the Monitoring and Controlling processes for Scope Management?

    A.   First Validate Scope, then Control Scope.

    B.   First Control Scope, then Validate Scope.

    C.   Both happen simultaneously.

    D.   There is not enough information to decide.

30. You are working on a construction project. You, your team, and your senior manager all feel that the work is complete. However, one of your important clients disagrees, and feels that one deliverable is not acceptable. What is the BEST way to handle this conflict?

    A.   Consult the contract and follow its claims administration procedure.

    B.   Renegotiate the contract.

    C.   File a lawsuit to force the stakeholder to accept the deliverable.

    D.   Terminate the contract and follow any termination procedure in the contract.

31. One way the Close Project or Phase process differs from Close Procurement is:

    A.   Procurement closure involves verification that all work and deliverables are acceptable, whereas Close Project or Phase does not.

    B.   Close Project or Phase is only a subset of Close Procurement.

    C.   Procurement closure means verifying that the project is complete or terminated; Close Project or Phase is the process of tying up all of the activities for every management process group.

    D.   Procurement closure is performed by the seller; Close Project or Phase is performed by the buyer.

32. Which of the following contracts has the MOST risk for the buyer?

    A.   Cost plus fixed fee (CPFF)

    B.   Time and materials (T&M)

    C.   Cost plus award fee (CPAF)

    D.   Fixed price (FP)

Exam Questions

33. You are managing a software project. During a walkthrough of newly implemented functionality, your team shows you a new feature that they have added to help make the workflow in the product easier for your client. The client didn't ask for the feature, but it does look like it will make the product easier to use. The team developed it on their own time because they wanted to make the client happy. You know this change would never have made it through change control. What is this an example of?

    A. Gold plating

    B. Scope creep

    C. Alternatives analysis

    D. Schedule variance

34. While identifying risks for a new construction project, you discover that a chemical you are using on your building cannot be applied in rainy conditions. You also learn that your project will be ready for the chemical application around the time when most of the rainfall happens in this part of the country. Since the project can't be delayed until after the rainy season and you need to make sure the building gets the chemical coating, you decide that your team will just have to allow enough time in the schedule for nonworked rain days.

This is an example of which strategy?

    A. Mitigate

    B. Exploit

    C. Accept

    D. Transfer

35. You are managing a construction project. During your risk identification interviews, you learn that there has been a string of construction site thefts over the past few months in the area where you will be building your project. The team agrees that it's unlikely that people will be able to steal from your site. Even if thieves could get around your security, it's even more unlikely that your project will lose a significant amount of material if a theft does occur. You decide to monitor the risk from time to time to be sure that it continues to have a low probability and impact. Where do you record the risk so that you don't lose track of it?

    A. In a trigger

    B. On a watchlist

    C. In the Probability and Impact matrix

    D. In the Monte Carlo analysis report

Exam Questions

36. Which of the following is NOT a characteristic of the Project Management plan?

- A. Collection of subsidiary plans
- B. Formal, written communication
- C. A bar chart that shows the order of tasks and their resource assignments
- D. Must be approved by project sponsor

37. You are developing the project scope statement for a new project. Which of the following is NOT part of creating a project scope statement?

- A. Validate Scope
- B. Using the project charter
- C. Alternatives identification
- D. Obtaining plan approval

38. Which of the following is NOT a tool or technique of Estimate Costs?

- A. Bottom-up
- B. Parametric
- C. Cost aggregation
- D. Analogous

39. You're managing a construction project to install several hundred air conditioner panels in a new office building. Every floor has identical panels. The customer, a construction contracting company, has provided specifications for the installations. The team is using a process to install and verify each panel. As the team completes each panel, your team's quality control inspector measures it and adds the data point to a control chart. You examine the control chart, and discover that the process is out of control and you need to take a close look at it immediately. Which of the following BEST describes what you found on the control chart?

- A. At least seven consecutive measurements are either above or below the mean but within the control limits.
- B. At least one point is inside of the control limits.
- C. At least seven consecutive measurements are inside of the control limits.
- D. At least one point is above or below the mean.

40. A senior manager is presenting to your client's board of directors about your project. As part of managing communications, you must deliver status updates and other project materials to him via overnight mail. If the materials do not arrive tomorrow, your company will miss a major contract deadline and you will lose future business from this important client. The deadline cannot be negotiated. The team worked right up to the last minute in order to give you the files. Due to a traffic jam, you are running late and the overnight delivery company will close in five minutes. You can only make it if you drive over the speed limit. Which of the following is correct?

- A. You must drive over the speed limit so that you can save the client relationship.
- B. You must stay within the speed limit, even if you lose the client.

Exam Questions

    C.   Update the organizational process assets to reflect the change.

    D.   You can use the earned value metrics to show that the SPI is over 1, meaning the project is not late.

41. You are planning a project that uses the same team as a project that is currently being performed by your company. What should you consult to find information about when those people will be available for your project?

    A.   The project schedule for your project

    B.   The project manager for the project that the team is working on

    C.   The Staffing Management plan for the project that the team is working on

    D.   The Communications Management plan for your project

42. When is the BEST time to have project kickoff meetings?

    A.   At the beginning of the project

    B.   When each deliverable is created

    C.   At the start of each phase

    D.   When the Communications Management plan is approved

43. Joe is a project manager on an industrial design project. He has found a pattern of defects occurring in all of his projects over the past few years, and he thinks there might be a problem in the process his company is using that is causing it. He uses Ishikawa diagrams to come up with the root cause for this trend over projects so that he can make recommendations for process changes to avoid this problem in the future. What process is he doing?

    A.   Plan Quality Management

    B.   Perform Quality Assurance

    C.   Control Quality

    D.   Perform Qualitative Risk Analysis

44. You are a project manager for a software project. Your team buys a component for a web page, but they run into defects when they use it. Those defects slow your progress down considerably. Fixing the bugs in the component will double your development schedule, and building your own component will take even longer. You work with your team to evaluate the cost and impact of all of your options and recommend hiring developers at the company that built the component to help you address problems in it. That will cost more, but it will reduce your delay by a month. What is your next step?

    A.   Fix the component.

    B.   Write up the change request and take it to the change control board.

    C.   Start Plan Procurements so you can get the contract ready for the vendor.

    D.   Change the scope baseline to include your recommendation.

Exam Questions

45. The terms of union contracts are considered _____ in your project plan.

    A. Assumptions

    B. Constraints

    C. Requirements

    D. Collective bargaining agreements

46. A change has occurred on your project. You've documented the change, filled out a change request, and submitted that request to the change control board (CCB). What's the NEXT thing that must happen on the project?

    A. A senior manager decides whether or not to make the change and informs the project management team of the decision.

    B. The project manager informs the CCB whether or not to approve the change.

    C. Stakeholders on the CCB use expert judgment to evaluate the requested change for approval.

    D. The project manager meets with the team to analyze the impact of the change on the project's time, scope, and cost.

47. You are managing a design project. You find that bringing all of your team members into a single room to work increases their communication, and helps build a sense of community. This is referred to as a:

    A. War room

    B. Virtual team

    C. Socially active team

    D. Common area

48. You are a project manager on a construction project. You have just prepared an RFP to send around to electrical contractors. You get a call from your uncle who owns an electrical contracting company. He wants to bid on your project. You know he's done good work before, and it may be a good fit for your company. How do you proceed?

    A. You disclose the conflict of interest to your company, and disqualify your uncle's company.

    B. You disclose the conflict of interest to your company, and make the selection based on objective criteria.

    C. You disclose the conflict of interest to your company, and provide your uncle with information that the other bidders don't have so that he has a better chance of winning the contract.

    D. You do not disclose the conflict of interest, and give your uncle the bid.

49. You are managing a software project. You are partway through the project, and your team has just delivered a preliminary version of part of the software. Your team gives a demonstration to the project sponsor and key stakeholders. Later, the sponsor informs you that there is an important client who will be using the software your team is building, and whose needs are not being met. As a result, you must now make a large and expensive change to accommodate that client. What is the BEST explanation for this?

Exam Questions

A. The sponsor is being unreasonable.

B. Stakeholder analysis was not performed adequately.

C. The team made a serious mistake and you need to use punishment power to correct it.

D. You do not have enough budget to perform the project.

50. Which of the following is NOT an input of the Close Project or Phase process?

A. Project Management plan

B. Project management methodology

C. Accepted deliverables

D. Organizational process assets

51. The scope baseline consists of:

A. The Scope Management plan, the project scope statement, and the WBS

B. The Scope Management plan, requirements documents, and the WBS

C. The Scope Management plan, the WBS, and the WBS dictionary

D. The project scope statement, the WBS, and the WBS dictionary

52. You are managing a construction project that is currently being initiated. You met with the sponsors, and have started to work on identifying stakeholders. You've documented several key stakeholders and identified their needs. Before you can finish initiating the project, your company guidelines require that you make a rough order of magnitude estimate of both time and cost, so that the sponsor can allocate the final budget.

What's the range of a rough order of magnitude (ROM) estimate?

A. −10% to +10%

B. −25% to +75%

C. −50% to +100%

D. −100% to +200%

53. Which of the following processes is in the Initiating process group?

A. Develop Project Charter

B. Develop Project Management Plan

C. Define Scope

D. Define Activities

Exam Questions

54. Mary is a project manager at a consulting company. The company regularly builds teams to create products for clients. When the product is delivered, the team is dissolved and assigned to other projects. What kind of organization is she working for?

    A. Weak matrix
    B. Projectized
    C. Functional
    D. Strong matrix

55. An important part of performing stakeholder analysis is documenting quantifiable expectations. Which of the following expectations is quantifiable?

    A. The project must improve customer satisfaction.
    B. The project should be higher quality.
    C. The project must yield a 15% reduction in part cost.
    D. All stakeholders' needs must be satisfied.

56. At the close of your project, you measure the customer satisfaction and find that some customer needs were not fully met. Your supervisor asks you what steps you took on your project to improve customer satisfaction. Which subsidiary plan would you consult to determine this information?

    A. Quality Management plan
    B. Communications Management plan
    C. Staffing Management plan
    D. Risk Management plan

57. Customer satisfaction should be measured at the end of the project to maintain long-term relationships. Which of the following is NOT always an aspect of customer satisfaction?

    A. The product meets its stated and unstated requirements.
    B. The project is profitable.
    C. The product is high quality.
    D. The customer's needs are met.

58. Dave is the project manager for a construction team that is building a gazebo. When the project first started, he met with the stakeholders to define the scope. The sponsors mentioned that the gazebo is a really important part of their daughter's wedding ceremony that was planned for seven months from then. In fact, they said that if the gazebo couldn't be completed in seven months, it wouldn't be worth it for them to even start the project. Dave wrote down the seven-month deadline to put in his project scope statement. In which section of the document did the deadline appear?

    A. Project deliverables
    B. Project objectives

Exam Questions

C.   Project constraints

D.   Project assumptions

59. Which of the following is a "hygiene factor" under Herzberg's Motivation-Hygiene Theory?

A.   Recognition for excellent work

B.   Self-actualization

C.   Good relations with coworkers and managers

D.   Clean clothing

60. A bar chart showing the number and type of resources you need throughout your project is called a(n) _____.

A.   Organizational chart

B.   Resource schedule

C.   Resource histogram

D.   Staffing timetable

61. You have identified an opportunity to potentially increase the project's value. Which of the following is an example of enhancing that opportunity?

A.   By forming a partnership with another company, you will increase the project's value for both companies.

B.   By taking additional actions, you increase the potential reward without reducing its probability.

C.   By taking out insurance, you can reduce potential costs to the project.

D.   By documenting the opportunity in the register, you can keep track of it and ensure it gets exploited.

62. Which of the following best describes the Plan-Do-Check-Act cycle?

A.   Invented by Joseph Juran, it's a way of tracking how soon defects are found in your process.

B.   Invented by Walter Shewhart and popularized by W. E. Deming, it's a method of making small changes and measuring the impact before you make wholesale changes to a process.

C.   Made popular by Phillip Crosby in the 1980s, it's a way of measuring your product versus its requirements.

D.   It means that you plan your project, then do it, then test it, and then release it.

Exam Questions

63. You are developing the project charter for a new project. Which of the following is NOT part of the enterprise environmental factors?

    A.   Lessons learned from previous projects

    B.   Knowledge of which departments in your company typically work on projects

    C.   The work authorization system

    D.   Government and industry standards that affect your project

64. You are managing a construction project to install new door frames in an office building. You planned on spending $12,500 on the project, but your costs are higher than expected, and now you're afraid that your project is spending too much money. What number tells you the difference between the amount of money you planned on spending and what you've actually spent so far on the project?

    A.   AC

    B.   SV

    C.   CV

    D.   VAC

65. Tom is the project manager on a construction project. Midway through his project, he realizes that there's a problem with the lumber they've been using in a few rooms and they're going to have to tear down some of the work they've done and rebuild. One of his team members suggests that the defect isn't bad enough to cause all of that rework. Tom says that he's worked on a project that made this same mistake before and they ended up having to redo the work when inspectors looked at the house. He convinces the team member that it's probably better to fix it now than later. What kind of power is he using to make the decision?

    A.   Legitimate

    B.   Expert

    C.   Referent

    D.   Reward

66. Which of the following is NOT one of the most common sources of project conflict?

    A.   Schedules

    B.   Priorities

    C.   Resources

    D.   Costs

67. Your company's quality assurance department has performed a quality audit on your project. They have found that your team has implemented something inefficiently, and that could lead to defects. What's the NEXT thing that should happen on your project?

Exam Questions

A. You work with the quality department to implement a change to the way your team does their work.

B. You document recommended corrective actions and submit them to the change control board.

C. You add the results of the audit to the lessons learned.

D. You meet with the manager of the quality assurance department to figure out the root cause of the problem.

68. A junior project manager at your company does not know how to perform earned value analysis. You spend a weekend with him to teach him how to do this. This is an example of:

A. Coaching

B. Fraternizing, and should be discouraged

C. Unpaid overtime

D. Giving access to proprietary information, and should be reported to PMI

69. Your client has terminated your project before it is complete. Which of the following is true?

A. You must stop all work and release the team immediately.

B. You must work with the team to document the lessons learned.

C. You must keep the team working on the project to give your senior management time to talk to the client.

D. You must update the Project Management plan to reflect this change.

70. When you look at a control chart that measures defects in the product produced by your project, you find that seven values are showing up below the mean on the chart. What should you do?

A. Look into the process that is being measured; there's probably a problem there.

B. Ignore the anomaly; this is the rule of seven, so statistically the data doesn't matter.

C. This indicates that the mean is too high.

D. You should adjust your lower control limit—the values indicate a problem with where the limits have been set.

71. A project manager uses a facilitator to gather opinions from experts anonymously. What risk identification tool or technique is being performed?

A. Brainstorming

B. Delphi technique

C. Interviews

D. SWOT analysis

Exam Questions

72. You have just been authorized to manage a new project for your company. Which of the following BEST describes your first action?

    A.   Create the work breakdown structure.

    B.   Develop the Project Management plan.

    C.   Start working on the project charter.

    D.   Figure out who has a stake in the project.

73. You're the project manager on a software project that is planning out various approaches to technical work. There's a 20% chance that a component you are going to license will be difficult to integrate and cost $3,000 in rework and delays. There's also a 40% chance that the component will save $10,000 in time and effort that would have been used to build the component from scratch. What's the EMV for these two possibilities?

    A. $13,000

    B. $7,000

    C. $3,400

    D.   –$600

74. You are managing a software engineering project when two team members come to you with a conflict. The lead developer has identified an important project risk: you have a subcontractor that may not deliver on time. Another developer doesn't believe that the risk is likely to happen; however, you consult the lessons learned from previous projects and discover that subcontractors failed to deliver their work on two previous projects. You decide that the risk is too big; you terminate the contract with the subcontractor, and instead hire additional developers to build the component. Both team members agree that this has eliminated the risk. Which of the following BEST describes this scenario?

    A.   Transferrence

    B.   Mitigation

    C.   Avoidance

    D.   Acceptance

75. Which of the following BEST describes the contents of a WBS dictionary entry?

    A.   The definition of the work package including its net present value

    B.   Work package ID and name, statement of work, required resources, and Monte Carlo analysis

    C.   Work package ID and name, statement of work, risk register, earned value calculation, scheduled complete date, and cost

    D.   Work package ID and name, statement of work, responsible organization, schedule milestones, quality requirements, code of account identifier, required resources, cost estimate

Exam Questions

76. Two of your project team members approach you with a conflict that they are having with each other over the technical approach to their work. One of the two people is very aggressive, and tries to get you to make a decision quickly. The other team member is quiet, and seems less willing to talk about the issue. The conflict is starting to cause delays, and you need to reach a decision quickly. What's the BEST approach to solving this conflict?

- A. Tell the team members that they need to work this out quickly, because otherwise the project could face delays.
- B. Since it's a technical problem, tell the team members that they should take it to the functional manager.
- C. Confront the issue, even though one team member is hesitant.
- D. Escalate the issue to your manager.

77. Tom is a project manager on an industrial design project. He is always watching when his team members come into the office, when they take their breaks, and when they leave. He periodically walks around the office to be sure that everyone is doing work when they are at their desks and he insists that he make every project decision, even minor ones. What kind of manager is he?

- A. Theory X
- B. Theory Y
- C. Cost cutter
- D. Effective

78. You are the project manager on a construction project. As you're planning out the work your team will do, you divide up all of the work into work packages and create a WBS that shows how they fit into categories. For each one of the work packages, you write down details such as initial estimates and information about what account it should be billed against. Where do you store all of that information?

- A. Scope Management plan
- B. WBS
- C. WBS dictionary
- D. Project scope statement

79. You are managing a project with an EV of $15,000, PV of $12,000, and AC of $11,000. How would you BEST describe this project?

- A. The project is ahead of schedule and within its budget.
- B. The project is behind schedule and within its budget.
- C. The project is ahead of schedule and over its budget.
- D. The project is behind schedule and over its budget.

Exam Questions

80. You are using a Pareto chart to examine the defects that have been found during an inspection of your product. Which process are you performing?

    A.   Perform Quality Assurance

    B.   Plan Quality Management

    C.   Control Quality

    D.   Validate Scope

81. A project manager is faced with two team members who have conflicting opinions. One team member explains her side of the conflict. The other team member responds by saying, "I know you'll never really listen to my side, so let's just go with her opinion and get back to work." This is an example of:

    A.   Withdrawal

    B.   Compromise

    C.   Smoothing

    D.   Forcing

82. Complete the following sentence: "The later a defect is found, _____ ."

    A.   the easier it is to find

    B.   the more expensive it is to repair

    C.   the less important it is to the product

    D.   the faster it is to repair

83. Your top team member has performed extremely well, and you want to reward her. She knows that you don't have enough money in the budget to give her a bonus, so she approaches you and requests an extra day off, even though she is out of vacation days. She asks if she can take one of her sick days, even though the company doesn't allow that. Which of the following is correct?

    A.   You should give her the time off, because McClelland's Achievement Theory states that people need achievement, power, and affiliation to be motivated.

    B.   You should give her the time off, because Expectancy Theory says that you need to give people an expectation of a reward in order to motivate them.

    C.   You should give her the time off because a Theory Y manager trusts the team.

    D.   You should not give her the time off.

84. Which of the following is NOT a tool or technique of Perform Qualitative Risk Analysis?

    A.   Risk urgency assessment

    B.   Expected monetary value analysis

    C.   Probability and Impact matrix

    D.   Risk categorization

Exam Questions

85. While creating a project charter, you discover a new project management software tool has come onto the market. You spend the weekend taking an online tutorial to learn about it. This is an example of:

    A. An assigned project manager using authority and responsibility

    B. Not paying for copyrighted software

    C. Contributing to the Project Management Body of Knowledge

    D. Enhancing personal professional competence

86. Which of the following is NOT a tool in Identify Risks?

    A. Brainstorming

    B. Risk urgency assessment

    C. Delphi technique

    D. SWOT analysis

87. You are managing a software engineering project when two team members come to you with a conflict. The lead developer has identified an important project risk: you have a subcontractor that may not deliver on time. The team estimates that there is a 40% chance that the subcontractor will fail to deliver. If that happens, it will cost an additional $15,250 to pay your engineers to rewrite the work, and the delay will cost the company $20,000 in lost business. Another team member points out an opportunity to save money an another area to offset the risk: if an existing component can be adapted, it will save the project $4,500 in engineering costs. There is a 65% probability that the team can take advantage of that opportunity. What is the expected monetary value (EMV) of these two things?

    A. –$14,100

    B. $6,100

    C. –$11,175

    D. $39, 750

88. Your project team has completed the project work. All of the following must be done before the project can be closed EXCEPT:

    A. Ensure that the schedule baseline has been updated.

    B. Get formal acceptance of the deliverables from the customer.

    C. Make sure the scope of the project was completed.

    D. Verify that the product acceptance criteria have been met.

89. During procurement closure, a procurement audit includes all of the following EXCEPT:

    A. Reviewing the contract terms to ensure that they have all been met

    B. Identifying successes and failures that should be recognized

    C. Documenting lessons learned

    D. Using the payment system to process consideration as per the terms of the contract

Exam Questions

90. You are reviewing performance goals to figure out how much bonus to pay your team members. What document would you consult to find your team's bonus plan?

    A.   The reward and recognition plan

    B.   The Staffing Management plan

    C.   The Human Resource Management plan

    D.   The project's budget

91. You're managing a construction project to install several hundred air conditioner panels in a new office building. You have completed 350 panels out of a total of 900 planned panels, but according to your schedule you should have completed 400 of them. Your company's contract states that you'll be paid a fixed price of $75 per panel. You've spent $45,000 so far on the project. Which of the following BEST describes your situation?

    A.   The CPI is .813, which means your project is currently over your budget.

    B.   The CV is –$4,350, which means your project is currently over your budget.

    C.   The TCPI is 1.833, which is the minumum CPI you need to stay within budget.

    D.   The SPI is .84, which means your project is behind schedule.

92. Your team has identified a risk with some of the chemicals you are using on your highway construction project. It is really difficult to mix them just right and, based on past projects, you've figured out that there's a high probability that about 14% of the chemical supply will be lost in mixing problems. You decide to buy an extra 15% of the chemicals up front so that you will be prepared for those losses and your project won't be delayed. Which response strategy are you using?

    A.   Avoid

    B.   Accept

    C.   Mitigate

    D.   Transfer

93. A project manager is planning the staffing levels that will be needed through the course of her project. She figures out the number of people that will be needed in each role over time and displays that information in a chart as part of her Staffing Management plan. What is that chart called?

    A.   Gantt chart

    B.   RACI matrix

    C.   Organization chart

    D.   Resource histogram

Exam Questions

94. Which of the following is NOT part of a typical change control system?

- A. Approval
- B. Change control board
- C. Project management information system
- D. Stakeholder analysis

95. A notice sent to a subcontractor about the contract is an example of which kind of communication?

- A. Informal verbal
- B. Formal written
- C. Formal verbal
- D. Informal written

96. You need to determine when to release resources from your project. Which part of the Staffing Management plan will be most useful for this?

- A. Resource histogram
- B. Safety procedures
- C. Recognition and rewards
- D. Training needs

97. Which of the following is NOT a type of communication?

- A. Formal written
- B. Paralingual
- C. Nonverbal
- D. Noise

98. A company is about to begin work on a large construction project to build four new buildings for a bank that wants to open new branches. The sponsor is writing a project charter. She recalls that a previous project the company performed for another bank ran over budget because the team had underestimated the effort required to install the reinforced walls in the vault. The previous project manager had documented the details of the lessons learned from this project. Where should the sponsor look for these lessons learned?

- A. The project records management system
- B. The company's organizational process assets
- C. The project's work performance information
- D. The project's performance reports

Exam Questions

99. The customer has reviewed the deliverables of a project and finds that they are acceptable, and must now communicate that acceptance to the project manager. Which form of communication is appropriate?

    A.   Informal written

    B.   Informal verbal

    C.   Formal written

    D.   Formal verbal

100. Which of the following is NOT found in a project charter?

    A.   The summary budget

    B.   High-level requirements

    C.   Procedures for managing changes to contracts

    D.   Responsibility and name of the person authorized to manage the project

101. Which of the following is NOT a project constraint?

    A.   Cost

    B.   Resources

    C.   Procurements

    D.   Scope

102. What is a risk owner?

    A.   The person who monitors the watchlist that contains the risk

    B.   The person who meets with stakeholders to explain the risk

    C.   The person who makes a risk happen

    D.   The person who is responsible for the response plan for the risk

103. You are managing a software engineering project when two team members come to you with a conflict. The lead developer has identified an important project risk: you have a subcontractor that may not deliver on time. Another developer doesn't believe that the risk is likely to happen; however, you consult the lessons learned from previous projects and discover that subcontractors failed to deliver their work on two previous projects. The lead developer suggests that you have two team members take three weeks to research the component being built by the subcontractor, and come up with some initial work that you can fall back on in case that subcontractor does not deliver. You decide to follow the lead developer's advice over the objections of the other team member. Which of the following BEST describes this scenario?

    A.   Transferrence

    B.   Mitigation

    C.   Avoidance

    D.   Acceptance

Exam Questions

104. You are managing a project with AC = $25,100, ETC = $45,600, VAC = –$2,600, BAC = $90,000, and EAC = $92,100. Your sponsor asks you to forecast how much money you expect to spend on the remainder of the project. Which is the BEST estimate to use for this forecast?

    A. $45,600

    B. $87,400

    C. $90,000

    D. $92,100

105. Which is the BEST description of project scope?

    A.   All of the features and deliverables your project will deliver

    B.   All of the products your project will make

    C.   All of the people involved in your project

    D.   All of the work you will do to build the product

106. Given this portion of the network diagram to the right, what's the LF of activity F?

    A.   10

    B.   11

    C.   16

    D.   17

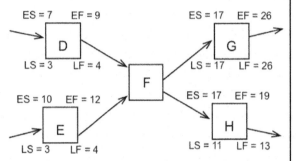

107. A team member is showing up late to work and leaving early, and it is affecting the project. The project manager decides that the team member must be reprimanded. Which of the following is the BEST way to handle this situation?

    A.   In a one-on-one meeting with the team member

    B.   At the next team meeting

    C.   In a private meeting with the team member and his functional manager

    D.   Over email

108. You are managing an accounting project when a new CFO is hired at your company. He'll be affected by all accounting projects in your company. What's the BEST thing for you to do?

    A.   Show him the project charter so that he knows that you are in charge of the project

    B.   Work with him to understand the current requirements and determine if he has new ones to add to the project

    C.   Keep working on the project and get his feedback when he can review the finished product

    D.   Add him to the communications plan

Exam Questions

109. You're managing an industrial design project. Your project is currently in the Initiating phase. The project charter has been created, and you are working on identifying the stakeholders. Which of the following is NOT something that you should do?

    A.   Review lessons learned from prior projects

    B.   Perform a stakeholder analysis

    C.   Review procurement documents

    D.   Create the change control system

110. Alberto is the project manager of a software implementation project. His company has made an organization-wide decision to move to a new accounting and human resources software package. He has read that some projects to implement the same package have resulted in the loss of personnel data when they tried to import it into the new system. He backs up the data so that it could be restored in the event of such a problem but also buys insurance to cover the cost of keying in the data manually if the implementation doesn't work. Which response strategies are Alberto using?

    A.   Mitigating and accepting

    B.   Mitigating and avoiding

    C.   Mitigating and transferring

    D.   Mitigating and sharing

111. Rekha is a project manager on a large construction project. Late in the project, her client demands a big change. She assesses the impact of the change and tells the client how much time and money it will cost. But the client says that he doesn't have the time or budget to allow the change. What's the BEST way for Rekha to handle this situation?

    A.   Have her senior managers meet with the client to explain the situation.

    B.   Hold a meeting with the client to figure out why he's asking for the change.

    C.   Do nothing; she's the project manager, so she sets the rules.

    D.   Have the client find more money for the budget.

112. You are conducting a status meeting and monitoring your risk register when you discover a risk that remains even after you implement all of your response strategies. What kind of risk is this and what should you do about it?

    A.   It's a secondary risk; you don't need to worry about it.

    B.   It's a residual risk; you need to plan a response strategy for it.

    C.   It's a residual risk; you don't need to plan a response strategy for it because you've already implemented all of the risk responses you can plan for.

    D.   It's a contingency reserve. You should only use it if the first risk occurs.

Exam Questions

113. Rekha is a project manager on a large construction project. Late in the project, her client demands a big change. She assesses the impact of the change and tells the client how much time and money it will cost. But the client won't allow any change to the schedule and won't pay anything more for change. Rekha explains that the proposed change is well outside the documented scope of the original work. The client tells Rekha that he doesn't care what was in the original scope and that she needs to implement the change with no impact to schedule or budget. What conflict resolution technique is the client using?

   A. Confronting

   B. Withdrawal

   C. Smoothing

   D. Forcing

114. Which of the following is NOT a tool of the Define Scope process?

   A. Facilitated workshops

   B. Constrained optimization

   C. Alternatives identification

   D. Product analysis

115. Which of the following BEST describes the records management system?

   A. A system to store contracts and project records for future project managers to reference

   B. A library that stores the lessons learned for past projects

   C. A filing system to store paid invoices

   D. A system to store human resource records, salary information, and work performance history

116. A project manager is negotiating with a contractor. Neither has a good idea of how long the project will take, or how much the materials will cost. Which contract type is MOST appropriate for this project?

   A. Cost plus fixed fee (CPFF)

   B. Time and materials (T&M)

   C. Cost plus award fee (CPAF)

   D. Firm fixed price (FFP)

117. Which is NOT an example of cost of quality?

   A. The cost of inspecting your product to be sure that it meets requirements

   B. The cost of reviewing documents used to produce your product to be sure that they do not have defects

   C. The cost of training your team on techniques that will help them avoid defects

   D. The cost of contracting another company to build part of the product

Exam Questions

118. Which of the following is NOT a stakeholder?

    A.   A project team member

    B.   An attorney from your company's competitor

    C.   A representative from your project team's union

    D.   The project sponsors

119. What is the main output of the Define Scope process?

    A.   Requirements documentation

    B.   Scope definition

    C.   Scope dictionary

    D.   Project scope statement

120. Paul is a project manager for an industrial design project. The project has a 60% chance of making the company $230,000 over the next year. It has a 40% chance of costing the company $150,000. What's the project's EMV?

    A. $138,000

    B. $60,000

    C. $78,000

    D. $230,000

121. The project manager for a construction project discovers that the local city council will vote on a zoning change that would open up a new neighborhood to commercial building. She contacts other construction companies in the area that would benefit from the change to ask them to attend the council meeting in order to convince the city council to vote for the change. A "Yes" vote will benefit all of the companies. This is an example of which risk response strategy?

    A.   Mitigate

    B.   Share

    C.   Exploit

    D.   Enhance

122. You are acquiring a project team, and as part of that process you must interview new project managers for your company. One of the candidates claims to be a PMP certified project manager, but you discover that she has never taken the PMP exam. What is the BEST way to handle this situation?

    A.   Consult the Human Resource Management plan to see if PMP certification is a resource requirement.

    B.   Report the person to your manager.

    C.   Report the person to PMI.

    D.   Call the police.

Exam Questions

123. Which of the following types of power is the most effective in leading teams?

- A. Expert
- B. Referent
- C. Reward
- D. Punishment

124. You are being hired to manage a highway construction project for a contractor working for Smith County. The sponsor is a project officer who works for the Smith County municipal government. You have three separate teams working all three shifts, with a separate foreman for each team. Each team has members from two different unions, and each union has its own representative. Who is the BEST person to approve the project charter?

- A. The project manager
- B. The Smith County project officer
- C. The team foreman
- D. The two union representatives

125. Which of the following is NOT an input of the Control Procurements process?

- A. Work performance data
- B. Procurement Management plan
- C. Agreements
- D. Procurement documents

126. While working on the WBS for a large project that spans multiple teams, you meet with a fellow project manager to discuss his portion of the work. He tells you in confidence that he lied about having a PMP certification, and never actually passed the exam. What is the BEST way to handle this situation?

- A. Report the person to PMI.
- B. Report the person to his manager.
- C. Ask him to tell the truth to his manager.
- D. Do nothing because you were told this in confidence.

127. Your team has recommended a change to the Validate Scope process. What's the first thing you should do?

- A. Implement the change.
- B. Analyze the change versus the Project Management plan to see what its impact will be.
- C. Write up a change request.
- D. Tell your team that the process has already been decided and they should follow it.

Exam Questions

128. As you are executing your project, you are constantly checking your risk register to be sure that you have planned responses for all of your risks. At one team status meeting, you find that a lower-priority risk has suddenly become more likely. Where do you keep information about low-priority risks?

- A. Triggers
- B. The watchlist
- C. Risk Management plan
- D. Qualitative analysis documents

129. Which of the following are NOT all examples of project documents?

- A. Scope baseline, project funding requirements, stakeholder requirements
- B. Activity list, stakeholder register, teaming agreements
- C. Forecasts, risk register, quality metrics
- D. Basis of estimates, resource requirements, statement of work

130. You're managing a project with a schedule performance index (SPI) of 1.07 and a cost performance index (CPI) of 0.94. How would you BEST describe this project?

- A. The project is ahead of schedule and within its budget.
- B. The project is behind schedule and within its budget.
- C. The project is ahead of schedule and over its budget.
- D. The project is behind schedule and over its budget.

131. Which of the following is the BEST example of a reward system?

- A. The team member who works the hardest will receive $1,000.
- B. Everyone will get a bonus of $500 if the project meets its quality goals, $500 if it meets its budget goals, and $600 if it comes in on time.
- C. The five team members who put in the most hours will get a trip to Disneyland.
- D. The team will only get a bonus if the project comes in 50% under budget, schedule, and quality metric goals; even though the team leads know this goal is unrealistic, they agree that it will motivate the team to work harder.

132. There are 17 people on a project. How many lines of communication are there?

- A. 136
- B. 105
- C. 112
- D. 68

133. Which of the following project selection methods is NOT a comparative approach (or benefit measurement model)?

- A. Linear programming
- B. Murder boards

Exam Questions

    C.   Benefit-to-cost ratios

    D.   Peer review

134. You're managing a team of project managers, and as part of developing the team you hold a PMP study group so that you and your coworkers can work together to study for the exam. One person recently took and passed the exam, and has offered to give you all of the questions he can remember. How should you respond?

    A.   Accept the questions, but to be fair to everyone in the study group you should make sure everyone gets a copy.

    B.   Refuse the offer, but encourage other study group members to make up their own minds.

    C.   Refuse the offer, and report the person to your manager.

    D.   Refuse the offer, and report the person to PMI.

135. The project charter is typically approved by the project sponsor, although some charters can be approved by key stakeholders instead. Which of the following BEST describes the role of the project sponsor on the project?

    A.   The sponsor manages the project.

    B.   The sponsor provides funding for the project.

    C.   The sponsor verifies that all of the work was completed.

    D.   The sponsor negotiates all contracts.

136. Two of your project team members approach you with a conflict that they are having with each other over the technical approach to their work. One of the two people is very aggressive, and tries to get you to make a decision quickly. The other team member is quiet, and seems less willing to talk about the issue. The conflict is starting to cause delays, and you need to reach a decision quickly. You spend the weekend studying conflict resolution techniques, which is an example of:

    A.   Contributing to the Project Management Body of Knowledge

    B.   Maslow's Hierarchy of Needs

    C.   Enhancing personal professional competence

    D.   Confronting (or problem solving)

137. You are managing a software project. The stakeholders have found a few requirements that were missed in the initial project scope statement. You put the change requests through change control and they are approved, so you need to update the scope statement to include the new work. Where can you find the most updated version of the scope statement?

    A.   In the configuration management system

    B.   In the document repository

    C.   In the Project Management plan

    D.   In the Communications Manaement plan

Exam Questions

138. After a status meeting, one of your team members, John, pulls you aside and tells you that he was insulted by a comment from another team member. He felt that the comment was racist. You meet with the team member who made the comment, Suzanne, but she says that the other team member's performance has been very poor. She has never made a comment like this before at the company. You review the records, and see that she is correct—he has consistently delivered lower-quality work than any other team member. What is the BEST way to handle this situation?

    A.   At the next team meeting, reprimand John for his poor performance and Suzanne for the racist comment.

    B.   Get John additional help for his poor performance.

    C.   Reprimand Suzanne in private for her racist comment, and follow any company policies for reporting racism among employees.

    D.   Suzanne has never had this problem before, so she should be given another chance.

139. You're managing a project that is currently executing. You're evaluating the work being performed by constantly measuring the project performance, and recommending changes, repairs, and corrections where necessary. What process are you performing?

    A.   Integrated Change Control

    B.   Monitor and Control Project Work

    C.   Control Scope

    D.   Communications Management

140. Which of the following helps you identify the root cause of 80% of the defects in your project using the 80/20 rule?

    A.   Scatter chart

    B.   Control chart

    C.   Cause-and-effect diagram

    D.   Pareto chart

141. As you complete each deliverable for your project, you check that it is correct along with your stakeholders and sponsors. Which process are you performing?

    A.   Define Scope

    B.   Define Activities

    C.   Validate Scope

    D.   Control Scope

142. You are a project manager on a large military contract that involves 7 subcontractor companies and a total of 1,253 team members, 752 stakeholders and sponsors, and 14 project managers (including you). You need to get a handle on the communications channels, because otherwise your project will devolve into chaos. How many potential channels of communication are there on this project?

Exam Questions

A. 2,019

B. 91

C. 2,037,171

D. No way to determine

143. Joe is an excellent programmer. He was promoted to the role of project manager because he understands technology better than anyone else in the company. Unfortunately, he is having trouble doing the project management job and his projects are failing. What is this an example of?

A. Gold plating

B. Halo effect

C. Pre-assignment

D. Ground rules

144. Which of the following describes the contents of a Staffing Management plan?

A. Organizational chart, training needs, estimated labor cost, and release criteria

B. Sponsor, organizational chart, validate scope plan, and schedule

C. RACI matrix, organizational chart, performance improvement plan, and budget

D. Resource histogram, training needs, recognition and rewards, and resource calendar

145. A member of your seven-person project team comes up with a controversial idea. Two others agree that it should be pursued. Two members of the team are opposed to it, while the remaining two have a different approach. You end up going with the controversial idea. This is an example of:

A. Unanimity

B. Majority

C. Plurality

D. Dictatorship

146. You have been hired by a contractor who wants you to manage a construction project for one of his clients. The project team has been working for two months, and is 35% done with the job. Two of your team members come to you with a conflict about how to handle the ongoing maintenance for a piece of equipment. You know that they can safely ignore the problem for a while, and you're concerned that if your project falls behind schedule before next week's stakeholder meeting, it will cause problems in the future. You tell the two team members that the problem really isn't as bad as they think it is, and if they take a few days to cool off about it you'll help them with a solution. This approach to conflict resolution is known as:

A. Withdrawal

B. Compromise

C. Smoothing

D. Forcing

Exam Questions

147. Which of the following is NOT true about the project charter?

  A.   It formally assigns the project manager.

  B.   It is always created by the project manager.

  C.   It contains external constraints and assumptions.

  D.   Includes a high-level milestone schedule.

148. You're managing a construction project to install several hundred air conditioner panels in a new office building. Every floor has identical panels. The customer, a construction contracting company, has provided specifications for the installations. The team is using a process to install and verify each panel. As the team completes each panel, your team's quality control inspector measures it and identifies defects. The root cause of each defect is identified. He keeps track of each step in this process and marks each one complete as it is finished. Which is the BEST tool to use for this?

  A.   Control chart

  B.   Fishbone diagram

  C.   Checksheets

  D.   Pareto chart

149. Which risk analysis tool is used to model your risks on a computer to show random probabilities?

  A.   Computerized risk audit

  B.   Monte Carlo analysis

  C.   EMV analysis

  D.   Delphi technique

150. You are a project manager for a software project. As you are defining the scope of the work you need to do, you sit down with all of the project's stakeholders and record all of the requirements you can get from them. Which of the following is NOT a valid requirement from stakeholder analysis?

  A.   The work the team does must be better than they did on their last project.

  B.   There can be no more than 5% schedule variance on the project.

  C.   The quality of the product must fit within organizational metrics for software quality.

  D.   The budget must be within 10% of our projected cost.

151. There have been several rounds of layoffs at your company. Now your project team is worried about their job security, and you've noticed that their performance has decreased significantly because of it. This is predicted by which motivational theory?

  A.   McGregor's Theory of X and Y

  B.   Maslow's Hierarchy of Needs

  C.   McClelland's Achievement Theory

  D.   Herzberg's Motivation-Hygiene Theory

Exam Questions

152. Which of the following is NOT included in a cost of quality calculation?

    A.   Team members' time spent finding and repairing defects

    B.   Quality managers' time spent writing quality standards

    C.   Project managers' time spent creating the Project Management plan

    D.   Team members' time spent reviewing specifications, plans, and other documents

153. Which of the following contracts has the MOST risk for the seller?

    A.   Cost plus fixed fee (CPFF)

    B.   Time and materials (T&M)

    C.   Cost plus percentage of costs (CPPC)

    D.   Fixed price (FP)

154. You are managing an industrial design project for an important client. Two of your team members have a disagreement on project priorities. One person wants to do certain activities first, while the other feels they should be left until the end of the project. You work with both people to forge a compromise where those activities are neither first nor last, but instead done in the middle of the project. Nobody is particularly unhappy with this solution. Another name for a compromise is a:

    A.   Win-win solution

    B.   Win-lose solution

    C.   Lose-lose solution

    D.   Standoff solution

155. A project manager on a construction project includes a line item in the budget for insurance for the equipment and job site. This is an example of:

    A.   Transference

    B.   Mitigation

    C.   Avoidance

    D.   Acceptance

156. You are talking to experts and gathering independent estimates for your contract. Which of the following BEST describes what you are doing?

    A.   Plan Procurement Management

    B.   Conduct Procurements

    C.   Control Procurements

    D.   Close Procurements

Exam Questions

157. A project manager is creating a report of the final status of a closed project to the stakeholders. Which of the following is NOT used in a final project report to communicate the status of a project?

    A.   Variance information

    B.   Lessons learned

    C.   Scope baseline

    D.   Status of deliverables

158. You are the project manager for a railroad construction project. Your sponsor has asked you for a forecast for the cost of project completion. Which of the following is the BEST metric to use for forecasting?

    A.   EV and AC

    B.   SV and CV

    C.   ETC and VAC

    D.   SPI and CPI

159. Information about the project must be distributed to all stakeholders. Which of the following process outputs is used to report the status and cost of project activities?

    A.   Work performance data

    B.   Issue logs

    C.   Status reports

    D.   Project records

160. A company uses a management technique that employs quality assurance techniques to continuously improve all processes. This is called:

    A.   Just-in-time management

    B.   Kaizen

    C.   Ishikawa diagrams

    D.   Inspection

161. You are initiating a project that has a virtual team. Half of your team members will be located in another country, where they are working for a subcontractor. The subcontractor's team members speak a different dialect of English than your team does. After a conference call to review the project charter, two of your team members make jokes about the way your subcontractor's team members speak. What is the BEST way to handle this situation?

    A.   Correct the team members individually, and hold a training session for your team to help remove communications barriers.

    B.   Immediately correct the two people in front of the rest of the team.

    C.   Report the team members to senior management and recommend that they be punished.

    D.   Remove noise from the communication by contacting the subcontractor and requesting that the team adjust the way they speak.

Exam Questions

162. Mike is a project manager for an IT technology implementation project. He is using an Ishikawa diagram to figure out what could cause potential risks on his project. Which process is he doing?

    A.   Identify Risks

    B.   Perform Qualitative Risk Analysis

    C.   Control Quality

    D.   Plan Risk Responses

163. Amit is the manager of a software project. His client has agreed on a project scope statement at the beginning of the project, but whenever the client verifies deliverables, she comes up with features that she would like to add into the product. Amit is working with the client to find what requirements were missed in the planning stages of the project and how to plan better in the future. What is the BEST description of his project's current situation?

    A.   Gold plating

    B.   Scope creep

    C.   Alternatives analysis

    D.   Schedule variance

164. You are executing a project, and as part of developing the project team you create materials for a PMP training seminar. Afterward, you decide to offer it to all other project managers in your company to help them obtain enough hours to qualify to take the PMP exam. This is an example of:

    A.   Organizational process assets

    B.   Contributing to the Project Management Body of Knowledge

    C.   Cheating, which should be reported to PMI

    D.   Lessons learned

165. You are managing a construction project to install wiring in an office building. While verifying that the scope of the work has been completed, you discover that one of your team members has taken a box of cable from the job site so he can wire his attic. What is the BEST way to respond?

    A.   Do nothing.

    B.   Report the team member to your manager.

    C.   Report the team member to PMI.

    D.   Call the police.

166. You are working with potential sponsors to determine which project your company will pursue. Based on the benefit-to-cost (BCR) ratios, which of the following four projects should you recommend?

    A.   Project A has a BCR of 5:2

    B.   Project B has a BCR of 5:4

Exam Questions

  C. Project C has a BCR of 3:1

  D. Project D has a BCR of 2:1

167. What are the five kinds of power?

  A. Legitimate, expert, reward, political, and bargaining

  B. Legitimate, expert, reward, political, and punishment

  C. Legitimate, expert, reward, economic, and bargaining

  D. Legitimate, expert, reward, referent, and punishment

168. You're managing a software project. Your team has discovered a problem, and as a result you've requested a change. The change will cost the project an extra three weeks, but without it several stakeholders might have problems with the final product. What's the NEXT thing that you should do?

  A. Instruct the team to make the change.

  B. Call a meeting with each stakeholder to figure out whether or not to make the change.

  C. Document the change and its impact, and put it through the change control system.

  D. Don't make the change because it will delay the project.

169. Which of the following is NOT part of the Close Project or Phase process?

  A. Making sure that all exit criteria have been met.

  B. Obtaining formal acceptance of all deliverables from all stakeholders.

  C. Moving the project's deliverables to the next phase or into production.

  D. Writing down lessons learned.

170. Which of the following are valid ways to break down the work in a WBS?

  A. By risk or quality metric

  B. By product feature or unit of work

  C. By project phase or project deliverable

  D. By charge code or initial estimate

171. Which of the following is an output of Direct and Manage Project Work?

  A. Work performance data

  B. Statement of work

  C. Issue log

  D. Agreements

172. You are working on a construction project. You, your team, and your senior manager all feel that the work is complete. Your stakeholders have communicated their final acceptance of the project. You are now meeting with your team to update the organizational process assets with a record of knowledge gained about the project to help future project managers with their projects. This is BEST described as:

Exam Questions

A. Lessons learned

B. Project records

C. Project management information system (PMIS)

D. Work performance information

173. Which of the following tools is used to document the lessons that were learned when the contract was administered?

A. Quality audit

B. Buyer-conducted performance review

C. Contract review

D. Procurement audit

174. You are managing a project to build a new wing onto a local school building over a summer break. One night, the school and your construction site are destroyed by a tornado. Your client demands that you continue work despite the disaster, but you consult the contract, and find a clause that states that you are not responsible for any more work. This is referred to as:

A. A force majeure clause

B. An "act of God" clause

C. Mitigation

D. An ex parte communication

175. A project manager is running into problems with the team. People are repeatedly running into trouble over seemingly small problems: who takes notes at meetings, what dress is appropriate for the office, who people need to notify when they take a day off. The problems started out small, but as more people run into more problems the situation is rapidly escalating. This situation is most likely caused by a lack of:

A. Sensitivity training

B. Common courtesy

C. A reward system

D. Ground rules

176. Which of the following BEST describes when you perform the Monitor and Control Project Work process?

A. Continuously throughout the project

B. As soon as every deliverable is completed

C. At scheduled milestones or intervals during the project

D. At the end of every project phase

Exam Questions

177. Which of the following is NOT an output of Monitor and Control Project Work?

    A.   Project Management plan updates

    B.   Change request status updates

    C.   Change requests

    D.   Project document updates

178. You are the project manger of a software project. Two developers, Bill and Alfredo, are having an argument about how to implement a feature. Bill thinks that it's more important that the project get done quickly, so he's suggesting that you reuse some work that's been done on a previous project to get started. Alfredo thinks that that work doesn't apply to this project and will just waste time. Bill is almost always right about these things and he's very influential on the team, so it's important that you keep him happy. What should you do?

    A.   Since you do want to get the project done quickly, you side with Bill.

    B.   Side with Alfredo; it could end up taking longer in the end.

    C.   Call a meeting to hear both sides of the situation and decide in favor of the solution that is best supported by objective evidence.

    D.   Call a meeting in private with Bill to hear more about his position.

179. Which of the following is NOT an example of a deliverable?

    A.   Project Management plan

    B.   Project schedule

    C.   Work breakdown structure

    D.   Parametric estimation

180. Which is NOT an input to the Create WBS process?

    A.   Project scope statement

    B.   Organizational process assets

    C.   Requirements documentation

    D.   WBS dictionary

181. Which of the following is the correct order of actions that you take during the Closing processes?

    A.   Get formal acceptance, release the team, write lessons learned, close the contract

    B.   Write lessons learned, release the team, get formal acceptance, close the contract

    C.   Get formal acceptance, write lessons learned, release the team, close the contract

    D.   Get formal acceptance, close the contract, write lessons learned, release the team

Exam Questions

182. At the beginning of the project, you hold a meeting with all of the stakeholders in your project in order to figure out how everyone will communicate as the work goes on. Which of the following terms best describes that meeting?

- A. Qualitative analysis
- B. Status meeting
- C. Communication plan meeting
- D. Kickoff meeting

183. You are managing a project with 23 team members and 6 key stakeholders. Two team members identify a problem with the current approach. Addressing that problem will require changes to the project plan and its subsidiary plans. One of the stakeholders previously indicated that any delays are unacceptable, and your team members tell you that it's possible the change could cause the team to miss at least one critical deadline. What is the BEST way to deal with this situation?

- A. Analyze the impact that the change will have on the work to be done, the schedule, and the budget.
- B. Deny the change because any delays are unacceptable.
- C. Gather consensus among the team that you should make the change before approaching the stakeholders, so that they can see the team supports making the change.
- D. Make the change to the project plan and subsidiary plans, and ask the team to implement the change.

184. You have been asked to select from three projects. Project A has a net present value of $54,750 and will take six months to complete. Project B has a net present value of $85,100 and will take two years to complete. Project C has a net present value of $15,000 and a benefit-cost ratio of 5:2. Which project should you choose?

- A. Project A
- B. Project B
- C. Project C
- D. There is not enough information to decide.

185. Which of the following is a defect?

- A. A mistake made by a team member on the job
- B. A change that the team needs to make in how they do the work
- C. A Project Management plan that does not meet its requirements
- D. A change request that's been rejected by the change control board

Exam Questions

186. A project manager is faced with two team members who have conflicting opinions. One team member explains her side of the conflict, and presents a possible solution. But before the other team member starts to explain his side of things, the project manager says, "I've heard enough, and I've decided to go with the solution I've heard." This is an example of:

    A.   Withdrawal

    B.   Compromise

    C.   Smoothing

    D.   Forcing

187. As you determine the acceptance criteria, constraints, and assumptions for the project you record them in which document?

    A.   Project Management plan

    B.   Project scope statement

    C.   Project charter

    D.   Communications Management plan

188. You've been hired by a large consulting firm to evaluate a software project for them. You have access to the CPI and EV for the project, but not the AC. The CPI is .92, and the EV is $172,500. How much money has actually been spent on the project?

    A. $158,700

    B. $172,500

    C. $187,500

    D.   There is not enough information to calculate the actual cost.

189. Approved changes are implemented in which process?

    A.   Direct and Manage Project Work

    B.   Monitor and Control Project Work

    C.   Perform Integrated Change Control

    D.   Develop Project Management Plan

190. You've been hired by a large consulting firm to lead an accounting project. You determine the needs of the project and divide the work up into work packages so that you can show how all of it fits into categories. What are you creating?

    A.   A WBS

    B.   A schedule

    C.   A project scope statement

    D.   A contract

Exam Questions

191. Over half of conflicts on projects are caused by:

- A. Bad habits, defects, technology
- B. Resources, priorities, schedules
- C. Budget, carelessness, personalities
- D. Technology, money, personalities

192. Which of the following is NOT a tool or technique of Control Risks?

- A. Bringing in an outside party to review your risk response strategies
- B. Revisiting your risk register to review and reassess risks
- C. Using earned value analysis to find variances that point to potential project problems
- D. Gathering information about how the work is being performed

193. You work for a consulting company and your team has implemented an approved scope change on your project. You need to inform your client that the change has been made. What's the best form of communication to use for this?

- A. Formal verbal
- B. Formal written
- C. Informal written
- D. Informal verbal

194. Which of the following is NOT typically found in a project charter?

- A. Project requirements
- B. Authorization for a project manager to work on a project
- C. Work packages decomposed into activities
- D. An initial set of schedule milestones

195. What are the strategies for dealing with positive risks?

- A. Avoid, mitigate, transfer, accept
- B. Transfer, mitigate, avoid, exploit
- C. Exploit, share, enhance, accept
- D. Mitigate, enhance, exploit, accept

196. In which process do you create the risk breakdown structure?

- A. Identify Risks
- B. Plan Risk Responses
- C. Perform Qualitative Risk Analysis
- D. Plan Risk Management

Exam Questions

197. Your project just completed, and one of your subcontractors has sent you floor seats to the next big hockey game to thank you for your business. What is the BEST way to respond?

    A.   Thank the subcontractor, but do not give him preference in the next RFP.

    B.   Thank the subcontractor, but politely refuse the gift.

    C.   Ask for tickets for the entire team, so that it is fair to everyone.

    D.   Report the subcontractor to PMI.

198. A project manager discovers that a project problem has occurred. The problem was never discussed during risk planning activities or added to the risk register, and it will now cost the project money. What is the BEST response?

    A.   Don't take any action, just accept that there's a problem that the team did not plan for.

    B.   Stop all project activity and approach senior management for advice.

    C.   Add the risk to the risk register and gather information about its probability and impact.

    D.   Use the management reserve to cover the costs of the problem.

199. You are managing a large construction project that's been broken down into subprojects (or phases). Each of these subprojects is scheduled to take between three and six months to complete. At the end of each sub-project, you plan to go through the closing processes and document lessons learned. Which of the following BEST describes what you must do at the beginning of each subproject or phase?

    A.   Make sure you don't involve the team, to avoid introducing too much project management overhead.

    B.   Identify the stakeholders.

    C.   Use the earned value technique to decide whether or not to finish the project.

    D.   Release all resources from the project and contact sellers to renegotiate all contracts.

200. Which of the following is NOT an output of the Control Stakeholder Engagement process?

    A.   Change requests

    B.   Deliverables

    C.   Updates to the Project Management plan

    D.   Organizational process asset updates

Before you look at the answers...

Before you find out how you did on the exam, here are a few
ideas to help make the material stick to your brain. Remember,
once you look through the answers, you can use these tips to
help you review anything you missed.

*This is especially useful
for conflict resolution
questions—the ones where
you're presented with a
disagreement between
two people and asked how
you'd handle it.*

❶ Don't get caught up in the question.

If you find yourself a little confsued about a question, the first thing you
should do is try to figure out exactly what it is the question is asking. It's
easy to get bogged down in the details, especially if the question is really
wordy. Sometimes you need to read a question more than once. The
first time you read it, ask yourself, "What's this question *really* about?"

❷ Try this stuff out on your job.

Everything you're learning about for the PMP exam is really practical.
If you're actively working on projects, then there's a really good chance
that some of the ideas you're learning about can be applied to your job.
Take a few minutes and think about how you'd use these things to make
your projects go more smoothly.

❸ Write your own questions.

*When you write your own question, you
do a few things:*

*● You reinforce the idea and make
it stick to your brain.*

*● You think about how questions are
structured.*

*● By thinking of a real-world
scenario where the concept is used, you
put the idea in context and learn how
to apply it.*

And all that helps you recall it better!

Is there a concept that you're just not getting? One of the best ways that
you can make it stick to your brain is to write your own question about
it! We included Question Clinic exercises in *Head First PMP* to help you
learn how to write questions like the ones you'll find on the exam.

❹ Get some help!

Join the **free PMP study forums** at the Head First
Labs website. That's a great place to ask questions and
find other people who are also studying for the exam.
Visit **http://www.headfirstlabs.com/** and click on
Forums to join.

Exam ~~Questions~~ Answers

1. Answer: C

Just because the change will help the project's timeline doesn't mean that it will be an overall benefit to the project. It's important to check how the project will impact the other two constraints as part of your change request. Once you know all the facts about the change, the change control board can make an informed decision about how to proceed.

2. Answer: A

Work performance data is what you're doing when you look at the work that the team is performing in order to determine whether the project is ahead or behind schedule. A really good way to do that is to use schedule variance (SV) and schedule performance index (SPI) calculations.

3. Answer: B

Quality audits are when your company reviews your project to make sure that you are following all of the processes in your company correctly. They are a tool of the Perform Quality Assurance process.

4. Answer: A

When a change has been approved, you always need to update the baseline and then implement the change. That way, you will be sure to track your performance versus new scope and schedule expectations and not the old ones.

5. Answer: A

Translating your communications into a different language so that they can be transmitted to someone else is an example of encoding.

6. Answer: A

The scope baseline contains the WBS and project scope statement, so you'll find constraints and assumptions there, too!

The project Scope Management plan is a really important tool in your project. It tells you exactly how you'll create the project scope, define the WBS, verify that the work has been done, and make changes to the scope. But it doesn't tell you about specific assumptions that you and the team have made, or constraints on your project. To find those, you should look in the requirements documentation and the project scope statement.

Exam ~~Questions~~ Answers

7. Answer: A

Stakeholder analysis is one of the tools and techniques of the Identify Stakeholders process. And that shouldn't really be a surprise. After all, the goal of stakeholder analysis is to write down the needs of your stakeholders. Identify Stakeholders is the only process in the Stakeholder Management knowledge area that is part of the Initiating process group.

8. Answer: C

It's not enough that the project manager approves of the Project Management plan; it needs to be approved by all of the stakeholders in the project. Everyone on the team should feel comfortable with the processes that are going to be used to do the work.

9. Answer: B

The most expensive defects are the ones introduced when the product is being designed. This is a little counterintuitive at first, but it really makes sense once you think about how projects are run. If your team introduces a defect into a product while it's being assembled, then they have to go back and fix it. But if there's a flaw in the design, then you have to halt production and go back and figure out all the things that flaw affected. You may have to order new parts, reassemble components, and maybe even go back and redesign the product from the ground up.

> That's why your Quality Management processes are so focused on reviewing EVERY deliverable—not just the final product, but all of the components, designs, and specifications, too.

10. Answer: A

This question is asking you to create a forecast using estimate to complete (ETC), which uses CPI to project how much money is likely to be spent for the rest of the project. The first step is to plug the numbers into the formula EAC = BAC / CPI, which yields EAC = $80,000 / .95 = $84,210. That's how much money you're likely to spend on the project. Now you can figure out ETC = EAC − AC = $84,210 − $25,000 = $59,210.

Answers

Exam ~~Questions~~

11. Answer: B

Decomposition is the main tool for creating the WBS. It just means breaking the work down into smaller and smaller pieces based on how your company does the work until it is small enough to categorize and organize hierarchically.

12. Answer: C

Quality Management is all about making sure that the product you are building conforms to your customer's requirements. If you have done a good job of gathering and understanding those requirements, all of the measurements you take on your project should help you see if what you are building will make your clients satisfied in the end.

13. Answer: D

The Communications Management plan defines all of the processes that will be used for communication on the project.

14. Answer: D

The work authorization system is a part of your company's enterprise environmental factors, and it's generally part of any change control system. It defines how work is assigned to people. If work needs to be approved by specific managers, the work authorization system will make sure that the right people are notified when a staff member's work assignments change.

15. Answer: A

There are a few really important ethical issues in this question. Your senior management was clear about the rules: go with the lowest bidder. And that's what you should do. But on top of that, you shouldn't choose your sellers based on perks that you'll get—that's called a bribe. And you should always refuse bribes.

16. Answer: C

The RACI matrix shows roles and responsibilities on your project. RACI stands for Responsible, Accountable, Consulted, Informed. Some people on your project will be responsible for activities; others might be accountable for them. The RACI matrix is a table that shows people and how they relate to the work that is being done.

Exam Questions Answers

17. Answer: B

Some questions on the exam might ask you about how to operate in another country. In this case, the question is about whether or not something is a bribe. Clearly, if it's a bribe, you can't pay it. But is it? If a payment to a government official (or anyone else) is customary, then it's not a bribe. You should go ahead and pay the police—as long as it's acceptable and legal in that country.

(Sometimes a bribe isn't money. Sometimes it's not 100% clear if something even is a bribe. But if you see an exam question where you're getting any reward for doing your normal job, make sure you treat that reward as a bribe—and refuse it!)

18. Answer: D

Confronting means figuring out the cause of the problem and fixing it. That's the best way to be sure that the right decision is made.

19. Answer: D

The most important part of the Control Quality process is that your team has to inspect each deliverable in order to verify that it meets its requirements. So what do you need to do that? Well, obviously you need the deliverables! And quality checklists are really useful too, because they help you inspect each deliverable. You need work performance data, because that tells you how well the team is doing the job. But validated changes aren't an input—they're the output!

(If you're wondering why the defect repairs are recommended rather than approved, it's because those defect repairs still need to go through change control! After all, there are some defects that are just not worth repairing, but only the stakeholders on the change control board can determine which ones are worth it and which ones aren't.)

20. Answer: D

You can't do anything about the problem until you understand it. You should meet with the client to get a better understanding of what went wrong and why the product is not meeting their needs.

21. Answer: C

Your risk register is one of the most important project management tools that you have—that's why you review it and go over your risks at every meeting. Any time you come across a new risk, the first thing you should do is document it in the risk register. It's really easy to lose track of risks, especially when you're running a big project. By adding every risk to the register, you make sure that you don't forget about any of them. So once you've identified the risk, what's the next step? You analyze the impact and probability of the risk! That's what the Perform Qualitative Risk Analysis process is for. You shouldn't take any other action until you've analyzed the risk. The reason is that it might turn out that the risk is very unlikely, and there might be another risk with a higher probability and larger impact that deserves your attention.

Answers

Exam Questions

22. Answer: D

Tom is using a template. As your company completes projects, the documents created along the
way are stored in an organizational process asset library. The WBSes from those past projects
can be a great way to be sure that you are thinking of all of the work that you will need to do
from the very beginning. Your project will never match the old WBS exactly, but there could
be work packages listed there that you might not have thought of on your own but really are
necessary in your project.

While a template is definitely one of your organization's process assets, it's NOT a tool or technique of the Create WBS process. It's an input!

23. Answer: A

The project charter does several important things: it lays out the project requirements, describes
an initial summary milestone schedule, documents the business case, and identifies initial
risks, assumptions, and constraints. But most importantly, a project charter identifies the project
manager, and assigns him or her the authority necessary to get the job done.

24. Answer: B

You need to figure out the root cause of the problem if you are going to find a lasting solution
to it. The best choice is to meet with the manager and understand why he offered the team
members to you and what you can do to work together to find the right people for your team. It's
possible that he has some information about those staff members that make them a good fit
after all.

25. Answer: C

The point of total assumption is the point at which the seller assumes the costs. In a firm fixed-
price contract, this is the point where the costs have gotten so large that the seller basically runs
out of money from the contract and has to start paying the costs.

26. Answer: C

All project reports must be communicated as formal written documents. Not only that, but
anything that has to do with a contract *definitely* needs to be formal written.

Answers

Exam ~~Questions~~

27. Answer: A

As a certified project management professional, it's your duty to respect copyrights. Purchased software is copyrighted, and you cannot use it without a license. Ever. If you don't have the budget to buy it, you can't use it.

> When a question says that you don't have enough money in the budget to do something that will keep your project from being late, then your project will be late. That's why time and cost are two of the constraints you have to deal with.

28. Answer: B

If you want to evaluate how the project is doing with respect to the schedule and budget, you need to calculate CPI and SPI. The first step is to write down the information you have so far: BAC = $450,000, planned % complete = 45%, actual % complete = 40%, and AC = $165,000. Now you can calculate PV = BAC x planned % complete = $450,000 x 45% = $202,500. And you can calculate EV = BAC x actual % complete = $450,000 x 40% = $180,000. Now you have the information you need to calculate CPI and SPI. CPI = EV / AC = $180,000 / $165,000 = 1.09, which is above 1.0—so your project is within its budget. And you can calculate SPI = EV / PV = $180,000 / $202,500 = .89, which is below 1.0—so your project is behind schedule.

29. Answer: D

Sometimes Validate Scope happens before Control Scope, and sometimes it happens afterward—and sometimes it happens both before AND afterward. That actually makes a lot of sense when you look at what those two processes do, and how they interact with each other. You always perform some Validate Scope activities at the end of your project, because you need to verify that the last deliverable produced includes all of the work laid out for it in the scope statement. Most projects will almost certainly have gone through Control Scope before then. So it might seem like Control Scope always happens before Validate Scope. But you don't just perform Validate Scope at the end—you actually do it after every deliverable is created, to make sure that all the work for that deliverable was done. Not only that, but sometimes Validate Scope fails because your team didn't do all of the work that was needed—that's why change requests are an output of Validate Scope. And if those changes include scope changes, then your project will end up going through Control Scope again— possibly for the first time in the project, if this is the first scope change you've had to make. So Control Scope can happen before Validate Scope, but it can also happen afterward as well. That's why there's no prescribed order for those two processes: they can happen in any order.

30. Answer: A

> Answer D is wrong because you can't just terminate a contract, since it's legally binding. But if a contract does eventually get terminated early during claims administration, you do have to follow any termination procedures in the contract.

When there's a dispute between a buyer and a seller, that's called a claim. Most contracts have some language that explains exactly how claims should be resolved—and since it's in the contract, it's legally binding, and both the buyer and seller need to follow it. Usually it's not an option to renegotiate a contract, especially at the end of the project after the work is complete, and lawsuits should only be filed if there are absolutely, positively no other options.

Answers

Exam Questions

31. Answer: C

When you're performing the (Close Procurements) process, you're closing out work done by a seller for a contract. To do that, you do a few things: you verify that all of the work and deliverables are acceptable, you finalize any open claims, and in case of early termination, you follow the termination clause in the contract. On the other hand, when you're performing the (Close Project or Phase) process, you're finalizing all of the various activities that you do across all of the process groups, and you're also verifying that the work and deliverables are complete.

Understanding the difference between these two things can really help you on the exam!

32. Answer: B

Of all of the contract types listed in the question, the Time and materials (T&M) contract is the riskiest kind of contract for the buyer, because if the cost of the materials gets really high then they're passed along to the buyer—and the seller doesn't have any incentive to keep them down! (It's true that cost plus award fee (CPAF) could involve paying an additional fee to the seller, but that fee is based entirely on the buyer's subjective evaluation of the seller's performance, which lowers the risk.)

Unplanned work done by the team is always gold plating, even if it makes the client happy. But if the client never asked for it, it's not scope creep because the project's planned scope never changed.

33. Answer: A

Gold plating is when you or your team add more work to the project that was not requested by the sponsor or client. It is always a bad idea to gold-plate a project because the impact is sometimes not immediately known. Sometimes, a feature that might seem really useful to your team is actually a detriment to the client. Gold-plated features can also introduce bugs that slow down later development.

34. Answer: C

This is an example of accepting a risk. The team can't do anything about the weather, so the project manager has accepted the fact that they could end up being delayed by it.

35. Answer: B

A watchlist is where you keep risks that don't have a high enough probability or impact to make it into the risk register but still need to be monitored. By recording the risk in a watchlist, you will have a reminder to check to be sure that circumstances haven't changed as your project goes on. That should give you enough time to come up with a risk response strategy if circumstances change as time goes on.

Exam Questions

36. Answer: C

The Project Management plan is not a bar chart (or a Gantt chart). It's the collection of all of the planning documents you create through all of the knowledge areas within the five process groups. It describes how your project will handle all of the activities associated with your project work.

37. Answer: A

Validate Scope is the Monitoring and Controlling process for the Scope Management knowledge area. It doesn't have anything to do with planning out the scope of the project—you do it as you complete each project phase to make sure that your team has completed all of the project work.

38. Answer: C

Cost aggregation is used to build your budget, but it is not a tool for cost estimation. Bottom-up, parametric, and analogous estimation techniques are used for both cost and time estimates.

39. Answer: A

A control chart is a really valuable tool for visualizing how a process is doing over time. By taking one measurement after another and plotting them on a line chart, you can get a lot of great information about the process. Every control chart has three important lines on it: the mean (or the average of all data points), an upper control limit, and a lower control limit. There's an important rule called the rule of seven that helps you interpret control charts. That rule tells you that if you find seven consecutive measurements that are on the same side of the mean, there's something wrong. That's because it's extremely unlikely for seven measurements like that to occur—it's much more likely that there's a problem with your process. If you can figure out an improvement to fix that, you'll have a lot fewer defects to repair later!

40. Answer: B

The PMP Code of Professional Conduct states that you must follow every law, no matter how trivial. Any time you see a question that asks about breaking a law, your answer should always be the choice that doesn't break it—no manner how minor the infraction, and how serious the consequences.

Answers

Exam ~~Questions~~

41. Answer: C

The Staffing Management plan tells you everything that you need to know about when resources will be released from a project. Since the team you need for your project is currently on another project, that project's Staffing Management plan will tell you when they will be released from that project and available for yours.

42. Answer: C

If your project is broken up into phases, you should have a kickoff meeting at the start of each phase. You use that meeting to talk about lessons learned from past projects and establish the way people will communicate as the project work goes on.

43. Answer: B

Joe is doing root-cause analysis on process problems: that's Perform Quality Assurance. Remember, Control Quality is when you are trying to find problems in your work products through inspection. Perform Quality Assurance is when you are looking at the way your process affects the quality of the work you are doing.

44. Answer: B

Once you've figured out the impact of the change to your schedule, budget, and scope, the next step is to take the change request to the change control board. If they approve your recommendation, then the request will be approved and you can update your baseline and implement the change.

45. Answer: B

When you work with a union, then the union contract can have an impact on your project. That means you need to consider the union itself a stakeholder, and when you do your planning you need to make sure any union rules and agreements are considered as constraints.

46. Answer: C

A change control board (or CCB) is a group of people that approves or rejects changes. It usually includes the sponsor, which makes sense because the sponsor is the one funding the project. It's not the project manager's job to tell the CCB whether or not to approve a change—they use their

Exam ~~Questions~~ Answers

expert judgment to figure out whether or not the change is valuable. It *is* the project manager's job to make sure the impact of the change on the triple constraint (time, scope, and cost) is evaluated, but that impact analysis should happen *before* the change request is sent to the CCB.

47. Answer: A

Colocation means that you have all of your team located in the same room. When you do this, you can increase communication and help them build a sense of community. Sometimes the room the colocated team meets in is called a war room.

When bidders are competing for a contract, you must make sure they all have the same information so that no one bidder is given an unfair advantage. That's why a bidder conference is a great tool—it gives all bidders access to the same information.

48. Answer: B

Any time there's a conflict of interest, it's your duty to disclose it to your company. After that, you should always proceed based on your company's policies. If there are no specific policies about that, then make sure that the conflict does not affect your decisions.

49. Answer: B

Stakeholder analysis means talking to the stakeholders and figuring out their needs, and it's something that you do when you're defining the project scope. If there's an important client who has needs that your project is supposed to fulfill, that client is always a stakeholder. And if your project is not meeting that client's needs, then you didn't do a good enough job when you were performing stakeholder analysis!

50. Answer: B

The project management methodology describes the process (or lifecycle) that you use to manage your project. It really doesn't have anything to do with closing a project or phase. The other three answers, however, do! You need the Project Management plan to give you the procedure for closing the project phase. You need the accepted deliverables to verify that they're complete. And you need your organizational process assets for lessons learned and closure guidelines.

51. Answer: D

The scope baseline is made up of the project scope statement and the WBS and the WBS dictionary. The WBS dictionary is considered a supporting document to the WBS, so if the WBS were to change, then the dictionary would, too.

Answers

Exam ~~Questions~~

52. Answer: B

A rough order of magnitude (ROM) estimate is an estimate that is very rough. According to the *PMBOK Guide*, you should expect a ROM estimate to be anywhere from –25% to +75% of the actual result. That means that if your ROM estimate for a project is six months, then you should expect the actual project to be anywhere from three months to nine months.

53. Answer: A

It's pretty easy to remember which processes are in the Initiating group, because there are only two of them! But more importantly, it's useful to know what you need to do when you initiate a project. First you need to create the project charter (by performing the Develop Project Charter process), which authorizes the project manager to do the work. And then you need to identify your stakeholders (by performing the Identify Stakeholders process), which helps you understand who needs your project done and what interest they have in it.

54. Answer: B

Mary is working for a projectized organization. In those companies, the project manager has authority over the team as well as the project.

55. Answer: C

It's very hard to figure out whether or not your project is successful unless you can measure that success. That's why you need to come up with goals that have numbers attached to them—which is what *quantifiable* means. Of all four answers, only answer C has a goal that you can actually measure.

56. Answer: A

Customer satisfaction is an important part of modern quality management. Remember, customer satisfaction is about making sure that the people who are paying for the end product are happy with what they get. But the way that you make sure that your customers are happy is by meeting their needs—and you do that by ensuring the product the team builds meets the customer's requirements. That's what Quality Management is all about, and it's an important reason that you do Quality Management.

Answers

Exam Questions

57. Answer: B

Customers can be satisfied even when a project is not profitable—customer satisfaction isn't always about money. Rather, customer satisfaction is about making sure that the people who are paying for the end product are happy with what they get. When the team gathers requirements for the specification, they try to write down all of the things that the customers want in the product so that you know how to make them happy. Some requirements can be left unstated, too. Those are the ones that are implied by the customer's explicit needs. In the end, if you fulfill all of your requirements, your customers should be satisfied.

58. Answer: C

Since the project absolutely must be completed in seven months for it to be worth doing, the deadline is a constraint. It must be met for the project to be considered successful.

59. Answer: C

Herzberg's Motivation-Hygiene Theory states that people need things like good working conditions, a satisfying personal life, and good relations with the boss and coworkers—these are called "hygiene factors." Until people have them, they generally don't care about "motivation factors" like achievement, recognition, personal growth, or career advancement.

60. Answer: C

The resource histogram is a bar chart that shows your staffing needs over time. If you need more testers in the end of the project than you do while you're building a product, for example, you can forecast how many you will need and what their skill level needs to be from the beginning. That way, you'll be sure that they're available when you need them.

61. Answer: B

There are four things you can do with any opportunity. You can exploit it by making sure you do everything you can to take advantage of it. You can share it by working with another company in a way that gives you a win-win situation. You can enhance it by figuring out a way to increase its value. Or, if there's no way to take advantage of it, you can just accept it and move on. In this case, taking additional actions that will increase the potential reward is enhancing the opportunity.

Answers

Exam ~~Questions~~

62. Answer: B

The Plan-Do-Check-Act cycle is a way of making small improvements and testing their impact before you make a change to the process as a whole. It comes from W. Edwards Deming's work in process improvement, which popularized the cycle that was originally invented by Walter Shewhart in the 1930s.

Just remember, lessons learned are your most important organizational process assets.

63. Answer: A

Lessons learned are part of the organizational process assets, not enterprise environmental factors. Your company's enterprise environmental factors tell you about how your company typically does business—like how your company's departments are structured, and the regulatory and industry environment your company operates in. An important enterprise environmental factor that you'll run across when you're planning a project is the work authorization system. That's your company's system to determine who is supposed to be working on what, and when the work should get done.

64. Answer: C

The cost variance (CV) is the difference between the amount of money you planned on spending and the total that you've spent so far. This should make sense—if your CV is negative, it means that you've blown your budget.

65. Answer: B

Tom is using expert power. Since he's been through this problem before, his team is more likely to accept his authority. Expert power is the best form of power to use when making project decisions. The team will respect decisions that are based on experience and expertise.

66. Answer: D

It's important to know that resources, schedules, and priorities cause 50% of project problems and conflicts. Sure, it's important for the PMP exam. But even more importantly, if you're trying to confront a problem by looking for the root cause of a conflict, the odds are that you'll find that cause in one of those three areas!

Answers

Exam ~~Questions~~

Answers

67. Answer: B

Quality audits are when your company reviews your project to see if you are following its processes. The point is to figure out if there are ways to help you be more effective by finding the stuff you are doing on your project that is inefficient or that causes defects. When you find those problem areas, you recommend corrective actions to fix them.

Any time you create recommended corrective actions, they go through change control.

68. Answer: A

Coaching is an important interpersonal skill for any project manager to have. Any time you do coaching, mentoring, training, or anything else to help others learn about project management, you're not only helping your team member, you're also contributing to the Project Management Body of Knowledge.

69. Answer: B

Even if a project is shut down before the work is completed, you still need to document the lessons learned and add them to the organizational process assets. In fact, if a project is terminated early, that's probably the best time to do that! When a project goes seriously wrong, then there are always important lessons that you can learn—even if it wasn't your fault!

70. Answer: A

Seven values on one side of the mean in a control chart indicate a problem with the process that is being measured.

71. Answer: B

The Delphi technique is a way to get opinions and ideas from experts. This is a technique that uses a facilitator who uses questionnaires to ask experts about important project risks. They take those answers and circulate them—but each expert is kept anonymous so he or she can give honest feedback.

72. Answer: D

Take a look at the answers to this question. What do you see? A list of processes—"Create WBS," "Develop Project Management Plan," "Develop Project Charter," and "Identify Stakeholders." Your job is to figure out which of these processes comes next. So what clues do you have to tell you where you are in the project lifecycle? Well, you've just been authorized to manage a new project. Since the project charter is what authorizes a project manager to work on a project, it means that the Develop Project Charter process has just been performed. So which of the processes in the list comes next? The other Initiating process: Identify Stakeholders.

Answers

Exam Questions

73. Answer: C

The expected monetary value (or EMV) of the problems integrating the component is the probability (20%) times the cost ($3,000), but don't forget that since it's a risk, that number should be negative. So its EMV is 20% x $3,000 = –$600. The savings from not having to build the component from scratch is an opportunity. It has an EMV of 40% x $10,000 = $4,000. Add them up and you get –$600 + $4,000 = $3,400.

74. Answer: C

The best thing that you can do with a risk is avoid it—if you can prevent it from happening, it definitely won't hurt your project. The easiest way to avoid a risk is to cut it out of your project entirely; in this case, getting rid of the subcontractor avoids the risk.

Sometimes avoiding one risk can lead to another. It's possible that there was a reason that you went with the subcontractor in the first place, and now you've exposed the project to a different risk! That's why Risk Management is so important.

75. Answer: D

The WBS dictionary always corresponds to an entry in the WBS by name and work package ID. So that's the easiest way to cross-reference the two. The statement of work describes the work that will be done. The responsible organization is the team or department who will do it. Schedule milestones are any set dates that will affect the work. The quality requirements describe how we will know if the work has been done properly. The resource and cost estimates are just a list of how many people will be needed to do the work and how much it will cost. Answer A couldn't be right because net present value doesn't have anything to do with individual work packages. The other options mention earned value and Monte Carlo analysis, which have nothing to do with Scope Management either.

76. Answer: C

The best way to resolve any problem is to confront the issue—because *confronting* means figuring out the source of the problem and then resolving the root cause of the conflict. Any time you have an opportunity to confront the problem, you should do it. Remember, one of the most important things that a project manager does is make sure that team conflicts get resolved. Sometimes questions are worded so that the word *confronting* sounds negative. Even when it is, it's still the best approach to resolving conflicts!

77. Answer: A

Tom is a Theory X manager. He believes that employees need to be watched all of the time and that all of his team members are selfish and unmotivated.

Exam Questions — ~~Answers~~

78. Answer: C

The WBS dictionary is the companion document to the WBS. It gives all the details that you know about each work package in the WBS, including estimates and billing information.

79. Answer: A

This is a calculation question that's asking you to use SPI and CPI to evaluate your project. Luckily, it's easy to do that! First calculate SPI = EV / PV = $15,000 / $12,000 = 1.25—so your project is ahead of schedule. Then calculate CPI = EV / AC = $15,000 / $11,000 = 1.36—so your project is within its budget.

80. Answer: C

Whenever you use any of the seven basic tools of quality to examine the results of an inspection of your product, you are in Control Quality. If you were examining the process your company uses to build multiple projects, you would be in Perform Quality Assurance.

81. Answer: A

Withdrawal happens when someone gives up and walks away from the problem, usually because he's frustrated or disgusted. If you see a team member doing this, it's a warning sign that something's wrong.

82. Answer: B

The reason we work to do quality planning up front is that it is most expensive to deal with problems if you find them late in the project. The best case is when you never inject the defects in the first place; then it doesn't cost anything to deal with them. Prevention is always better than inspection.

83. Answer: D

You must always follow your company's policy—it's your ethical duty as a project manager. You should find some other way to reward her that is not against your company's rules.

Answers

Exam ~~Questions~~

---

84. Answer: B

Perform Qualitative Risk Analysis is all about figuring out prioritizing each risk, and figuring out its probability and impact. It's an important part of risk planning. But it's not about coming up with specific numbers! That's what Perform Quantitative Risk Analysis is for—and EMV analysis is part of quantitative (not qualitative) analysis, because it's where you assign numeric values to risks.

Remember, quantitative means numbers and qualitative means judgments!

85. Answer: D

The PMP Code of Professional Conduct tells us that an important part of any project manager's career is enhancing personal professional competence. This means increasing your knowledge and applying it so that you can improve your ability to manage projects.

86. Answer: B

A risk urgency assessment is a tool of Perform Qualitative Risk Analysis. Identify Risks is all about finding risks. Perform Qualitative Risk Analysis is about ranking them based on what your team thinks their impact and probability will be for your project.

Brainstorming and Delphi technique are both part of information-gathering techniques, which is one of the tools and techniques of Identify Risks.

87. Answer: C

To calculate the expected monetary value (EMV) of a set of risks and opportunities, multiply each probability by its total cost and add them together. In this question, the cost of the risk is –$15,250 + –$20,000 = –$35,250, so its EMV is 40% x –$35,250 = -$14,100. The value of the opportunity is $4,500 and its probability is 65%, so its EMV is 65% x $4,500 = $2,925. So the total EMV for the two is –$14,100 + $2,925 = –$11,175.

Don't forget that the cost of a risk is negative, and the cost of an opportunity is positive.

88. Answer: A

Before you can close your project, there are a few things you need to do. Remember the acceptance criteria in the scope statement? Well, those criteria need to be met. And you need to get formal written acceptance from the customer. And every work item in the WBS needs to be completed.

Until the customer accepts the final product, your project isn't done!

89. Answer: D

Once you've closed out a procurement, it's important to conduct a procurement audit. This is

---

Answers

Exam Questions

where you go over everything that happened on the project to figure out the lessons learned, and look for anything that went right or wrong. However, consideration—or payment—is not part of an audit (unless there was a problem processing or paying it).

That's why the payment system is one of the Administer Contracts tools and techniques, and not part of Close Procurements—you can't close out the contract until it's been paid.

90. Answer: B

The Staffing Management plan includes a "Reward and Recognition" section that describes how you'll reward your team for good performance. It also contains training requirements and release criteria.

There's no such thing as a "Reward and Recognition plan" in the PMBOK Guide.

91. Answer: C

It's pretty obvious just from a quick glance at the numbers that this project is in trouble. The total budget for completion is 900 panels x $75 per panel = $67,500, and your actual costs are already $45,000. But if you look at all the answers, every one of them could potentially be correct: you know that you're pretty far below budget, so the CPI will be below 1 and the CV will be negative. And since you should have completed 400 panels, you're behind schedule, so you know your SPI will also be below 1. So which of the answers is right? There's only one way to find out: do the calculations. Actual % complete is 350 ÷ 900 = 38.9% so PV = $67,500 x .389 = $26,258. Planned % complete is 400 ÷ 900 = 44.4%, so PV = $67,500 x .444 = $29,970. CPI = $26,258 ÷ $45,000 = .584, and CV = $26,258 − $45,000 = −$18,742, so neither of those numbers matches the answers for A and B. And for answer D, SPI = $26,258 ÷ $29,970 = .876, which doesn't match, either. But for answer C, TCPI = ($67,500 − $26,258) ÷ ($67,500 − $45,000) = 1.833, which does match the answer. Answer C is correct.

92. Answer: C

By buying the extra chemical stock, you are mitigating the risk.

93. Answer: D

A resource histogram is just a way to visualize the number of people in each role that you will need on your project as time goes on. Once you have figured out your schedule and the order of activities, you figure out how many people it's going to take to do the work and plot that out over time. Then you have a good idea of what the staffing needs of your project will be.

Stakeholder analysis is important, but it's not part of change control.

94. Answer: D

Change control is how you deal with changes to your Project Management plan. And a change control system is the set of procedures that lets you make those changes in an organized way. A typical change control system includes a change control board, utilizes a project management information system, and ends with either approval or rejection.

Answers

Exam ~~Questions~~

95. Answer: B

Any time you have any communication having to do with the contract, it's always formal written communication.

96. Answer: A

One of the most important elements of the Staffing Management plan is the timetable, which tells you who will work on what, and when they will be released from the project. One of the most common ways of showing the timetable is the resource histogram (or staffing histogram). That timetable will let you know exactly when you plan to release your project resources.

97. Answer: D

While noise can interfere with communication, it's not a communication type.

98. Answer: B

Lessons learned from past projects are always part of a company's organizational process assets, and are usually stored in a process asset library. The other three answers are important project tools, but they're not where you find lessons learned.

All the answers to that question sounded good, right? Just remember, lessons learned are your most important organizational process assets!

99. Answer: C

Once your project team is done with the work, it's time to check the deliverables against the scope statement, WBS, and Scope Management plan. If your deliverables have everything in those documents, then they should be acceptable to stakeholders. When all of the deliverables in the scope are done to their satisfaction, then you're done with the project! What comes next? Formal acceptance, which means you have written confirmation from the stakeholders that the deliverables match the requirements and the Project Management plan. Since this communication is a project document, it's formal written communication.

100. Answer: C

The procedure for managing changes to a contract is found in the Contract Management plan. The other three answers are all things you typically find in a project charter.

Answers

Exam ~~Questions~~

101. Answer: C

The most important project constraints that you'll see on the exam are scope, quality, schedule, budget, resources, and risk. Any change to one of those constraints affects the others. It's important to balance all of these constraints throughout your project.

102. Answer: D

Every risk should have a risk owner listed in the register. That person is responsible for keeping the response plan up to date and making sure the right actions are taken if the risk does occur.

103. Answer: B

Risk mitigation means taking some sort of action that will cause a risk, if it materializes, to do as little damage to your project as possible. Having team members spend time doing work to prepare for the risk is a good example of risk mitigation.

104. Answer: A

Sometimes you don't need to do any calculations when you run across a question like this. The question asked you which number to use for a forecast of how much money you expect to spend on the rest of the project. Well, isn't that the definition of ETC? Since you were given the value of ETC, you could just use that number!

105. Answer: D

Product scope means the features and functions of the product or service being built. Project scope means the work that's needed to build the product.

106. Answer: A

It's just easy to calculate the late finish (LF) of an activity in a network diagram. Look at the following activity, take its LS (late start), and subtract one. If there's more than one following activity, use the one with the lowest LS. So for activity F in the question, the following activities are G, with an LS of 17, and H, with an LS of 11. So the LF of F is 11 – 1 = 10.

Answers

Exam ~~Questions~~

107. Answer: A

Punishment power is exactly what it sounds like—you correct a team member for poor behavior. Always remember to do this one-on-one, in person, and in private! Punishing someone in front of peers or superiors is extremely embarrassing, and will be really counterproductive.

Punishment isn't usually the best way to handle a situation, but if it's the only option, make sure you do it right.

108. Answer: B

Since the CFO is affected by your project, that means he's a stakeholder. The best thing you can do in this situation is get the new stakeholder's opinion incorporated in the project up front. It's important that all of the project stakeholders understand the needs and objectives that the project is meant to address. The worst case is to have the stakeholder's opinion incorporated at the end of the project—that could mean a lot of rework or even an entirely unacceptable product.

109. Answer: D

An important part of identifying stakeholders is reviewing lessons learned from prior projects (because they may help you identify stakeholder issues early), performing stakeholder analysis (which often involves a power/interest grid), and reviewing procurement documents (because a contract often brings extra stakeholders with it). However, you don't create the change control system in the Initiating phase—that's something that you do as part of your project planning activities.

110. Answer: C

The project manager is mitigating the risk by backing up the data so that it doesn't get lost. He is transferring it to the insurance company by insuring the company for the cost of rekeying the information.

111. Answer: B

This project is not in good shape. The client has needs that aren't being met, but there may not be enough time or money to meet them. What's the project manager going to do? Well, the first thing that you should do any time you have a problem is try to figure out what's causing it. All of the other answers involved taking some sort of action, and you should never take action until you've figured out the root cause of the problem.

Answers

Exam ~~Questions~~

112. Answer: C

Residual risks are risks that remain even after you have planned for and implemented all of your risk response strategies. They don't need any further analysis because you have already planned the most complete response strategy you know in dealing with the risk that came before them.

113. Answer: D

The client is trying to command Rekha to do what he says even though she has good reasons for not doing it. He isn't working to solve the problem, he's just forcing the resolution to go his way.

114. Answer: B

Constrained Optimization doesn't have anything to do with Define Scope—it's a kind of benefit selection method. The other answers are all tools of the Define Scope process.

115. Answer: A

The records management system is one of the tools that you use in the Close Procurements process. It's what you use to store your contracts and any related documents, so that future project managers can refer to them in future projects.

116. Answer: B

Time and materials (T&M) contracts are used in labor contracts. In a T&M contract, the seller pays a rate for each of the people working on the team plus their material costs. The "time" part means that the buyer pays a fixed rate for labor—usually a certain number of dollars per hour. And the "materials" part means that the buyer also pays for materials, equipment, office space, administrative overhead costs, and anything else that has to be paid for.

117. Answer: D

Any activity that helps you find, prevent, or fix defects in your product is included in the cost of quality. The activities you do to build the product don't count toward that number.

Exam ~~Questions~~ Answers

118. Answer: B

A stakeholder is anyone who is affected by the cost, time, or scope of your project. And that includes unions—if you have team members who are in a union, then you always need to consider that union as a stakeholder and make sure its needs are met. However, you don't need to consider the needs of your company's competitors.

119. Answer: D

The project scope statement defines the scope of work for the project. It's where everyone comes to a common understanding about the work that needs to be accomplished on the project.

120. Answer: C

$230,000 x 0.70 = $ 138,000 savings, and $150,000 x 0.40 = –$60,000 expenses. Add them together and you get $78,000.

When you calculate EMV, anything that saves your project money is counted as positive, and anything that costs it money is negative. Multiply each by the probability and add them together.

121. Answer: B

The project manager is asking the other companies to help her make this opportunity happen and they can all share in its benefits.

122. Answer: C

If you discover that someone claims to have the PMP credential but is not actually certified, you must contact PMI immediately so that it can take action.

123. Answer: A

The most effective type of power for a project manager is Expert power. That's when your team respects you because they know that you know what you are talking about.

Exam ~~Questions~~ Answers

124. Answer: B

Since the Smith County project officer is the sponsor, he's the person who is best suited to sign the charter. A project charter is typically approved and signed by the sponsor. Some projects are approved by key stakeholders, but they are never approved by project managers (since the project manager is only granted authority once the project is signed) or team members.

125. Answer: B

Control Procurements is the Monitoring and Controlling process for Procurement Management. It's when you run into a change that has to be made to a specific contract. You use work performance information to determine how the contract is going, and the contract and procurement documents to see exactly what everyone's on the hook for. But you don't actually see the Procurement Management plan as an input to Control Procurements.

126. Answer: A

If you discover that someone claims to have the PMP credential but is not actually certified, you must contact PMI immediately so that it can take action.

127. Answer: B

You may get a question on the exam that asks what to do when you encounter a change. You always begin dealing with change by consulting the Project Management plan.

128. Answer: B

Sometimes you'll find that some risks have obviously low probability and impact, so you won't put them in your register. Instead, you can add them to a watchlist, which is just a list of risks that you don't want to forget about, but you don't need to track as closely. You'll check your watchlist from time to time to keep an eye on things.

129. Answer: A

Everything listed in each of the answers is a project document…except for the scope baseline. The baselines and subplans are all part of the Project Management plan, so they don't fall under the heading of "project documents."

Answers

Exam ~~Questions~~

---

130. Answer: C

When you're looking at CPI and SPI numbers, remember: lower = loser. If your CPI is below 1.0, then your project is over its budget. If the SPI is below 1.0, then the project is behind schedule. In this case, the project is ahead of schedule, since its SPI is above 1.0. But it's over its budget, because it's got a CPI that's below 1.0.

131. Answer: B

The key to a good bonus system is that it must be achievable and motivate everyone in the team to work toward it. If you are only rewarding one team member or a few people in the group, the rest of the team will not be motivated. Also, making the goals too aggressive can actually de-motivate people.

132. Answer: A

The formula for lines of communication is n x (n–1) / 2. So the answer to this one is (17 x 16) / 2 = 136.

133. Answer: A

There are two kinds of project selection methods. Benefit measurement models, or comparative methods, are used to compare the benefits and features of projects. Mathematical models use complex formulas to determine which project has the most value to the company. You should get familiar with some of the more common comparative approaches to project selection, like murder boards, benefit-to-cost ratios, and peer reviews.

134. Answer: D

If you find out that someone is cheating on the PMP exam by distributing questions that are on it, you must report that person to PMI immediately. If that person is a PMP-certified project manager, he will be stripped of his certification.

135. Answer: B

The project sponsor is the person (or people) that pays for the project. Sometimes this means the sponsor directly provides funding; other times, it means the sponsor is the person who signs the organizational approval to assign resources. Either way, you can usually tell who the sponsor is by finding the person who can approve or deny the budget.

136. Answer: C

An important part of any project manager's career is enhancing personal professional competence. This means increasing your knowledge and applying it so that you can improve your ability to manage projects.

The "document repository" sounded good, but you won't find that term anywhere in the PMBOK Guide. Watch out for made-up terms on the exam!

137. Answer: A

The configuration management system is there to be sure that everybody on the team has the most updated version of all of the project documents. Whenever a project document is changed, it is checked into the configuration management system so that everyone knows where to go to get the right one.

138. Answer: C

Project managers must have a "zero tolerance" policy on racist remarks, or any other cultural insensitivity. If there is an incident involving racism, sexism, or any other kind of discrimination, your top priority is to correct that. Every company has a policy that guides how you handle this kind of situation, so a question involving racism will usually involve the company's policy or HR department.

139. Answer: B

An important part of making sure that your project goes well is keeping an eye on the work, and that's what the Monitor and Control Project work process is for. It's where you constantly evaluate the work being done, and any time you see a problem you recommend changes, defect repairs, and preventive and corrective actions.

The bars in the Pareto chart show the number of defects in each category, with a line overlaid that shows the percentage of the total defects found.

140. Answer: D

Pareto charts plot out the frequency of defects and sort them in descending order. The right axis on the chart shows the cumulative percentage. This helps you figure out which root cause is responsible for the largest number of defects. The 80/20 rule states that 80% of defects are caused by 20% of the root causes you can identify. So if you do something about that small number of causes, you can have a big impact on your project.

Answers

Exam ~~Questions~~

141. Answer: C

You need to make sure that what you're delivering matches what you wrote down in the scope statement. That way, the team never delivers the wrong product to the customer. As you complete each deliverable, you work with the stakeholders and the sponsor to make sure that you did the right work.

142. Answer: C

Even though the numbers are large, this is a simple application of the channels of communication formula: # lines = n x (n – 1) ÷ 2. There is a total of 1,253 + 752 + 14 = 2,019 people. So the number of channels is 2019 x 2018 ÷ 2 = 2,037,171. That's a pretty staggering number, but it's realistic for a large project—and it's why Communications Management is such an important part of a project manager's toolbox.

143. Answer: B

The halo effect is when you put someone in a position he can't handle, just because he's good at another job. Just because Joe is a great programmer, that doesn't mean he'll be a good project manager.

144. Answer: D

The Staffing Management plan always includes a resource histogram, so that should be your first clue about which one of these answers is right. The resource histogram shows what kind of resource is needed through each week of your project and how many staff members you need. When planning out your staffing needs, you need take into account the training it will take to get them up to speed as well as the kinds of incentives you are going to offer for a job well done. Resource calendars are important too; they'll tell you when your staff members will be available. You need to think about what staff members need to get done before they are released to work on other projects.

Exam ~~Questions~~

145. Answer: C

Plurality is an example of a group decision-making technique in which a decision can be made by the largest block of people in the group, even if they don't have a 50% majority.

146. Answer: C

Smoothing is minimizing the problem, and it can help cool people off while you figure out how to solve it. But it's only a temporary fix, and does not really address the root cause of the conflict.

147. Answer: B

The project charter is often created without the project manager's involvement. Sometimes it is handed to the project manager by the sponsor or high-level manager.

148. Answer: C

Checksheets are a great way to keep a tally of any quality-related activities you need to repeat while controlling quality or performing quality assurance on your project.

149. Answer: B

Monte Carlo analysis is a way of seeing what could happen to your project if probability and impact values changed randomly.

150. Answer: A

Saying that the work must be "better" is subjective. Requirements gathered in stakeholder analysis need to be quantifiable. That way, the team has a goal they can shoot for and you can always tell how close to or far from it you are.

~~Exam Questions~~ Answers

151. Answer: B

Maslow's Hierarchy of Needs says that people have needs, and until the lower ones (like acceptance on the team, job safety, or job security) are satisfied, they won't even begin to think about the higher ones (fulfilling their potential and making a contribution).

Cost of quality doesn't include the time the project manager spends putting together the Project Management plan—except for the time spent on the quality portions!

152. Answer: C

Cost of quality is what you get when you add up the cost of all of the prevention and inspection activities you are going to do on your project. It doesn't just include the testing. It includes any time spent writing standards, reviewing documents, meeting to analyze the root causes of defects, reworking to fix the defects once they're found by the team—absolutely everything you do to ensure quality on the project.

153. Answer: D

A fixed-price (FP) contract means that the buyer pays one amount regardless of how much it costs the seller to do the work. A fixed-price contract only makes sense in cases where the scope is very well known. If there are any changes to the amount of work to be done, the seller doesn't get paid any more to do it.

So if the costs get really high, then the buyer has to swallow them.

154. Answer: C

A lot of people think compromise is a great way to handle conflicts. But any time there's a compromise, it means that everyone needs to give up something. That's why compromise is often called a lose-lose solution. It's always better to confront the problem and fix the root cause of the conflict. You should only force people to compromise if that's the only option.

Insurance is just a contract that you use to pay a company to take on some of your risk.

155. Answer: A

One effective way to deal with a risk is to pay someone else to accept it for you. This is called transference. The most common way to do this is to buy insurance.

156. Answer: B

When you're working with procurements, independent estimates is one of the tools and techniques of the Conduct Procurements process. It certainly sounds a lot like something you'd

Answers

Exam ~~Questions~~

do while planning out your procurements. Don't forget that the Conduct Procurements process involves finding sellers as well as carrying out the work to complete the contract. That's why you use things like bidder conferences and qualified seller lists in Conduct Procurements.

157. Answer: C

The scope baseline is not a particularly useful thing once a project's done. A baseline is what you use to measure any changes to the project—whenever there's a change, you always want to compare it against the baseline. But once the project is done, the baseline isn't necessary anymore.

158. Answer: C

Forecasting is a cost monitoring tool that helps you predict how much more money you'll need to spend on the project. So which of the cost metrics would you use to do that? There are two useful numbers that you can use for forecasting. One of them is called estimate to complete (ETC), which tells you how much more money you'll probably spend on your project. And the other one, variance at completion (VAC), predicts what your variance will be when the project is done.

Did you think the answer was "Status reports"? You generally won't see that as a valid answer on the exam. The PMBOK Guide is clear on this: a PM's job is to plan the work and control the project, not just gather and report status.

159. Answer: A

You create one of the most important outputs of your entire project when the team is doing the project work. Work performance data tells you the status of each deliverable in the project, what the team's accomplished, and all of the information you need to know in order to figure out how your project's going. But you're not the only one who needs this—your team members and stakeholders need to know what's going on, so they can adjust their work and correct problems early on.

160. Answer: B

Kaizen is a Japanese word that means "improvement"—and it's also a management technique that helps your company use problem solving to constantly find new ways to improve. Kaizen focuses on making small improvements and measuring their impact. It's is a philosophy that guides management, rather than a particular way of doing quality assurance.

Ishikawa diagrams—or fishbone diagrams—are an important tool that's used in Kaizen.

Answers

— Exam ~~Questions~~ —

161. Answer: A

The PMP Code of Professional Conduct requires cultural sensitivity to others. It's unacceptable to belittle people based on how they speak, the way they dress, or any other aspect of their cultural background. If you see a member of your team doing this, it's your responsibility to do what's necessary to correct the behavior and prevent it from happening in the future.

Don't assume that every time you see a fishbone diagram the question is talking about Control Quality.

162. Answer: A

Diagramming techniques (including Ishikawa diagrams and flowcharts) are a tool of the Identify Risks process. You use them to find the root cause of defects in Quality Management processes, but they can also be useful in finding the risks that can lead to trouble in Risk Management.

163. Answer: B

The project's scope is changing every time the client is asked to verify the product—that's scope creep. The best way to avoid that is to be sure that the project scope statement that is written in the planning stages of the project is understood and agreed to by everyone on the project. Scope changes should never come late in the project; that's when they cost the most and will jeopardize the team's ability to deliver.

164. Answer: B

Any time you hold a seminar, give a talk, write an article, or help others learn about project management, you're contributing to the Project Management Body of Knowledge.

This is an important part of every PMP-certified project manager's career!

165. Answer: D

If you discover that someone has broken the law, it is your duty to call the authorities and report that person. You need to do this, even if it seems like the offense is minor.

166. Answer: C

When you're asked to use benefit-to-cost (BCR) ratios to select a project, always choose the project with the highest BCR because that's the project that gives you the most benefit for the least cost. An easy way to do it is to divide: Project A has a BCR of 5:2, and 5 / 2 is 2.5. Do that with all four projects, and you find that project C has the highest BCR.

Exam ~~Questions~~ Answers

167. Answer: D

Legitimate power is the kind of power you have when you tell someone who reports to you to do something. Expert power is when your opinion carries weight because people know that you know what you're talking about. Reward power is when you promise a reward for doing as you ask. Referent power is when people do what you say because of your association with somebody else. Punishment power is when people do what you say because they are afraid of the consequences.

168. Answer: C

Every change request needs to be evaluated to determine whether or not it should be made. That's what we do in the Perform Integrated Change Control process—every change is analyzed to determine its impact. It's then documented as a change request and put into the change control system. That's where the stakeholders on the CCB determine if the change should be made.

169. Answer: B

By the time the Close Project or Phase process happens, you should have already gotten formal written acceptance for the deliverables. That's what the Validate Scope process is for, and you verify that formal acceptance in the Close Project or Phase process.

170. Answer: C

The WBS work packages can be displayed by project phase or by project deliverable. It depends on how your company needs to see the work organized. If you use the same phased lifecycle for all projects, it can be easier to show all of the work as it breaks down within each phase. If you have various teams depending on the deliverables your team will produce, it can make sense to break the work down by project deliverable.

171. Answer: A

The two main outputs of Direct and Manage Project Work are deliverables and work performance data. Work performance data is a name for all of the actual data that comes from the work your team is doing.

Exam ~~Questions~~ Answers

172. Answer: A

Lessons learned are some of your most important organizational process assets. At the end of every project, you sit down with the project team and write down everything you learned about the project. This includes both positive and negative things. That way, when you or another project manager in your company plans the next project, you can take advantage of the lessons you learned on this one.

It's really important that you work with the team to write down the lessons you've learned, because they have a lot of insight into what went right and wrong on the project.

173. Answer: D

Once you've closed out a contract, it's important to conduct a procurement audit. This is where you go over everything that happened on the project to figure out the lessons learned, and look for anything that went right or wrong.

174. Answer: A

"Force majeure" is a kind of clause that you'll see in a contract. It says that if something like a war, riot, or natural disaster happens, you're excused from the terms of the contract.

175. Answer: D

Ground rules help you prevent problems between team members, and let you establish working conditions that everyone on the team can live with. You set up the ground rules for a project to help guide people in their interactions with each other. Make sure you discuss the ground rules with the team during the kickoff meeting!

176. Answer: A

One of the most important things that you do as a project manager is to constantly monitor the project for changes, and take the appropriate action whenever you make a change. But changes don't happen on any sort of schedule—if they did, it would make project management a whole lot easier! That means you need to continuously monitor your project to figure out whether or not its plans and scope need to change.

You can think of Change Requests as what you get when someone finds a problem and needs to make a change. Once you've figured out whether or not to do that change (in Perform Integrated Change Control), you give the person an update on its status.

177. Answer: B

This question is basically asking you the difference between change requests and change request status updates. Change request status updates are outputs from Perform Integrated Change Control, not Monitor and Control Project Work.

Answers

Exam ~~Questions~~

178. Answer: C

You can't know the answer to technical questions as well as your team. So, while it's important to understand both sides of the issue, your job is to make sure that problems are confronted and fairly evaluated.

179. Answer: D

Parametric estimation is a tool for creating estimates. It's not a deliverable.

180. Answer: D

The WBS dictionary is an output of the Create WBS process. It is created along with the WBS and gives all of the details about each work package in the WBS.

The team always needs to help you document the lessons learned for the project.

181. Answer: C

This question isn't hard if you remember one really important fact: you need your team's help when you're writing the lessons learned. That's why you can't release the team until the lessons learned are documented and added to the organizational process assets. Also, the last thing you do on the project is close the contract. The reason for this is that you don't want to have to wait for payment before releasing the team, because most contracts have payment terms that allow for some period of time before full payment is required.

182. Answer: D

The kickoff meeting gets all of the stakeholders together to explain how communication will go. That way, everyone knows who to talk to if things go wrong or they run into any questions.

183. Answer: A

Not every change needs to be made. Before you make any change, you always need to evaluate its impact on the triple constraint—time, cost, and scope—and how those changes will affect the quality of the deliverables. Until you analyze that impact, there's no way to know whether or not it makes sense to make the change.

Answers

Exam ~~Questions~~

---

184. Answer: B

The idea behind net present value (or NPV) is that you can compare potential projects by figuring out how much each one is worth to your company right now. You figure out a project's NPV out by coming up with how much the project is worth, and then subtracting how much it will cost. If you're asked to choose between projects and given the NPV of each of them, choose the one with the biggest NPV. That means you're choosing the one with the most value!

185. Answer: C

It's easy to get change, defects, and corrective actions mixed up—they're all words that sound suspiciously similar! Just remember: a defect is any deliverable that does not meet its requirements. A defect is *not* always caused by a mistake—defects can come from lots of sources, and team members' errors only cause some defects. For example, plenty of defects are caused by equipment problems.

> Don't forget that the Project Management plan itself is a deliverable! That means that it can have defects, too—a lot of companies have specific standards and requirements that every project plan must meet. And if a defect is found in the plan after the work has started, then you need to go through change control to repair it!

186. Answer: D

Forcing means putting your foot down and making a decision. One person wins, one person loses, and that's the end of that.

187. Answer: B

The project scope statement is where you figure out exactly what your stakeholders need, and turn those needs into exactly what work the team will do to give them a great product. Any constraints or assumptions that need to be made to determine the work need to be written down in the scope statement as well.

188. Answer: C

You can figure out the actual cost that was spent on a project, even if all you're given are some of the project metrics. In this case, if you only have CPI and EV, you can figure out the AC by writing down the formula that has all three of them: $CPI = EV / AC$. Now flip the formula around: $AC = EV / CPI = \$172,500 / .92 = \$187,500$.

Answers

Exam ~~Questions~~

189. Answer: A

Changes are found in Monitor and Control Project Work; they are approved in Perform Integrated Change Control, and implemented in Direct and Manage Project Execution. When you are monitoring and controlling the project work, you are always looking for changes that might need to be made to your plan and assessing their impact. Then you present those changes to the change control board for approval. If they approve, you implement them in the Direct and Manage Project Execution process—that's where all the work gets done.

190. Answer: A

A work breakdown structure is the best way to visualize all of the work that will be done on your project. It divides all of the work up into work packages and shows how it fits into higher-level categories. By looking at the WBS, you can communicate to other people just how much work is involved in your project.

191. Answer: B

Over half of the conflicts on projects come from resources, priorities, and schedules. It can be tough to get resources assigned to projects, especially if they have skills that are in high demand. Sometimes multiple projects (and even roles within projects) are vying to get top priority. Finally, you probably don't need to think too hard to remember a conflict about schedules on a project you've worked on—many projects start with overly aggressive deadlines that cause conflicts from the very beginning.

192. Answer: D

When do you gather work performance information? You do it when you're reporting on the performance of the team—that's why it's a tool and technique of the Control Communications process. But that's not something you do during Control Risks—work performance information is an input to that process, which means it needs to be gathered *before* you start controlling your risks.

Some questions on the PMP exam will describe tools or techniques rather than using their names. A question might say "bringing in an outside party to review your risk response strategies" instead of "risk audit." You're actually asked about concepts you've learned, not just about a bunch of things you've memorized.

Answers

Exam ~~Questions~~

193. Answer: B

You should always use formal written communication when you are communicating with clients about changes in your project.

194. Answer: C

The project charter is created long before you start identifying work packages and activities. Those things are done as part of the project planning, which happens only after the project charter is completed.

195. Answer: C

Positive risks are opportunities that could happen on your project. The strategies for dealing with them are all about making sure that your project is in a position to take advantage of them or at least share in them with other projects if possible.

196. Answer: D

The RBS is part of the Risk Management plan. It's structured very similarly to an WBS. The RBS helps you to see how risks fit into categories so you can organize your risk analysis and response planning.

197. Answer: B

The PMP Code of Professional Conduct says that you're not allowed to accept any kind of gift, not even if it's after the project has finished. That would be the same thing as taking a bribe.

198. Answer: D

This is a tough situation for any project manager. You've got a problem that's happened, and you didn't plan for it. Now it's going to cost you money. What do you do? Well, you can't just accept it and move on—that's only something you do with risks that have no other option. You have options with a problem that happens during your project. And you can't just go to the boss, because you're the project manager and it's your job to figure out what to do. There's no use in doing risk planning, because you already know the probability (100%) and impact (the cost of fixing the problem). So what do you do?

Exam Questions Answers

That's where your reserve comes in. There are two kinds of reserves: a contingency reserve and a management reserve. The contingency reserve is what you use for "known unknowns"—you use it to pay for risks that you've planned for. But this situation isn't like that. That's why you tap into the management reserve. That's the money in the budget you set aside for "unknown unknowns"—problems that you didn't plan for but which came up anyway.

199. Answer: B

When you have a project that's broken up into subprojects or phases, it's important that you perform the Initiating processes at the beginning of the project. Answer B is the one that best describes something that happens during the processes in the Initiating group—performing the Identify Stakeholders process.

200. Answer: B

This question looks hard, but it's actually pretty easy if you remember that Control Stakeholder Engagement is just an ordinary Monitoring and Controlling process—it's the one for the Stakeholder Management knowledge area. Once you know that, it's easy to pick out the output that doesn't fit! When you're handling a change in a Monitoring and Controlling process, you update your project plan and organizational process assets, and you request changes. But you don't create deliverables.

So how'd you do?

PMI uses a scoring system called the "Modified Agnoff Technique" (which it explains in the PMP Handbook, available for download from their website), which makes it a little hard to predict exactly how you'll do. But if you're scoring in the 80% to 90% range on this exam, then you're in really good shape.

The End

Index

Symbols

A

C

E

N

negative risk
  EMV on 606
  strategies for 583
negative stakeholders 49, 57, 666, 672
negotiation
  in Acquire Project Team process 470
  in Develop Project Team process 477
Net present value (NPV) 346–347, 354–355
network diagrams
  creating 267
  critical path in
    backward pass 296
    finding float for activities 288–291
    using Critical Path Method 285–287
    using early start and early finish 293, 297–299
    using forward pass method 295
    using late start and late finish 294
  finding float for activities in 288–291, 301
  in Define Activities process 248
networking, in Plan Human Resource Management process 464
neutral role, stakeholders in 672–674
noise, in communication model 512–513
nominal group techniques 181, 186, 419
nonverbal communication
  in Manage Communications process 522–523
  vs. paralingual communication 536
norming, as stage of team development 482, 492
NPV (Net present value) 346–347, 354–355

O

observation
  in Manage Project Team process 486
  using 182
Open Issues section, in project management plan 123
operational work
  about 47
  vs. projects 48
operations management, function of 26

operations (processes)
  anatomy of 79–81
  combining 82
  in process groups 76–78, 81
  test practice question answers on 728
  test practice questions on 710–711
  vs. projects 17–18
operations teams, as stakeholders 26
opportunity cost 346–347
organizational groups, as stakeholders 50
organizational process assets
  about 81
  as project input 115
  in Acquire Project Team process 470
  in Close Procurements process 646
  in Close Project or Phase process 140–141
  in Communications Management process 518
  in Conduct Procurements process 634–635, 636
  in Control Communications process 528, 536
  in Control Costs process 359–360
  in Control Procurements process 641
  in Control Quality process 424
  in Control Schedule process 311–312
  in Create WBS process 197
  in Define Activities process 252
  in Define Scope process 187
  in Develop Project Charter process 108, 117
  in Develop Schedule process 280
  in Direct and Manage Project Work process 126
  in Estimate Activity Durations process 272–274
  in Estimate Activity Resources process 269
  in Estimate Costs process 342
  in Identify Risks process 560
  in Identify Stakeholders process 664
  in Manage Communications process 520
  in Manage Stakeholder Engagement process 679
  in Monitor and Control Project Work process 132
  in Perform Integrated Change Control process 133
  in Perform Qualitative Risk Analysis process 566
  in Perform Quantitative Risk Analysis process 573
  in Plan Cost Management process 340
  in Plan Human Resource Management process 464–466
  in Plan Procurement Management process 620
  in Plan Quality Management process 417
  in Plan Risk Management process 554
  in Plan Schedule Management process in 250

Have it your way.

O'Reilly eBooks

- Lifetime access to the book when you buy through oreilly.com
- Provided in up to four, DRM-free file formats, for use on the devices of your choice: PDF, .epub, Kindle-compatible .mobi, and Android .apk
- Fully searchable, with copy-and-paste, and print functionality
- We also alert you when we've updated the files with corrections and additions.

oreilly.com/ebooks/

Safari Books Online

- Access the contents and quickly search over 7000 books on technology, business, and certification guides
- Learn from expert video tutorials, and explore thousands of hours of video on technology and design topics
- Download whole books or chapters in PDF format, at no extra cost, to print or read on the go
- Early access to books as they're being written
- Interact directly with authors of upcoming books
- Save up to 35% on O'Reilly print books

See the complete Safari Library at safari.oreilly.com

Get even more for your money.

Join the O'Reilly Community, and register the O'Reilly books you own. It's free, and you'll get:

- $4.99 ebook upgrade offer
- 40% upgrade offer on O'Reilly print books
- Membership discounts on books and events
- Free lifetime updates to ebooks and videos
- Multiple ebook formats, DRM FREE
- Participation in the O'Reilly community
- Newsletters
- Account management
- 100% Satisfaction Guarantee

Signing up is easy:

1. Go to: oreilly.com/go/register
2. Create an O'Reilly login.
3. Provide your address.
4. Register your books.

Note: English-language books only

To order books online:
oreilly.com/store

For questions about products or an order:
orders@oreilly.com

To sign up to get topic-specific email announcements and/or news about upcoming books, conferences, special offers, and new technologies:
elists@oreilly.com

For technical questions about book content:
booktech@oreilly.com

To submit new book proposals to our editors:
proposals@oreilly.com

O'Reilly books are available in multiple DRM-free ebook formats. For more information:
oreilly.com/ebooks

O'REILLY®

Spreading the knowledge of innovators   oreilly.com

CPSIA information can be obtained at www.ICGtesting.com
Printed in the USA
BVOW09n0600040315

390189BV00004B/8/P